Communications
in Computer and Information Science 1194

Commenced Publication in 2007
Founding and Former Series Editors:
Phoebe Chen, Alfredo Cuzzocrea, Xiaoyong Du, Orhun Kara, Ting Liu,
Krishna M. Sivalingam, Dominik Ślęzak, Takashi Washio, Xiaokang Yang,
and Junsong Yuan

More information about this series at http://www.springer.com/series/7899

Miguel Botto-Tobar · Marcelo Zambrano Vizuete ·
Pablo Torres-Carrión · Sergio Montes León ·
Guillermo Pizarro Vásquez ·
Benjamin Durakovic (Eds.)

Applied Technologies

First International Conference, ICAT 2019
Quito, Ecuador, December 3–5, 2019
Proceedings, Part II

 Springer

Editors
Miguel Botto-Tobar ⓘ
Eindhoven University of Technology
Eindhoven, The Netherlands

Marcelo Zambrano Vizuete ⓘ
Universidad Técnica del Norte
Ibarra, Ecuador

Pablo Torres-Carrión ⓘ
Universidad Técnica Particular de Loja
Loja, Ecuador

Sergio Montes León ⓘ
Universidad de las Fuerzas Armadas (ESPE)
Quito, Ecuador

Guillermo Pizarro Vásquez ⓘ
Universidad Politécnica Salesiana
Guayaquil, Ecuador

Benjamin Durakovic ⓘ
International University of Sarajevo
Sarajevo, Bosnia and Herzegovina

ISSN 1865-0929   ISSN 1865-0937   (electronic)
Communications in Computer and Information Science
ISBN 978-3-030-42519-7   ISBN 978-3-030-42520-3   (eBook)
https://doi.org/10.1007/978-3-030-42520-3

This Springer imprint is published by the registered company Springer Nature Switzerland AG
The registered company address is: Gewerbestrasse 11, 6330 Cham, Switzerland

Preface

The First International Conference on Applied Technologies (ICAT 2019) was held on the main campus of the Universidad de las Fuerzas Armadas (ESPE), in Quito, Ecuador, during December 3–5, 2019, and was jointly organized by the Universidad de las Fuerzas Armadas (ESPE), the Universidad Técnica Particular de Loja, and the Universidad Técnica del Norte, in collaboration with GDEON. The ICAT series aims to bring together top researchers and practitioners working in different domains in the field of computer science to exchange their expertise and discuss the perspectives of development and collaboration. The content of this three-volume set is related to the following subjects: technology trends, computing, intelligent systems, machine vision, security, communication, electronics, e-learning, e-government, and e-participation.

ICAT 2019 received 328 English submissions written by 586 authors from 23 different countries. All these papers were peer-reviewed by the ICAT 2019 Program Committee consisting of 191 high-quality researchers. To assure a high quality and thoughtful review process, we assigned each paper to at least three reviewers. Based on these reviews, 124 full papers were accepted, resulting in an acceptance rate of 38%, which was within our goal of less than 40%.

We would like to express our sincere gratitude to the invited speakers for their inspirational talks, to the authors for submitting their work to this conference, and the reviewers for sharing their experience during the selection process.

December 2019

Miguel Botto-Tobar
Marcelo Zambrano Vizuete
Pablo Torres-Carrión
Sergio Montes León
Guillermo Pizarro Vásquez
Benjamin Durakovic

Organization

General Chair

Miguel Botto-Tobar   Eindhoven University of Technology, The Netherlands

Organizing Committee

Miguel Botto-Tobar	Eindhoven University of Technology, The Netherlands
Marcelo Zambrano Vizuete	Universidad Técnica del Norte, Ecuador
Pablo Torres-Carrión	Universidad Técnica Particular de Loja, Ecuador
Sergio Montes León	Universidad de las Fuerzas Armadas (ESPE), Ecuador, and Universidad Rey Juan Carlos, Spain
Guillermo Pizarro	Universidad Politécnica Salesiana, Ecuador
Benjamin Durakovic	International University of Sarajevo, Bosnia and Herzegovina
Jose Bucheli Andrade	Universidad de las Fuerzas Armadas (ESPE), Ecuador

Steering Committee

Miguel Botto-Tobar	Eindhoven University of Technology, The Netherlands
Ángela Díaz Cadena	Universitat de Valencia, Spain

Publication Chair

Miguel Botto-Tobar   Eindhoven University of Technology, The Netherlands

Program Chairs

Technology Trends

Jean Michel Clairand	Universidad de Las Américas, Ecuador
Miguel Botto-Tobar	Eindhoven University of Technology, The Netherlands
Hernán Montes León	Universidad Rey Juan Carlos, Spain

Computing

Miguel Zúñiga Prieto	Universidad de Cuenca, Ecuador
Lohana Lema Moreira	Universidad de Especialidades Espíritu Santo (UEES), Ecuador

Intelligent Systems

Janeth Chicaiza	Universidad Técnica Particular de Loja, Ecuador
Pablo Torres-Carrión	Universidad Técnica Particular de Loja, Ecuador
Guillermo Pizarro Vásquez	Universidad Politécnica Salesiana, Ecuador

Machine Vision

Julian Galindo	LIG-IIHM, France
Erick Cuenca	Université de Montpellier, France
Jorge Luis Pérez Medina	Universidad de Las Américas, Ecuador

Security

Luis Urquiza-Aguiar	Escuela Politécnica Nacional, Ecuador
Joffre León-Acurio	Universidad Técnica de Babahoyo, Ecuador

Communication

Nathaly Verónica Orozco Garzón	Universidad de Las Américas, Ecuador
Óscar Zambrano Vizuete	Universidad Técnica del Norte, Ecuador
Pablo Palacios Jativa	Universidad de Chile, Chile
Henry Ramiro Carvajal Mora	Universidad de Las Américas, Ecuador

Electronics

Ana Zambrano Vizuete	Escuela Politécnica Nacional (EPN), Ecuador
David Rivas	Universidad de las Fuerzas Armadas (ESPE), Ecuador

e-Learning

Verónica Falconí Ausay	Universidad de Las Américas, Ecuador
Doris Macías Mendoza	Universitat Politécnica de Valencia, Spain

e-Business

Angela Díaz Cadena	Universitat de Valencia, Spain
Oscar León Granizo	Universidad de Guayaquil, Ecuador
Praxedes Montiel Díaz	CIDEPRO, Ecuador

e-Government and e-Participation

Vicente Merchán Rodríguez	Universidad de las Fuerzas Armadas (ESPE), Ecuador
Alex Santamaría Philco	Universidad Laica Eloy Alfaro de Manabí, Ecuador

Program Committee

A. Bonci	Marche Polytechnic University, Italy
Ahmed Lateef Khalaf	Al-Mamoun University College, Iraq
Aiko Yamashita	Oslo Metropolitan University, Norway
Alejandro Donaire	Queensland University of Technology, Australia
Alejandro Ramos Nolazco	Instituto Tecnólogico y de Estudios Superiores Monterrey, Mexico
Alex Cazañas	The University of Queensland, Australia

Alex Santamaria Philco	Universitat Politècnica de València, Spain
Alfonso Guijarro Rodriguez	University of Guayaquil, Ecuador
Allan Avendaño Sudario	Escuela Superior Politécnica del Litoral (ESPOL), Ecuador
Alexandra González Eras	Universidad Politécnica de Madrid, Spain
Ana Núñez Ávila	Universitat Politècnica de València, Spain
Ana Zambrano	Escuela Politécnica Nacional (EPN), Ecuador
Andres Carrera Rivera	The University of Melbourne, Australia
Andres Cueva Costales	The University of Melbourne, Australia
Andrés Robles Durazno	Edinburg Napier University, UK
Andrés Vargas Gonzalez	Syracuse University, USA
Angel Cuenca Ortega	Universitat Politècnica de València, Spain
Ángela Díaz Cadena	Universitat de València, Spain
Angelo Trotta	University of Bologna, Italy
Antonio Gómez Exposito	University of Sevilla, Spain
Aras Can Onal	Tobb University Economics and Technology, Turkey
Arian Bahrami	University of Tehran, Iran
Benoît Macq	Université Catholique de Louvain, Belgium
Bernhard Hitpass	Universidad Federico Santa María, Chile
Bin Lin	Università della Svizzera italiana (USI), Switzerland
Carlos Saavedra	Escuela Superior Politécnica del Litoral (ESPOL), Ecuador
Catriona Kennedy	The University of Manchester, UK
César Ayabaca Sarria	Escuela Politécnica Nacional (EPN), Ecuador
Cesar Azurdia Meza	University of Chile, Chile
Christian León Paliz	Université de Neuchâtel, Switzerland
Chrysovalantou Ziogou	Chemical Process and Energy Resources Institute, Greece
Cristian Zambrano Vega	Universidad de Málaga, Spain, and Universidad Técnica Estatal de Quevedo, Ecuador
Cristiano Premebida	Loughborough University, ISR-UC, UK
Daniel Magües Martinez	Universidad Autónoma de Madrid, Spain
Danilo Jaramillo Hurtado	Universidad Politécnica de Madrid, Spain
Darío Piccirilli	Universidad Nacional de La Plata, Argentina
Darsana Josyula	Bowie State University, USA
David Benavides Cuevas	Universidad de Sevilla, Spain
David Blanes	Universitat Politècnica de València, Spain
David Ojeda	Universidad Técnica del Norte, Ecuador
David Rivera Espín	The University of Melbourne, Australia
Denis Efimov	Inria, France
Diego Barragán Guerrero	Universidad Técnica Particular de Loja (UTPL), Ecuador
Diego Peluffo-Ordoñez	Yachay Tech, Ecuador
Dimitris Chrysostomou	Aalborg University, Denmark
Domingo Biel	Universitat Politècnica de Catalunya, Spain
Doris Macías Mendoza	Universitat Politècnica de València, Spain

Jorge Charco Aguirre   Universitat Politècnica de València, Spain
Jorge Eterovic   Universidad Nacional de La Matanza, Argentina
Jorge Gómez Gómez   Universidad de Córdoba, Colombia
Juan Corrales   Institut Universitaire de France et SIGMA Clermont,
                                  France
Juan Romero Arguello   The University of Manchester, UK
Julián Andrés Galindo   Université Grenoble Alpes, France
Julian Galindo   Inria, France
Julio Albuja Sánchez   James Cook University, Australia
Kelly Garces   Universidad de Los Andes, Colombia
Kester Quist-Aphetsi   Center for Research, Information, Technology
                                  and Advanced Computing, Ghana
Korkut Bekiroglu   SUNY Polytechnic Institute, USA
Kunde Yang   Northwestern Polytechnic University, China
Lina Ochoa   CWI, The Netherlands
Lohana Lema Moreira   Universidad de Especialidades Espíritu Santo (UEES),
                                  Ecuador
Lorena Guachi Guachi   Yachay Tech, Ecuador
Lorena Montoya Freire   Aalto University, Finland
Lorenzo Cevallos Torres   Universidad de Guayaquil, Ecuador
Luis Galárraga   Inria, France
Luis Martinez   Universitat Rovira i Virgili, Spain
Luis Urquiza-Aguiar   Escuela Politécnica Nacional (EPN), Ecuador
Maikel Leyva Vazquez   Universidad de Guayaquil, Ecuador
Manuel Sucunuta   Universidad Técnica Particular de Loja (UTPL),
                                  Ecuador
Marcela Ruiz   Utrecht University, The Netherlands
Marcelo Zambrano Vizuete   Universidad Técnica del Norte, Ecuador
María José Escalante   University of Michigan, USA
  Guevara
María Reátegui Rojas   University of Quebec, Canada
Mariela Tapia-Leon   University of Guayaquil, Ecuador
Marija Seder   University of Zagreb, Croatia
Mario Gonzalez Rodríguez   Universidad de las Américas, Ecuador
Marisa Daniela Panizzi   Universidad Tecnológica Nacional Aire, Argentina
Marius Giergiel   KRiM AGH, Poland
Markus Schuckert   Hong Kong Polytechnic University, Hong Kong
Matus Pleva   Technical University of Kosice, Slovakia
Mauricio Verano Merino   Technische Universiteit Eindhoven, The Netherlands
Mayken Espinoza-Andaluz   Escuela Superior Politécnica del Litoral (ESPOL),
                                  Ecuador
Miguel Botto-Tobar   Eindhoven University of Technology, The Netherlands
Miguel Fornell   Escuela Superior Politécnica del Litoral (ESPOL),
                                  Ecuador
Miguel Gonzalez Cagigal   Universidad de Sevilla, Spain
Miguel Murillo   Universidad Autónoma de Baja California, Mexico

Miguel Zuñiga Prieto	Universidad de Cuenca, Ecuador
Milton Román-Cañizares	Universidad de las Américas, Ecuador
Mohamed Kamel	Military Technical College, Egypt
Mohammad Al-Mashhadani	Al-Maarif University College, Iraq
Mohammad Amin	Illinois Institute of Technology, USA
Monica Baquerizo Anastacio	Universidad de Guayaquil, Ecuador
Muneeb Ul Hassan	Swinburne University of Technology, Australia
Nam Yang	Eindhoven University of Technology, The Netherlands
Nathalie Mitton	Inria, France
Nathaly Orozco	Universidad de las Américas, Ecuador
Nayeth Solórzano Alcívar	Escuela Superior Politécnica del Litoral (ESPOL), Ecuador, and Griffith University, Australia
Noor Zaman	King Faisal University, Saudi Arabia
Omar S. Gómez	Escuela Superior Politécnica del Chimborazo (ESPOCH), Ecuador
Óscar León Granizo	Universidad de Guayaquil, Ecuador
Oswaldo Lopez Santos	Universidad de Ibagué, Colombia
Pablo Lupera	Escuela Politécnica Nacional, Ecuador
Pablo Ordoñez Ordoñez	Universidad Politécnica de Madrid, Spain
Pablo Palacios	Universidad de Chile, Chile
Pablo Torres-Carrión	Universidad Técnica Particular de Loja (UTPL), Ecuador
Patricia Ludeña González	Universidad Técnica Particular de Loja (UTPL), Ecuador
Paúl Mejía	Universidad de las Fuerzas Armadas (ESPE), Ecuador
Paulo Batista	CIDEHUS.UÉ, Portugal
Paulo Chiliguano	Queen Mary University of London, UK
Paulo Guerra Terán	Universidad de las Américas, Ecuador
Pedro Neto	University of Coimbra, Portugal
Praveen Damacharla	Purdue University Northwest, USA
Priscila Cedillo	Universidad de Cuenca, Ecuador
Radu-Emil Precup	Politehnica University of Timisoara, Romania
Ramin Yousefi	Islamic Azad University, Iran
René Guamán Quinche	Universidad de los Paises Vascos, Spain
Ricardo Martins	University of Coimbra, Portugal
Richard Ramirez Anormaliza	Universitat Politècnica de Catalunya, Spain
Richard Rivera	IMDEA Software Institute, Spain
Richard Stern	Carnegie Mellon University, USA
Rijo Jackson Tom	SRM University, India
Roberto Murphy	University of Colorado Denver, USA
Roberto Sabatini	RMIT University, Australia
Rodolfo Alfredo Bertone	Universidad Nacional de La Plata, Argentina
Rodrigo Barba	Universidad Técnica Particular de Loja (UTPL), Ecuador

Rodrigo Saraguro Bravo	Universitat Politècnica de València, Spain
Ronald Barriga Díaz	Universidad de Guayaquil, Ecuador
Ronnie Guerra	Pontificia Universidad Católica del Perú, Peru
Ruben Rumipamba-Zambrano	Universitat Politècnica de Catalanya, Spain
Saeed Rafee Nekoo	Universidad de Sevilla, Spain
Saleh Mobayen	University of Zanjan, Iran
Samiha Fadloun	Université de Montpellier, France
Sergio Montes León	Universidad de las Fuerzas Armadas (ESPE), Ecuador
Stefanos Gritzalis	University of the Aegean, Greece
Syed Manzoor Qasim	King Abdulaziz City for Science and Technology, Saudi Arabia
Tatiana Mayorga	Universidad de las Fuerzas Armadas (ESPE), Ecuador
Tenreiro Machado	Polytechnic of Porto, Portugal
Thomas Sjögren	Swedish Defence Research Agency (FOI), Sweden
Tiago Curi	Federal University of Santa Catarina, Brazil
Tony T. Luo	A*STAR, Singapore
Trung Duong	Queen's University Belfast, UK
Vanessa Jurado Vite	Universidad Politécnica Salesiana, Ecuador
Waldo Orellana	Universitat de València, Spain
Washington Velasquez Vargas	Universidad Politécnica de Madrid, Spain
Wayne Staats	Sandia National Labs, USA
Willian Zamora	Universidad Laíca Eloy Alfaro de Manabí, Ecuador
Yessenia Cabrera Maldonado	University of Cuenca, Ecuador
Yerferson Torres Berru	Universidad de Salamanca, Spain, and Instituto Tecnológico Loja, Ecuador
Zhanyu Ma	Beijing University of Posts and Telecommunications, China

Organizing Institutions

Sponsoring Institutions

UNIVERSIDAD DE CUENCA
DEPARTAMENTO DE CIENCIAS
DE LA COMPUTACIÓN
FACULTAD DE INGENIERÍA

Collaborators

Contents – Part II

Machine Vision

Intelligent Systems

Evaluation of WhatsApp to Promote Collaborative Learning in the Use of Software in University Professionals

William Montalvo[1] , Fernando Ibarra-Torres[2] ,
Marcelo V. Garcia[2,3](✉) , and Valeria Barona-Pico[2]

[1] Universidad Politecnica Salesiana, UPS, 170146 Quito, Ecuador
wmontalvo@ups.edu.ec
[2] Universidad Tecnica de Ambato, UTA, 180103 Ambato, Ecuador
{of.ibarra,mv.garcia,va.barona}@uta.edu.ec
[3] University of Basque Country, UPV/EHU, 48013 Bilbao, Spain
mgarcia294@ehu.eus

Abstract. In this article we discuss how users of Mobile Instant Messaging (MMI) (WhatsApp) applications collaborate in environments that promote knowledge networks. For this study the researcher analyzed the interaction of users added to a group in WhatsApp, which is based on concepts of connectivism, communication, collaborative learning and knowledge creation. This study deals with the granting of knowledge on a specific computer subject and its transmission by all the members of the group, facilitating the work for the instructor and increasing the users' compression capacity. Each user can become an instructor for the new members of the group. Strengthening collaborative learning processes. Part of the study is the use of the questionnaire tool that collected information from group members, reception capacity, age and difficulties with MMI.

Keywords: Knowledge · Collaborative learning · Mobile Instant Messaging · Smartphones · Knowledge transmission · Creation of new knowledge

1 Introduction

With the rise and combination of the internet and learning, the way in which human beings work with the transmission of knowledge has changed more and more [1]. The emergence of new methodologies, tools and software seek to facilitate the acquisition and transmission of knowledge, changing the way of learning and avoiding dependence on a single instructor; diminishing the importance of socio-cultural aspects such as age, sex, gender, level of education or others [2].

The emergence of new technologies seeks to facilitate and support the learning process [3], and new collaborative techniques for the knowledge acquisition process are increasing [4].

According to the Household Panel of the National Commission on Markets and Competition (CNMC), by the end of 2017 the daily use of instant messaging services,

© Springer Nature Switzerland AG 2020
M. Botto-Tobar et al. (Eds.): ICAT 2019, CCIS 1194, pp. 3–12, 2020.
https://doi.org/10.1007/978-3-030-42520-3_1

such as WhatsApp, almost doubles that of mobile and fixed calls. Instant messaging is used by six out of 10 people a day (60%), a figure far higher than the daily use of calls made from mobiles (24%), landlines (12%) or online (4%). SMS have become almost obsolete and are rarely used. Almost 60% of people in the study never send them (CNMC, 2017).

To [5], study how the use of instant messaging provides flexibility in communication between employees, improving the coordination and logistics of medical staff in caring for their patients. Other studies advocate direct and indirect improvements in work outcomes and performance when using instant messaging services at work [6, 7].

This seeks to analyze learning and how the parts act to reinforce it [8]. According to the literature [9], today we can identify three types of learning:

- Formal learning: occurs in structured environments such as educational institutions or companies. It uses validations to verify acquired knowledge [10].
- Non-formal learning: are those obtained in defined environments but that were not initially designed for that.
- Informal learning: it is obtained daily in the different activities and spheres. It is disorganized and unstructured learning.

The combination of these types of learning with new methods and/or methodologies of communication that appear, become vehicles to promote the transmission of knowledge that can build a society with more prepared individuals [11].

Multimedia messaging applications are perfectly compatible with learning communities [12], where through the proposition of a topic and a brief explanation conversations are developed, the first content can be shared freely with the whole group, which in the first instance gives rise to interaction between users and most importantly the transmission of knowledge between people.

In a literature analysis [13, 14], in the interactions as a product of the topic proposal one can very well acquire additional knowledge and improve the learning experience. From this point of view, it is feasible to cite the theory of connectivism [13] (online or offline) which identify that the learning process is enriched by the time of connection of the members of the group.

This article deals precisely with these last evidences, the propagation of information and learning between users in a certain group and their interaction in the event of queries or doubts that arise.

Publications such as [15, 16], focus more on collaborative learning between college and university students, rather than on university professionals who currently use WhatsApp to share information and knowledge. It is therefore necessary to analyze how university professionals use this technological tool for learning.

The platform used for the study is WhatsApp: which is an Internet messaging application. Currently owned by the company Facebook, having millions of users worldwide. Indirectly it becomes a tool for informal learning, based on connectivity.

In this way, the objectives of this article can be defined as follows:

1. Analysis of responses in conversations and interaction among group members.
2. Determine learning patterns in the use of multimedia messaging applications.

3. Evaluate the type of learning and transmission of knowledge between users, in order to study what types of learning take place in these environments.

The goals, and the rest of the content, concepts and definitions necessary for development are discussed in the sections: Materials and Methods, Results, Discussions and Conclusions.

2 Materials and Methods

2.1 Materials

WhatsApp. It has become one of the most popular applications for exchanging messages. It has been increasing its features allowing the exchange of multimedia content and even the making of voice and video calls [17]. It is intended to take advantage of these features to use them in favor of the diffusion, transmission of knowledge in a certain group. The use of this tool allows a constant interaction between the members of the group, increasing the capacity to transmit knowledge indistinctly of the schedule.

Knowledge is generated with the publication of a topic or indication and the participants or members of the group retransmit or add content allowing the creation of a cooperative model.

The study contemplates a data analysis of 18 months and approximately 75000 messages, using the indicated tool and the opportunities they offer in the field of learning. Depending on the characteristics of the WhatsApp tool, members of the group are placed with knowledgeable messages in the area of information technology in three days a day, Ecuadorian time:

- 8:00 am
- 14:00 pm
- 19:00 pm

In each of the sessions, different data on the use of institutional software was presented, allowing the members of the group to access and follow up on all the contents. The percentage of connection and revision of the exposed information varies according to the schedule in which it is published, in Fig. 1 the percentage of connections according to the schedule is offered. Of the 82 members of the group, 54 people reviewed the information from the first schedule until before the next publication, 67 people reviewed the information from the second schedule before the next publication, and 82 people reviewed the information published in the third schedule before the next publication.

Figure 2 shows the percentage of users who connect to the software and interact with each other, retransmitting the knowledge provided and providing constant feedback between members. In Fig. 3 we can show that only 9 users (10.97%) of the total needed the group moderator to instruct them again, the rest of users (89.02%) were able to receive the knowledge at the right time and/or through the active collaboration of the other members of the group, constant feedback between them. In other words, the members of the group become channels of communication. Now in this context we can mention that for users with more experience in their respective positions it is easier to capture, analyze and process information to turn it into knowledge.

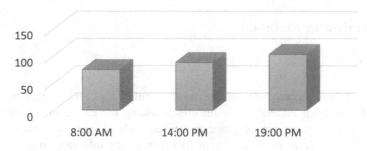

Fig. 1. Review of information in published timetables.

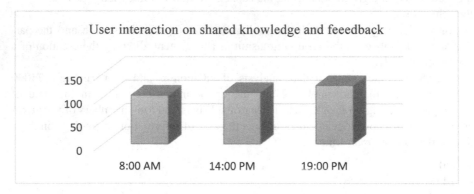

Fig. 2. Statistics on the interaction of user's members of the group sharing knowledge to peers

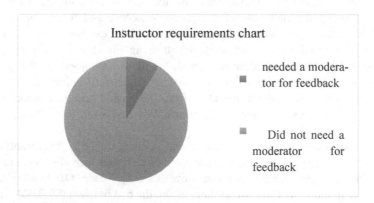

Fig. 3. Instructor need analysis. Once the knowledge is shared, the most experienced users provide feedback to users.

Multimedia Messaging. It is considered as the most evolving means to collaborate with the creation and transmission of knowledge. They play an important role in the learning process [18]. Allowing the creation of learning networks, for our study we have designed a case study to understand concepts in a practical way [17].

The topics proposed and exposed in the conversations, open forms of discussion and acquisition of knowledge from the non-formal point of view.

2.2   Method

The interaction of users in the group and the information contents is necessary to establish correct channels of information and retrieval [8]. In our case, between the members with their respective university positions and the reception and feedback to other users of the same group.

The methods used can be summarized as follows:

- The group's messages were exported from the WhatsApp application by sending them by mail and then downloading them to the PC in a TXT format.
- To filter the information and be able to analyze it, we use a spreadsheet tool.
- To obtain extra data and contribute to our analysis, we use the free app Analyzer for WhatsApp and access certain visual statistical features [11].

Additionally, the research is also observational and descriptive of the habits and behaviours of the individuals in the group. The technique used is quantitative. It was conducted to members of the Technical University of Ambato (Ambato, Ecuador) during the last months of 2017, every month of 2018 and the first quarter of 2019.

A total of 78 surveys were conducted and answered. After purifying the sample, eight of them were discarded when contradictions were detected in the answers or unanswered questions. The final sample is made up of 82 university professionals, of whom 35 are women and 47 are men.

Since this is a survey that adapts personal interview techniques, the answers correspond to the self-perception of the users about their frequency and purposes of use, the usefulness of its possible implementation in the academic work environment.

3   Results

3.1   Results of the Questionnaire

We have proceeded to use the questionnaire completed by users members of the group, which contains questions related to interaction, reception, transmission and feedback to help with learning. 102 users, of which 98% completed the questionnaire.

Figures 4, 5, 6, 7 and 8 show the data retrieved through the questionnaires and the segmentation of responses by age and position of the users. The Likert scale is used (1–5 with no null value).

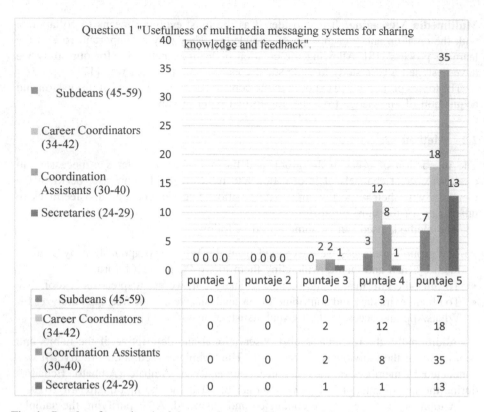

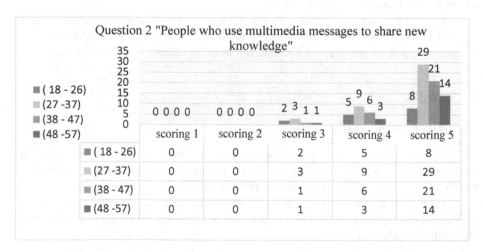

Fig. 4. Results of question 1 of the questionnaire. The use and feasibility of feedback among users are evaluated. Grouped by position or hierarchy and age.

Fig. 5. Results from question 2 of the questionnaire, use frequently to share new knowledge and constant feedback to other users

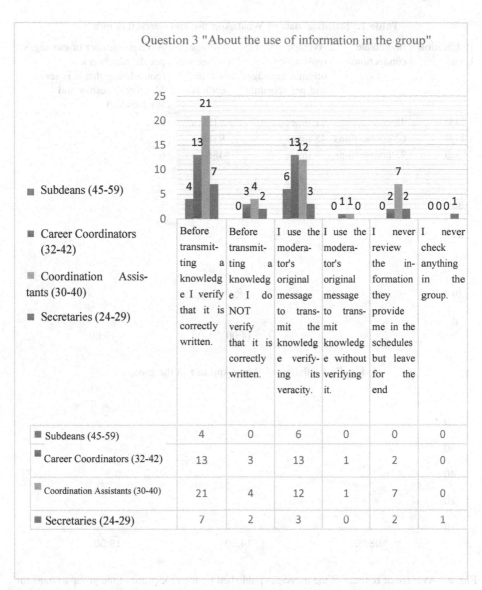

	Before transmitting a knowledge I verify that it is correctly written.	Before transmitting a knowledge I do NOT verify that it is correctly written.	I use the moderator's original message to transmit the knowledge verifying its veracity.	I use the moderator's original message to transmit knowledge without verifying it.	I never review the information they provide me in the schedules but leave for the end	I never check anything in the group.
▪ Subdeans (45-59)	4	0	6	0	0	0
▪ Career Coordinators (32-42)	13	3	13	1	2	0
▪ Coordination Assistants (30-40)	21	4	12	1	7	0
▪ Secretaries (24-29)	7	2	3	0	2	1

Fig. 6. Results of question 3 of the questionnaire, on the use of information in the group. Responses grouped by position.

3.2   Results Obtained in WhatsApp Statistics

In order to obtain results from the use of WhatsApp and review of the information entered, the information was filtered with the help of the Analyzer for WhatsApp tool and manual records.

Table 1 generates a report with the data obtained. Additionally, this information was evaluated to check connection time, transmission, collaboration and feedback.

Table 1. Statistical data of WhatsApp use and interaction in it

Publication hours	Immediate connections	Average replications per original message and per schedule	Average connection time in seconds	Average number of messages per day/day/week (considering that it is sent Monday, Tuesday and Wednesday)
8:00	45 connections	31 times	6120 s	6
14:00	62 connections	45 times	9180 s	7
19:00	73 connections	59 times	34884 s	6

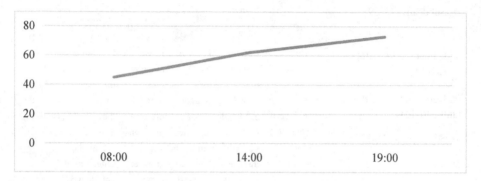

Fig. 7. Evolution of WhatsApp use in the group

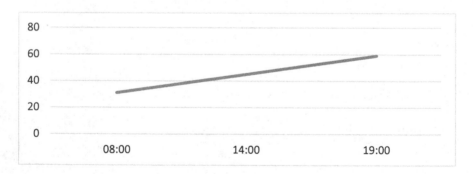

Fig. 8. Average of replicas of the messages published by the moderator of the group to transmit knowledge

4 Discussions and Conclusions

WhatsApp and other multimedia messaging tools represent a good opportunity for collaboration in learning and teaching new knowledge. They also provide useful features for formal and non-formal learning by allowing us to collect content and aspects specific to knowledge communities. Giving feedback on the content in the group and increasing the possibilities of opening channels of knowledge transmission.

WhatsApp is possibly the most notorious communication trend of our time and revolutionized written communication around the world. Regardless of age, this tool is constantly used to share information. The study shows that academic staff are highly dependent on instant messaging applications such as the one studied. Around 85% use the application more in the afternoon and evening. 50% of the staff read or wrote a WhatsApp in the group during snack hours.

From the point of view of the moderator of the group, the schedules of use is worrying, since it can affect in a direct way two aspects: the feeding and the rest, reducing the concentration and therefore the performance of the professionals when returning to their corresponding activities. It can also indirectly affect health, as it generates stress and anxiety, which accentuate this decline in work performance.

As a result of this work, we can deduce that the use of this tool by university professionals, can contribute to generate knowledge communities, as long as its use was appropriate. Being university professionals, the risk of sharing erroneous information decreases due to the degree of maturity of the people, and the capacity of assimilation increases. This study allows us to identify the hours of most use of the tool and the ability of people to retain knowledge and then retransmit it. People in middle and junior jobs use the tool the most and pass on knowledge.

References

1. García-Peñalvo, F.J., Seoane-Pardo, A.M.: Una revisión actualizada del concepto de eLearning. Décimo Aniversario. Educ. Knowl. Soc. **16**, 119–144 (2015)
2. Dodero, J.M., et al.: Development of e-learning solutions: different approaches, a common mission. IEEE Revista Iberoamericana de Tecnologias del Aprendizaje **9**, 72–80 (2014)
3. Berlanga, A.J., García-Peñalvo, F.J.: Learning design in adaptive educational hypermedia systems. JUCS – J. Univ. Comput. Sci. (2008). http://www.jucs.org/doi?doi=10.3217/jucs-014-22-3627. Accessed 4 Dec 2019
4. Kravets, A., Shcherbakov, M., Kultsova, M., Shabalina, O. (eds.): The Contribution of Gamification on User Engagement in Fully Online Course. Creativity in Intelligent Technologies and Data Science. Springer, Cham (2015). https://doi.org/10.1007/978-3-319-23766-4_56. Accessed 4 Dec 2019
5. Iversen, T.B., Melby, L., Toussaint, P.: Instant messaging at the hospital: supporting articulation work? Int. J. Med. Inform. **82**, 753–761 (2013)
6. Garcia, C.A., Caiza, G., Naranjo, J.E., Ortiz, A., Garcia, M.V.: An approach of training virtual environment for teaching electro-pneumatic systems. IFAC-PapersOnLine **52**, 278–284 (2019)
7. Garcia, C.A., Naranjo, J.E., Ortiz, A., Garcia, M.V.: An approach of virtual reality environment for technicians training in upstream sector. IFAC-PapersOnLine **52**, 285–291 (2019)
8. Salaverría, R.: Aproximación al concepto de multimedia desde los planos comunicativo e instrumental, pp. 383–395. ESMP (2001)
9. González-González, C., Blanco-Izquierdo, F.: Designing social videogames for educational uses. Comput. Educ. **58**, 250–262 (2012)
10. Mwakapina, J.W., Mhandeni, A.S., Nyinondi, S.: WhatsApp mobile tool in second language learning: opportunities, potentials and challenges in higher education settings in Tanzania. Int. J. Engl. Lang. Educ. **4**, 70 (2016)

11. Ngaleka, A., Uys, W.: m-Learning with WhatsApp: a conversation analysis, pp. 282–2900. Academic Conferences Ltd., Cape Town (2013)
12. Surendeleg, G., Tudevdagva, U., Kim, Y.S.: The contribution of gamification on user engagement in fully online course. In: Kravets, A., Shcherbakov, M., Kultsova, M., Shabalina, O. (eds.) CIT&DS 2015. CCIS, vol. 535, pp. 710–719. Springer, Cham (2015). https://doi.org/10.1007/978-3-319-23766-4_56
13. Awada, G.: Effect of WhatsApp on critique writing proficiency and perceptions toward learning. Cogent Educ. **3**, 1264173 (2016)
14. Barhoumi, C.: The effectiveness of WhatsApp mobile learning activities guided by activity theory on students' knowledge management. Contemp. Educ. Technol. **6**, 221–238 (2015)
15. Tulika, B., Dhananjay, J.: A study of students' experiences of mobile learning. Global J. Hum.-Soc. Sci. **14**, 26–33 (2014)
16. Brusilovsky, P., Millán, E.: User models for adaptive hypermedia and adaptive educational systems. In: Brusilovsky, P., Kobsa, A., Nejdl, W. (eds.) The Adaptive Web. LNCS, vol. 4321, pp. 3–53. Springer, Heidelberg (2007). https://doi.org/10.1007/978-3-540-72079-9_1. Accessed 4 Dec 2019
17. Karnouskos S., et al.: Experiences in integrating Internet of Things and cloud services with the robot operating system. In: 2017 IEEE 15th International Conference on Industrial Informatics (INDIN), pp. 1084–1089. IEEE, Emden (2017). http://ieeexplore.ieee.org/document/8104924/. Accessed 4 Dec 2019
18. Hassan, A.Q.A., Ahmed, S.S.: The impact of WhatsApp on learners' achievement: a case study of English language majors at King Khalid University. Int. J. Engl. Lang. Educ. **6**, 69 (2018)

Deep Learning for Weather Classification from a Meteorological Device

Nayely Galicia[1], Eddy Sánchez-Delacruz[1](✉)(iD), Rajesh R. Biswal[2],
Carlos Nakase[1], José Mejía[3], and David Lara[1]

[1] Instituto Tecnológico Superior de Misantla, Veracruz, Mexico
eddsacx@gmail.com
[2] Escuela de Ingeniería y Ciencias, Tecnológico de Monterrey, Nuevo Leon, Mexico
[3] Universidad Autónoma de Ciudad Juárez, Chihuahua, Mexico

Abstract. In order to realize climate prediction or weather forecast, there exist large and expensive meteorological stations which monitor various weather variables, such as temperature, wind speed, humidity, etc. However, these prediction systems are deployed to monitoring urban zones or large population areas. Therefore, predictions for communities far from urbanization are in the practice, imprecise. Currently, these inaccurate predictions for the climate changes, affect the agriculture in several areas, due to inadequate planning by the farmer, which is based on *a priori* knowledge that the inhabitants have with the experience from the observation of the climate behavior, action that is highly imprecise and unpredictable. Therefore, in this work, for a more accurate weather forecast in rural regions, a portable low-cost meteorological device is proposed, which using suitable sensors measures and record the weather variables such as temperature, humidity of the air, luminosity, rain, humidity, and atmospheric pressure. Using these weather data, a classification of the target class set {*rainy, cloudy, sunny*}, is made based on the parameters obtained through the device. Then, using a combinations of assembled algorithms with deep learning, optimum results are obtained with the following classifiers: *MultiClassClassifier+Multilayer perceptron*, using the sampling criteria 2/3-1/3, cross validation and representative sample. The classification results are comparable and competitive, with respect to those reported in the state-of-the-art, and stands out by distinguishing the target classes with a high degree of precision.

Keywords: Deep learning · Weather prediction · Meteorological device

1 Introduction

Despite of the existence of different systems and devices to know the weather forecast, these are not usually available for the agricultural community, where the climate is different according with the geographical area. Also, in rural zones, the agricultural community lacks of internet connectivity or sophisticated mobile

© Springer Nature Switzerland AG 2020
M. Botto-Tobar et al. (Eds.): ICAT 2019, CCIS 1194, pp. 13–25, 2020.
https://doi.org/10.1007/978-3-030-42520-3_2

devices to have a timely weather forecast. The meteorological agencies make use of information captured by satellites, combined with the measurement data from largo and expensive equipment installed in different parts of the country. Climate forecasting is generally carried out to cover densely populated/urbanized zones (entities) or regions (municipalities), with the disadvantage of biasing for some rural regions where diverse climatic and topographic parameters converge particular that impact on their environment. In general, the farmer plans the agricultural activities such as sowing, cleaning and harvesting depending on the environmental factors. In this work, assembled classifiers are implemented having deep learning as a basic algorithm. Unlike robust climate prediction systems, these classifier matching strategies perform well with a minimum of local historical information to characterize the multi-class environmental condition {*rainy, cloudy, sunny*}. The classifiers are trained by the weather data variables registered by the device, such as humidity, height above sea level, solar brightness and atmospheric pressure. The results and their respective validation give a classification very close to the optimal value.

2   Previous Works

The literature is very limited with respect to works that describe the relevant technological contributions about the design of architectures of the devices with models to predict the climate. Among the most significant works, it can be found the publication of Flamengo *et al.* [1], which explains a methodology used in models for meteorological stations in Argentina. Their research was aimed to associate the factors that alter the pressure and temperature of the seas, with statistical methods, dividing the area by regions and labeling them according to the rainfall presents to determine its predictors: dry, very dry, normal, humid and very humid. The authors justify their research based on the effects of climate change on agricultural activities. Other interesting work, is presented by Cruz [2], where with the Earth and Atmosphere Observation Group (GOTA) a set of meteorological predictions was done using with the application in the short term of numerical methods. GOTA has a website to publish the weather forecasts results. The purpose of the GOTA project was to elaborate an user interface *app* that allow download and visualize relevant data from the site with a simple and friendly operation for the user. The aforementioned *app* shows the predictions maps, the different climate variants and the temporary variable alteration that are of utmost importance for a precise location. Farfn and Barrantes [3] highlight that changes in the climate state is affecting agricultural and livestock producers in the region of Chiquinquir, Boyac, Colombia, that is why, that the study of the weather forecast is very important for the field sector. The authors propose a monitoring system composed of hardware and software, with the ability to collect information for the variables of humidity, pressure, temperature, heat, wind, among others that allows to deduce the climatic changes of the region. With the data acquisition of the values of each variable, they generate reports with the possible climatic changes, as well as their predictions. Krishnamurthi et al. [4], present the development of an automated system to monitor the weather. The meteorological variables (temperature and humidity) are calculated in function of the pressure differences in the air between the different areas of interest. The authors implemented a device based on Arduino microcontroller module, and embedded systems that uses different sensors, then from a historical database with updated register, the results of prediction were shown graphically.

Kedia [5] designed a mechanism to improve accuracy for a weather station installed on top of urban buildings. This unit registers measurements from sensors for different climatic aspects, for real time weather prediction.

3   Materials and Method

3.1   Requirements

The system was designed taking as main architecture the master-slaves topology, were a computer (with Intel Core i3-2370M CPU at 2.4 GHz, using 6 GB DDR3 1333 MHz of RAM and 320 GB HDD for data storage, with Microsoft Windows 7 Home Basic 64-bits as OS) was used as master. The software design and programming editors used to develop the software were as follows: Fritzing to design and emulate the electrical circuit of the device (topology of the sensor network). The communication between the Mega2560 card and the sensors was codified in Arduino Software IDE v.1.8.3. For acquiring and registering data the following sensors were used:

- DHT11 Temperature and Humidity Sensor, with temperature range of 0 to 50 Celsius degrees and Humidity range of 20–90% RH;
- BMP180 barometric sensor with $I2C$ interface, measuring range from 300 to 1100 hPa with and accuracy of 0.02 hPa;
- YL-83 Rain Sensor as a drop or rain detector;
- Light Dependent Resistor (LDR) for light measurement; and
- FC-28 Soil Moisture Sensor which measures moisture percentages with a 1024 resolution.

These sensors communicate with an Arduino Mega2560 card using digital ports. The power supply consisted of a set of power battery bank of 2600 mAh, complemented with an emergency or replacement unit of 1800 mAh. A type B micro USB cable was used to connect the Arduino Mega2560 card with the power supply. The circuit prototype was elaborated in a protoboard and an acrylic box was used to protect the entire circuit. Also, a microSD adapter module was connected to the system for data storage. The information obtained from the device was compared with the information from the weather forecasts of the National Meteorological System (NMS) and the websites meteored.mx and Yahoo!. The register of which was taken by means of a screenshot on a Lanix Ilium LT500 smartphone , ARM Cortex-A7/1100 MHz, Android 5.1 Lollipop 32-bits, RAM 1 GB and internal storage of 8 GB. The experiments of combination of assembled classifiers and the deep learning algorithm were implemented in Waikato Environment for Knowledge Analysis (WEKA) [13].

3.2 Method

The design, construction and validation of the system was done in six steps, including a process of analysis and re-engineering, either in hardware or software, that are described briefly here:

1. Analysis of requirements: The problem is treated and analysed to select the tools to be used: hardware and software.
2. Prototype: The device is designed and built taking into consideration the operation mode of sensors and other parameters.
3. Data acquisition and data collection: Measurements in field are donde using the device and they are stored.
4. Data analysis and comparison: the values obtained are compared with the predictions of climate predictions obtained from the NMS and the websites (meteored.mx and Yahoo!) to know the performance of the device and approximation with the available weather information.
5. Experimentation: It is the most important stage, where assembled classifiers are combined with the deep learning algorithm *Multilayer perceptron*. For this, three sampling criteria are applied: 2/3-1/3, cross validation and representative sample.
6. Analysis and validation of the results: The measurements are checked with the percentage of correctly classified instances and validated with confusion matrices, to know where each target class is confused.

4 Development and Experiments

4.1 Architecture Design

The electronic architecture of the device is shown in Fig. 1, where the circuit was implemented in simulation to validate its correct operation, avoiding wrong connection in the real world components. The desired functions were programmed with a specified capture time, sampling frequency, storage and data types.

4.2 Dataset

The registered and stored data were the variables of temperature in degrees Celsius, humidity of the air, measured in percentage, luminosity, rain, humidity of the earth, atmospheric pressure in millibars. The data acquisition of these variables was done during 14 days, from 9am to 5pm, at a sample rate of 4 samples per second. During the acquisition period some technical failures related with false connections in the prototype were presented during its operation, this helped to detect the weaknesses to improve its structure and programming, and thus have a good quality device to reach the goal. A total 27888 records were obtained. Table 1 shows a fragment of the records.

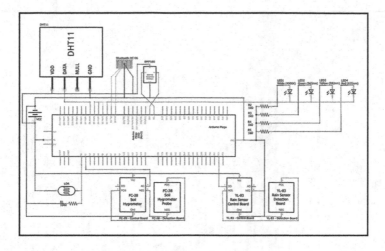

Fig. 1. Device topology

4.3   Data Comparison

Considering the variables *temperature*, *humidity*, *rain* and *climate*, a parallel comparison was done with the values registered in the forecasters SMN, Yahoo! and Meteored. Graphs were designed considering records of fourteen days; then, for each day the values are shown by bars in the following order: (i) device, (ii) NMS, (iii) Yahoo! and (iv) Meteored. For the *temperature* and *humidity* it is observed in the Figs. 2 and 3, that the values of the SMN, Yahoo! and Meteored do not correspond to the actual values obtained with the device *in situ*. What confirms the imprecise prediction in these forecasters.

For the variable *rain* (Fig. 4), it is observed that both the values registered by the device and those issued by the forecasters correspond on most cases except in intervals 6, 7, 11 and 12. On the other hand, the variable *clima*, corresponds in some days and not in others, see Fig. 5.

Table 1. Registers stored in a file with .csv format

Temperature (°C)	Humidity (%)	Light	Rain	Earth	Pressure (mbar)
12	83	424	0	441	995.29
12	82	471	0	544	995.33
12	82	415	0	495	995.24
13	81	413	0	494	995.27
12	82	633	0	676	995.24
13	81	361	0	442	995.27
13	80	414	0	488	995.23
13	80	412	0	486	995.22
13	80	412	0	485	995.27

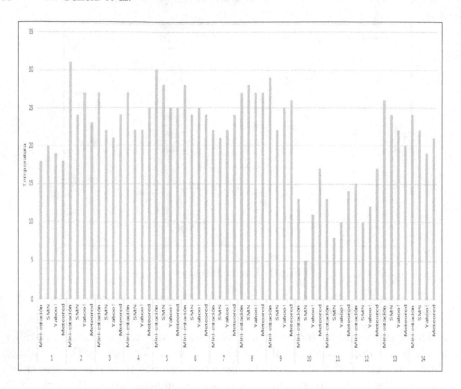

Fig. 2. Data comparison: *temperature*.

4.4 Classification Algorithms

The deep learning algorithm used was the *Multilayer perceptron*, available in WEKA as *Dl4jMlpClassifier* (see Algorithm 1). Among the deep learning algorithms, it is distinguished by being fast and efficient with a set of data of considerable dimensions, this is due to the fact that the hidden layers focus on few examples and some updates are made for each neuron [12]. In addition, it is simple to implement and, since it does not require any internal storage (such as derivatives, for example) or hyperbolic tangent calculation, it is computationally efficient [6].

The algorithm *Multilayer perceptron* was combined with the following assembled classifiers: *AdaBoostM1, AttributeSelectedClassifier, Bagging, ClassificationViaRegression, CostSensitiveClassifier, CVParameterSelection, FilteredClassifier, IterativeClassifierOptimizer, LogitBoost, MultiClassClassifier, MulticlassClassifierUpdateable, MultiScheme, RandomCommittee, RandomizableFilteredClassifier, RandomSubSpace, Stacking, Vote* y *WeighteInstancesHandlerWrapper*.

The best result, for the three sampling criteria, was generated with the combination: *MultiClassClassifier+Multilayer perceptron*.

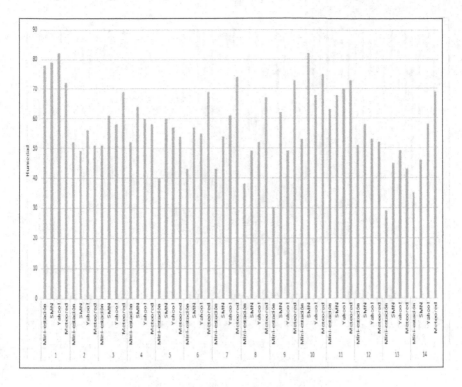

Fig. 3. Data comparison: *humidity*.

The assembled classifier *MultiClassClassifier* is useful for handling multi-class datasets with classifiers of two classes. This classifier is also capable of applying error correction output codes for greater accuracy [13]. The Algorithm 2 represents this assembled classifier.

4.5   Experiments

Each combination, in the three sampling criteria, was done with 10 iterations. The combinations resulted in different correct percentages of classification, see Table 2, where the highest results, for each sampling criterion were:

- 2/3-1/3: *Multilayer perceptron* and *MultiClassClassifier+Multilayer perceptron*, both with 99.82%.
- Cross validation: *Multilayer perceptron* with 99.82% and *MultiClassClassifier+Multilayer perceptron* with 99.85%.
- Representative sample: *MultiClassClassifier+Multilayer perceptron* with 99.74%.

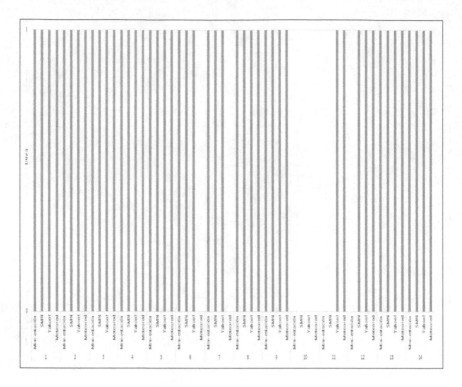

Fig. 4. Data comparison: *rain*.

4.6   Analysis of Results

For each sampling criterion, in relation to the classifications with the highest percentages, the confusion matrices were obtained, which are then presented from higher to lower percentage, for their respective analysis.

Confusion Matrix

- Representative sample: Implementing the algorithm *Multilayer perceptron* only, we observe its behavior in Table 3, where the correctly classified instances are shown, corresponding to the diagonal with the values 282, 37 and 59, which equals to 378 instances and equivalent to 99.74% obtained. Outside the diagonal, the values 0, 0, 1, 0, 0 and 0, are shown which indicate that only in the class *cloudy* there was a confusion, thus amounting to 0.26% of incorrect classification. It is important to note that for this case, a total of 378 instances were used in the set of tests, which is lower compared with the total number of instances used in cross validation and 2/3-1/3. It shows that with a minimum amount of historical information, the algorithm *Multilayer perceptron* classifies with greater precision, thus proving the representative sample to be a valid statistical measure for the calculation from a considerable population, as it is the case of our 27888 records in total.

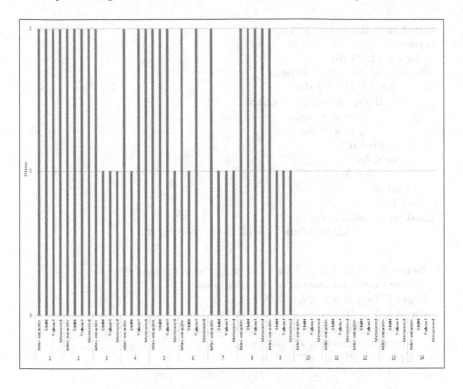

Fig. 5. Data comparison: *weather*.

- Cross validation: The results obtained by the implementation of the algorithm *Multilayer perceptron*, can be observed in Table 4, where the correctly classified instances are shown, which correspond to the diagonal with the values 20583, 2533 and 4723, thus adding up to 27839 instances and equals to 99.82%. Outside the diagonal, the values 21, 2, 26, 0, 0 and 0 are obtained, which indicate the confusions, in total add up to 49 instances and represent 0.18% of incorrect classification. On the other hand, by combining the algorithm *MultiClassClassifier+Multilayer perceptron*, we observe its behavior in Table 5 where the correctly classified instances are shown, and correspond to the diagonal with the values 20582, 2541 and 4723, which totals to 27846 instances and are equivalent to 99.85%. Outside the diagonal, the values 24, 0, 18, 0, 0 and 0, indicate the confusions, which adds up to 42 instances and represent the 0.15% of incorrect classification. We can observe, that the class *sunny* confuses 23 instances, while implementing the algorithm *Multilayer perceptron* without combination, that is, one less than those obtained while combining the classifiers. On the other hand, the cloud class is better classified by combining *MultiClassClassifier+Multilayer perceptron*, thereby confusing only 28 instances.
- 2/3-1/3: Table 6 shows the behavior on implementation of the algorithm *Multilayer perceptron*, where the correctly classified instances are shown, which

Input Initialize w and b to zero.
repeat
  for $l \in \{1...l\}$ **do**
    if $y_l(w \cdot x_l + b) \leq \beta$ **then**
      for $i \in \{1...N\}$ **do**
        if $\|v_i \cdot x_l + d_i\| \leq 1$ **then**
          $v_i \leftarrow v_i + \lambda y_l x_l$
          $d_i \leftarrow d_i + \lambda y_l$
        end if
      end for
      $b \leftarrow b + \lambda y_l$
    end if
  end for
until termination criterion

Algorithm 1: Multilayer perceptron.

Begin $Z = \{z_1, z_2, ..., z_n\}$ with $z_i = (x_i, y_i)$ as a training set.
M, the maximum number of classifications.
Begin $F^0(x_i) = 0$ y $p(x_i) = 1/2$, i=1,...,n
for $m = 1, ...M$; **do**
  $w_i = p(x_i)(1_p(x_i))$, $z_i = (y_i)/p(x_i) \ (1-p(x_i))$.
  Adjust the function $F^m(x)$; z_i to x_i
  Update $F^{(m)}(x_i) = F^{(m-1)}$ y
  $p(x_i) = exp(F^{(m)}(x_i)/(1 + exp(F^{(m)}(x_i))$
end for

Algorithm 2: MultiClassClassifier.

correspond to the diagonal with the values 6885, 839 and 1555, adding a total of 9279 instances and equivalent to 99.82%. The values 15, 0, 2, 0, 0 and 0, outside the diagonal indicates the confusions, which in total add up to 17 instances and represent the 0.18% of incorrect classification. The results of combining the algorithm *MultiClassClassifier+Multilayer perceptron*, can be observed in Table 7, where the correctly classified instances are shown, which correspond to the diagonal with the values 6884, 840 and 1555, and add up to a total of 9279 instances and equal to 99.82%. Outside the diagonal, the values 16, 0, 1, 0, 0 and 0, which indicate the confusions, add up to 17 instances in total and represent the 0.18% of incorrect classification. Although in both cases the percentages of correct classification are equal, the confusion matrix allows us to observe a better performance when implementing the algorithm *Multilayer perceptron* without combination, given that in the classes *sunny* and *cloudy*, the number of confusions is decreased.

Table 2. Results of the combination of algorithms.

Assembled classifier	MLP*	2/3 − 1/3	C. V.**	R. S.***
		99.82	**99.82**	99.47
AdaBoostM1			97.24	
AttributeSelectedClassifier		98.00	98.31	97.10
Bagging			73.89	74.41
ClassificationViaRegression		74.22	73.89	74.41
CostSensitiveClassifier				
CVParameterSelection		16.73%	73.89%	74.41%
FilteredClassifier		99.02%	99.22%	98.94%
IterativeClassifierOptimizer				
LogiBoost		M. Ins	M. Ins	M. Ins.
MultiClassClassifier		**99.82%**	**99.85%**	**99.74%**
MulticlassClassifierUpdateable		Error	Error	Error
MultiScheme				
RandomCommittee		M. Ins	M. Ins	74.41%
RandomizableFilteredClassifier		96.62%	94.80%	98.15%
RandomSubSpace		99.10%	98.96%	98.94%
Stacking				
Vote				
WeighteInstancesHandlerWrapper		16.73%	73.89%	74.41%

* MLP = Multilayer perceptron.
** V. C. = Cross validation.
*** M.R. = Representative sample.

Table 3. *MultiClassClassifier+Multilayer perceptron.*

a	b	c	Classified as
282	0	0	a = sunny
1	37	0	b = cloudy
0	0	59	c = cloudy

Table 4. *Multilayer perceptron.*

a	b	c	Classified as
20583	21	2	a = sunny
26	2533	0	b = cloudy
0	0	4723	c = rainy

Table 5. *MultiClassClassifier+Multilayer perceptron.*

a	b	c	Classified as
20582	24	0	a = sunny
18	2541	0	b = cloudy
0	0	4723	c = rainy

Table 6. *Multilayer perceptron.*

a	b	c	Classified as
6885	15	0	a = sunny
2	839	0	b = cloudy
0	0	1555	c = rainy

Table 7. *MultiClassClassifier+Multilayer perceptron.*

a	b	c	Classified as
6884	16	0	a = sunny
1	840	0	b = cloudy
0	0	1555	c = rainy

5 Conclusions and Future Work

Because of the real time, sensitivity of the sensors, and the capturing of records *insitu*; the measurements of the proposed device, have better precision for measure climatological parameters than the general-purpose forecasters. On the other hand, and as a core part of this research, we found that it is possible to implement a deep learning algorithm, based on a minimum of historical information on climatological variables, to forecast with high confidence the weather class: cloudy, rainy, sunny. In this study a highly competitive result were obtained with the proposed scheme. As future works of this project it is proposed:

- To add other sensors to the device for considering other variables, and thus have the option to study the behavior of the best ranked classifiers with specific variables, i.e., making a selection of relevant attributes by means of algorithms that have been established for data sets like those in this paper, namely: CFS, Chi Squared, Consistency, Information Gain and Symmetrical Uncertainty [7–11].
- Having identified the best classifiers, in this study, their implementation in a new set of data that considers the four seasons of the year. Currently this field work is in process.
- Design a support system for decision making (or *app*), considering comparative statistics useful to users in rural areas.

- create a network of several capture devices, strategically distributed to be able to implement a warning system before a possible abrupt climate change, in the *app*.
- The transfer of technology to the end user, that is, to the farmer of rural regions, as a final phase of this project, for the application perspective.

References

1. Flamengo, E., Rebella, C.M., Carballo, S., Rodrguez, R.O.: Methodology of seasonal rainfall forecast in Argentine regions. Argentina de agrometeorologa, pp. 133–141 (2002)
2. Cruz, E.: Desarrollo de una aplicacin para visualizacin de predicciones meteorolgicas en dispositivos mviles (2015)
3. Farfan, J.H.R., Barrantes, M.F.T.: Sistema prototipo de captura y anlisis de datos meteorolgicos, para generacin de alertas y reportes de pronostico climtico acertado en el municipio de Chiquinquira. Matices tecnolgicos 6 (2017)
4. Krishnamurthi, K., Thapa, S., Kothari, L., Prakash, A.: Arduino based weather monitoring system. Int. J. Eng. Res. Gen. Sci. **3**, 452–458 (2015)
5. Kedia, P.: Localised weather monitoring system. Int. J. Eng. Res. Gen. Sci. **4**, 315–322 (2016)
6. Collobert, R., Bengio, S.: Links between perceptrons, MLPs and SVMs. In: Proceedings of the Twenty-first International Conference on Machine Learning, pp. 23. ACM (2004)
7. Hernández-Torruco, J., Canul-Reich, J., Frausto-Solís, J., Méndez-Castillo, J.J.: Feature selection for better identification of subtypes of Guillain-Barr syndrome. Comput. Math. Methods Med. **2014**, 1–9 (2014)
8. Liu, H., Li, J., Wong, L.: A comparative study on feature selection and classification methods using gene expression profiles and proteomic patterns. Genome Inf. **13**, 51–60 (2002)
9. Liu, Y., Schumann, M.: Data mining feature selection for credit scoring models. J. Oper. Res. Soc. **56**(9), 1099–1108 (2005)
10. Pitt, E., Nayak, R.: The use of various data mining and feature selection methods in the analysis of a population survey dataset. In: Proceedings of the 2nd International Workshop on Integrating Artificial Intelligence and Data Mining, vol. 84, pp. 83–93. Australian Computer Society (2007)
11. Zheng, Z., Wu, X., Srihari, R.: Feature selection for text categorization on imbalanced data. ACM SIGKDD Explor. Newslett. **6**(1), 80–89 (2004)
12. Goodfellow, I., Bengio, Y., Courville, A.: Deep Learning. The MIT Press, Cambridge (2017)
13. Witten, I.H., Frank, E., Hall, M.A., Pal, C.J.: Data Mining: Practical Machine Learning Tools and Techniques. Morgan Kaufmannm, Burlington (2016)

Resource Management Strategy in Case of Disaster Based on Queuing Theory

Darin Mosquera[3], Edwin Rivas[3], and Luis Alejandro Arias[1,2(✉)]

[1] Universidad ECCI, Bogotá, Colombia
lincarias@yahoo.com
[2] Autónoma de Colombia, Bogotá, Colombia
[3] Faculty of Engineering, Universidad Distrital Francisco José de Caldas,
Bogotá, Colombia
jdmosquera@correo.udistrital.edu.co,
erivast@udistrital.edu.co

Abstract. In the present article, the main needs of collection centers and immediate care facilities in case of disasters are analyzed. A model is proposed for the services provided by these collection centers based on queuing theory, including an assessment of the arrival rates and service capacities, waiting times before being treated or receiving no healthcare service. A management algorithm is proposed that allows changes in real time of the system dynamics so it can adjust to queuing models with different features in order to carry out an effective help for system users. This reduces the service time and integral attention of the people affected by a disaster in favor of rapid recovery from a psychological and social point of view.

Keywords: Disaster attention centers · Resource management · Basic needs · Queuing theory · System dynamics

1 Introduction

The vulnerability of humankind in case of a natural disaster is undeniable. The impact in a society of such an event in terms of the human losses and financial costs associated with the destruction of the infrastructure is considerable in spite of the prevention and mitigation strategies proposed from a technological development standpoint (Laframboise and Loko 2012). Additionally, the management of the attention given to the survivors during the initial moments after the disaster implies a high degree of susceptibility of the population, which deserves a special check-up to avoid the escalation of risks and affectations even higher than those of the disaster itself (Estrat 2018). The immediate needs, especially after a disaster, are hard to assess due to their diversity and variability. Some of the basic needs of immediate care are: primary healthcare, psychological care, provision of food, clothes and shelter, information on missing relatives, review of civil work and potentially vulnerable infrastructure, management of contagious diseases, prevention of vandalism (Lucia and Leon 2014).

The scientific literary review on the disaster management subject show that there are different mechanisms and strategies aimed at streamlining the work of healthcare

M. Botto-Tobar et al. (Eds.): ICAT 2019, CCIS 1194, pp. 26–38, 2020.
https://doi.org/10.1007/978-3-030-42520-3_3

and relief agencies such as: rapid response teams, emergency military units, national coordination centers, special government units in combination with the national armies (Solinska-Nowak et al. 2018) (Kobayashi 2014) (Lucia and Leon 2014). Figure 1 shows some relevant references in terms of how some countries handle disasters. In this article, the management elements of disaster attention and collection centers (DACC) are proposed in order to handle the attention for survivors and other people affected by a natural disaster. The DACC should be managed so that the resources and overall attention are optimized in favor of providing a service in the shortest time possible and channel the resources towards the basic needs that may eventually require higher demand or priority within the community to be cared for.

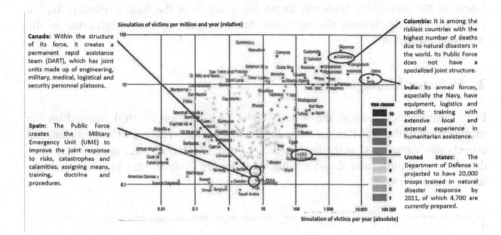

Fig. 1. Actions of the public force for disaster management on a global scale. Source: (Lucia and Leon 2014)

This work is part of the doctoral proposal called "Energy Resilient Management System in a disaster situation", which is under development by GCEM research groups (Electromagnetic Compatibility Group) and ORION (Telematics Research Group), attached to the Francisco José de Caldas District University Colombia. In the doctoral proposal one of the main motivations is to make an analysis of the real situations that cause natural phenomena due to their unpredictability and the randomness underlying when managing resources for your attention. When a phenomenon occurs and reaches a considerable force level, impacting a community, region or city, it becomes a natural disaster and affects the architecture of a city especially infrastructure Criticisms such as those related to energy and the management of food and health resources. Before said situation is necessary to activate, manage and execute plans for the administration of resources so dynamic in such a way as to maximize the provision of the most urgent services (health, food and energy among others).

This paper proposes a dynamic management of queue theory based on what community requests affected know and activate the management of resources tailored to the needs that go presenting. If in the course of time the demand for requests increases, the provision of the service will be adapted to improve the contingent situation.

The present article is organized as follows: general aspects of disaster management strategies used worldwide are described in Sect. 2, with a particular emphasis in the structural organization of Colombia. This section also includes the fundamental precepts of queuing theory and the applicability of resource coordination models against emergency situations. Section 3 describes the algorithm to be implemented for the dynamic management of changes in queue types for the service of emergency response as well as the parameters and variables to consider in the modelling process. Section 4 shows an illustrative case study of the current management strategy adopted in disaster response and collection centers in Colombia and the management strategy proposed based on the algorithm. Section 5 shows the analysis of the results. Finally, Sect. 6 gives a set of conclusions and recommendations regarding the potential use of the proposed algorithm.

2 Disaster Management

For global entities such as the World Health Organization (WHO) focused on the management of the activities for the reduction of disasters. There are three fundamental aspects that correspond to the three phases of the 'Disaster Cycle': disaster response, disaster preparation, and disaster mitigation. Figure 2 shows the proposed disaster management cycle.

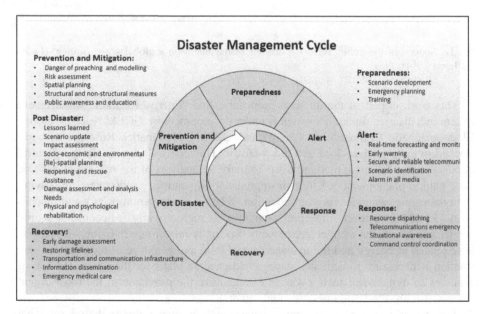

Fig. 2. Phases of the disaster cycle. Adapted from (Khan et al. 2010).

The phases concerning the response and recovery stages once the disaster has taken place are the focus of attention of this article. It specifically encompasses the management, coordination and dispatch of resources on both the human and the technical aspect, in order to respond to the basic needs of the affected population in an effective and integral manner in the shortest time possible.

The main idea is to assess the management of attention centers and their performance as well as measure the management both qualitative and quantitatively based on preestablished levels taken from desirable references. These are fixed by each government based on their experiences with disaster management (de Desarrollo 2015) (Herrera Cepeda 2018) (Estrat 2018).

3   Queuing Theory Applied to Disaster Management

Queuing theory has been widely used in the assessment and optimization of management process on an industrial level, in order to prioritize high-productivity indexes related with the opportune attention of the clients and users of a system (Singer et al. 2008). However, the applicability of queuing theory in scenarios other than the productive one, which is the case of disaster response and organized dispatch of resources, can be seen as a valuable tool focused on relevance objectives such as the preservation of life and the improvement of quality conditions for the surviving population of a disaster (Sandoval and Voss 2016). The basic structure of a queue system includes an input source, represented by the surviving people compared to their corresponding requests that seek to fulfill their basic needs (Varela 2010). These people enter the system, which is the DACC in this context, and join the queue. In certain moments, the members of the queue are chosen to offer the requested service in terms of a set of rules, known as the discipline of the queue. At this point, it is important to mention that the discipline of the queues for disaster management is always of the Dynamic type. Although the strategy naturally comes from a FCFS (First Come First Served) type, it is necessary to consider some basic needs such as the emergency attention for wounded people in grave state. When life is at imminent danger, the queue must migrate to a PQ (Priority Queuing) type (Asif and El-Alfy 2005).

After the selection, people are attended through a service mechanism that in case of a disaster focuses on supplying the most urgent needs of each individual, seeking the highest effectiveness in terms of the total coverage of the need and the reduction of the service time (Chatzigeorgiou and Stephanides 2006) (Lisboa et al. 2002). After the service has been provided, the user must leave the queue system. A subjacent specificity in the application of queuing theory in disaster management stems the multiplicity of requests of a single user in proportion to the different needs to be satisfied, which is why the service disciplines must be dynamic in the DAAC. A basic illustration of the queuing model can be seen in Fig. 3.

The global queuing model chosen for the DAAC is the M/M/s/k corresponding to a system with multiple servers (s > 1) where the service rate depends on the number of clients in the system. By breaking down the notation, the elements are now described:

M/M are the distributions of the arrival and service times that have been defined as the exponential type. s corresponds to the number of services provided by the DAAC.

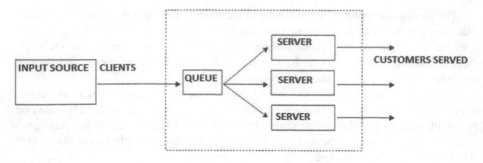

Fig. 3. Queuing system model

The parameter k corresponds to the capacity of the system. Although the contingency plans in case of emergencies stipulate in their protocols some standards regarding the number of people that can be attended in the healthcare centers, international experience shows that often the capacity of said places is overwhelmed. Hence, the optimization of resources and redirectioning to other centers are typical management strategies in said situations (Tapia et al. 2017).

Some parameters to keep in mind are:

λ corresponds to the arrival rate
μ corresponds to the service rate
$\rho = \frac{\lambda}{s\mu}$ is the overall utilization factor that corresponds to the proportion of expected occupation time of the servers

If $\rho < 1$ then the system is stabilized. Otherwise, the number of people in the system increases endlessly.

L is the expected value of people in the system
L_q is the expected number of people in queue
W is the average waiting time of the system
p_n is the probability that n people are in the system (in stationary state)

Each server has the resources needed to attend to the basic needs proposed at the moment. Then, the analysis performs each task (or basic need) in a differentiated manner, hence assessing the attention per task as if the queuing model had a single server: M/M/1. This could be migrated towards a queuing model M/M/s.

$\rho_i = \frac{\lambda_i}{\mu_i}$ is the utilization factor of each basic need that must be fulfilled.

4   Proposed Algorithm for Management of the Disaster Attention and Collection Center (DACC)

With the purpose of improving the attention in DACCs, a management model is proposed based on the queuing theory, by choosing an initial notation of the M/M/k type with multiple sources of request arrival (λ_i) and multiple servers with attention capacities (μ_i). For the DAAC, it is proposed to carry out an initial evaluation of the performance for the differential attention given to each type of basic need demanded by the users (survivors) that reach the DACC. Afterwards, the global performance of the evaluation is carried out for the entire DACC. The details of the algorithm phases are shown in Fig. 4.

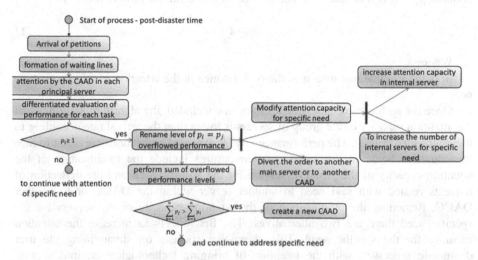

Fig. 4. Algorithm proposed for the management of the DACC (CAAD)

The management algorithm begins with the initial phase, sometime after the disaster takes place, when the DACC is installed. The disaster response services in different countries seek to make the installation time as short as possible, yet the variability of this time can change in terms of the location of the disaster, the state of the access roads, the preparation times of the staff and their level of preparation, among other factors (Capacci and Mangano 2014) (Cepeda 2018).

Once the DACC has been installed, the surviving population starts to arrive with the basic needs to be satisfied. The arrival rates λ_i are known by their stochasticity, causing them to be often characterized by probability density functions such as the Poisson function, Markov chains, Montecarlo and exponential probability density functions, among others ("Normal, Binomial, Poisson Distributions" 2014) (Zonouz 2012) (Jung et al. 2013).

The attention is provided by the teams of servers of the DACC, associated with a groups of specialized people and teams corresponding, for example, to doctors, nurses, paramedics, vein-focused crisis equipment, artificial respirators and others. The preparation of resources for the teams of servers may vary and it is understandable that for some types of needs, such as locating a missing relative or psychological care, the human resource may prevail over the technological resource (Farra et al. 2015) (Altay and Green III 2006). The attention capacity of the different server teams is characterized by the parameter μ_i.

Once the attention processes have begun, a periodic evaluation of the performance levels of each team of servers is proposed after each hour. The initial evaluation time is the first 12 h, which is considered as the critical time in terms of disaster management (Administration 2015). Such performance evaluation must be carried out in a differentiated manner for each basic need that is being handled in the DACC. Therefore, the variable $\rho_i = \frac{\lambda_i}{\mu_i}$ is included. The comparison is based on the formulation (1)

$$\rho_i \geq 1 \tag{1}$$

Where:

ρ_i is the variable that measures the performance in the attention of a basic specific need

Once the performance assessment has been concluded, the algorithm identifies that the attention capacity of the group of servers is lower than the arrival rate according to the established criteria. The performance level is renamed and actions are taken to face the situation. Some of the initial countermeasures include the modification of the attention capacity of the group of servers for the specific need and the deflection of requests related with said need to another server within the DACC or to another DACC. Regarding the modification of the attention capacity of the servers for the specific need there are two alternatives. The first one is to increase the attention capacity for the specific need. This alternative consists on streamlining the user diagnostic processes, with the intention of bringing technological equipment with lower response times, as well as personalized staff with more skill and experience to satisfy the specific need at hand. The other alternative increases the number of internal servers of a main server (s_{ji}) that meet the specific need whose performance level has been exceeded.

The proposal of decision management reviews the resources available in the DACC always seeking the best attention in time and quality for the affected population. The proposed management algorithm does a global assessment of the performance indicators that have been overflowed in each main server. In case of having several basic needs with levels above the established performance criteria, the call or creation of an additional main server in the DACC is recommended. A similar assessment is carried out for all servers and, in case their capacity is exceeded, then the creation of an additional DACC is proposed near the existing DACC.

5 Testing

Initially, the DACC behavior was tested by assuming that the M/M/s/k model has lower arrival rates than the attention capacity of the servers for a value k = 600 people in the queue of the DACC and s = 5 servers which is the estimated attention that the DACC would manage.

Table 1 shows the results obtained of the global system behavior and differentiated by the basic needs to satisfy, keeping a value μ = 10 per minute under different global arrival rates differentiated in one of the servers (server 1).

Table 1. Global behavior of the DACC in terms of attention and differentiated for each task for one of the servers (main server 1) in scenario 1: $\lambda = 40$, $s = 5$, $\mu = 10$

$\mu = 10$ people/ min, s=5 servers				
Arrival rates, λ (people/min)	Average number of clients in the system, L	Average number of clients in line (queue), Lq	Global use of the DACC servers, ρ	Average time within the system, W (sec)
40	7,0779	2,2165	0,8	0

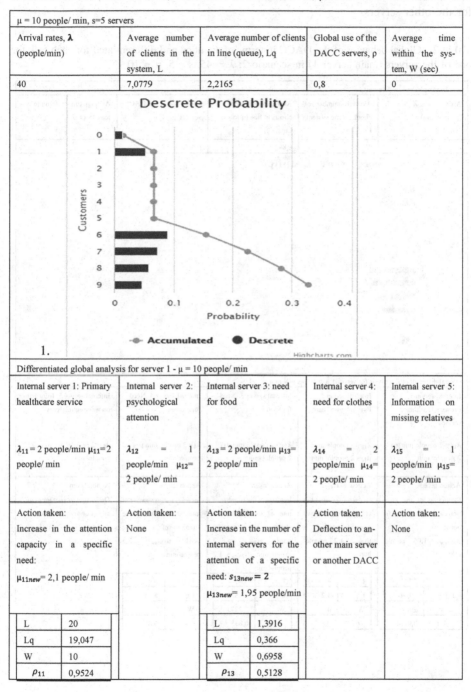

1.

Differentiated global analysis for server 1 - μ = 10 people/ min				
Internal server 1: Primary healthcare service	Internal server 2: psychological attention	Internal server 3: need for food	Internal server 4: need for clothes	Internal server 5: Information on missing relatives
λ_{11} = 2 people/min μ_{11}=2 people/ min	λ_{12} = 1 people/min μ_{12}= 2 people/ min	λ_{13}= 2 people/min μ_{13}= 2 people/ min	λ_{14} = 2 people/min μ_{14}= 2 people/ min	λ_{15} = 1 people/min μ_{15}= 2 people/ min
Action taken: Increase in the attention capacity in a specific need: μ_{11new}= 2,1 people/ min	Action taken: None	Action taken: Increase in the number of internal servers for the attention of a specific need: s_{13new} = 2 μ_{13new}= 1,95 people/min	Action taken: Deflection to another main server or another DACC	Action taken: None

L	20
Lq	19,047
W	10
ρ_{11}	0,9524

L	1,3916
Lq	0,366
W	0,6958
ρ_{13}	0,5128

Tables 2 and 3 show the results for global increases of the arrival rates and the actions taken during the internal management of server 1. Similar tests were carried out for the other servers.

Table 2. Global behavior of the DACC in terms of attention and differentiated for each task for one of the servers (main server 1) in scenario 2: $\lambda = 45$, $s = 5$, $\mu = 10$

$\mu = 10$ people/ min, s=5 servers				
Arrival rates, λ (people/min)	Average number of clients in the system, L	Average number of clients in line (queue), Lq	Global use of the DACC servers, ρ	Average time within the system, W (sec)
45	11,777	6,8624	0,9	0

Descrete Probability

Global differentiated analysis for server 1 $\mu = 10$ people/ min				
Internal server 1: Primary Healthcare service	Internal server 2: Psychological care	Internal server 3: Need for food	Internal server 4: Need for clothes	Internal server 5: Information on missing relatives
$\lambda_{11} = 1$ people/min $\mu_{11} = 2$ people/ min	$\lambda_{12} = 1$ people/min $\mu_{12} = 2$ people/ min	$\lambda_{13} = 2$ people/min $\mu_{13} = 2$ people/ min	$\lambda_4 = 3$ people/min $\mu_{14} = 2$ people/ min	$\lambda_5 = 2$ people/min $\mu_{15} = 2$ people/ min
Action taken: Reduction of the attention capacity in a specific need: $\mu_{11new} = 1{,}05$ people/ min	Action taken: Reduction of the attention capacity in a specific need: $\mu_{12new} = 1{,}25$ people/ min	Action taken: Increase in the number of internal servers for the attention of a specific need: $s_{13new} = 2$ people/min $\mu_{13new} = 1{,}2$ people/ min	Action taken: Increase in the number of internal servers for the attention of a specific need: $\mu_{14new} = 3{,}5$ people/min	Action taken: Deflection to another main server or another DACC

L	20		L	4		L	5,45		L	6
Lq	19,04		Lq	3,2		Lq	3,7879		Lq	5,1429
W	20		W	4		W	2,7273		W	2
ρ_{11}	0,9524		ρ_{12}	0,8		ρ_{13}	0,833		ρ_{14}	0,8571

Table 3. Global behavior of the DACC in terms of attention and differentiated for each task for one of the servers (main server 1) in scenario 3: $\lambda = 49,9$, s = 5, $\mu = 10$

$\mu = 10$ people/ min, s=5 servers				
Arrival rates, λ (people/min)	Average number of clients in the system, L	Average number of clients in line (queue), Lq	Global use of the DACC servers, ρ	Average time within the system, W (sec)
9,9	42,87	37,87	0,998	0

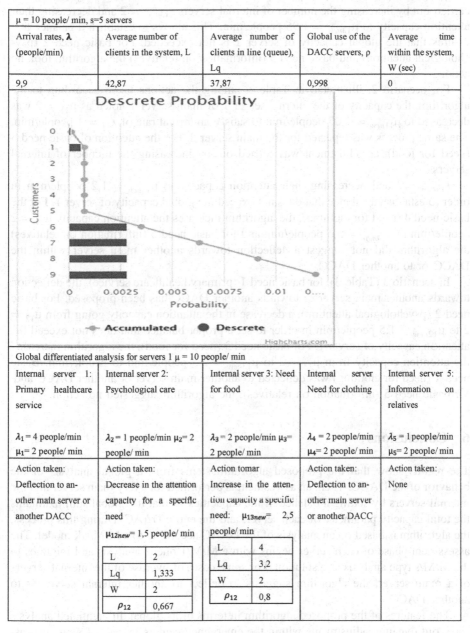

Global differentiated analysis for servers 1 $\mu = 10$ people/ min				
Internal server 1: Primary healthcare service	Internal server 2: Psychological care	Internal server 3: Need for food	Internal server Need for clothing	Internal server 5: Information on relatives
$\lambda_1 = 4$ people/min $\mu_1 = 2$ people/ min	$\lambda_2 = 1$ people/min $\mu_2 = 2$ people/ min	$\lambda_3 = 2$ people/min $\mu_3 = 2$ people/ min	$\lambda_4 = 2$ people/min $\mu_4 = 2$ people/ min	$\lambda_5 = 1$ people/min $\mu_5 = 2$ people/ min
Action taken: Deflection to another main server or another DACC	Action taken: Decrease in the attention capacity for a specific need $\mu_{12new} = 1,5$ people/ min	Action tomar Increase in the attention capacity a specific need: $\mu_{13new} = 2,5$ people/ min	Action taken: Deflection to another main server or another DACC	Action taken: None

For internal server 2:

L	2
Lq	1,333
W	2
ρ_{12}	0,667

For internal server 3:

L	4
Lq	3,2
W	2
ρ_{12}	0,8

The analysis of the results in Table 1 show that, among the actions taken according to the algorithm, the internal capacity of server 1 in main server 1, noted as $\mu_{11} = 2$

people/min was increased to $\mu_{11new} = 2,1$ people/min in order to satisfy an arrival rate of $\lambda_{11} = 2$ people/min. To handle the basic need 3 (need for food) an adjustment was carried out by increasing the number of internal servers $s_{13new} = 2$ and decreasing their attention capacity to $\mu_{13new} = 1,95$ people/min. This allows to establish a balance that assures that the global capacity of server 1 is not exceeded. For basic need 2 (psychological attention) and basic need 5 (information on relatives) the algorithm took no action.

For scenario 2, illustrated in Table 2, among the actions taken according to the algorithm, the capacity of the internal server 1 of main server 1 noted as $\mu_{11} = 2$ was decreased to $\mu_{11new} = 1,05$ people/min to satisfy an arrival rate of $\lambda_{11} = 1$ people/min. The same process was repeated for the main server 2. For the attention of basic need 3 (need for food), an adjustment was carried out by increasing the number of internal servers

$s_{13new} = 2$ and decreasing their attention capacity to $\mu_{13new} = 1,2$ people/min in order to establish a balance that does not exceed the global capacity of server 1. For the basic need 4 (need for clothing), the algorithm increases the attention capacity $\mu_{14} = 2$ people/min to $\mu_{14new} = 3,5$ people/min and for basic need 5 (information on relatives) the algorithm did not suggest a deflection towards another main server within the DACC or to another DACC.

In scenario 3 (Table 3), for basic need 1 (primary healthcare service), the deflection towards another main server or towards another DACC has been proposed. For basic need 2 (psychological attention), a decrease in the attention capacity going from $\mu_{12} = 2$ to $\mu_{12new} = 1,5$ people/min in order to set a proper balance that does not exceed the attention capacity of server 1. For basic need 3 (need for food), the algorithm increased the attention capacity from $\mu_{13} = 2$ to $\mu_{13new} = 2,5$ people/min to people/min. Basic need 4 (need for clothing) was deflected to another main server or another DACC and, for basic need 5 (information on relatives), the algorithm suggested no action.

6 Conclusions

The results show that the proposed algorithm stems from a global analysis of the behavior of the DACC. Then, an assessment is performed for every main server and the internal servers by setting a balance of the capacities so that the global sum maintains the total capacity parameter of each server, and the entire DAAC. During this process, the algorithm is based on an analysis of the DAAC according to a M/M/s/k model. The assessment phase of each server begins with a M/M/1 queue analysis and migrates to the M/M/s type analysis as a solution. For some cases of overuse of the internal servers of a main server, the algorithm suggests to deflect to another internal server or to another DACC.

The features of the proposed algorithm in terms of its global differentiated analysis carry out dynamic adjustments within the emergency centers in case of natural disasters, where the adequate management of existing resources is prioritized in order to efficiently cover the basic needs of the population.

In a disaster situation, it must be taken into account that the demand versus the service time is so great that it is necessary to increase the speed of response and adjust the dynamics of the way services are provided in order to ensure a timely response. In this sense, the present work proposes a dynamic way to meet the needs, improving the attention and service of each of the people affected.

References

U.S. Fire Administration: Operational lessons learned in disaster response, June 2015

Altay, N., Green III, W.G.: Interfaces with other disciplines: OR/MS research in disaster operations management. Eur. J. Oper. Res. **175**, 475–493 (2006). https://doi.org/10.1016/j.ejor.2005.05.016

Asif, H., El-Alfy, E.-S.: Performance evaluation of queuing disciplines for multi-class traffic using OPNET simulator. In: 7th WSEAS International Conference on Mathematical Methods and Computational Techniques in Electrical Engineering, Sofia, 27–29 October 2005, pp. 1–6 (2005)

Capacci, A., Mangano, S.: As catástrofes naturais. Rev. Colombiana de Geografia **24**(2), 35–51 (2014). https://doi.org/10.15446/rcdg.v24n2.50206

Herrera Cepeda, G.: Gestión del riesgo y atención de desastres con Profesionales Oficiales de la Reserva del Ejército colombiano. Revista Científica General José María Córdova **16**(22), 1–20 (2018). ISSN 1900-6586

Chatzigeorgiou, A., Stephanides, G.: Evaluation of a queuing theory and systems modeling course based on UML (2006). https://doi.org/10.1109/ICALT.2006.1652486

Estrat, S.: Informe de riesgos mundiales 2018. 13. a edición (2018)

Farra, S.L., Miller, E.T., Hodgson, E.: Virtual reality disaster training: translation to practice. Nurse Educ. Pract. **15**(1), 53–57 (2015). https://doi.org/10.1016/j.nepr.2013.08.017

de Desarrollo, B.I.: Indicadores de Riesgo de Desastre y de Gestión de Riesgos (2015)

Jung, J., et al.: Monte Carlo analysis of plug-in hybrid vehicles and distributed energy resource growth with residential energy storage in Michigan. Appl. Energy **108**, 218–235 (2013). https://doi.org/10.1016/j.apenergy.2013.03.033

Khan, H., Vasilescu, L.G., Khan, A.: Disaster management cycle – a theoretical approach. University of Science and Technology, Pakistan (2010)

Kobayashi, M.: Experience of infrastructure damage caused by the Great East Japan Earthquake and countermeasures against future disasters. IEEE Commun. Mag. **52**(3), 23–29 (2014). https://doi.org/10.1109/MCOM.2014.6766080

Laframboise, N., Loko, B.: Natural disasters: mitigating impact, managing risks. International Monetary Fund (2012)

Lisboa, C., Santana, M., Carlucci, R., Lankalapalli, N., De Carvalho, S.: Performance evaluation of complex systems: queuing statecharts approach. In: Proceedings of I Workshop on Performance of Computational Systems (WPerformance)/Brazilian Computing Conference (SBC), Florianòpolis, SC, Brazil, July 2002 (2002)

Lucia, E., Leon, G.: Criminalidad Derivada De Desastres Naturales: Propuesta Para La Generación De Políticas Públicas Ervyn Norza Céspedes, Elba Lucia Granados Leon, Vanesa Sarmiento Dussán, Dayana Fonseca Hernández, Giovanni Torres Guman (2014)

Normal, binomial, poisson distributions (2014)

Sandoval, V., Voss, M.: Disaster governance and vulnerability: the case of Chile, December 2016. https://doi.org/10.17645/pag.v4i4.743

Singer, M., Donoso, P., Scheller-Wolf, A.: Una introducción a la teoría de colas aplicada a la gestión de servicios. Rev. ABANTE **11**(2), 93–120 (2008)

Solinska-Nowak, A., et al.: An overview of serious games for disaster risk management – prospects and limitations for informing actions to arrest increasing risk. Int. J. Disaster Risk Reduct. **31**, 1013–1029 (2018). https://doi.org/10.1016/j.ijdrr.2018.09.001

Tapia, E., Reddy, E., Oros, L.: Retos e incertidumbres en la predicción y prevención del riesgo sísmico. Rev. de Ingeniería Sismica **96**, 66–87 (2017)

Varela, J.E.: 2 Teoría de colas o líneas de espera, pp. 17–33 (2010)

Zonouz, S.: A fuzzy Markov model for scalable reliability analysis of advanced metering infrastructure. In: Innovative Smart Grid Technologies (ISGT), pp. 1–5 (2012). https://doi.org/10.1109/ISGT.2012.6175770

Application of Techniques Based on Artificial Intelligence for Predicting the Consumption of Drugs and Substances. A Systematic Mapping Review

Pablo Torres-Carrión$^{(\boxtimes)}$ ⓘ, Ruth Reátegui, Priscila Valdiviezo, Byron Bustamante, and Silvia Vaca

Universidad Técnica Particular de Loja, San Cayetano Alto S/N, Loja, Ecuador
{pvtorres, rmreategui, pmvaldiviezo, bfbustamante, slvaca}@utpl.edu.ec

Abstract. The consumption of alcohol, drugs and substances constitutes one of the most serious public health problems worldwide, in particular due to the social consequences related to violence, abandonment of studies, family disintegration and the great problem of drug trafficking that is strengthened at the increase consumption. In addition, Artificial Intelligence is consolidating as an area of interdisciplinary knowledge, helping with the agile relationship of variables and indicators, which facilitates the discovery of behavioral patterns of human and material entities. From this perspective, the present investigation is proposed, which details universal indicators regarding the research carried out in the conjunction of these areas of knowledge. A systematic search has been carried out of scientific articles of journals indexed in the Scopus and WoS databases. There is an exhibition organized by subareas, countries, researchers, problems raised, regarding scientific production to facilitate possible future research in these areas.

Keywords: Consumption of drugs · Artificial intelligence · Systematic mapping review · Prediction models

1 Introduction

The consumption of alcohol, drugs and substances constitutes one of the most serious public health problems worldwide, in particular due to the social consequences related to violence, abandonment of studies, family disintegration and the great problem of drug trafficking that is strengthened at the increase consumption [1]. In Latin American countries, studies have reported a prevalence of drug use ranging from 65% in Colombia to 18.8% in El Salvador [2]. As the statistics indicate, alcohol continues to position itself as the drug of choice for university students, using it more frequently and intensely than the rest of the other drugs combined, resulting in more than 200 health conditions, among diseases and disorders, which are associated with excessive alcohol consumption [3]. For its part, artificial intelligence (AI) has positioned itself in the

---

The original version of this chapter was revised: the names and sequence of the authors have been corrected as "Pablo Torres-Carrión, Ruth Reátegui, Priscila Valdiviezo, Byron Bustamante and Silvia Vaca". The correction to this chapter is available at https://doi.org/10.1007/978-3-030-42520-3_52

© Springer Nature Switzerland AG 2020, corrected publication 2020
M. Botto-Tobar et al. (Eds.): ICAT 2019, CCIS 1194, pp. 39–52, 2020.
https://doi.org/10.1007/978-3-030-42520-3_4

academy as a cross-cutting science to enhance the development of other sciences [4], being medicine and psychology part of them. With this precedent, and considering that drugs, alcohol and other substances constitute a social problem, and it has been proposed as an objective to know from the universality of science, the investigations that have been carried out in this field, and with the application of artificial intelligence techniques, which include data mining and machine learning.

1.1 Theoretical Context of Fields of Knowledge

Scientific literature has been reported that psychosocial variables such as perceived stress [5], depression [6], loneliness [7], psychological inflexibility [8], impulsivity [9], personality [10] and life satisfaction [11] are good predictors of the onset, maintenance and relapse of alcohol [12, 13], tobacco [14] and other chemical substances [15, 16]. The psychosocial variables, such as mental health and substance use variables are briefly exposed through referential psychological tests.

Artificial Intelligence Techniques
Artificial intelligence (AI) optimizes the processing and analysis of large amounts of information from emerging algorithms that detect patterns that allow predictive models to be established. Previous studies have already used AI in the health field [16, 17] to predict the desirable characteristics of new drugs [18] or the response to certain drugs [19]. In the context of the drug problem, AI offers tools for the analysis of large volumes of information associated with consumption, including sociodemographic, psychosocial and health variables, which are complexly related to each other. Unlike classical statistical analyzes or models, artificial intelligence allows us to generate prediction models, in our case for the problematic or controlled consumption of any drug, based on the autonomous, dynamic, continuous and incremental processing of the available data, in our case based on information on drug use, sociodemographic variables, and health, using techniques for the selection, processing and analysis of data known as "data mining" [20]. The analysis of sociodemographic, psychosocial and health variables associated with drug use through AI, allows the detection of patterns associated with drug abuse or controlled substance use; from these, it is possible to design more effective prevention programs and social policies to address such consumption, as well as more efficient in terms of cost/benefit analysis, since the predictive models themselves are updated with the new information available.

In this investigation several AI techniques and algorithms are considered. Some concerning Big Data, which include Data Structures, Data Analytics and methods such as clustering, Association Rules, Regression, Classification, Time Series Analysis and Text Analysis; and tools like as MapReduce, Hadoop, Data Base Analytic, End-game, among others [21]. In the Machine Learning (ML) field, we have selected: supervised and unsupervised learning, Batch and Online, Instance-Based Learning and Model-Based Learning. The main algorithms, classified as supervised, are reviewed: k-Nearest Neighbors, Linear Regression, Logistic Regression, Support Vector Machines (SVMs), Decision Trees and Random Forests, Neural networks; and no supervised: Clustering, such as k-Means, Hierarchical Cluster Analysis (HCA), Expectation Maximization; Visualization and dimensionality reduction, such as Principal Component Analysis

(PCA), Kernel PCA, Locally-Linear Embedding (LLE), t-distributed Stochastic Neighbor Embedding (t-SNE); and of Association rule learning, as they are Apriori and Eclat [22]. These two areas of artificial intelligence summarize the main techniques and algorithms used for the design and production of intelligent computer models and solutions.

1.2 Previous Literature Reviews

In the search process, applying the Torres-Carrion method [23], we have selected the investigations of the types of literature review, survey or meta-analysis that have similar objectives or with a high degree of relation to those exposed in this proposal. or the similarity, in addition to the search results according to the script obtained, the research questions that are presented in the methodology section, and the problem presented in the initial part of the introduction are considered. All research is more than five years old and does not answer the research questions proposed in this study (Table 1).

Table 1. Related reviews

Article	Analysis	#Papers
Study protocol for a systematic review of evidence for digital interventions for comorbid excessive drinking and depression in community-dwelling populations [24]	Manuscript focuses on protocol to alcohol consumption and depression as a consequence, taking as its premise the effectiveness of digital interventions to reduce excessive consumption. All documents published until 2019 are searched in the MEDLINE, The Coch-rane Library, CENTRAL, CINAHL, PsycINFO, ERIC and SCI databases	Does not specify
Applications of machine learning in addiction studies: a systematic review [25]	This systematic review se aplica en aplicaciones de ML methods in general addiction research. Se detalla artículos que refieren a drugs addiction (N = 14, 82.4%), que incluyen: smoking (N = 4), alcohol drinking (N = 3), cocaine (N = 4), opioid (N = 1), and multiple substances (N = 2); other addiction are gambling (N = 2) and internet gaming (N = 1). Se busca en las bases de datos MEDLINE, EMBASE and the Cochrane Database	17
Automatable algorithms to identify nonmedical opioid use using electronic data: a systematic review [26]	In the manuscript, they consider the utility, the validation attempts and the application of algorithms to detect the use of non-medical opioids. Indexed articles are searched in PubMed and Embase. Of the 15 resulting algorithms, 11 were developed via regression modeling, and 4 used natural language processing, data mining, audit analysis, or factor analysis	15

Section two details the applied methodology, research questions, inclusion criteria and semantic structure of the search. In the next section, the results of each research question are presented. Finally, in the last section, conclusions are exposed.

2   Methodology

For the systematic search, the methodology of Torres-Carrion [23] is applied, which divides the process into three phases: Planning, Conducting and Reporting the review. The third part is complemented from [27].

2.1   Research Questions

Six research questions are proposed, adapted from [27] to these areas of knowledge. Table 2 details the variables that allow detailed answers.

Table 2.   Research questions

Question	Indicators
RQ1: How many studies are in the WOS and Scopus databases from 2015 to 2019?	• Number of documents in Scopus • Number of documents in WoS • Number of duplicated documents • Number of open access documents • Quartile of documents
RQ2: Who are the authors of the most cited articles?	• Most cited authors • Most cited articles • Relationship between co-authors • Association between authors and citation
RQ3: What is the geographical distribution of the authors?	• Countries where the authors are from. • Cooperation relationship for research between countries
RQ4: What are the journals with more publications on this line of research?	• List of Journals with more than two publications • Organization of journal by quartile
RQ5: In what contexts are these studies developed?	• Psychosocial factors • Mental health and substance use variables • Artificial intelligence techniques
RQ6: What are the main topics addressed in this line of research?	Cluster of main topics

2.2   Inclusion, Exclusion and Quality Criteria

The criteria for filtering the documents in the search are detailed in Table 3. It details those characteristics that could generate noise in the results, or that due to their publication time have lost relevance. The quality criteria require the expertise of the reviewer when making a prior reading of the abstract of the document, and if necessary a revision of the document to corroborate their contribution to the study.

Table 3. Inclusion and exclusion criteria

Criteria	Inclusion	Exclusion
Theoretical field	• Drugs, tobacco and other substances consumer • Artificial Intelligence (ML, BD)	• Studies analyzed with other statistical or Psychological methodology • Other interaction technologies
Databases	Web of Science (WOS) o Scopus	Google Scholar and other index files in WoS o Scopus
Document type	Article, review, book chapters, conference paper	Speech documents, editorial, Note, ESCI or other
Year	2015–2019	Before 2015 or in press
Area	Computer Science, Psychology	

2.3 Semantic Search Structure

This process is the basis for the proper selection of articles according to the areas of knowledge, and following the methodology [23], the "mentefacto conceptual" is applied for the organization of the area of knowledge, its classification, the discrimination of related areas and detail and characterization of the central concept, which in our case is drug use; This concept is complementary to the search objectives, artificial intelligence. As shown in Table 4, from the thesaurus of synonyms, the search script is generated using conjunction and disjunction operators, as well as word proximity operators (W/n and $NEAR/n$). The script is organized in four levels: in the first level the consumption of drugs, tobacco, alcohol and substances is exposed; the second level corresponds to artificial intelligence and its techniques; the third level corresponds to the union of two research areas (L1 + L2); in the fourth level applies revision process with the quality, inclusion and exclusion protocols; and the fifth level to the combination of the two databases (WoS and Scopus), with 21 repeated articles.

Table 4. Semantic search structure

Level	Thesaurus	SCOPUS script	Scopus	WOS
L1	**Consumption of drugs, tobacco, alcohol and substances**			
	• drug • tobacco • alcohol • substance **use**, consumer, addiction, dependence, abuse	((**drug** OR narcotic OR opiate OR dope OR cocaine OR stuff) OR (**tobacco** OR cigarette OR smoke OR cigar) OR (**alcohol** OR liquor OR beer OR wine OR cocktail) OR **substance**) W/4 (**use** OR consum* OR addicti* OR dependence OR abuse)	395,867	440,629

(*continued*)

Table 4. (*continued*)

Level	Thesaurus	SCOPUS script	Scopus	WOS
L2	**Artificial Intelligence**			
	• Artificial intelligence • Machine learning • Data mining • Natural Language Processing • Fuzzy Logic • Expert System • Neural Networks	(**ai** OR ((artificial* OR bot OR robot OR machine OR simulat* OR agent OR predict* OR algorithm OR system OR sensor OR comput*) W/2 (intelligen* OR smart* OR able* OR skill OR intellect OR perception)) OR ((**machine** OR deep OR reinforcement OR supervised OR unsupervised) W/2 **learn***) OR ((**data** OR text OR video OR imag*) W/2 **mining**) OR (natural W/2 languaje W/2 processing) OR (fuzzy W/2 logic) OR (expert W/2 system) OR (neural W/2 networks))	1,571,963	420,189
L3	L1 AND L2		2,158	3.535
L4	**Review Protocol**			
	(Last 5 years) from 2015		994	1,813
	(Research Areas) Computer Science, Psychology		305	279
	Article, book chapters, conference paper		277	270
	Journal		144	270
	Quality criteria		113	135
L5	Combination of results in Scopus and WOS (**repeated = 21**)			**227**

3   Results

As a result of the systematic search, 113 documents were obtained from Scopus and 135 from WoS; 21 documents are indexed in both databases, leaving a result of 227 documents, which are input to answer the six research questions. The list of documents is available for review on the web and accessed from the link: http://bit.ly/MR_Drugs_AI.

3.1   RQ1: How Many Studies are in the WOS and Scopus Databases from 2015 to 2019?

41% of the resulting documents are indexed in Scopus, 50% in WoS and 9% in both databases. 70% of the documents are in journals of the first quartile, and 23% in the second quartile. It is important to remember that only articles published in the journal have been considered for this study, as specified in the inclusion and exclusion criteria;

this is intended to analyze only mature research, with relevant results. The detail of these results is explained in Table 4 of the semantic search structure (Fig 1).

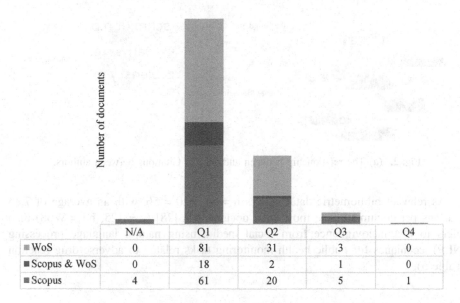

	N/A	Q1	Q2	Q3	Q4
■ WoS	0	81	31	3	0
■ Scopus & WoS	0	18	2	1	0
■ Scopus	4	61	20	5	1

Fig. 1. Articles by data base and quartile

3.2 RQ2: What are the Papers with More Cites?

The documents have been written by a total of 520 authors, with at least 20 authors listed in 2 articles, and 3 authors in a maximum of 5 articles. In Fig. 2(a). the relationship between authors (co-authorship) is explained, with Barlow (documents d = 4) and Novins (d = 2) in the first cluster, and Goklish (d = 3), Lanzerlere-Hinton (d = 2), Suttle (d = 2) and others in the second cluster; to establish the relationship, the standardization method of Association strength in VOSviewer is applied. In Fig. 2(b). 2 clusters are shown, where size represents the number of citations, and links the relationship in citations between authors; Schmid (d = 6, citations c = 62), Farris (d = 5, c = 60) and Zvolensky (d = 5, c = 60), which belong to the first cluster stand out. This explains the relationship between co-authorship, number of documents published and the number of citations, grouped into clusters. The two figures are not directly related, considering different parameters.

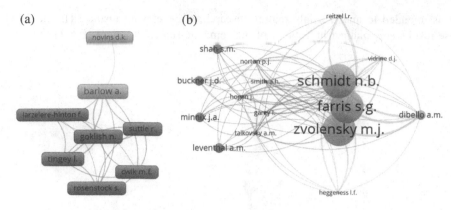

Fig. 2. (a) The relationship between authors. (b) Citations between authors.

As relevant bibliometric data, indices h = 21, h10 = 56, with an average of 7.85 citations per document. The most cited document is [28] (c = 125, bd = WoS) than refers to Pharmacovigilance from social media, using natural language processing (NLP) techniques for public health monitoring tasks related to adverse drug reaction (Table 5).

Table 5. Documents and citations

Number of cites	Papers
>100	217, 220
51–100	198, 206
41–50	205, 132, 168, 225
31–40	210
21–30	164, 177, 226, 215, 130, 152, 181, 175, 171, 172, 207, 218, 221, 224
11–20	125, 195, 199, 203, 103, 188, 196, 157, 202, 59, 96, 114, 211, 227, 113, 155, 185, 74, 106, 151, 77, 85, 118, 121, 176, 209, 81, 166, 219
6–10	75, 216, 187, 119, 129, 134, 141, 149, 165, 208, 212, 38, 41, 91, 98, 102, 123, 133, 162, 180, 60, 92, 112, 136, 40, 156, 179
5	33, 65, 69, 95, 117, 135, 139, 146, 163, 178, 182, 183, 200, 214
4	58, 72, 84, 87, 99, 120, 142, 154, 159, 161, 170, 173, 190
3	61, 67, 71, 137, 147, 150, 192, 197
2	18, 25, 31, 43, 78, 83, 86, 88, 90, 110, 127, 128, 138, 167, 184, 189, 193, 204, 223
1	3, 6, 7, 17, 27, 28, 40, 49, 51, 54, 62, 64, 73, 82, 94, 100, 105, 108, 109, 116, 124, 126, 143, 144, 153, 160, 169, 174, 186, 191, 194, 213
0	60 remaining articles

3.3   RQ3. What is the Geographical Distribution of the Authors?

For each of the 27 countries, the total strength of the co-authorship links with other countries was selected and organized in a cluster according to the relationship between co-authors from various countries. In the Fig. 3(a) are visible three clusters: The first (green) cluster is led by United Kingdom, and complemented by India, Colombia and France; the second (blue) led by Spain, with collaborations from Chile, Israel and Italy; and finally the largest group is in the third cluster (red), led by the USA, followed by China, South Korea, Germany, Cyprus, Malaysia and Finland; the third group does not include any Latin American country.

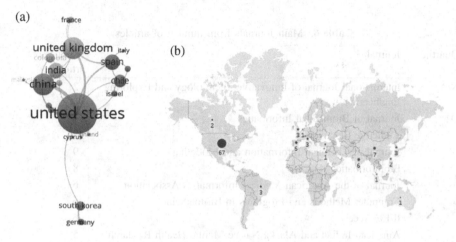

Fig. 3. (a) Association between authors country. (b) Documents by country (Color figure online)

For the geographical distribution of the authors, the first author of the publication was considered. United States is the country with the highest number of publications (59%), followed by China (6.3%), India (5.2%) and Spain (4.4%). Brazil (2.7%) is the only South American country on the map, in contrast to what is shown in Fig. 3(a) where several South American countries appear, but as co-authors, not as the first author.

3.4   RQ4: What are the Journals with More Publications on This Line of Research?

All journals with at least three scientific articles have been considered. The list according to quartiles has been made with the data of Scopus, available in Scimago Journal & Country Rank[1]. There are eighteen journals of first quartile with at least 3 documents, and six of second quartile. The Journal of Biomedical Informatics (h = 83;

---

[1] https://www.scimagojr.com/ (revised in September 2019).

SJR = 1.2; IF = 2.95; total cites = 7.431) has the highest number of documents on the subject (n = 17; 7.4%); As detailed in its web portal, this journal is open access, from the Elsevier publishing house; and, publishes articles motivated by applications in the biomedical sciences, related to new methodologies and techniques that have general applicability and that form the basis for the evolving science of biomedical informatics. Other important journals are Addictive Behaviors (h = 114; SJR = 1.42; IF = 2.963; total cites = 12.833) and Journal of Chemical Information and Modeling (h = 142; SJR = 1.45; IF = 3.966; total cites = 16.352), both with a complementary focus on computational methods, related to the scope of this study, on machine learning and data mining techniques (Table 6).

Table 6. Main Journals from number of articles

Quartile	Journal	Number of articles
N/A	International Journal of Innovative Technology and Exploring Engineering	2
Q1	Journal of Biomedical Informatics	17
	Addictive Behaviors	12
	Journal of Chemical Information and Modeling	9
	Bioinformatics	8
	Journal of the American Medical Informatics Association	6
	Computer Methods and Programs in Biomedicine	6
	IEEE Access	6
	American Indian and Alaska Native Mental Health Research	5
	BMC Bioinformatics	5
	Journal of Substance Abuse Treatment	5
	Journal of Affective Disorders	5
	International Journal of Medical Informatics	4
	Neurocomputing	4
	American Journal of Drug and Alcohol Abuse	3
	Artificial Intelligence in Medicine	3
	Expert Systems with Applications	3
	Frontiers in Human Neuroscience	3
	Journal of Cheminformatics	3
Q2	Molecular Informatics	5
	Journal of Bioinformatics and Computational Biology	4
	Journal of Medical Systems	4
	Computers in Biology and Medicine	3
	Journal of Computational Biology	3
	Journal of Computer-Aided Molecular Design	3

3.5   RQ5: In What Contexts are These Studies Developed?

In the context of the investigation, three main areas of knowledge are identified: Artificial Intelligence Techniques (blue), Tobacco, drugs and other substances (orange); and psychological profile of the consumer (red). In the first group, Machine Learning (ML) and Data Mining stand out, and as algorithm types: classification and prediction; Natural language processing is shown thanks to sustained studies in the analysis of medical documents. The consumption of alcohol and tobacco has been mainly investigated, with less relevance in other drugs and substances. As for the consumer profile, there is no greater evidence to study the topics presented in Sect. 1.1 of theoretical context, except for Psychological inflexibility (Fig 4).

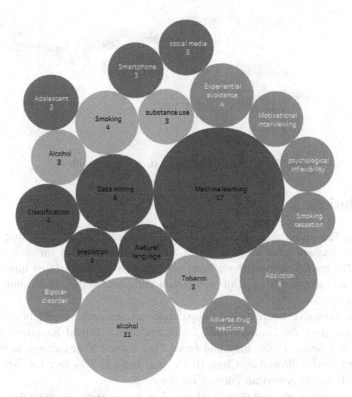

Fig. 4. Frequency of articles by research context (Color figure online)

3.6   RQ6: What are the Main Topics Addressed in This Line of Research?

A total of 438 keywords are obtained, of which 47 are repeated at least 1 time. In the Fig. 5 the total strength of the co-occurrence links with other keywords was calculated; the keywords with the greatest total link strength was selected. Four clusters are obtained by applying the Fractionalization method (min = 4) in VOSviewer; the first cluster (yellow) has as reference "data mining" (n = 7) and related health services; the second (network) refers to "machine learning" (n = 17) and the various techniques and algorithms used in the investigations; the third (blue) groups "alcohol" (n = 14), "smoking" (n = 5) and other keywords related to substance use and population; Finally, the fourth cluster (green) relates "tobacco" (n = 5) and health variables, such as "addiction", "psychological inflexibility", "substantive abuse", among others. Special interest is generated by the fourth group for the relationship between tobacco and health variables, as an emerging fact of study, as well as the yellow and red clusters, for their grouping around AI techniques.

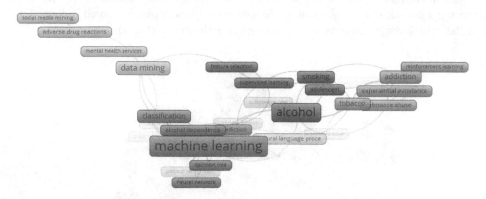

Fig. 5. Main topics of study. (Color figure online)

4 Conclusions

- There are few publications that refer to the investigation of tobacco, drug and other substance use applying methodologies and techniques based on artificial intelligence. The results obtained have been published in journals of first quartile (70%) and second quartile (23%), with high bibliometric data in the results (index h = 21, h10 = 56).
- There is extensive international cooperation for research, which brings together researchers from the United States, China and India, United Kingdom, India and France; and Spain, Chile, Italy and Israel. The United States shows supremacy in research (59%), followed by China (6.3%) and India (5.2%) very far. Brazil (2.7%) is the only South American country on the map.
- The consumption of alcohol and tobacco has been mainly investigated, with less relevance in other drugs and substances, being relevant in the consumer profile Psychological inflexibility and smoking cessation. As AI techniques, Machine Learning and Data mining have been mainly applied, and as algorithm types, classification and prediction.

References

1. United Nations Office on Drugs and Crime: World Drug report, Vienna, Austria (2019)
2. Herrera Rodríguez, A., et al.: Policonsumo simultáneo de drogas en estudiantes de facultades de ciencias de la salud/ciencias médicas en siete universidades de cinco países de América Latina y un país del Caribe: implicaciones de género, legales y sociales (2012)
3. United Nations Office on Drugs and Crime: World Drug Report, Vienna, Austria (2017)
4. Szolovits, P.: Artificial Intelligence in Medicine. Routledge, Abingdon (2019)
5. Cohen, S., Kamarck, T., Mermelstein, R.: A global measure of perceived stress. J. Health Soc. Behav. **24**(4), 385–396 (1983)
6. Kroenke, K., Spitzer, R.L., Williams, J.B.W.: The PHQ-9: validity of a brief depression severity measure. J. Gen. Intern. Med. **16**, 606–613 (2001)
7. Hughes, M.E., Waite, L.J., Hawkley, L.C., Cacioppo, J.T.: A short scale for measuring loneliness in large surveys: results from two population-based studies. Res. Aging. **26**, 655–672 (2004)
8. Ruiz, F.J., Luciano, C., Cangas, A.J., Beltrán, I.: Measuring experiential avoidance and psychological inflexibility: the Spanish version of the acceptance and action questionnaire-II. Psicothema **25**, 123–129 (2013)
9. Steinberg, L., Sharp, C., Stanford, M.S., Tharp, A.T.: New tricks for an old measure: the development of the barratt impulsiveness scale-brief (BIS-Brief). Psychol. Assess. **25**, 216 (2013)
10. Renau, V., Oberst, U., Gosling, S., Rusiñol, J., Chamarro, A.: Translation and validation of the ten-item-personality inventory into Spanish and Catalan. Aloma Rev. Psicol. Ciències l'Educació i l'Esport. **31**, 85–97 (2013)
11. Diener, E.D., Emmons, R.A., Larsen, R.J., Griffin, S.: The satisfaction with life scale. J. Pers. Assess. **49**, 71–75 (1985)
12. Babor, T.F., Higgins-Biddle, J.C., Saunders, J.B., Monteiro, M.G.: Audit: The Alcohol Use Disorders Identification Test: Guidelines for Use in Primary Health Care. World Health Organization, Geneva (2001)
13. Saunders, J.B., Aasland, O.G., Babor, T.F., De la Fuente, J.R., Grant, M.: Development of the alcohol use disorders identification test (AUDIT): WHO collaborative project on early detection of persons with harmful alcohol consumption-II. Addiction **88**, 791–804 (1993)
14. Heatherton, T.F., Kozlowski, L.T., Frecker, R.C., Fagerstrom, K.: The Fagerström test for nicotine dependence: a revision of the Fagerstrom Tolerance Questionnaire. Br. J. Addict. **86**(9), 1119–1127 (1991)
15. Ruisoto, P., Cacho, R., López-Goñi, J.J., Vaca, S., Jiménez, M.: Prevalence and profile of alcohol consumption among university students in Ecuador. Gac. Sanit. **30**, 370–374 (2016)
16. Iavindrasana, J., Cohen, G., Depeursinge, A., Müller, H., Meyer, R., Geissbuhler, A.: Clinical data mining: a review. Yearb. Med. Inform. **18**, 121–133 (2009)
17. Hamet, P., Tremblay, J.: Artif. Intell. Med. Metab. Exp. **69**, S36–S40 (2017)
18. Kumar, R., Sharma, A., Haris Siddiqui, M., Kumar Tiwari, R.: Prediction of metabolism of drugs using artificial intelligence: how far have we reached? Curr. Drug Metab. **17**, 129–141 (2016)
19. Chakradhar, S.: Predictable response: finding optimal drugs and doses using artificial intelligence. Nat. Med. **23**, 1244–1247 (2017)
20. Jothi, N., Husain, W.: Data mining in healthcare–a review. Procedia Comput. Sci. **72**, 306–313 (2015)
21. Long, C.: Data Science and Big Data Analytics. Discovering, Analyzing, Visualizing and Presenting Data. Wiley, Indianapolis (2015)

22. Géron, A.: Hands-on Machine Learning with Scikit-Learn and Tensorflow: Concepts, Tools, and Techniques to Build Intelligent Systems. O'Reilly Media, Inc., Sebastopol (2017)
23. Torres-Carrión, P., González-González, C., Aciar, S., Rodríguez-Morales, G.: Methodology for systematic literature review applied to engineering and education. In: EDUCON2018 – IEEE Global Engineering Education Conference, Santa Cruz de Tenerife – España. IEEE Xplore Digital Library (2018)
24. Schulte, B., Kaner, E.F.S., Beyer, F., Schmidt, C.S., O'Donnell, A.: Study protocol for a systematic review of evidence for digital interventions for comorbid excessive drinking and depression in community-dwelling populations. BMJ Open 9(10), e031503 (2019)
25. Mak, K.K., Lee, K., Park, C.: Applications of machine learning in addiction studies: a systematic review. Psychiatry Res. 275, 53–60 (2019)
26. Canan, C., Polinski, J.M., Alexander, G.C., Kowal, M.K., Brennan, T.A., Shrank, W.H.: Automatable algorithms to identify nonmedical opioid use using electronic data: a systematic review. J. Am. Med. Inform. Assoc. 24, 1204–1210 (2017)
27. García-González, A., Ramírez-Montoya, M.-S.: Systematic mapping of scientific production on open innovation (2015–2018): opportunities for sustainable training environments. Sustainability 11, 1–15 (2019)
28. Nikfarjam, A., Sarker, A., O'Connor, K., Ginn, R., Gonzalez, G.: Pharmacovigilance from social media: mining adverse drug reaction mentions using sequence labeling with word embedding cluster features. J. Am. Med. Inform. Assoc. 22, 671–681 (2015)

Generating Individual Gait Kinetic Patterns Using Machine Learning

César Bouças[1]([✉]), João P. Ferreira[1,2], A. Paulo Coimbra[1],
Manuel M. Crisóstomo[1], and Paulo A. S. Mendes[1]

[1] Institute of Systems and Robotics, Department of Electrical
and Computer Engineering, University of Coimbra, Coimbra, Portugal
cesar.boucas@isr.uc.pt
[2] Department of Electrical Engineering, Superior Institute of Engineering
of Coimbra, Coimbra, Portugal
ferreira@isec.pt

Abstract. In this study, data of 42 healthy individuals walking over
a treadmill was used to train and test a neural network that produced
individual kinetic patterns of gait cycle as output for a set of atomic
features (gender, age, mass, height and gait speed) used as input. The
proposed method implements a 3-layer feedforward architecture capable
to produce the 3D gait patterns of ankle, knee and hip moment at once,
with an average root mean squared error (RMSE) of 7% and average correlation coefficient (ρ) of 0.94 with respect to the ground truth patterns
of the test set. The presented strategy may be used to support individual
gait clinical analysis as an alternative to the use of the normal literature
pattern that do not take into account the specific characteristics of the
patients.

Keywords: Human gait kinetics · Time series generation · Machine
Learning

1 Introduction

The study of human gait covers the way people use the movement of limbs to
perform terrestrial locomotion. Gait analysis is one of the most active research
fields in Biomechanics and have a broad range of applications, such as pathology
detection [14], rehabilitation [16], prosthesis design, biometric identification [3]
and bipedal robotic locomotion.

Clinical gait analysis methods aims to provide an objective record that quantifies the magnitude of patients deviations from normal gait. Whereas a set of

The Fundação para a Ciência e Tecnologia (FCT) and COMPETE 2020 program
are gratefully acknowledged for funding this work with PTDC/EEI-AUT/5141/2014
(Automatic Adaptation of a Humanoid Robot Gait to Different Floor-Robot Friction
Coefficients).

pathology-related gait disorders are identified and can be used to support diagnosis and target treatments.

However, the human gait characteristics are unique to each person and can be clearly differentiated from gait patterns of other individuals [10]. In fact, human gait has been used to successfully perform biometric identification with high performance levels [3]. Therefore, the determination of specific gait patterns of a certain individual may support clinicians and researchers in the individualisation of their analyses, diagnoses and interventions. The idea is to compare to the deviations from the patient's expected gait pattern instead of comparing the deviation from the normal literature pattern. This is related to Personalized Medicine: the concept that managing a patient's health should be based on the individual patient's specific characteristics, including age, gender, height and weight.

Previews works showed that such biometric characteristics in conjunction with walking speed are sufficient to generate personalized patterns of human gait. Special attention is given to the walking speed, since it directly affects the kinematic, kinetic and other human gait characteristics.

The work of [15] takes walking speed, gender, age and body mass index as input to several regression models that outputs syntethic sagittal gait kinematics with root mean squared error (RMSE) < 3%. Moreover, [12] used a generalized regression neural network (GRNN) that receive gait parameters and anthropometric data as input and generate Fourier coefficients as output. Inverse Fourier transforms is applied to obtain the final knee joint angle curve with $\rho > 0.98$. And [7] uses computational intelligence techniques to generate the human knee joint angle walking pattern in the sagittal plane.

While kinematics is the study of movements without reference to the forces that cause motion, kinetics is the study of forces that cause motion. Both having direct implications for joint functions of human gait.

In the context of gait kinetics generation, a wavelet neural network was proposed to predict joint moments based on ground reaction forces (GRFs) and electromyography (EMG) signals [2], results showed a normalized root mean squared error (NRMSE) < 10% and $\rho > 0.94$. In addition, [9] used a neural network to predict kinetic knee extensor and flexor torque based on age, gender, height, body mass, EMG signals, joint position and joint velocity. And [6] proposed to use a three-layer perceptron to predict the external knee adduction moment, based on force plate data and anthropometric measurements.

The present work uses just biometric features in conjunction with gait speed to generate 3-dimensional components of knee, hip and ankle joint moments. Concretely, a feedforward neural network is designed to generate nine time series corresponding to an approximation for the measured gait cycle patterns of abduction, adduction, rotation, flexion and extension of lower limbs. The input to the network is the gender, age, mass, height and gait speed of individuals.

2   Material and Method

In order to build a personalized solution capable of producing gait pattern moments that vary with the input, we propose to use data-driven approaches to learn directly from training data rather than from knowledge provided by gait analysis experts.

As usual in a Machine Learning strategy, our goal is to learn a function $f(x) = y$. In our case, x represents a person walking at a specific gait speed and y is the expected healthy gait cycle kinetic pattern of x. The function f is to be learnt from examples. The method can be resumed in three steps: dataset acquisition, baseline definition and finally obtaining an f from the dataset that performs better than the baseline.

2.1   Dataset

Data used in this work was obtained from a public dataset from the Laboratory of Biomechanics and Motor Control at the Federal University of ABC, Brazil [8]. Whereas data about individual participants were acquired through 3D motion-capture system, force plates and instrumented treadmill.

The volunteer participants were 42 healthy persons: 24 males and 18 females. They walked for 90 s at eight different gait-speeds. The last 30 s of each trial were recorded and processed.

Figure 1 illustrates the ensemble average gait cycles of ankle abduction and adduction obtained for each of the trials of a female participant.

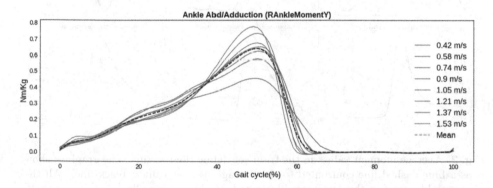

Fig. 1. Trials data of a female subject (Age = 24, Mass = 48 kg, Height = 158 cm) at several gait speeds (m/s) and the mean values. Each curve corresponds to the ensemble average gait cycle values obtained from a specific trial.

Data is separated among raw and text files, the latter contains the time-normalized gait cycles ensemble averages (101 time-normalized points) for each participant's treadmill and overground trials measurements. Sample frequencies of the kinematics and kinetics data differ and the number of gait trials is not the same across all participants.

The present work makes use of the kinetics treadmill data, the biometric data of the participants and the data corresponding to the average gait speeds of the trials, extracted from the metadata file.

2.2 Baseline

Given the train set containing trials data, the mean values per variable was calculated and taken as baseline predictions. This is equivalent to define an estimator that always outputs the same average moment curves no matter the input features.

Figure 2 shows average values of the joint moments present in train and test set trials.

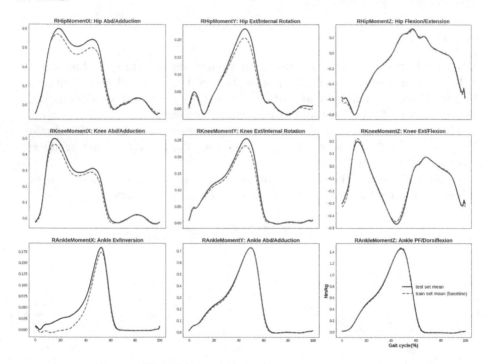

Fig. 2. Average moment values in the train set define the baseline. The graphics show it as a blue dashed line confronted to the average test set values (black line). All the nine variables used in this study are illustrated. (Color figure online)

Such strategy produces a strong baseline since it maintains a similarity with respect to the ground truth series. However, it suffers from the lack of specificity as it does not vary appropriately with the input features.

2.3 Input and Output

Let's define a supervised learning strategy where f is learned from examples $(x_i, y_i)_{i=0..|N|}$ and:

- x_i is a walking person represented by a set of atomic features: gender, age, mass, height, and gait speed.
- y_i is a multivariate time series with 101 steps and 9 variables representing the 3-dimensional hip, knee, and ankle joint moments of a gait cycle. Table 1 shows the variables names and its description.
- $|N| = 328$ is the size of the dataset used in the experiment. It corresponds to the number of files, each one containing ensemble data of the joint moments of a complete gait cycle at certain gait speed.

Table 1. Output variables.

Joint	Component	Name	Description
Right hip	X	RHipMomentX	Hip Abd/Adduction
Right hip	Y	RHipMomentY	Hip Ext/Internal Rotation
Right hip	Z	RHipMomentZ	Hip Flexion/Extension
Right knee	X	RKneeMomentX	Knee Abd/Adduction
Right knee	Y	RKneeMomentY	Knee Ext/Internal Rotation
Right knee	Z	RKneeMomentZ	Knee Extension/Flexion
Right ankle	X	RAnkleMomentX	Ankle Ev/Inversion
Right ankle	Y	RAnkleMomentY	Ankle Abd/Adduction
Right ankle	Z	RAnkleMomentZ	Ankle PF/Dorsiflexion

2.4 Data Preparation

Data corresponding to x_i was extracted from the metadata and y_i was extracted from the ASCII files containing ensemble averages. Little pre-processing work was required since the released data have been already time-normalized and ensembled to reflect a complete full gait cycle for each trial.

In order to split the dataset into **train** and **test** sets, two strategies were considered: split the data by subject, and to split it by trial. Empirical observation showed that splitting the data by trials leads to a better accuracy. But that could be an indication of overfitting since the data of all subjects could appear during training phase, letting the model learn about all subject's gait pattern before the test phase.

Thus, the data were divided by making a random selection of 28 subjects and its ensemble files to **train** the model. The remaining 14 subjects and it's trial files were held out hidden from the train phase in order to evaluate the model only at the **test** phase. Then, at test phase, the trained model is confronted with trials of completely unseen subjects.

2.5 Feedforward Neural Network

A multivariate time series generator was designed as a neural network, whereas the input x_i is first transformed by a hidden layer with 20 sigmoid neurons (units), then by a linear (identity) layer with as many units as the number of steps of the time series multiplied by the number of variables, that is 101×9.

The network was implemented using Keras [4] with TensorFlow backend [1]. Weights for all layers were initialized to a random distribution and biases initialized to 0. Figure 3 illustrates the architecture: a fully connected feedforward neural network with three layers.

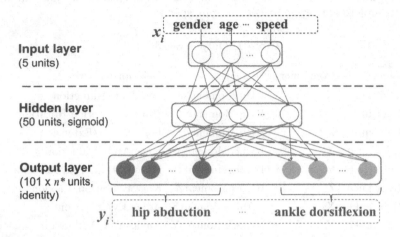

Fig. 3. The neural network architecture with the number of neurons/units of each layer and its activation functions. Units are represented by circles and connections are represented by arrows. The network is fully connected and each arrow points forward to the propagation direction. The n^* represents the number of variables of the time series in y_i, thus $n^* = 9$ and $|y_i| = 101 \times 9$ given that each series contains 101 timesteps.

This architecture is attractive due its ability to generate multivariate time series for each input features while maintaining a single model.

The hyperparameters tuned at training were the number of hidden units and it's activation function. Whereas the value of 20 hidden units with sigmoid activation function demonstrated to be sufficient to reach the best average results at training phase.

2.6   Train

The goal of training is to obtain from examples, a model capable to predict time series output as similar as possible to the ground truth (golden) values in the dataset.

The network was trained through 2000 epochs of backpropagation using the AdaMax algorithm [11] and mean squared error (MSE) as loss function. It was configured to early stop when the model stays more than 100 epochs with no improvement over the loss function.

We selected as hyper-parameters the number of units in the hidden layer and their activation function. In order to calibrate these parameters we used empirical evaluation, and the number of hidden units and dropout rate that presented the best performance was respectively 20 and sigmoid. Table 2 resume such empirical evaluation.

Table 2. Hyper-parameters values. Sigmoid, radial basis and hyperbolic tangent activation functions were tested.

Parameter	Tested values	Best
Number of hidden units	5 to 32	**20**
Activation function of hidden units	sigmoid, radial, tanh	**sigmoid**

Training in the complete set was performed using a virtual machine with 1 CPU and 12 GB of memory.

2.7 Test

After the model f is obtained through training, that is, the neural network parameters has been calibrated, generating the gait pattern of an unseen feature set values x' is done by a simple call to the trained model: $f(x') = \hat{y}$. Where $\hat{y}$ is the patterns generated by the model.

Therefore, the test phase is done by iterating through each example of the test set, i.e., each pair $(x'_j, y'_j)_{j=0..|M|}$, where $|M|$ is the test set size. At each iteration the predicted time series is obtained $f(x'_j) = \hat{y}_j$ and stored in order to later be compared to the golden time series (y'_j) using evaluation metrics to measure the error between the predicted and the ground truth series: $y - \hat{y}$.

2.8 Evaluation Metrics

Root mean squared error (RMSE), dynamic time warping (DTW) and correlation coefficient (ρ) were used to measure the similarity of the two series: the original ground truth golden series $(y = s_1, s_2, ..., s_{101})$ and the series predicted by the model $(\hat{y} = \hat{s}_1, \hat{s}_2, ..., \hat{s}_{101})$.

The RMSE measures the root average of the squares of the errors and its calculated as:

$$RMSE(y, \hat{y}) = \sqrt{MSE(y, \hat{y})} = \sqrt{\frac{1}{101} \sum_{i=1}^{101} (s_i - \hat{s}_i)^2} \tag{1}$$

DTW is computed as the Euclidean distance between the time series aligned [18], i.e., if P is the optimal alignment path:

$$DTW(y, \hat{y}) = \sqrt{\sum_{(i,j) \in P} (s_i - \hat{s}_i)^2} \tag{2}$$

Pearson correlation coefficient ρ can be expressed in terms of covariance (cov) and standard deviation (σ):

$$\rho(y, \hat{y}) = \frac{cov(y, \hat{y})}{\sigma_y \sigma_{\hat{y}}} \tag{3}$$

By using these three metrics, we capture diverse aspects of the similarity between predicted and ground truth series. The goal is to minimize RMSE and DTW while maximizing the value of ρ.

3   Results

For each example present in the test set, the input features were used to predict the joint curves using the trained FFNN. Evaluation metrics were calculated and averaged separately for each joint moment variable. Comparing the predicted values with respect to the ground truth values. Table 3 shows the obtained mean values of the metrics calculated for the FFNN model as well for the baseline predictions.

Table 3. Metrics obtained from comparing the predicted series with the test set ground truth series averaged among all test trials.

Method	RMSE	DTW	ρ
FFNN	**0.074**	**0.387**	**0.945**
Baseline	0.098	0.566	0.904

A detailed sumarization of the average FFNN predictions quality is shown in the Table 4.

In order to illustrate the FFNN's superior ability to produce personalized patterns when compared to the baseline, one subject (age of 33, 75 Kg of mass and 179 cm tall) was randomly selected. From this subject, 3 variables related to the hip, knee and ankle moments were extracted from a specifc trail (gait speed of 0.68 m/s). We compared the ground truth series of this trail with the baseline and with the FFNN model prediction. Figure 4 shows a visual comparation and Table 5 confirms the visual perception: the metrics of the FFNN prediction were better than the metrics of the baseline.

All the data [8] as well as the source code of the experiments are open and publicly available[1].

---

[1] github.com/cesarboucas/gait.

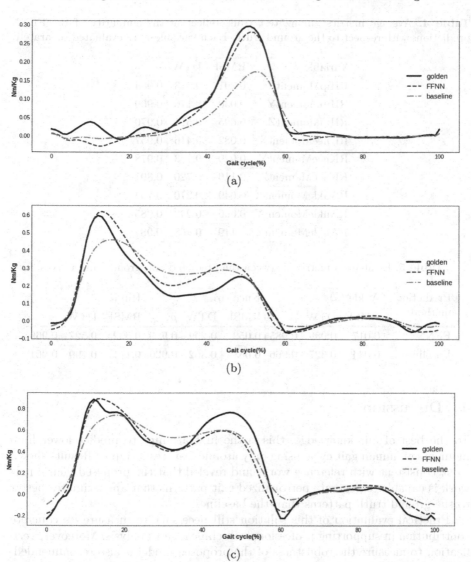

Fig. 4. Groud truth (golden) series, the baseline series and the series predicted by the FFNN of a specific trail. Three variables were selected to illustrate a measurement taken from the ankle (a), knee (b) and hip (c). In all cases, the FFNN prediction turned out to be a better approximation to the golden series than the baseline.

Table 4. Average metrics among test trails reflecting the similarity of the FFNN predictions with respect to the ground truth. Each variable were evaluated separately.

Variable	RMSE	DTW	ρ
RHipMomentX	0.091	0.513	0.984
RHipMomentY	0.030	0.176	0.960
RHipMomentZ	0.095	0.511	0.970
RKneeMomentX	0.082	0.416	0.957
RKneeMomentY	0.029	0.163	0.978
RKneeMomentZ	0.116	0.720	0.891
RAnkleMomentX	0.049	0.270	0.794
RAnkleMomentY	0.056	0.242	0.985
RAnkleMomentZ	0.119	0.468	0.984

Table 5. Evaluation metrics between the predicted and groud truth curves.

Prediction method	Ankle (a)			Knee (b)			Hip (c)		
	RMSE	DTW	ρ	RMSE	DTW	ρ	RMSE	DTW	ρ
FFNN	0.017	0.089	0.985	0.052	0.299	0.980	0.092	0.527	0.969
Baseline	0.044	0.327	0.966	0.068	0.362	0.926	0.111	0.799	0.961

4   Discussion

To the best of our knowledge, this is the first attempt to predict lower limb moments of human gait cycle using only atomic features as input. Results shown to be consistent with reference works and reveled that the proposed neural network is capable to generate personalized gait patterns that approximates better to the ground truth patterns than the baseline.

Practical evaluation of this solution still necessary to measure its concrete contribution in supporting professionals in clinical gait analysis. Moreover, verification to measure the robustness of the proposed model is also recommended. In fact, to provide robust solutions for the presented and similar tasks remains challenging given the small size of available gait datasets, as well the little informative character of the atomic inputs.

As an option to enrich the input, other measured gait characteristics such as ground reaction forces (GRF) could be incorporated into it. And to address the lack of datasets with proper size, data from diverse open resources could be combined to form an augmented base. A difficult task since each dataset is strongly dependent to its laboratory specificity and measurement conditions [13,17].

Actually, given that improving the access to data has become vital in the progression of many studies, notably in fields relating to human biomechanics, as further work, we advocate to the definition of a benchmark human gait dataset.

Following standards defined by initiatives such as [19] and using synthetic data obtained with the use of the proposed model.

Such synthetic gait patterns can be used for example, in data augmentation strategies. That could be accomplished by defining a classification task, e.g., classify a pattern between male and female gait, then make use of the Train on Synthetic, Test on Real (TSTR) strategy as proposed by [5].

Using this method, the quality of the Time Series Generator is measured by the accuracy gain that is achieved when it's generated data is used to augment the training set for a classifier.

5   Conclusion

This work showed that five biometric features and gait speed are sufficient to generate personalized knee, ankle and hip gait cycle moment patterns. The evaluation procedure revealed that the multivariate series produced as output by our method is an approximation of the ground truth curves with superior quality and personalization potential than the baseline strategy.

The proposed method used the right side joint moments of human gait on treadmill. To perform the same analysis for the left joints of humans walking overground is analog and can be an immediate continuation to the present work. This can lead to insights about kinetic implications of the use of the hands on the treadmill. Since in some trials the participants walked while hanging onto the handrails [8].

Finally, given the simplicity of the proposed architecture, accuracy improvements may be reached in several ways. For example, by using hyperparameter optimization methods, slicing strategies as well as designing a more sophisticated network architecture. Such as the Generative Adversarial Networks (GANs), that has achieved state-of-the-art performances in several time series generation tasks recently [5].

References

1. Abadi, M., et al.: TensorFlow: large-scale machine learning on heterogeneous systems (2015). https://www.tensorflow.org/. Software available from tensorflow.org
2. Ardestani, M.M., et al.: Human lower extremity joint moment prediction: a wavelet neural network approach. Expert Syst. Appl. **41**(9), 4422–4433 (2014)
3. Chao, H., He, Y., Zhang, J., Feng, J.: Gaitset: regarding gait as a set for cross-view gait recognition. arXiv preprint arXiv:1811.06186 (2018)
4. Chollet, F., et al.: Keras (2015). https://keras.io
5. Esteban, C., Hyland, S.L., Rätsch, G.: Real-valued (medical) time series generation with recurrent conditional gans. arXiv preprint arXiv:1706.02633 (2017)
6. Favre, J., Hayoz, M., Erhart-Hledik, J.C., Andriacchi, T.P.: A neural network model to predict knee adduction moment during walking based on ground reaction force and anthropometric measurements. J. Biomech. **45**(4), 692–698 (2012)
7. Ferreira, J.P., Vieira, A., Ferreira, P., Crisostomo, M., Coimbra, A.P.: Human knee joint walking pattern generation using computational intelligence techniques. Neural Comput. Appl. **30**(6), 1701–1713 (2018)

8. Fukuchi, C.A., Fukuchi, R.K., Duarte, M.: A public dataset of overground and treadmill walking kinematics and kinetics in healthy individuals. PeerJ **6**, e4640 (2018)
9. Hahn, M.E.: Feasibility of estimating isokinetic knee torque using a neural network model. J. Biomech. **40**(5), 1107–1114 (2007)
10. Horst, F., Lapuschkin, S., Samek, W., Müller, K.R., Schöllhorn, W.I.: Explaining the unique nature of individual gait patterns with deep learning. Sci. Rep. **9**(1), 2391 (2019)
11. Kingma, D.P., Ba, J.: Adam: a method for stochastic optimization. arXiv preprint arXiv:1412.6980 (2014)
12. Luu, T.P., Low, K., Qu, X., Lim, H., Hoon, K.: An individual-specific gait pattern prediction model based on generalized regression neural networks. Gait Posture **39**(1), 443–448 (2014)
13. Majernik, J.: Normative human gait databases-use own one or any available. Stat. Res. Lett. **2**(3), 69–74 (2013)
14. Mehrizi, R., Peng, X., Zhang, S., Liao, R., Li, K.: Automatic health problem detection from gait videos using deep neural networks. arXiv preprint arXiv:1906.01480 (2019)
15. Moissenet, F., Leboeuf, F., Armand, S.: Lower limb sagittal gait kinematics can be predicted based on walking speed, gender, age and BMI. Sci. Rep. **9**(1), 9510 (2019)
16. Morone, G., et al.: Robot-assisted gait training for stroke patients: current state of the art and perspectives of robotics. Neuropsychiatr. Dis. Treat. **13**, 1303 (2017)
17. Whittle, M.W.: An Introduction to Gait Analysis. Butterworth-Heinemann, Oxford (2007)
18. Sakoe, H., Chiba, S.: Dynamic programming algorithm optimization for spoken word recognition. IEEE Trans. Acoust. Speech Sig. Process. **26**(1), 43–49 (1978)
19. Salathé, M., Wiegand, T., Wenzel, M.: Focus group on artificial intelligence for health. arXiv preprint arXiv:1809.04797 (2018)

Knee Injured Recovery Analysis Using Extreme Learning Machine

João P. Ferreira[1,2](✉) ⓘD, Bernardete Ribeiro[3] ⓘD,
Alexandra Vieira[2,3], A. Paulo Coimbra[2] ⓘD,
Manuel M. Crisóstomo[2] ⓘD, César Bouças[2], Tao Liu[4],
and João Páscoa Pinheiro[5]

[1] Department of Electrical Engineering,
Superior Institute of Engineering of Coimbra, Coimbra, Portugal
ferreira@isec.pt
[2] ISR - Department of Electrical and Computer Engineering,
University of Coimbra, Coimbra, Portugal
[3] CISUC - Department of Informatics Engineering,
University of Coimbra, Coimbra, Portugal
[4] State Key Laboratory of Fluid Power and Mechatronic Systems,
School of Mechanical Engineering,
Zhejiang University, Hangzhou 310027, China
[5] Rehabilitation Medicine Department,
Coimbra University Hospital, Coimbra, Portugal

Abstract. The physiotherapists analyse gait patterns to recognize normal and pathological gait movements. The gait patterns are affected by the characteristics of the individual (gender, age, weight and height) and the walking speed. In this paper, a gait analysis system to evaluate the severity of gait pathology is proposed. The Machine Learning (ML) algorithm can generate reference knee patterns for specific individuals. Gait index are used to compare the patterns generated by the ELM and patterns of the patients who suffered a surgical knee reconstruction. Two gait index are compared: The Gait Variable Score (GVS) and the Global Index (G_{Index}) developed by the authors. The G_{Index} classified 7 patients as not recovery, corroborating with the opinion of physiotherapists, while the GVS only classified 2 as not recovered. The proposed gait analysis system using the Extreme Learning Machine (ELM) and the G_{Index} can be useful tool for physiotherapy team in the gait pathology diagnosis and evaluation of future pathologies.

Keywords: MLP · ELM · MSVR · Knee pattern · GVS

1 Introduction

The human gait is a cyclic pattern composed by a set of complex movements which are executed by several parts of the body [1]. The gait cycle, repeated over the person's gait, begins with the first touch of the heel on the floor and ends when the same heel touch on the floor by the second time [2].

© Springer Nature Switzerland AG 2020
M. Botto-Tobar et al. (Eds.): ICAT 2019, CCIS 1194, pp. 65–79, 2020.
https://doi.org/10.1007/978-3-030-42520-3_6

The analysis of the human gait can be applied in different fields such as physiotherapy and identification of people for forensics [3] and security purposes [4]. The physiotherapy area analysis the gait limitations caused by sports activities, and musculoskeletal and neurological diseases, such as Parkinson disease [5].

In gait analysis, the physiotherapist compares the knee pattern of the patient with a reference healthy knee pattern to recognize normal and pathological gait patterns [3]. Normally, the reference healthy knee pattern is a literature knee pattern which describes the knee pattern of people without any type of gait disease. However, this comparison is not specific for each patient who has personal characteristics (gender, age, weight, height and gait speed) [6]. The personal characteristics affect the knee joint pattern and the literature pattern does not take them into account. The ML algorithms were been used to generate reference knee pattern for an individual with specific characteristics [7], which improves gait analysis' results.

A visual comparison between the patient's pattern and the pattern generated by the ML algorithm can result in a weak classification of the gait severity. The gait pathology severity can be quantified using gait indices which compare knee patterns using multivariate statistic methods. Gait measures have received increasing attention in the evaluation of patients with knee osteoarthritis (OA). Comprehending gait parameters is an essential requirement for studying the causes of knee disorders [40]. OA has a significant impact on quality of life (QOL) [41].

In this paper, an innovative gait analysis system is proposed. The system uses specific reference knee patterns generated by ML algorithms and gait indices to analyze the gait limitation. After training and testing three ML algorithms (MLP, ELM and MSVR) with gait data of people without gait limitations (healthy people), the authors select the ML algorithm with best results in generation of gait patterns, resulting the ELM as a best [21]. In second part of the work, it is created a G_{Index} to measure the difference between real, generated by the selected ELM and healthy people knee patterns. After the G_{Index} results are used by the developed RDI to evaluate the recovery stage of patients submitted to a ligamentoplasty surgery about four years ago.

The proposed system can be a great help in rehabilitation processes providing a personalized analyze for each patient.

This paper is organized as follows. Section 2.2 presents gait indices for gait analysis. Section 2.3 is dedicated to the methodology where data collection and comparison of patterns processes are described. Section 3 shows the excrimental results of the proposed gait system. In Sect. 4 is described the evaluation of the experimental results. Section 5 presents the concluding remarks and the future work plan.

2 Material and Method

In gait analysis, physiotherapists use measure parameters such as angles of joints movements and pressure distribution under feet [8]. The available systems to collect gait parameters are video tracking [9], force plates [10], treadmills [11], insoles [12] and instrumented shoes [13]. The presents work starts by collecting the angles of joints movements using the low vision system composed by a treadmill, two cameras and several passive marks. The vision system is described in Subsect. 2.3.

The state-of-the-art of ML algorithms and gait indices for gait analysis is presented in Subsects. 2.1 and 2.2, respectively.

2.1 ML Algorithms for Gait Analysis

ML algorithms can extract important information from a large gait dataset [14]. Rani and Arumugam [15] shows that the ELM is better to classify the children abnormal gait than the Support Vector Regression. Luu et al. [7] proposed a General Regression Neural Network (GRNN) able to successfully generates healthy gait patterns. Artificial Neural Network [16] realizes the human gait classification into six different stages. Prakash et al. [17] presented a multi-model technique (Random Forest, Linear Discriminant Analysis and Support Vector Machine) with subject identification rate of 93.33%.

In a previews work Ferreira *et al.* [21] performed the comparison between the artificial neural network (ANN), extreme learning machine (ELM) and multi-output support vector regression (MSVR), and this study suggests that the ELM is the best ML algorithm to generate the knee joint angle pattern for each gender. This ELM is used in this presented work.

2.2 Indices for Gait Analysis

The gait index with major clinical acceptance is the Gillette Gait Index (GGI) [18, 19]. The GGI presents some limitations, such as: necessity of data of about 40 normal people, dependence of the reference data diversity, complex calculation with results of difficult interpretation. The Gait Deviation Index (GDI) was developed to overcome the GGI limitations [19].

The Gait Profile Score (GPS) is identical to the GDI, but with an easy form of calculation [19]. The GPS consists of the Root Mean Square (RMS) of N kinematics variables [20]:

$$GPS = \frac{\sum_{i=1}^{N} GVS_i^2}{N} \tag{1}$$

where GVS_i means each kinematic variable of the Gait Variable Score index and is calculated as the RMS of a single variable [20]:

$$GVS_i = \frac{\sum_{t=1}^{T} (x_{i,t} - \bar{x}_{i,t}^{ref})^2}{T} \tag{2}$$

where T is the number of gait cycle divisions, $x_{i,t}$ is the value of variable i in point t of the real knee pattern, and $ref_{i,t}$ is the mean value of variable i in point t of the reference knee pattern [20].

The GPS is proportional to the patient pathology, i.e. a higher GPS indicates a worse pathology severity [20].

2.3   Data Collection

The development work follows an incremental strategy which begins with the performance's study of three ML algorithms, already presented in the work [21] and ends
with the development of gait indices.

The vision system developed by Ferreira et al. [22] is used to collect the gait
patterns. The vision system is composed by a treadmill, two cameras and 20 passive
marks. Each camera positioned in one side of the treadmill is characterised by a CMOS
640 × 480 (VGA) sensor, maximum of 30 frames/sec. and USB 2.0 interface. The
alignment and the calibration of the system were done according to P. Ferreira et al.
[22]. The passive marks were positioned in the joints of each side of the individual, as
represented for the right side in Fig. 1: shoulder, elbow, wrist, pelvis, leg, knee, ankle,
heel and two extremities of the finger toes. After the measurement of the distances
between the marks to know the depth was realized the pelvis calibration. The pelvis
calibration gives the pelvis position taking into account the relationship between leg,
knee and pelvis marks. The pelvis mark, represented in Fig. 1 with triangle was
removed before the gait record because it would be occluded by upper limb of the
individual.

The vision system was used to collect data from two groups of volunteers, one with
individuals with healthy gaits and other with patients who were subjected to a ligamentoplasty (surgical reconstruction) about four years ago, after suffering a rupture of
the anterior cruciate ligament of the knee. The healthy group, used in [21] was composed by 14 females and 11 males, and the patients group by 10 males. The low
number of specimen used is acceptable for a prototype system. Tables 1 and 2 present
physical characteristics intervals, mean and standard deviation of the healthy and
patients groups, respectively.

Fig. 1. Position of passive marks during the gait record

Each individual walked five different speeds on the treadmill. The value of the
speeds corresponds to 25%, 50%, 75%, 100% and 125% of the comfortable speed. The
comfortable speed is the velocity appropriated for an individual with a specific gender,

age and weight, as proposed by Bohannon [23]. The objective of the work is the development of a method to analysis the gait of individual with gait pathologies which gives priority to the study of the lower gait speeds. The selection of maximum and minimum speeds was limited by the vision system characteristics: minimum treadmill speed of 0.28 m/s and maximum speed for a good accuracy 1.53 m/s.

Table 1. Characteristics of healthy group

Gender	Characteristic	Minimum	Maximum	Mean	Standard deviation
Female	Height (m)	1.59	1.69	1.64	0.03
	Age (years)	18	58	33	16
	Weight (kg)	47	90	65	13
Male	Height (m)	1.69	1.90	1.78	0.07
	Age (years)	19	55	35	13
	Weight (kg)	58	120	81	19

Table 2. Characteristics of patients group

Gender	Characteristics	Minimum	Maximum	Mean	Standard deviation
Male	Height (m)	1.69	1.89	1.79	0.06
	Age (years)	20	38	30.10	5.36
	Weight (kg)	74	111	86.30	11.55

This work only used the data of the knee joint for realize the gait analysis. The processing of the data were done according to Ferreira et al. [22].

2.4 Generation of Knee Joint Pattern Using ELM

Input and output matrices, for the ELM, grouped the processed knee joint patterns in function of the gender, i.e. each gender has an input and an output matrices. Women and men present different gait pattern because they walk with different styles and present different physical characteristics (such as height and weight).

The knee dominance (right or left) was not considered in the data division due to two reasons:

- It is expected that healthy individuals present similar right and left knee patterns which was proved by the collected data. The difference measured between the maximum value of the left and right knee is small (about 0.04 rad). Furthermore, this difference is close to the vision system calibration error value (0.035 rad).
- The reference literature for knee joint pattern [24] used in this work as comparison base also join the right and left knee patterns.

The input matrices had one line for each physical characteristic (height, age and weight) and one line for the gait speed. The ELM only generate accurate patterns for input features with values within the range limited by the input matrices, described in

Table 1 for each gender. The output matrices had 100 lines (% of gait cycle) and the same number of columns than the input matrices. The Fig. 2 present a generated knee pattern of a female (22 years old, height of 1.60 m and weight of 59 kg.) for 5 different gait speeds [21] and literature knee pattern [24].

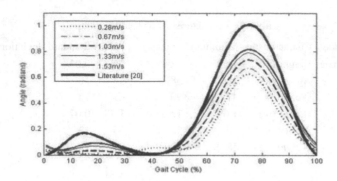

Fig. 2. Knee patterns of five gait speeds generated by the selected ELM for a female characteristics, and literature knee pattern [24].

2.5 Patterns Comparison

A method to measure and compare the severity of the patients pathology was developed. The developed method is composed by three phases as illustrated in Fig. 3: (a) Calculation of indices of the knee joint pattern; (b) Calculation of G_{Index} and GVS; (c) Calculation of RDI using G_{Index} and GVS.

The characteristics (gender, age, weight, height and gait speed) and the processed real knee patterns of the patients are the inputs of the method.

1. Calculation of indices of the knee joint pattern
This phase begins with the generation of the reference knee patterns to the patient, by the ELM of the correspondent gender which was selected in the first part of the work [21]. All the patients included in this study can be analysed by the method because their characteristics represented in Table 2 are within the range of the characteristics of the healthy group (Table 1). The data of the health group is the base of the development of the ELM and the base of the reference gait indices presented above.

After, the maximum of the generated patterns and the real patterns are aligned in function of the knee: patterns of the right knee are shifted to 75% of the gait cycle and patterns of the left knee are shifted to 25% of the gait cycle. This alignment in function of the knee helps the following calculation and representation of the results.

The graphical and maths analysis functions use the aligned patterns to calculate 7 indices (I_C) for each pattern and 4 indices (I_{BC}). The 7 I_C described in Table 3 are indices used by publications in the area [25–27]. The 4 I_{BC}, defined in Table 3, compare the generated and the real patterns.

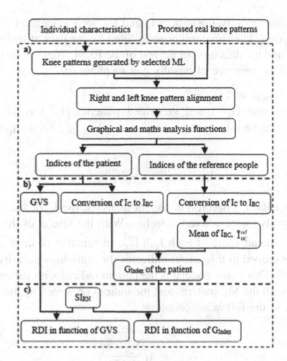

Fig. 3. Flowchart of the three phases (A, B and C) of the method to calculates the RDI in function of GVS and G_{Index}

Table 3. Pattern indices: I_C and I_{BC}

I_C	Definition
A_{0-50}	Area under knee pattern between 0–50% of gait cycle
A_{50-100}	Area under knee pattern between 50–100% of gait cycle
Amp_{0-50}	Maximum amplitude between 0–50% of gait cycle
Amp_{50-100}	Maximum amplitude between 50–100% of gait cycle
PSD	Power spectral density, which represents the Discrete Fourier Transformer (DFT) area of the pattern
D_{area}	Area of the derivative of the knee pattern in function of the knee pattern
D_{centre}	Distance between coordinate system's origin and area centre of the derivative of knee pattern in function of his own
I_{BC}	Definition
RI	Reference index measures the angle between x-axis and linear regression line of reference and real patterns. RI = π/4 rad means that real and reference patterns are equal. RI ≠ π/4 rad can indicate some limitation in the gait
R^2	Coefficient of determination of the linear regression between the real and reference patterns
DTW	Dynamic Time Warping measures the similarity between the real and reference patterns. DTW = 0 indicates that the two patterns are equal [28]
GVS	GVS is the RMS of knee joint angle

Before the patient analysis, the indices I_C and I_{BC} were calculated for a subgroup of the men healthy group. The subgroup is constituted by the data of two men which was used by the ELM, and the data used in the test of the ELM. These indices of the healthy group will be used as reference of healthy gait in phase (c).

2. Calculation of G_{Index} and GVS

The comparison of real $\left(I_C^{real}\right)$ and generated patterns $\left(I_C^{ML}\right)$ requires indices that measure the difference between patterns, here named as I_{BC}. So the I_C were converted to I_{BC} as

$$I_{BC} = \left|\left(I_C^{real}\right)_i - \left(I_C^{ML}\right)_i\right| \tag{3}$$

where i represents each I_C. Before the patient analysis, the I_C of the subgroup of the men healthy group also were converted to I_{BC}. With the results of the healthy people were calculated the mean value of each I_{BC}, $\overline{I}_{BC}^{ref}$, in function of knee and test speed.

The G_{Index}, developed in this work, indicates the pathology severity in function of knee and gait speed. This index compares the patient indices with the expected indices, estimated for him with the ML patterns and the indices presented by the healthy people. The G_{Index} results of the following equation:

$$G_{Index} = \sqrt{\frac{\sum_{i=1}^{10} x_i^2}{10}} \tag{4}$$

where x_i is calculated as

$$x_i = \frac{(I_{BC})_i - (\overline{I}_{BC}^{Ref})_i}{(\overline{I}_{BC}^{Ref})_i} = \frac{(I_{BC})_i}{(\overline{I}_{BC}^{Ref})_i} - 1 \tag{5}$$

where i represent each one of the 10 I_{BC}, calculated above. The index R^2 was not in Eq. (5), his value supports G_{Index} result, if $R^2 \geq 90\%$ the G_{Index} is viable. The G_{Index} result is not viable due to problem in data collection or low generation capacity of the selected ML. G_{Index} value can be equal or higher than zero, where is value increases with level of the gait pathology severity.

3. Calculation of RDI

The Symmetry Index measures the symmetry between the movement of right and left knee. According to Gouwanda [29], x-axis and y-axis represent the patterns of right and left knee, respectively. Ideally right and left knee must describe the same movement ($SI_{ideal} = \pi/4$ rad) however this does not occur due to the dominance of one leg [30]. The SI is important because, during recovery, the leg's dominance can be changed, for example to protect the injured knee.

In this work, the SI normalized (SI_N) is obtain through Eq. (6). The SI_N measures the deviation between the SI value of the patient ($SI_{patient}$) and the SI_{ideal}. The SI_N also indicates the dominant knee: $SI_N > \pi/2$ dominance of left knee an $SI_N < \pi/2$ dominance of right knee.

$$SI_N = \left(SI_{patient} - SI_{ideal}\right)/SI_{ideal} \qquad (6)$$

Graphic representations of GVS and G_{Index} in function of SI_N are used to classify the gait of the patients as normal or pathological. These graphics include a reference area that represents a normal gait. The severity of the gait pathology increases with the distance to reference area. The positive x-axis limit of the reference area is the higher absolute value obtained for the subgroup of the healthy group in Eq. (6). The negative limit is the negative value of the positive limit. The minimum y-axis value of the reference area is zero (ideal value), and the maximum value is the highest value of GVS and G_{Index} obtained in healthy subgroup.

The RDI is the module of the vector between ideal and patient indices, for each knee and gait speed. The ideal index is represented in the graphic origin. The RDI value increases with the ideal distance.

The RDI was used to realize a comparative analysis between the recovered patients and between the not recovered patients. For the patients classified as recovered, the RDI is the sum of vector's magnitude of each patient. For the not recovered patients the RDI only results from the sum of vector's magnitude of the points represented outside the reference area, which are the indices with more importance in the patient recovery.

3 Results

The analysis of the proposed methods takes into account their experimental results. Comparing between real and generated patterns of patients using the ELM.

The healthy subgroup, dominant of right knee, presents a SI = 0.77 ± 0.06 rad. The standard deviation value obtained is within the values showed for healthy people [31, 32]. The results of this group defines the limits of the reference area where the maximum SI_N, G_{Index} and GVS values are 0.18, 0.97 and 1.21 rad, respectively.

Figures 4 and 5 show the GVS and G_{Index} results, respectively, of the patients in function of the SI_N. From this results were excluded results without an accurate pattern generated by the ELM. Regarding graphics symbols: colors represent different patients, named in legend by 'P'; symbols represent different gait speeds, where S1, S2, S3, S4, S5 match to the following velocities 0.36, 0.78, 1.06, 1.33, 1.28 m.s^{-1}, respectively; right ('R') and left ('L') knees are represented by filled and unfilled symbols, respectively.

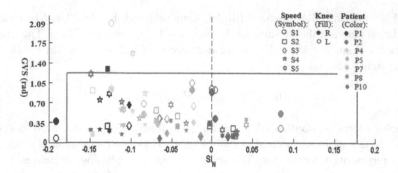

Fig. 4. GVS results of patients in function of SI_N

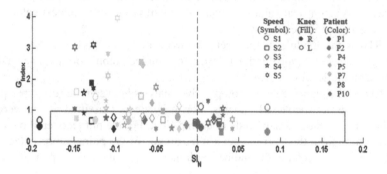

Fig. 5. GIndex results of patients in function of SI_N

In Table 4 is presented the patients' knee subjected to the ligamentoplasty (in brackets is presented patient's knee with problems before suffering the rupture of the anterior cruciate ligament of the knee) and the knee with limitations according GVS and G_{Index}.

Table 4. Patients' knee injured and knee classify as injured by GVS and G_{INDEX}

Patient	Injured knee	GVS	G_{Index}
P1	L	R and L	R and L
P2	R	–	R and L
P4	L (R)	L	L
P5	L	–	R
P7	L (R)	–	R and L
P8	R	–	L
P10	L	–	R and L

Table 5 shows the results of the RDI in function of the gait stage of the patients: recovery or not recovery. The RDI uses GVS and G_{Index} indices.

Table 5. RDI of the patients based in GVS and G_{INDEX}, in function of their gait state

Recovery patients	RDI		Not recovery patients	RDI	
	GVS (rad)	G_{Index}		GVS (rad)	G_{Index}
P1	–	–	P1	1.73	9.50
P2	5.21	–	P2	–	4.08
P4	–	–	P4	3.65	9.19
P5	4.56	–	P5	–	2.24
P7	4.53	–	P7	–	1.81
P8	2.83	–	P8	–	2.86
P10	2.55	–	P10	–	3.50

In ascending order of pathology severity (RDI), the GVS presents the recovered patients as P10 < P8 < P7 < P5 < P2; and the not recovered as P1 < P4. The G_{Index} organized the patients by the following ascending order: P7 < P5 < P8 < P10 < P2 < P4 < P1.

4 Discussion

The authors propose an analysis system for gait pathology diagnosis and evaluation of future pathologies. The analysis system realizes two processes: generation of a reference knee joint pattern for an individual with specific characteristics using the ELM and evaluation of the severity of the gait pathology.

The visual analysis of the patients during the data collection on the treadmill indicated that all the patients have a normal gait. However the results represented in Fig. 4 and in Fig. 5 show the existence of limitations in the gait of patients, specifying the gait speed and the knee that contribute for this classification. GVS (Fig. 4) classifies two patients as not recovery and G_{Index} (Fig. 5) classifies seven patients as not recovery, corroborating with the opinion of physiotherapists. Some years, after the ligamentoplasty, is possible that the patients are still not recovered, as the patients of article [33] who still present limitations in gait two years after the ligamentoplasty.

Both indices indicate P1 and P4 as the patients with more limitations in gait (Table 5). The difference between the number of patients classified as not recovery shows that the G_{Index} can detect more limitations, in gait, than GVS, once taking in account more indices in his calculation.

The majority of patients classified as not recovery shows gait limitations at low and high gait speeds as represented in Fig. 5. This information can be used by the rehabilitation team to direct the recovery treatment.

The GVS and the G_{Index} indicate that some patients have limitations in the knee which wasn't subjected to the ligamentoplasty (Table 4). This is possible if the patient used in excess the opposite knee with the objective to protect the injured knee. This excess of use of the opposite knee can cause limitations on it.

The good results and the high detection ability of the G_{Index} make it a useful gait index.

5 Conclusions

In this manuscript, it is proposed a gait analysis system composed by two parts: generation of specific reference knee patterns and evaluation of the severity of gait pathology.

The use of the knee patterns generated by the ELM as reference of healthy gait can provides an objective gait pathology analysis instead of a subjective one resulted of the use of a literature pattern which is generic for people with different characteristics.

In the evaluation of the severity of the gait pathology is compared the results of two gait index: GVS and G_{Index}. These gait index compares the patterns of patients submitted to a ligamentoplasty and the patterns generated by the selected ML algorithm, ELM. The detection capacity of gait limitations of the G_{Index} outperforms the GVS. The G_{Index} classifies 7 patients as not recovery, corroborating with the opinion of physiotherapists, and only classifies 2 patients as not recovery. The ordination of the recovered and the not recovered patients allows to compare the patients' pathology severity of the gait. This ordination shows consistency between the results of the two gait index. The patients classified as recovery by the GVS are defined by the G_{Index} as patients with minor severity of pathology and the both gait index classify the P1 and P4 as the patients with worse gait severity.

The gait analysis system proposed by the authors can be a useful tool for physiotherapy team in gait pathology diagnosis and evaluation of future pathologies.

The future work includes the increase of database and the addiction of data collected with an instrumented shoe registered by the team in national patent entitled "Instrumented shoe for gait analysis" [34].

In opposition to the ELM presented in this paper, the deep algorithms are robust against the overfitting problems and can automatically extract high-level feature hierarchies from high-dimensional data [35]. Some publications in the area of human gait analysis [36–38] show that the deep learning algorithms outperform the shallow algorithms. The studies [4] and [39] use the DBN (Deep Belief Networks) as a deep learning algorithm for gait analysis. So, the authors intend to develop a new ML algorithm constituted by unsupervised DBN layers to extract the features of the data and a supervised ELM on top layer.

Additionally, the authors intend to develop a contextual adaptation model to evaluate future gait pathologies. This model will use: gait indices; muscle strength test and analysis of gait pathology achieved by medical team; and DBN-ELM approach with data from the more frequent gait pathologies.

Acknowledgments. The Fundação para a Ciência e Tecnologia (FCT) is gratefully acknowledged for funding this work with the grants SFRH/BD/132408/2017 and PTDC/EEI-AUT/5141/2014 (Automatic Adaptation of a Humanoid Robot Gait to Different Floor-Robot Friction Coefficients). The authors also acknowledge the COMPETE 2020 program for the financial support with the PTDC/EEI-AUT/5141/2014.

Compliance with Ethical Standards
Conflict of Interest. All authors declare that they have no conflict of interest.

References

1. Sousa, A.S.P.: Controlo Postural em Marcha Humana: Análise Multifactorial. Ph.D. thesis, Faculdade de Engenharia da Universidade do Porto, Porto, Portugal (2010)
2. Araújo, A.G.N., Andrade, L.M., De Barros, R.M.L.: System for kinematical analysis of the human gait based on videogrammetry. Fisioter e Pesqui **11**(1), 3–10 (2005)
3. Muro-de-la-Herran, A., García-Zapirain, B., Méndez-Zorrilla, A.: Gait analysis methods: an overview of wearable and non-wearable systems, highlighting clinical applications. Sensors **14**(2), 3362–3394 (2014)
4. Nair, B.M., Kendricks, K.D.: Deep network for analyzing gait patterns in low resolution video towards threat identification. Electron. Imaging **2016**(11), 1–8 (2016)
5. Hannink, J., Kautz, T., Pasluosta, C.F., Klucken, J., Eskofier, B.M.: Sensor-based gait parameter extraction with deep convolutional neural networks. IEEE J. Biomed. Health Inf. **21**(1), 85–93 (2017)
6. Yun, Y., Kim, H.-C., Shin, S.Y., Lee, J., Deshpande, A.D., Kim, C.: Statistical method for prediction of gait kinematics with Gaussian process regression. J. Biomech. **47**(1), 186–192 (2014)
7. Luu, T.P., Low, K.H., Qu, X., Lim, H.B., Hoon, K.H.: An individual-specific gait pattern prediction model based on generalized regression neural networks. Gait Posture **39**(1), 443–448 (2014)
8. Winter, D.A.: The Biomechanics and Motor Control of Human Movement, 4th edn. Wiley, Hoboken (2009)
9. Abbass, S.J., Abdulrahman, G.: Kinematic analysis of human gait cycle. NUCEJ **16**(2), 208–222 (2014)
10. Lincoln, L.S., Bamberg, S.J.M., Parsons, E., Salisbury, C., Wheeler, J.: An elastomeric insole for 3-axis ground reaction force measurement. In: Proceedings of the IEEE RAS EMBS International Conference on Biomedical Robotics and Biomechatronics, pp. 1512–1517 (2012)
11. Najafi, B., Khan, T., Wrobel, J.: Laboratory in a box: wearable sensors and its advantages for gait analysis. In: Proceedings of the International Conference of the IEEE Engineering in Medicine and Biology Society, EMBS, pp. 6507–6510 (2011)
12. Crea, S., Donati, M., De Rossi, S.M.M., Oddo, C.M., Vitiello, N.: A wireless flexible sensorized insole for gait analysis. Sensors **14**, 1073–1093 (2014)
13. Lind, R.F.: Wearable ground reaction force foot sensor. United States patent US 2014/0013862 A1, 16 January 2014
14. Ornetti, P., Maillefert, J.F., Laroche, D., Morisset, C., Dougados, M., Gossec, L.: Gait analysis as a quantifiable outcome measure in hip or knee osteoarthritis: a systematic review. Joint Bone Spine **77**(5), 421–425 (2010)
15. Rani, M.P., Arumugam, G.: Children abnormal GAIT classification using extreme learning machine. Glob. J. Comput. Sci. Technol. **10**(13), 66–72 (2010)
16. Kong, W., Saad, M.H., Hannan, M.A., Hussain, A.: Human Gait State Classification using Artificial Neural Network, pp. 0–4 (2014)
17. Prakash, C., Mittal, A., Tripathi, S., Kumar, R., Mittal, N.: A framework for human recognition using a multimodel gait analysis approach. In: International Conference on Computing, Communication and Automation (ICCCA 2016), pp. 1–5 (2016)
18. Schwartz, M.H., Rozumalski, A.: The gait deviation index: a new comprehensive index of gait pathology. Gait Posture **28**(3), 351–357 (2008)

19. Baker, R., et al.: The gait profile score and movement analysis profile. Gait Posture **30**(3), 265–269 (2009)
20. Celletti, C., et al.: Use of the gait profile score for the evaluation of patients with joint hypermobility syndrome/ehlers-danlos syndrome hypermobility type. Res. Dev. Disabil. **34** (11), 4280–4285 (2013)
21. Ferreira, J.P., Vieira, A., Ferreira, P., Crisóstomo, M., Coimbra, A.: Human knee joint walking pattern generation using computational intelligence techniques. Neural Comput. Appl. **30**(6), 1701–1713 (2018). https://doi.org/10.1007/s00521-018-3458-5
22. Ferreira, P.A., Ferreira, J.P., Crisóstomo, M., Coimbra, A.P.: Low cost vision system for human gait acquisition and characterization. In: IEEE International Conference on Industrial Engineering and Engineering Management, vol. 2016-Decem, pp. 291–295 (2016)
23. Bohannon, R.W.: Comfortable and maximum walking speed of adults aged 20–79 years: reference values and determinants. Age Ageing **26**(1), 15–19 (1997)
24. Darras, N.: Gait Analysis ADplot (2013). https://sites.google.com/site/gaitanalysisadplot/file-cabinet. Accessed 30 Dec 2018
25. Mostayed, A., Mynuddin, M., Mazumder, G., Kim, S., Park, S.J., Korea, S.: Abnormal Gait Detection Using Discrete Fourier Transform, vol. 3, no. 2, pp. 1–8 (2010)
26. Gabel, M., Gilad-Bachrach, R., Renshaw, E., Schuster, A.: Full body gait analysis with Kinect. In: Proceedings of the Annual International Conference of the IEEE Engineering in Medicine and Biology Society, EMBS, pp. 1964–1967 (2012)
27. Pietraszewski, B., Winiarski, S., Jaroszczuk, S.: Three-dimensional human gait trajectory – reference data for normal men. Acta Bioeng. Biomech. **14**(3), 9–16 (2012)
28. Müller, M.: Dynamic time warping. In: Müller, M. (ed.) Information Retrieval for Music and Motion, pp. 69–84. Springer, Heidelberg (2007). https://doi.org/10.1007/978-3-540-74048-3_4
29. Gouwanda, D.: Comparison of gait symmetry indicators in overground walking and treadmill walking using wireless gyroscopes. J. Mech. Med. Biol. **14**(01), 1450006 (2014)
30. Lathrop-Lambach, R.L., et al.: Evidence for joint moment asymmetry in healthy populations during gait. Gait Posture **40**(4), 526–531 (2013)
31. Herzog, W., Nigg, B.M., Read, L.J., Olsson, E.: Asymmetries in ground reaction force trajectorys in normal human gait. Med. Sci. Sport. Exerc. **21**(1), 110–114 (1989)
32. Bensoussan, L., Mesure, S., Viton, J.M., Delarque, A.: Kinematic and kinetic asymmetries in hemiplegic patients' gait initiation trajectorys. J. Rehabil. Med. **38**(5), 287–294 (2006)
33. Costa, L., et al.: Application of machine learning in postural control kinematics for the diagnosis of Alzheimer's disease. Comput. Intell. Neurosci. **2016**, 1–15 (2016)
34. Ferreira, J., et al.: Calçado instrumentado para análise da marcha. Patent PT 108143 A1. Internet, 4 January 2018. http://www.marcasepatentes.pt/files/collections/pt_PT/49/55/573/593/2016-07-11.pdf
35. Alsheikh, M.A., Selim, A., Niyato, D., Doyle, L., Lin, S., Tan, H.-P.: Deep activity recognition models with triaxial accelerometers. In: The Workshops of the Thirtieth AAAI Conference on Artificial Intelligence Artificial Intelligence Applied to Assistive Technologies and Smart Environments: Technical Report WS-16-01, pp. 8–13 (2015)
36. Ordóñez, F.J., Roggen, D.: Deep convolutional and LSTM recurrent neural networks for multimodal wearable activity recognition. Sensors **16**(1), 115 (2016)
37. Castanharo, R., Da Luz, B.S., Bitar, A.C., D'Elia, C.O., Castropil, W., Duarte, M.: Males still have limb asymmetries in multijoint movement tasks more than 2 years following anterior cruciate ligament reconstruction. J. Orthop. Sci. **16**(5), 531–535 (2011)

38. Wang, Y., Chen, Y., Bhuiyan, M.Z.A., Han, Y., Zhao, S., Li, J.: Gait-based human identification using acoustic sensor and deep neural network. Futur. Gener. Comput. Syst. **86**, 1228–1237 (2017)
39. Sukhbaatar, S., Makino, T., Aihara, K., Chikayama, T.: Robust generation of dynamical patterns in human motion by a deep belief nets. J. Mach. **20**, 231–246 (2011)
40. Zeng, W., Ma, L., Yuan, C., et al.: Artif. Intell. Rev. **52**, 449 (2019). https://doi.org/10.1007/s10462-018-9645-z
41. Han, A.B.S., Gellhorn, A.C.: Trajectories of quality of life and associated risk factors in patients with knee osteoarthritis. Am. J. Phys. Med. Rehabil. **97**(9), 620–627 (2018)

Driving Mode Estimation Model Based in Machine Learning Through PID's Signals Analysis Obtained From OBD II

Juan José Molina Campoverde[✉]

Automotive Mechanical Engineering: GIIT Transport Engineering
Research Group, Salesian Polytechnic University, Cuenca, Ecuador
jmolinac4@est.ups.edu.ec

Abstract. In this paper a driving mode estimation model based in machine learning architecture is presented. With the statistic method, Random Forest, the highest inference of driving variables is determined through the best attributes for a training model based in OBD II data. Engine sensors variables are obtained with the aim of explaining the behavior of the PID signals in relation to the driving mode of a person, according to specific consumption and engine performance, characterizing the signals behavior in relation to the different driving modes. The investigation consists of 4 power tests in the dynamometer bank at 25%, 50%, 75% and 100% throttle valve opening to determine the relationship between engine performance and normal vehicle circulation, through the engine most influential variables like MAP, TPS, VSS, Ax and each the transmission ratio infer in the fuel consumption study and engine performance. In this study Random Forest is used achieving an accuracy rate of 0.98905.

Keywords: Random forest · OBD II · Fuel consumption · Eco drive · K-means · S-Golay

1 Introduction

Facing the increase in environmental pollution caused by greenhouse gas emissions produced by combustion engines [1], electronic injection systems were implemented. Through the on-board diagnostic (OBD-II), available since 1994 [2], different variables of sensors are obtained, like longitudinal acceleration (Ax), Engine Speed (RPM), Vehicle Speed Sensor (VSS), Manifold Air Flow (MAF), Manifold Absolute Pressure (MAP) [3]. A methodology used to determine sensors behavior is carried out by driving modes study (Sport, Normal and ECO) where Lv et al. [4], highlight behaviors and interactions between the driver and the vehicle, these modes are identified according to driving styles by means of maneuvers [5], to estimate fuel consumption and polluting emissions [6]. Due to the average driver can increase efficiency by 10% simply by changing the driving mode [7], Hooker et al., in case of Eco drive mode, covers low consumption driving behaviors due to the accelerator pedal differential use [8], which influences in fuel economy and vehicle emissions to reduce its environmental impact [9]. In addition, an efficient driving mode is the main solution to reduce greenhouse

© Springer Nature Switzerland AG 2020
M. Botto-Tobar et al. (Eds.): ICAT 2019, CCIS 1194, pp. 80–91, 2020.
https://doi.org/10.1007/978-3-030-42520-3_7

gases emission produced by the vehicles [10]. According to Sivak et al., factors such as drivers decision respect to the vehicle maintenance or route selection influence in fuel economy [9], as well as making early gear changes avoid high speeds, reduce polluting emissions and fuel consumption [11, 12]. Lapuerta et al. [13], considers how it infers the altitude above sea level effect in internal combustion engines due to the air density that modifies its thermodynamic cycle and as consequence decreases the engine power, increases the polluting emissions and fuel consumption, considering Cuenca altitude, 2550 m.a.s.l. [14].

2 Materials and Methods

2.1 Experimental Overview

For the development of the present investigation, a review of previous researches was realized to propose an adequate data acquisition and analysis methodology. Zhou et al. [15], propose an optimization model of engine repair based on the classification of a Random Forest algorithm, which by adaptation and decision of more significant variables allows classifying engine real data, demonstrating that the more samples are trained the higher the classification accuracy. Considering the weight of applied variables can reach 79.1% of precision in the applied model. Damström et al. [16], propose a methodology that consists of finding the ideal number of groupings controlling similar characteristics between them by means of a cross correlation, but it has the disadvantage that it doesn't guarantee to find a global minimum which the approximation of cluster centroid changes its results. Based in the Corcoba et al. methodology [17], a diffuse logic system was used to simulate human behavior to estimate an efficient driver behavior by input and output variables, its output is rated with a classification between 0 to 1 depending how efficient the driving is, the proposal characterized the driving style by the data acquisition through telemetry. Pereira et al. [18], present a mobile application based on diffuse logic, through OBD-II data acquisition to characterize the driving efficiency by estimating the specific consumption with an accuracy of 85% in this model.

Oñate et al. [19], propose an intelligent system of erratic driving diagnosis through OBD-II and GPS data acquisition, which proposes reduce accidents by extracting driving signals, mathematical signal analysis and driving fault detection through Neural Networks, evaluating the model they guaranteed its viability and reliability. Chen et al. [20], present a semi – supervised classification method through a Support Vector Machine (SVM) algorithm, by data acquisition of velocity and engine speed data by OBD II from gasoline and diesel passengers vehicles with the vehicle identification number (VIN), it can classify the vehicles through a specific classification rule, whose purpose is applying to vehicles with internet connection and to improve estimation of CO_2 emissions. Corcoba et al. present an efficient driving model through OBD-II data acquisition by a smartphone which provides ecological driving suggestions to the driver for bad driving habits, with the aim of improve the driving style through the efficiency consumption by algorithms such as J48 [21], Naïve Bayes and Random Forest, where the last one allows a better control of random attributes for each node and was selected because its execution time is less than J48, especially when the attributes number or the training set is bigger with a classification certainty of 83.17 and 71.15% respectively.

Given the background, the following methodology is proposed (Fig. 1):

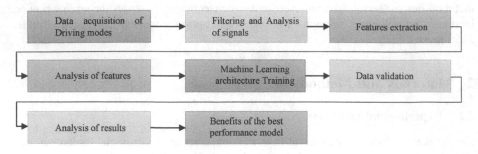

Fig. 1. Methodology

2.2   OBD-II Data Acquisition Though Data Logger Device

OBD-II data acquisition is realized by the data logger device, "Freematics ONE+", which allows to extract all the Electronic Control Unit (ECU) sensors data and register the information in a ".CSV" archive, it's saved in a micro SD for a later processing and research. Stored Variables are shown on Table 1.

Table 1.   Stored variables

Variables	Nomenclature	Units
Longitudinal acceleration	Acceleration	m/s^2
Vehicle speed sensor	VSS	m/s
Engine coolant temperature	ECT	°C
Intake air temperature	IAT	°C
Manifold absolute pressure	MAP	KPa
Oxygen sensor	O2	V
Engine speed	RPM	rpm
Throttle position sensor	TPS	%
Long term fuel trim	LTFT	%
Short term fuel trim	STFT	%

For the experiment development, some test were realized on the dynamometer bank MAHA LP 3000 based in the ISO 17359 2018 "Condition monitoring and diagnostics of machines" [22] to obtain engine power and specific consumption data according to quality criterions.

To characterize method based in Meseguer et al. [1–3], Dörr et al. propose analyze driving incidence factors, such as, input, study, noise and controllable variables. These variables are shown in Table 2.

Table 2. Incidence factors

Variables	Factors	Symbol	Unit
Input	Longitudinal acceleration	A	[m/s^2]
	Vehicle speed	VSS	[m/s]
	Manifold absolute pressure	MAP	[kPa]
	Engine speed	η	[rpm]
	Throttle position sensor	TPS	[%]
Controllable	Coolant temperature	T Coolant	[°C]
	Rolling temperature	T Rolling	[°C]
	Lubricant temperature	T Lubricant	[°C]
Noise	Oxygen sensor	O2	[V]
	Air temperature	T Air	[%]
	Fuel temperature	T Fuel	[°C]
	Atmospheric pressure	P Atm	[Pa]
	Relative humidity of air	H Rel	[%]
Study	Power	P	[kW]
	Specific consumption	C esp	[g/kWh]

2.3 Signal Filtering

Motor signals when acquiring data present noise due to the existence of incidence factors in the vehicle, for this reason, these signals go through a filter stage; this consists in a comparison between real signals versus those obtained from sensors, for this purpose the S-Golay filter [24] was applied in order to obtain clean signals for a correct interpretation.

3 Results

3.1 Experimentation Results

With 4 power test, the vehicle performance was characterized as the volumetric fuel consumption that a low engine speed it is low, then when speed engine is increased by the increased throttle opening, the volumetric fuel consumption gets a gradually increase due to a bigger amount of air – fuel mixture causing more engine power, (see Fig. 2).

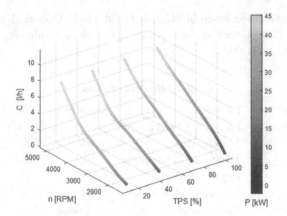

Fig. 2. Volumetric consumption

Relating obtained results of volumetric fuel consumption with engine power at 25%, 50%, 75% and 100% throttle valve opening, engine specific consumption is determined by (1):

$$\text{Equation: } C_{esp} = \frac{\rho \dot{v}}{P} \left[\frac{g}{kWh} \right] \tag{1}$$

Where:

$\dot{v} = Fuel\,Volumetric\,Flow\,(l/h)$
$\rho = Fuel\,Density\,(g/l)$
$P = Engine\,Power\,(kW)$.

The engine specific consumption depends directly of engine speed and the input air – fuel mixture regulated by the throttle valve opening angle, (see Fig. 3).

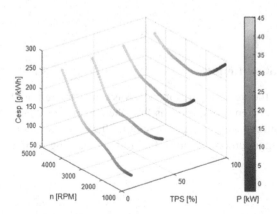

Fig. 3. Engine specific consumption.

3.2   Machine Learning Architectures Application

For obtained data application from dynamometer bank in real driving cycles a model that explain the relationship between engine performance and normal circulation vehicle is proposed, for later determine by means of an artificial neural network application the fuel consumption and finally to apply fuzzy logic to obtain the driving mode.

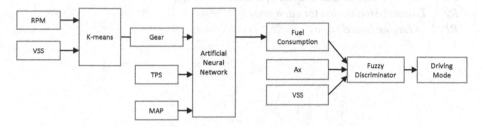

Fig. 4. Driving estimation model

Vehicle performance depends directly of transmissions ratios of each gear, so these values are obtained by the application of K-means method based in relation analysis between speed vehicle and speed engine (VSS/RPM) taken from a normal driving cycle, (see Fig. 4).

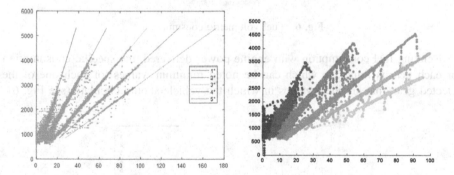

Fig. 5. K-means classification.

Where the following transmission ratios values are obtained (Table 3):

Table 3. Transmission ratios

Transmission ratios				
Rn	R1	R2	R3	R4	R5
0	0.0088	0.0137	0.0212	0.0299	0.0425

Fuel Volumetric Consumption depends directly of engine speed and the vehicle gear selected, as shown in (2) (see Fig. 5).

$$C_n \left[\frac{l}{h} \right] = R_n * RPM \tag{2}$$

Where:

$C_n = Fuel\ Volumetric\ Consumption\ for\ each\ gear\ (l/h)$
$R_n = Transmission\ Ration\ for\ each\ gear$
$RPM = Engine\ Speed\ (rpm)$

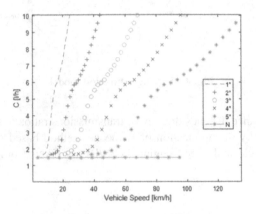

Fig. 6. Fuel volumetric consumption

Relating fuel consumption with engine power delivered, the specific consumption for each gear is found, in which can be noted minimum values for each one of the selected gear, that are ideal values in which the vehicle should circulate (see Fig. 6).

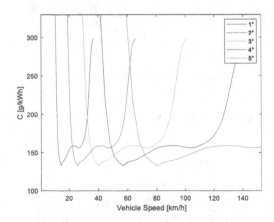

Fig. 7. Specific consumption for each gear.

Given the performance, maximum engine acceleration for each selected gear is determined, as shown in (3)

$$A_x = \frac{M_{nor} * \eta_i * \frac{R_C R_{Cn}}{re}}{m_{veh} * \gamma_m}$$ (3)

Where:

$A_x = Longitudinal\ Acceleration$
$\eta_i = Gear\ efficiency$
$M_{nor} = Normal\ Momentum$
$\frac{R_C R_{Cn}}{re} = R_n = Total\ Transmission\ Ratio$
$m_{veh} = Vehicle\ Mass$
$\gamma_m = Rotative\ Mass\ Coficient$
$n = Selected\ gear$
$\gamma_m = 1,05 + 0,0025 R_n^2$

Vehicle maximum acceleration for each gear depends on engine speed and due to transmission ratio of the vehicle velocity [25], as shown in Fig. 8.

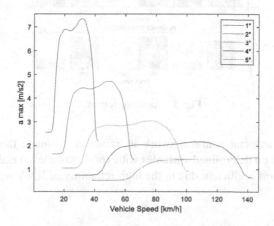

Fig. 8. Vehicle maximum acceleration for each gear

For an efficient driving gear changes should be done when specific consumption is relatively low according with Fig. 7. Table 4 shows values where fuel consumption is minimized.

Table 4. Specific consumption eco drive

Eco drive					
Gear change	Speed (km/h)	Rpm max	Previous RPM	Ax (m/s²)	Previous Ax (m/s²)
1–2	26,03	3400	1900	6,908	4,205
2–3	42,47	3100	2000	4,501	2,767
3–4	61,48	2900	2100	2,860	1,990
4–5	86,71	2900	2000	2,033	1,380

Random Forest algorithm is presented, which its main purpose is to find relationship between input and output study variables, to determine more important variables in fuel consumption according with driving characteristics. Through study attributes, Random Forest allows to classify real engines data, demonstrating that while more samples being trained the greater would be the classification accuracy [15], also it allows a better control of the random attributes number for each node with an execution time really efficient [21], proving that fuel consumption depends directly of MAP, TPS, VSS, Ax and the selected gear, but not from engine speed, because this factor is in close relationship with VSS and Ax, with a determination index of 0.98905, (see Fig. 9).

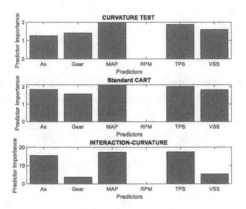

Fig. 9. Random forrest.

From this, an artificial neural network is trained to estimate the volumetric fuel consumption based on determined variables with 16661 samples to estimate the driving mode. The prediction is efficient due to the high reliability of RNA with a global index of 0.98905 (Table 5).

Table 5. ANN training

	Samples	Samples (%)	R
Training	11663	70	0.98886
Validation	2499	15	0.98878
Testing	2499	15	0.99026

The found settings (see Fig. 10):

With this model an efficient vehicle speed is estimated by VSS and acceleration is determined to evaluate change gear according to engine speed. As the engine speed increases, a gear change is required, with the aim of increase vehicle speed and avoid excessive fuel consumption so that when the gear change is made the engine revolutions fall until get the new gear, considering that if the change gear is not realized, fuel consumption gets increased.

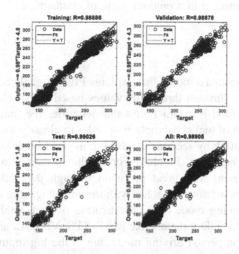

Fig. 10. Relationship between expected values and estimated values by ANN

By Predictor importance the strength of the relationship of the predictors and the PID data result is quantified by the import of the samples, the viability of the established model is determined according to its importance and the training given.

In this work a diffuse classifier such as Manami is proposed with a total of 43 rules, which has many input variables: speed, acceleration, volumetric fuel consumption of the vehicle and determines the driving mode between Sport, Normal and Eco. The classifier is applied to obtained data in real driving (see Fig. 11) during start-up the Sport driving mode is used meanwhile eco mode is used in decelerations and small displacements and for the rest of scenarios normal mode is frequently used.

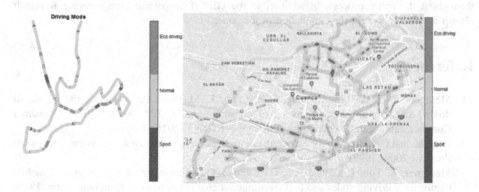

Fig. 11. Characterization of the driving mode [23]

Through the characterization of the driving modes it is determined that the Eco-driving mode is the most efficient due to the differentiated use of the acceleration pedal which establishes the specific consumption is low compared to the other modes by optimizing fuel consumption in a random route of conduction.

4   Conclusions

In this paper it was obtained a model that explains PID's signals behavior according to a person driving mode through thresholds values based on Machine Learning architecture by statistic method Random Forest that identifies the major importance variables on the estimation of fuel consumption in order to classify the driving modes in a random route, representing by sections the driving modes Eco drive, Normal and Sport according to driving behavior. Random Forest allows estimating sensors variables values that are unknown, with great precision through data set adjustments using characteristics of engine sensors variables, with the advantage that the model presents a minimum error for driving mode data classification of each vehicle with a determination index of 0.98905. PID's signals like MAP, TPS, VSS, Ax and the selected gear, directly influence in the person driving mode due to the big input mixture air – fuel, where highlights specific consumption, polluting emissions and vehicle performance considering that engine speed doesn't have a relationship with VSS and Ax, it doesn't infer in the model behavior. The application of K-means method based on the analysis of the relationship between vehicle speed and engine speed (VSS/RPM) though obtained data from the dynamometer bank establishes a model based on the specific consumption and engine performance for driving modes characterization under normal driving conditions.

This paper aims to facilitate the study of future projects for the determination of the estimation of the driving mode through Machine Learning architectures applied to the study of the analysis of polluting emissions.

Acknowledgment. To Mr. Nestor Diego Rivera Campoverde, for his direction and unconditional collaboration in the realization of the paper with his contributions and his knowledge throughout the entire process, in addition to the GIIT Transportation Engineering Research Group for his support for the completion of the paper.

References

1. Meseguer, J., Toh, C.K., Calafate, C.T., Cano, J., Manzoni, P.: Assessing the impact of driving behavior on instantaneous fuel consumption. In: 12th Annual IEEE Consumer Communications and Networking Conference (CCNC) (2015)
2. International Organization for Standardization, Road vehicles, Diagnostic systems, Keyword Protocol 2000 (1999)
3. Meseguer, J., Toh, C.K., Calafate, C., Cano, J.Y., Manzoni, P.: DrivingStyles: a mobile platform for driving styles and fuel consumption characterization. J. Commun. Netw. **19**(2), 162–168 (2017)

4. Lv, C., et al.: Cyber-physical system based optimization framework for intelligent powertrain control. SAE Int. J. Commer. Veh. **10**(1), 254–264 (2017)
5. Mohd, T.A.T., Hassan, M., Aris, I., Azura, C., Ibrahim, B.S.K.K.: Application of fuzzy logic in multi-mode driving for a battery electric vehicle energy management. Int. J. Adv. Sci. Eng. Inf. Technol. **7**, 284–290 (2017)
6. Dia, H., Panwai, S.: Impact of driving behavior on emissions and road network performance. In: IEEE International Conference on Data Science and Data Intensive Systems (2015)
7. Barkenbus, J.N.: Eco-driving: an overlooked climate change initiative. Energy Policy **38**(2), 762–769 (2010)
8. Hooker, J.N.: Optimal driving for single-vehicle fuel economy. Transp. Res. A Gen. **22**(3), 183–201 (1988)
9. Sivak, M., Schoettle, B.: Eco-driving: strategic, tactical, and operational decisions of the driver that influence vehicle fuel economy. Transp. Policy **22**, 96–99 (2012)
10. Andrieu, C., Saint Pierre, G.: Comparing effects of eco-driving training and simple advices on driving behavior. Procedia – Soc. Behav. Sci. **54**, 211–220 (2012)
11. Beckx, C., Panis, L.I., De Vlieger, I., Wets, G.: Influence of gear changing behavior on fuel-use and vehicular exhaust emissions. Highw. Urban Environ. **12**, 45–51 (2007)
12. McIlroy, R., Stanton, N., Godwin, L., Wood, A.: Encouraging eco-driving with visual, auditory, and vibrotactile stimuli. IEEE Trans. Hum.-Mach. Syst. **47**(5), 661–672 (2017)
13. Lapuerta, M., Armas, O., Agudelo, J., Sánchez, C.: Study of the altitude effect on internal combustion engine operation. Part 1: performance. Technol. Inf. **17**(5), 21–30 (2006)
14. Rivera, N., Chica, J., Zambrano, I., García, C.: Estudio del comportamiento de un motor ciclo otto de inyección electrónica respecto de la estequiometría de la mezcla y del adelanto al encendido para la ciudad de cuenca. Revista Politécnica **40**(1), 59–67 (2017)
15. Zhou, Y., Guo, J., Fu, L., Liang, T.: Research on aero-engine maintenance level decision based on improved artificial fish-swarm optimization random forest algorithm. In: 2018 International Conference on Sensing, Diagnostics, Prognostics, and Control (2018)
16. Damström, J., Gerlitz, C.: Classification of Power Consumption Patterns for Swedish Households Using K-means. K4. Power Consumption Analysis (2016)
17. Corcoba, V., Muñoz, M.: Eco-driving: energy saving based on driver behavior. In: XVI Jornadas de ARCA sobre Sistemas Cualitativos y sus Aplicaciones en Diagnosis, Robótica e Inteligencia Ambiental (JARCA) (2015)
18. Pereira, A., Alves, M., Macedo, H.: Vehicle driving analysis in regard to fuel consumption using fuzzy logic and OBD-II devices. In: 2016 8th Euro American Conference on Telematics and Information Systems (EATIS) (2016)
19. Oñate, J.A., Christian M. Quintero, G., Pérez, J.M.: Intelligent erratic driving diagnosis based on artificial neural networks. In: IEEE ANDESCON (2010)
20. Chen, S., Lin, R., Liu, W., Tsai, J.: The semi-supervised classification of petrol and diesel passenger cars based on OBD and support vector machine algorithm. In: 2017 International Conference on Orange Technologies (ICOT) (2017)
21. Corcoba, V., Muñoz, M.: Artemisa: using an android device as an eco-driving assistant. Cyber J.: Multidiscip. J. Sci. Technol. J. Sel. Areas Mechatron. (JMTC), June Edition, 3–7 (2011)
22. ISO 17359: Condition monitoring and diagnostics of machines (2018)
23. Google Maps. https://www.google.com/maps. Accessed 01 July 2019
24. Pang, C.K., et al.: Intelligent energy audit and machine management for energy-efficient manufacturing. In: IEEE 5th International Conference on Cybernetics and Intelligent Systems (CIS) (2011)
25. Aparicio, F., Vera, C., Díaz, V.: Teoría de los vehículos automóviles. Universidad Politécnica de Madrid, Madrid, pp. 279–285 (1995)

Deep Learning Methods in Natural Language Processing

Alexis Stalin Alulema Flores[1,2(✉)]

[1] University of New Mexico, Albuquerque, NM, USA
alulema@unm.edu
[2] Number8, Louisville, KY, USA
aalulema@number8.com

Abstract. The purpose of this paper is to make a concise description of the current deep learning methods for natural language processing (NLP) and discusses their advantages and disadvantages. The research further discusses the applicability of each deep learning method in the context of natural language processing. Additionally, a series of significant advances that have driven the processing, understanding, and generation of natural language are also discussed.

Keywords: Natural language processing (NLP) · Deep learning · Neural networks

1 Introduction

NLP is currently an active field, in which deep learning (DL) research has progressed substantially. However, while in other areas such as computer vision (CV), the accuracy to be obtained is relatively high, in NLP, there are still many logical and precision problems to be solved, which will have a profound impact on performance improvement obtained through the application of DL.

There are several practical everyday applications where DL-based NLP is applied, such as Apple's voice assistant. Cloud-based services from leading companies in NLP have created natural linguistic understanding-based applications, such as Microsoft LUIS, Google AutoML Natural Language, among others.

Languages that use ideographic characters, such as Chinese, require morphological analysis, a technique that considers a unique syntax and word separation.

NLP has several issues that need further research, but some of the limitations have been solved with the introduction of new models of neural networks, and specialized frameworks such as TensorFlow or PyTorch. Additional research in DL will result in new models with greater efficiency and performance. Some other desirable features of the new models are discussed later in this paper.

© Springer Nature Switzerland AG 2020
M. Botto-Tobar et al. (Eds.): ICAT 2019, CCIS 1194, pp. 92–107, 2020.
https://doi.org/10.1007/978-3-030-42520-3_8

2   Distributed Representations of Words

For text information to be processed by DL models, it is necessary to transform words into binary digits, which is the input format that the models can understand. Computers process information in binary digits (bits). The most basic representation of words is a **sparse representation** or **one-hot** encoded representation.

For example, if we have a vocabulary of 10,000 words:

$$V = [a, Aaron, \ldots, Zulu, \langle Unknown \rangle], 0 - index\,based$$

'a' is in index 0, 'Aaron' is in index 1, and so on. To represent a word within this vocabulary, the vector that represents it is 10,000 long, with the value 1 in the index representing the word within the dictionary, for example:

$$Dog(4,001) = [0,0,0,0,\ldots,1,\ldots,0,0]$$

However, there are a couple of drawbacks related to this representation:

- The size of the representation increases with the size of the vocabulary
- It does not provide any information about the correlation between words

A **distributed representation** is how we represent a word as some latent features that each correspond to some semantic or syntactic concept. For example (Table 1):

Table 1. Example of distributed representation.

Word	Features			
	Colorr	Daylight	...	Polarity
Black	−1.0	0.01	...	0.02
White	1.0	0.02	...	−0.01
Night	−0.97	0.97	...	0.01
Day	0.98	0.99	...	−0.02
Wonderful	0.01	0.05	...	0.94

This is the representation that we expect from the neural network, which will be trained with the input "one-hot encoded" words to get a weight for several features.

3   Word2Vec

Word2Vec [1] is an algorithm used to learn these distributed representations of words (word embeddings). It uses a 2-layered feed-forward neural network (FFNN) trained to reconstruct the linguistic context of words (Fig. 1).

```
The quick brown fox jumps over the lazy dog.
{(the, quick), (the, brown)}

The quick brown fox jumps over the lazy dog.
{(quick, the), (quick, brown), (quick, fox)}

The quick brown fox jumps over the lazy dog.
{(brown, the), (brown, quick), (brown, fox), (brown, jumps)}

The quick brown fox jumps over the lazy dog.
{(fox, quick), (fox, brown), (fox, jumps), (fox, over)}
```

Fig. 1. Word context pairs [2].

The core concept behind Word2Vec is, given a word, it seeks to predict the words that appear around this word in the sentences. Words that appear in similar contexts around a word are assumed to have a similar meaning and semantic relationship.

3.1   Skip-Gram Network Architecture

In this architecture, the input sentences are transformed into one-hot vectors and then passed into the hidden layer, which learns the context. Lastly, the hidden layer's output is passed to the output layer that predicts the probability that a word can appear in the sentence.

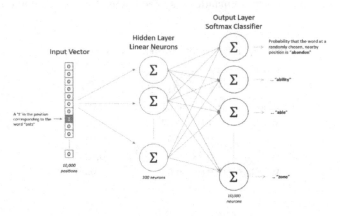

Fig. 2. Word context pairs [2].

3.2   Hidden Layer

From Fig. 2, the hidden layer has 300 nodes; each node represents a feature to learn for every word in the vocabulary. A matrix makes this layer of 300 neurons by 10,000 words, where each row corresponds to a single word in the vocabulary, and columns represent dimensions to learn for each concept. The layout is the concept behind **distributed representation**.

It is not clear which of these dimensions represents a specific concept. The number of dimensions is entirely arbitrary, and the designer chooses it according to a project's

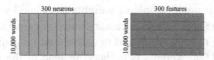

Fig. 3. Hidden layer weight matrix to word vector lookup table [2].

needs. If the input text is only financing data, then 50 dimensions could be sufficient. If the input text is literature, then approximately 500 dimensions are required.

This matrix is multiplied by the one-hot vector, and it selects the matrix row corresponding to the "1", as shown in the following example:

$$[0 \quad 0 \quad 1 \quad 0] \times \begin{bmatrix} 25 & 17 & 18 \\ 24 & 11 & 1 \\ 19 & 23 & 12 \\ 5 & 10 & 7 \end{bmatrix} = [19 \quad 23 \quad 12]$$

The hidden layer does not use the activation function. The input and output vectors are both one-hot vectors, but when the trained network is evaluated, the output vector is a probability distribution where each node is the index of a word in the vocabulary.

3.3 Word2Vec – Output Layer

Each node in the output layer represents a single word in the vocabulary. Each node in the hidden layer is connected to every node in the output layer. The output layer applies the softmax function in every node for each node in the hidden layer (Fig. 4).

Softmax summarizes this into a probability distribution where each node in the output layer obtains the probability of that word in the context.

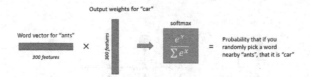

Fig. 4. Calculating the output neuron for word "car" [2].

3.4 Negative Sampling

From Fig. 3, there are $300 \times 10,000 = 3,000,000$ weights in the hidden layer, which makes it extremely slow to train since it needs a significant amount of training data while simultaneously avoiding over-fitting. However, given that it is predicting a single word at a time, it only has a unique "positive" output out of 10,000 possible outputs.

For efficiency, only a small sample of weights associated with "negative" samples is updated randomly. The approach is necessary since negative samples do not give the network the information to learn from, and negative sampling speeds up the training as only a fraction of 3,000,000 weights is updated instead of the entire data.

The desirable network output for a negative word is a "0". For negative sampling, a small number of negative words is selected to update the weights. If five negative words are chosen for a total of six output neurons, it updates a total of 6×300 weights.

4 Recurrent Neural Networks (RNN)

When dealing with text classification models, sequential data is often used with some aspects of temporal change. Typically, it is necessary to analyze a sequence of words, and the output can be a single value (e.g., sentiment classification), or another sequence (e.g., text summarization, language translation, entity recognition).

FFNNs have problems when trying to predict the next word after a specific word or words when the input text has the following characteristics:

- The length of each sentence is not fixed, and the amount of words available is vast
- Unexpected expressions and words, such as spelling errors, jargon, acronyms, etc.

The setback can be problematic for neural networks since the number of neurons in each layer, including the input layer, needs to be fixed, and the networks must have matching size for data entry. Additionally, the length of the input data is not fixed and can vary significantly, and hence it cannot be applied to the model without additional adaptations. For these reasons, FFNNs are limited in prediction, classification, or language generation tasks.

RNNs were created to solve the FFNNs limitations since the networks are linked with each other, thereby allowing consistency of information in the structure (Fig. 5).

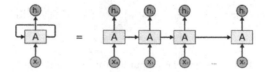

Fig. 5. Recurrent neural net [3].

Inside A, there is an FFNN, from X_0 to X_t are the words to input to the network (Fig. 6).

The outputs h_0 to h_{t-1} are feeding the next node, and h_0 to h_t are the outputs. Input vector X can be, for example, a 300-dimensional vector and the output vector h can be, for example, a 200-dimensional vector (hidden layer). This nodes have a simple structure, such as a single *tanh* layer.

This is effective for sequence models and predicting text at the end of the sequence because the state encodes information about all of the previous inputs.

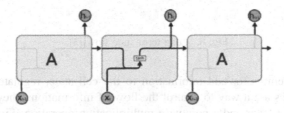

Fig. 6. Standard RNN [3].

4.1 Long Term Dependency Problem

It is incredibly robust for short sequences when data is long since the **vanishing gradient problem** appears. Once a prediction is made, it shows the final value, and then it backpropagates all of that information through the full network. The problem arises when a text is long since the content it can learn from the start of the sequence is almost zero.

Theoretically, the RNNs can handle "long-term dependencies," although, in practice, the RNNs cannot learn from expressions where the context does not allow to predict the next word accurately. For example, the phrase "fish swim in the ..." clearly suggests that the next word is "water," but phrases like "I studied in Italy, I learned ..." leave a vast range of options for choosing the next word.

4.2 Long Short Term Memory (LSTM) Networks

LSTMs are designed to solve the latest problem by remembering information for long periods. LSTMs keep track of 2 states: *long term state* and *short term state* (Fig. 7).

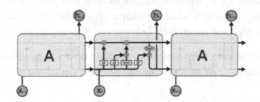

Fig. 7. Long short term memory [3].

Long term state flows along the top and can be updated very easily, as it is only two operations, multiplication, and addition (Fig. 8).

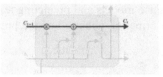

Fig. 8. Long term state flow [3].

LSTMs can remove or add information to the cell state, regulated by structures called gates. **Gates** are a way to control the flow of information. They comprehend a sigmoid neural net layer and a pointwise multiplication operation (Fig. 9).

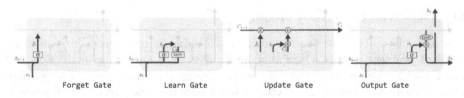

Fig. 9. LSTM gates [3].

Forget Gate. It outputs a value between 0 and 1. It represents *"having the short term state; what information should it forget?"*.

Learn Gate. This gate does two things: it concatenates the input and passes it to both sigmoid and *tanh* activation. It learns both i_t and $\tilde{C}_t$. $\tilde{C}_t$ is the candidate new state, and i_t is an adjustment factor of a dimension's relevancy. *Based on the sent information and the short term state, it finds the candidate's new long term state.*

Update Gate. Here we know what we want to forget, and we know what we want to learn. First, we multiply the previous long term state by the forget vector. *For every prior dimensional state, we determine the fraction of this that we need to forget.* Next, it sums: *"now we have forgotten what we wanted from the previous state, let's add it in the new information that we wanted to learn from this input."*

Output Gate. The new long term state is available; now we need to know *"what is the important information needed to know in the short term about the long term state."*

4.3   Gated Recurrent Unit (GRU)

This gate combines the forget and input gates into a single *"update gate."* It also merges the cell state and hidden state, and some other minor changes. The resulting model is more straightforward than LSTM models and has become increasingly popular (Fig. 10).

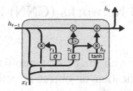

Fig. 10. Gated recurrent unit [3].

4.4 Dropout

Dropout is a method for regularizing a neural network, similar to L1 and L2. Its working mechanism is as the net is learning, it applies probabilistically sampling to determine the nodes that it is going to update to avoid becoming over-dependent on just a couple of nodes (Fig. 11).

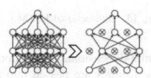

Fig. 11. Dropout.

4.5 Bidirectional LSTM (BiLSTM)

This RNN is the same as the LSTM, except that it has two LSTMs. S_0 reads the input sequence from the front to the back. The second LSTM S'_0 reads the same series in reverse. Finally, it concatenates both states together into the final output and passes it to the next layer. Even when it represents one layer [of BiLSTM], it is two LSTMs (Fig. 12).

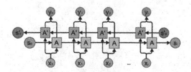

Fig. 12. Bidirectional LSTM [4].

The logic behind this is that even when LSTMs are more robust, like RNNs with long sequences and vanishing gradient problem solved, and it could work with languages where the order of the words sometimes keeps the sense of the sentence.

5 Convolutional Neural Networks (CNN) for Language Tasks

Problems that involve working with image or video data use the CV models, which typically requires image classification or object detection. The community has ultimately inspired the application of CV models to other domains, like NLP (Fig. 13).

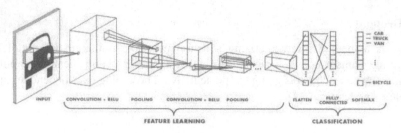

Fig. 13. Bidirectional LSTM [5].

CNN extracts features from 3-dimensional input through convolutional layers and pooling layers, and then these features are passed into a general multi-layer neural network. This preprocessing classifies data according to the following specifications:

- Separates the input data into several domains
- Extracts the features from the respective domains, such as borders and positions

5.1 Convolution Function

Convolution works define a "small window" (e.g., 2 × 2, 3 × 3). Now, the function takes this window and slides it across the image applying several filters to extract features. These filters are called **kernels**, and convolved images are called **feature maps**. Different features can be extracted by changing kernel values (Fig. 14).

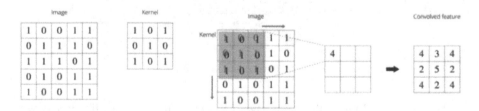

Fig. 14. Convolution [6].

-1	-1	-1		0.1	0.1	0.1
-1	8	-1		0.1	0.1	0.1
-1	-1	-1		0.1	0.1	0.1

Fig. 15. Examples of the kernel [6].

From Fig. 15, the left kernel extracts the edges of the image because it accentuates the color differences. On the other hand, the right kernel blurs the picture because it degrades the original values. Each kernel has its values and extracts particular features from the image. The number of feature maps and the number of kernels are always the same, which means if we have 20 kernels, we also have 20 feature maps (Fig. 16).

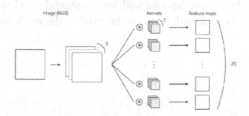

Fig. 16. Convolution in a multi-channel image [6].

When the image has multiple channels, kernels will be adapted separately for each channel. Therefore, for RGB with three channels, we have 60 convolved images first, composed of 20 mapped images for each of the three channels. All the convolved images initially from the same image will be combined into one feature map. As a result, there will be 20 feature maps. In other words, images are decomposed into different channeled data, and then combined into mixed-channeled images again.

The activation function will activate all the convolved values. Currently, the rectifier is the most popular function to use with CNN, but it can use the sigmoid function, the hyperbolic tangent, or any other activation functions available instead.

It is essential to mention that on CNN, it is not necessary to set these kernel values manually. Once initialized, CNN itself will learn the proper weights through the learning algorithm, which means parameters trained in a CNN are the weights of the kernel.

5.2 Pooling Function

This function does not train or learn by itself but downsamples images propagated by convolutional layers. It could be thought that it may lose some vital information from the data, but this is necessary to make the network keep its translation invariance. Max-Pooling and Average are the most common ways of downsampling (Fig. 17).

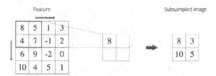

Fig. 17. Max-pooling [6].

5.3   Convolutional Neural Networks in NLP

Convolutions are used to reach a state, and after applying this function, the prediction task is done with an FFNN. The convolutions filter and transform an image to find the state that represents it, and the subsequent actions of the neural network depend on this state. Something similar is what should be done with text; it should start learning the sentence's state, and once this has been found, it should be used in the prediction task (Fig. 18).

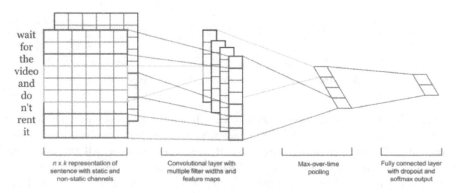

Fig. 18. CNN applied to text [7].

Images use 2-dimensional convolutions, but for text, we only use a 1-dimensional convolution. As such, for input matrices, we have each word on the rows going down the columns across the vector representation of words (8 rows, 300 columns). Each row will be a Word2Vec representation for the specific word, and then it needs to define a filter, but it is only a 1-dimensional filter; for example, the filter could be of size 2.

It should expand all of the columns and expand two rows (because of the filter of size 2), and perform the same convolutional function, except that it slides down and encapsulates two words at a time, then it does the convolution, and passes it forward. Using this analogy, we may find 10 of these filters, one of size 2, or 3, 4, etc. As such, the output from the different filters works like filters through an image. Finally, it applies max-pooling, and it unwraps the result to have a fully connected layer.

The reason we perform a 1-dimensional convolution is that, unlike an image where data lives in 2 dimensions, text lives in only one dimension. Columns represent words and considering that it is not necessary to know what those dimensions represent, it does not make sense to exclude some features by performing a 2-dimensional convolution.

Word2Vec learns general concepts about a word and how words relate to each other in a broad sense. Convolutions learn how these words relate to each other in a specific way, such as a feeling. CNN is used for many-to-one classification because it is not explicit but captures the generic meaning of the sentence.

6   Generalized Language Models (GLM)

A few years ago, machine learning experts started to realize that even when they can learn languages, they cannot transfer this general knowledge about embedded words to specific tasks. CV embraces transfer learning, and researchers have created several huge pre-trained models, like ImageNet and others, to classify and identify hundreds of thousands of classes of images. This is achieved by using the trained models into specific tasks by refining the end of the model to retain just the last few layers on the task's particular data. In general terms, transferring the model's information into a smaller task means leveraging much knowledge about edges, rotation, depth, among others.

NLP has embraced this concept as well, by learning the general knowledge about a language into a pre-trained model and using this model as a baseline to train a smaller model for a specific task.

The probability for the sentence is shown in the following equation [8] (Table 2):

$$P(S) = P(W_1, \ldots, W_n)$$
$$= \Pi_i P(W_i | W_1, \ldots, W_{i-1})$$
$$= P(W_1) * P(W_2 | W_1) * P(W_3) * P(W_1, W_2) * \ldots * P(W_n | W_1, \ldots, W_{n-1})$$

Table 2. Latest work on generalized learning models.

Name	Details	Date
ELMO	Embedded language model Author: *Allen Al/University of Washington*	March 2018
ULMFiT	Universal language model fine-tuning Author: *fast.ai*	May 2018
MultiFiT	Multi-lingual language model fine-tuning Author: *fast.ai*	Sep 2019
BERT	Bidirectional encoder representations from transformers Author: *Google AI*	Nov 2018
XLNet	Transformer XL Net Author: *Google Brain*	Jun 2019
RoBERTa	Robust optimized BERT pretraining approach Author: Facebook AI	Jul 2019,... Nov 2019
GPT/GPT-2	Generative pre-trained transformer Author: *OpenAI*	Jun 2018,... Aug 2019

6.1   ULMFiT [9]/MultiFiT [10]

MultiFiT is an extension of ULMFiT, designed to enable training and fine-tuning language models in their language of choice, although internally it works like ULMFiT (Fig. 19).

Fig. 19. ULMFiT process [9].

Step 1: GLM Pre-Training. This involves pre-training a GLM using an *AWD-LSTM* with vast data sources like Wikipedia, which provide as much knowledge as possible and train a model that tries to predict the next word given text input.

AWD-LSTM is like a regular LSTM but super regularized, which implies many dropouts in the embedding layer, and uses some optimization tricks.

Step 2: Refine GLM for Target Task. This involves fine-tuning the previous language model for a specific task. It starts with the pre-trained model and trains again on vocabulary for a particular task.

It uses **Discriminative Fine-Tuning**, which means that different learning rates are used for each layer since they capture different information because the model learns at different rates. We start by defining a learning rate, and this will be updated on every layer over the network, training a couple of layers at a time for a few epochs, and adjusting the learning rate as it continues.

Also, it uses **Slanted Triangular Learning Rates (STLR)**, which means that learning rates are first increased (starting with high learning rates), and then slightly decreased. The reason is that it begins learning quickly, but learning too fast usually results in issues, like converging, and so it has to reduce that learning rate to prevent the model from converging. Over time, it first increases the rate, and then gradually decreases it.

Step 3: Target Task Classification Training. This step uses the previous model as the embedding input and produces a state for each input sentence.

It consists of appending two FF layers and a softmax output layer to predict target labels. Internally, it uses **Concat Pooling**, which extracts max and the mean pooling over the history of hidden states and appends to the final state.

It also uses **Gradual Unfreeze**, which updates only a single GLM layer on each epoch during training. Freeze and Unfreeze in Deep Learning mean that weight on that layer is learnable. If something is frozen when the gradient descent is executing, then the weight is not going to be updated, and, if it is unfrozen, then this weight can be updated from what it has learned from the model.

6.2 BERT/GPT-2 – Transformer Model

BERT [11] and GPT [12] are similar models; both of them use the same approach of learning a Generalized Language Model and use supervised fine-tuning. These models use an **Attention Mechanism** and a **Transformer Model** instead of an RNN.

Attention Mechanism. Attention is an additional layer that you can put on top of an RNN. Its objective is "taking care" of the current information in the input sequence when making a prediction. A BiLSTM receives the source text, and then there is the attention distribution similar to the discrete probability distribution.

According to Fig. 20, when the prediction of the next word is in process, it cares about the word 'victorious', so the layer is trained to know this word based on the fact that it sees the word 'Germany', and it has to output the next word in this text summarization problem. The model has to focus on 'victorious,' and then, based on that attention, it can more effectively choose 'beat' as the next word.

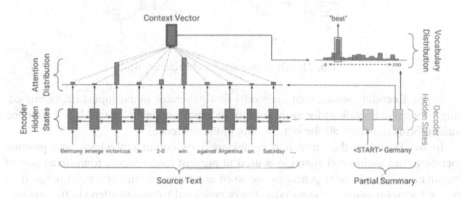

Fig. 20. Attention mechanism [13].

The attention mechanism is suitable in many-to-many RNN where we want to input something and then output a summary, or if we want to translate a language. Attention is a simple NN layer at the top that learns the distribution over the input sequences that require attention at the moment that the model is making the prediction.

Transformer Model [14]. The objective of this model is parallelizing the process by replacing recurrence with attention and encoding the symbol position in the sequence.

This model encodes each position and applies the attention mechanism to match two distant words of both the inputs and outputs, which can be parallelized, thus accelerating the training. It also reduces the number of sequential operations to relate two symbols from input/output sequences to a constant amount of processes. This model uses the **Multi-Head Attention Mechanism** that allows modeling dependencies without considering its distance in the input or output phrase, and the **Scaled Dot-Product Attention**, which makes a weighted sum of the values, where the weight assigned to each value is determined by the dot-product of the query with all the keys. In general terms, *"Multi-Head Attention is just several attention layers stacked in parallel, with different linear transformations of the same input."*

Considering that RNN or CNN is not present in this model, it should inject information about both relative and absolute position for the token in the sequence to the embeddings, and make use of the order of the sequence (Fig. 21).

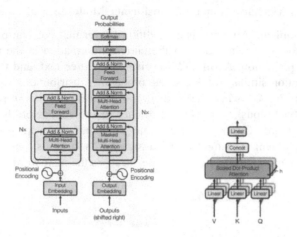

Fig. 21. The transformer architecture [14]

In the **Encoder**, we see that the model first generates initial inputs (input embedding + position encoding) for each word in the input sentence, for which self-attention adds information from all the other words in the sentence.

In the **Decoder**, the input is a combination of output embedding and position encoding, and multi-head attention is used to prevent future words from being part of the attention. The decoder generates one word at a time, and the first word is based on the final representation of the encoder. Every predicted word will attend to the previous words of the decoder at that layer and the final representation of the encoder.

The general idea of the Transformer Model is having multiple layers of the attention mechanism that is used to look back on time in the input sequence. It means that it *inputs and outputs the sentence* simultaneously, and what the attention learns from each output matters for the input.

7 Conclusions

NLP has been under exhaustive research and development, and although predictive models have greatly improved, there are still many technological challenges to overcome and achieve higher levels of accuracy for its various applications.

Similar works, like Sebastian Ruder's [15] or Elvis Saravia's [16], are pervasive and descriptive about the latest research in NLP. This document has focused on describing the latest and more remarkable works in Deep Learning methods only and merely explaining the core concepts.

GLMs and GPT-2, in particular, have been a significant boom among the community due to the improvements in related technologies that have allowed incorporating support technologies that enable training models with previously unimaginable amounts of data. In the past, these types of models were almost impossible to train without a sufficient amount of data. Now that we have techniques like ULMFiT, the amount of data you need is substantially less. Nevertheless, the performance of recent NLP models has substantially improved by using vast amounts of data, like XLNet or RoBERTa.

While new NLP models are under development, we will see innovations in Cloud Computing, specialized chips, and Deep Learning processes being executing into distributed environments, from different sources and clouds. However, the most critical improvements in NLP will be based on an increasingly deep understanding of the way humans build and understand language, which has allowed us to map behavior and the way we process linguistic information into cognitive models much more precise.

References

1. Word2vec, in Wikipedia. https://en.wikipedia.org/wiki/Word2vec
2. McCormick, C.: Word2vec Tutorial - The Skip-Gram Model. http://mccormickml.com/2016/04/19/word2vec-tutorial-the-skip-gram-model/
3. Olah, C.: Understanding LSTM Networks. https://colah.github.io/posts/2015-08-Understanding-LSTMs/
4. Olah, C.: Neural Networks, Types, and Functional Programming (2015). http://colah.github.io/posts/2015-09-NN-Types-FP/
5. Prabhu, R.: Understanding of Convolutional Neural Network (CNN) – Deep Learning (2018). https://medium.com/@RaghavPrabhu/understanding-of-convolutional-neural-network-cnn-deep-learning-99760835f148
6. Sugomori, Y., Kaluža, B., Soares, F.M., Souza, A.M.F.: Deep Learning - Practical Neural Networks with Java. Packt Publishing, Birmingham (2017). Chapter 4
7. Britz, D.: Understanding Convolutional Neural Network for NLP. http://www.wildml.com/2015/11/understanding-convolutional-neural-networks-for-nlp/
8. Jurafsky, D., Martin, J.H.: N-gram Language Models (2019). https://web.stanford.edu/~jurafsky/slp3/3.pdf
9. Howard, J., Ruder, S.: Introducing state of the art text classification with universal language models. http://nlp.fast.ai/classification/2018/05/15/introducing-ulmfit.html
10. Eisenschlos, J., Ruder, S., Czapla, P., Kardas, M.: Efficient multi-lingual language model fine-tuning. http://nlp.fast.ai/classification/2019/09/10/multifit.html
11. Devlin, J., Chang, M.-W., Lee, K., Toutanova, K.: BERT - Pre-training of Deep Bidirectional Transformers for Language Understanding. Google AI (2019)
12. Radford, A., Wu, J., Child, R., Luan, D., Amodei, D., Sutskever, I.: Language Models are Unsupervised Multitask Learners (2019). https://d4mucfpksywv.cloudfront.net/better-language-models/language-models.pdf
13. Abigail: Taming Recurrent Neural Networks for Better Summarization (2017). http://www.abigailsee.com/2017/04/16/taming-rnns-for-better-summarization.html
14. Thiruvengadam, A.: Transformer Architecture: Attention Is All You Need (2018). https://medium.com/@adityathiruvengadam/transformer-architecture-attention-is-all-you-need-aeccd9f50d09
15. Rudes, S.: Tracking Progress in Natural Language Processing (2019). https://nlpprogress.com/
16. Saravia, E.: Deep Learning for NLP, An Overview of Recent Trends (2018). https://medium.com/dair-ai/deep-learning-for-nlp-an-overview-of-recent-trends-d0d8f40a776d

Comparison of Attenuation Coefficient Estimation in High Intensity Focused Ultrasound Therapy for Cancer Treatment by Levenberg Marquardt and Gauss-Newton Methods

Laura de los Ríos Cárdenas[1] , Leonardo A. Bermeo Varón[1]([✉]) ,
and Wagner Coelho de Albuquerque Pereira[2]

[1] Department of Engineering, Universidad Santiago de Cali,
Street 5 no. 62-00, Cali, Colombia
leonardo.bermeo00@usc.edu.co
[2] Department of Biomedical Engineering,
Universidade Federal de Rio de Janeiro, Av. Horácio Macedo 2030,
Centro de Tecnologia, COPPE/UFRJ, Rio de Janeiro, Brazil

Abstract. Hyperthermia using High-Intensity Focused Ultrasound (HIFU) is an acoustic therapy used in clinical applications to destroy malignant tumors of bone, breast, brain, kidney, pancreas, prostate, rectum and testicle. This technique consists in increase the temperature in the tumor or the specific area, to achieve coagulative necrosis and immediate cell death. Although hyperthermia can cure cancer, it can also cause side effects and even damage healthy cells or tissues. Therefore, for having a successful treatment, it is important to monitor and observe what is the tissue behavior, as well as its changes, before, during and after the procedure. Mathematical models are tools that can be useful to simulate an adequate therapy by differentiating characteristics that will depend on each individual. An attenuation coefficient estimation for a forward model with a rectangular two-dimensional domain is presented, the estimation was made with simulated numerical data and simulated experimental data by Levenberg Marquardt and Gauss-Newton Methods. The results demonstrate that by identifying the attenuation coefficient of each patient. By estimating the attenuation coefficient parameter it is possible to predict the thermal responses of the tissue to be treated and, based on them, to plan an adequate cancer treatment by inducing heat by HIFU.

Keywords: Parameter estimation · High-Intensity Focused Ultrasound · Hyperthermia · Levenberg Marquardt · Gauss-Newton

1 Introduction

Cancer is the main disease causing the most deaths worldwide, it is due to that cancer cells can spread both in tissues and any organ, causing an increased risk in the treatment and most cases if it is not treated on time, the death of the patient who suffers.

© Springer Nature Switzerland AG 2020
M. Botto-Tobar et al. (Eds.): ICAT 2019, CCIS 1194, pp. 108–118, 2020.
https://doi.org/10.1007/978-3-030-42520-3_9

According to WHO 8.8 million deaths occurred in 2015 caused by any type of cancer, and it is estimated that, by 2030, 13.2 million per year will occur [1].

The medical sciences have had many technological advances, that allow applying optimal treatment to combat diseases such as cancer, avoiding surgical procedures or conventional treatments that in some cases they are not effective and affect the patient. Therefore, studies that improve existing therapies, are considered an important issue for researchers and patients, where the development of a less painful and more effective for cancer treatment, will provide a better-quality life.

Alternative therapy is the increase of temperature in the cancerous tissue that produces necrosis and achieves the destruction of the tumor [2], it consists in raising the temperature of the point to be treated at more than 38 °C, by induction of electromagnetic waves [3–10] and acoustic pressure [2, 11–17]. One of these techniques is High-Intensity Focused Ultrasound (HIFU), a non-invasive technique used as a cancer treatment by ablation or hyperthermia, the generated heat is highly localized and thus, it can destroy tumors such as bone, brain, breast, liver, pancreas, rectum, kidney, testes and prostate, in some cases without anesthesia and on an outpatient basis [12].

This type of technique is based on the use of an acoustic transducer to focus the ultrasound energy, which propagates through the tissue and is absorbed as heat energy in a specific zone. The ultrasonic transducer uses an acoustic lens which focuses multiple beams of ultrasound that intersect at a target, each beam passes through the tissue with little effect, but at the focal point where they converge the beams, there is an absorption effect of the acoustic energy in the tissue which may cause thermal effects useful for cancer treatment with HIFU. Thus, the characterization of the temperature field provided by transducers is important for predicting these effects [15, 16]

HIFU therapy needs to plan an optimal treatment that allows establishing parameters that vary from individual to individual. Mathematical models and optimization algorithms are tools that can be useful for planning and controlling these treatments.

Mathematical models allow to simulate HIFU therapy using an ultrasonic transducer applied in a specific area affected by a tumor, frequently and with a set time. The results have shown that with the simulations obtained it is possible to observe parameters and characteristics appropriate for an effective treatment, thus being able to monitor the effects before, during and after applying the HIFU in the patient without damaging the surrounding tissue [13, 18–20], One of the important parameters in HIFU therapy is the attenuation coefficient, due to the effect it has on the thermal response in biological tissues.

Some studies reveal that with an adequate estimation of the attenuation coefficient, it was possible to identify whether the tumor was benign or malignant, so the parameter is considered very important for different parametric studies that use ultrasound to apply to a tissue or skin [21–24].

Estimates of parameters are fundamentally an optimization problem, often can be found using non-statistical minimization algorithms. Levenberg Marquardt and Gauss-Newton methods are not statistical iterative algorithms used in different disciplines for data adjustment with least-squares [25–28]. These methods allow establishing the ideal parameters in a different process that allow knowing better the problem that is being covered from simulated and experimental data of the system.

In this paper, the estimation of parameters is performed with two algorithms, Levenberg Marquardt and Gauss-Newton methods, using the pressure acoustic and heat transfer model on a two-dimensional model. The main objective is to establish optimal therapy in the performing of the therapy. In general, the study consists of a sensitivity analysis that allows to observe the behavior of the system parameters with respect to the temperature profiles, through heat induction from an external source into the tissue, planning of the experiment and the parameters estimate considering experimental data and simulation of the physics problem, for establishing appropriate parameters that dependent of the individual who will be treated.

2 Methodology

The forward problem is implemented and solved in Comsol Multiphysics® 5.3 to find the temperature of a rectangular two-dimensional domain profile in two materials: (i) silicone, used for simulating muscle tissue due to present the similar thermosphysical properties and (ii) water used for simulating HIFU transducer.

The acoustic field is simulated by Eq. 1, where r and z are the radial and axial coordinates, ω is the angular frequency, m is the circumferential mode number to setting pressure acoustic equation, q is the inward heat flux of the material, p is the acoustic pressure, ρ_0 is the density and c is the speed of sound propagating through the material.

$$\frac{\partial}{\partial t}\left[-\frac{r}{\rho_0}\left(\frac{\partial p}{\partial r}-qr\right)\right]+\frac{\partial}{\partial z}\left[-\frac{r}{\rho_0}\left(\frac{\partial p}{\partial z}-qz\right)\right]-\frac{rp}{\rho_0}\left[\left(\frac{\omega}{c_0}\right)^2-\left(\frac{m}{r}\right)^2\right]=rQ \quad (1)$$

The initial acoustic pressure is calculated by Eq. 2, where Z is the material acoustic impedance and I is the acoustic intensity.

$$P=\sqrt{2ZI} \quad (2)$$

The temperature field is calculated by the Bioheat Equation (Eq. 3), where c_p is the specific heat, c_b is the blood specific heat, ρ is the density, ρ_b is the blood density, Q is the external heat source, Q_m is the metabolic heat source, ω_b is the blood perfusion rate, T_b is the blood temperature and k is the thermal conductivity [29].

$$\rho c_p \frac{\partial T}{\partial t}+\nabla\cdot(-k\nabla T)=\rho_b c_b \omega_b(T_b-T)+Q_m+Q \quad (3)$$

The external heat source is defined by Eq. 4, where the μ_0 is the amplitude of the particle velocity and α is the attenuation coefficient:

$$Q=2\alpha I=\alpha\rho c\mu_0^2 \quad (4)$$

For all cases examined, a rectangular two-dimensional axisymmetric domain with dimensions $L_r = 50$ mm and $L_z = 85$ mm was considered, with the ultrasonic transducer located at the bottom with dimensions $L_r = 12.5$ mm and $L_z = 2$ mm as shown in Fig. 1.

Table 1 shows the physical properties involved in the models considering the material properties of the domain (silicone) and the HIFU transducer properties (water) that vary depending on the region where it makes the analysis. The models were defined by using frequency analysis in pressure for the acoustics model and the time-dependent analysis in temperature. In the Bioheat transfer model, the thermal properties within the biological tissue were zero, because is simulated with a solid material (silicone).

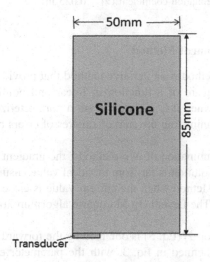

Fig. 1. A rectangular two-dimensional domain

Table 1. Physical properties of the domain [30]

Properties	
Nominal frequency (f)	1 MHz
Acoustic intensity (I)	0.5 W · cm^{-2}
Initial temperature (T$_0$)	36.8 °C
Cooling temperature (T$_c$)	24 °C
Time of exposure	200 s
Metabolic heat source (Q$_m$)	0 W · m^{-3}
Blood specific heat (c$_b$)	0 J/(kg · K)
Blood perfusion rate (ω_b)	0 s−1
Blood density (ρ_b)	0 kg · m^{-3}
Reference wave speed C$_{ref}$)	1,483 m · s^{-1}
Physical properties of the domain	
Density (ρ)	1,350 kg · m^{-3}
Thermal conductivity (k)	0.275 W · m^{-1}K^{-1}

(*continued*)

Table 1. (*continued*)

Heat capacity (c_p)	705 J $\cdot$ kg^{-1}K^{-1}
Speed of sound (c)	998 m $\cdot$ s^{-1}
Attenuation coefficient (α)	19.8 m^{-1}
Physical properties of the transducer	
Density (ρ)	1,000 kg $\cdot$ m^{-3}
Thermal conductivity (k)	0.6 W $\cdot$ m^{-1}K^{-1}
Heat capacity (c_p)	4,178 J $\cdot$ kg^{-1}K^{-1}
Speed of sound (c)	1,480 m $\cdot$ s^{-1}
Attenuation coefficient (α)	0.025 m^{-1}

2.1 Levenberg Marquardt Method

Levenberg Marquardt method is an iterative method that provides a numerical solution to find the minimum square of a function in linear and nonlinear problems. Least-squares problems arise when it is required to set a parameterized function to a set of measured data points, minimizing the sum of squares of errors between the data points and function [25].

This method is a combination of two methods: the gradient descent method, when the value of the current solution is far from the ideal value ensuring slow convergence, and the Gauss-Newton Method when the current value is close to the ideal presenting rapid convergence [25]. The Levenberg Marquardt algorithm implemented is described below.

The numerical solution Y_t (Eq. 5) is obtained by the forward problem solution at all times and all positions defined in Eq. 3, with the parameter exact values and 5,284 elements.

$$Y_t = Y_1, Y_2, \ldots, Y_n \tag{5}$$

The simulated experimental data is defined in Eq. 6, where σ is a Gaussian additive error to numerical solution with mean zero and standard deviation of 1%.

$$Y = Y_t + \sigma Y_t \tag{6}$$

The forward problem solution with initial guess of parameters (P^k), getting the temperatures vector (T_{Pk}) with 618 elements.

$$T_{p^k} = T_1, T_2, \ldots, T_n \tag{7}$$

Calculate the likelihood objective function S.

$$S_k = \left[Y - T_{p^k} \right]^T \left[Y - T_{p^k} \right] \tag{8}$$

Calculate the estimate with P^{k+1} (Eq. 9), where $\mu = 0.1$ is the dumpling factor W is the measurements error covariance matrix and Ω is the diagonal matrix of $J^T W J$.

$$P^{k+1} = P^k + [J^T W J + \mu \Omega]^{-1} J^T W [Y - T(P^k)] \tag{9}$$

Calculate the sensitivity matrix J formed by each sensitivity coefficient vector and ε is the disturbance (Eq. 10), to identify the parameters of the model that show variations when applying parametric changes of the system [31]:

$$J_{ij} = \frac{T_i(P_1, P_2, \ldots, P_j + \varepsilon P_j, \ldots, P_N) - T_i(P_1, P_2, \ldots, P_j - \varepsilon P_j, \ldots, P_N)}{2\varepsilon P_j} \tag{10}$$

Obtain the forward problem solution with a new P^{k+1}, and calculate the objective function S.

$$S_{k+1} = [Y - T_{p^{k+1}}]^T [Y - T_{p^{k+1}}] \tag{11}$$

If $S(P^{k+1}) > S(P^k)$, the μ value becomes 10μ.
If $S(P^{k+1}) < S(P^k)$, a new estimate is accepted, the μ value becomes 0.1μ.

Calculate the estimate with a new P^{k+1} (Eq. 9). Stopping criteria is described in the Eq. 12, where $\varepsilon_1 = 10^{-5}$.

$$\left\| P^{k+1} - P^k \right\| = \varepsilon_1 \tag{12}$$

2.2 Gauss-Newton Method

The second method implemented to the estimation was the Gauss-Newton method, the most fundamental iterative method, that provides a numeric solution to find the least-square of a function in linear problems [26]. It is a fast method due to locally converges quadratically and only when the parameters are near the ideal value. The Gauss-Newton algorithm implemented is described below.

Calculate the Eqs. 5–8, and the estimate with P^{k+1}, where J is the sensitivity matrix (Eq. 10) is described in the Eq. 13.

$$P^{k+1} = P^k + [J^T W J]^{-1} J^T W [Y - T(P^k)] \tag{13}$$

Stopping criteria is described in the Eq. 12, where $\varepsilon_1 = 10^{-5}$.

The numerical simulation was performed on a computer intel core i3 6th generation, CPU of 2.3 GHz and 8 GB of RAM, where the optimization algorithm was performed 50 times to calculate the mean, standard deviation and a 95% confidence interval.

3 Results and Discussions

Figure 2 is presented the temperature distribution in the domain at time $t = 200$ s for $f = 1$ MHz and $I = 0.5$W $\cdot$ cm^{-2} obtained with Comsol Multiphysics® 5.3. The result indicating an increase of the temperature of 13.9 °C, the maximum temperature is presented in the focused zone. The increasing values are appropriate for the hyperthermia treatment.

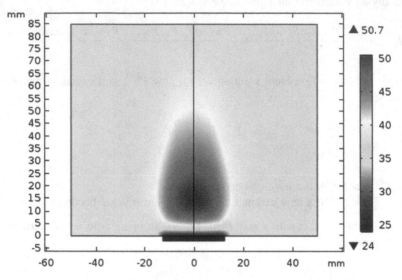

Fig. 2. Temperature distribution (°C) in the domain at time = 200 s.

Figure 3 shows the simulated temperature transient of 3 points with r and z coordinates in the domain (silicone). It is observed that the increase in temperature is time-dependent. These results are similar to those presented by [32].

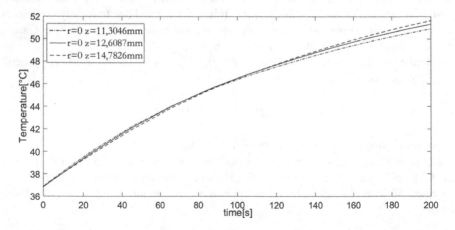

Fig. 3. Temperature transient in three (3) different location of the domain.

Considering the sensitivity analysis presented by [33], the estimate of attenuation coefficient (α) in the domain of the silicone was performed. The simulated experimental measures were obtained by forward problem solution through a mathematical model taking (Eq. 3) with the largest number of elements in the domain (5,284), with Gaussian error additive with mean zero and a standard deviation of 1%. Figure 4 presents the simulated experimental data and the exact temperature in the (0.00; 14.78). It is considered a unique point of simulated experimental measures for the estimation process.

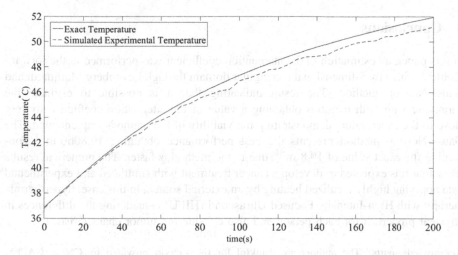

Fig. 4. Numerical and simulated experimental temperature transient in (0.00; 14.78).

Table 2 present the results of the parameter estimation (α) with both methods. Indicating the mean, the standard deviation, 95% confidential interval and run-time mean after performing the estimation for 50 repetitions of each algorithm. The result indicates that the estimated value of the parameter α with Levenberg Marquardt method had a confidence interval of 19.7489–19.8262. Similarly, with Gauss-Newton method the estimated value of the parameter α to the confidence interval was 19.7920–19.8200, indicating that the estimated values with both methods, approximately 95% of the performances of the algorithms and the exact value were in this range (19.8000).

These results showed that the attenuation coefficient (α) exerts a great influence on the dynamic temperature profiles of the model, as well as the effectiveness of the methods implemented to make the estimation.

The results can be establishing an appropriate parameter that differs from individual to individual, it is possible to perform optimal cancer therapy inducing heat with HIFU, allowing monitoring of treatment before, during and after applying it to the patient, ensuring a more effective therapy

Additionally, the Gauss-Newton presented in a shorter run-time due to fast convergence and more precision. The little variability of results shows the accuracy of the implemented methods for the estimate of parameters, allows to obtain values close to the exact value of the α from the simulated experimental value.

Table 2. Results of the estimation of the attenuation coefficient (α)

Methods	Levenberg Marquardt	Gauss-Newton
Exact α (m^{-1})	19.8000	19.8000
Mean (m^{-1})	19.7876	19.8060
Standard deviation (m^{-1})	0.1393	0.0504
Run-time (s)	4,789.2991	3,450.3230
Confidence interval (m^{-1})	19.7489–19.8262	19.7920–19.8200

4   Conclusions

In this paper, an estimation of the attenuation coefficient was performed in the forward problem with a two-dimensional rectangular domain through Levenberg Marquardt and Gauss-Newton method. The result indicating that it is possible to estimate the parameter with both methods obtaining a value of the attenuation coefficient to very close to the exact value, demonstrating the viability of the methods implemented. The Gauss-Newton method presents the best performance obtaining 19.8060 m^{-1} concerning the exact value of 19.8 m^{-1}, due to the method is faster. The numerical results show that it is expected to develop a cancer treatment with simulated and experimental data applying highly localized heating by an external source, in this case, hyperthermia therapy with High-Intensity Focused Ultrasound (HIFU) considering the differences in physical properties for each person and also considering tumor characteristics.

Acknowledgments. The authors are thankful for the support provided by CNPq, CAPES, FAPERJ (Brazil) and DGI of Universidad Santiago de Cali, Colombia, project No. 819- 621118-120.

Disclosure Statement. No potential conflict of interest was reported by the authors.

References

1. WHO: Cancer. www.who.int. Accessed 01 Oct 2019
2. Sengupta, S., Balla, V.K.: A review on the use of magnetic fields and ultrasound for non-invasive cancer treatment. J. Adv. Res. **14**, 97–111 (2018)
3. Gas, P., Miaskowski, A.: Specifying the ferrofluid parameters important from the viewpoint of magnetic fluid hyperthermia. In: Selected Problems of Electrical Engineering and Electronics, pp. 1–6 (2015)
4. Kurgan, E., Gas, P.: Treatment of tumors located in the human thigh using RF hyperthermia. Prz. Elektrotechniczny. **87**, 103–106 (2011)
5. Kurgan, E., Gas, P.: Analysis of electromagnetic heating in magnetic fluid deep hyperthermia. In: 7th International Conference Computational Problems of Electrical Engineering (CPEE), Sandomierz, Poland, pp. 1–4 (2016)
6. Gas, P.: Essential facts on the history of hyperthermia and their connections with electromedicine. Prz. Elektrotechniczny. **87**, 37–40 (2011)

7. Bermeo Varon, L.A., Orlande, H.R.B., Eliçabe, G.E.: Estimation of state variables in the hyperthermia therapy of cancer with heating imposed by radiofrequency electromagnetic waves. Int. J. Therm. Sci. **98**, 228–236 (2015)

8. Bermeo Varon, L.A., Orlande, H.R.B., Eliçabe, G.E.: Combined parameter and state estimation in the radiofrequency hyperthermia treatment of cancer. Heat Transf. Part A Appl. **70**, 581–594 (2016)

9. Lamien, B., Varon, L.A.B., Orlande, H.R.B., Elicabe, G.E.: State estimation in bioheat transfer: a comparison of particle filter algorithms. Int. J. Numer. Methods Heat Fluid Flow **27**, 615–638 (2017). https://doi.org/10.1108/HFF-03-2016-0118

10. Lamien, B., Orlande, H.R.B., Eliçabe, G.E.: Inverse problem in the hyperthermia therapy of cancer with laser heating and plasmonic nanoparticles. Inverse Probl. Sci. Eng. **25**(4), 608–631 (2016)

11. Grull, H., Langereis, S.: Hyperthermia-triggered drug delivery from temperature-sensitive liposomes using MRI-guided high intensity focused ultrasound. J. Control. Release. **161**, 317–327 (2012)

12. Ter Haar, G., Coussios, C.: High intensity focused ultrasound: physical principles and devices. Int. J. Hyperth. Off. J. Eur. Soc. Hyperthermic Oncol. North Am. Hyperth. Gr. **23**, 89–104 (2007)

13. Canney, M.S., Bailey, M.R., Crum, L.A., Khokhlova, V.A., Sapozhnikov, O.A.: Acoustic characterization of high intensity focused ultrasound fields: a combined measurement and modeling approach. J. Acoust. Soc. Am. **124**, 2406–2420 (2008)

14. Ellens, N.P.K., Hynynen, K.: High-intensity focused ultrasound for medical therapy. In: Gallego-Juárez, J.A., Graff, K.F. (eds.) Power Ultrasonics, pp. 661–693. Woodhead Publishing, Oxford (2015)

15. Vaezy, S., et al.: Real-time visualization of high-intensity focused ultrasound treatment using ultrasound imaging. Ultrasound Med. Biol. (2001). https://doi.org/10.1016/S0301-5629(00)00279-9

16. Zhou, Y.-F.: High intensity focused ultrasound in clinical tumor ablation. World J. Clin. Oncol. **2**, 8–27 (2011)

17. Al-Bataineh, O., Jenne, J., Huber, P.: Clinical and future applications of high intensity focused ultrasound in cancer. Cancer Treat. Rev. **38**, 346–353 (2012)

18. Teja, J.L., Vera, A., Leija, L.: Acoustic field comparison of high intensity focused ultrasound by using experimental characterization and finite element simulation. In: Comsol Conference, p. 7 (2013). Ch.Comsol.Com

19. de los Ríos, L., Varon, L.A.B., de Albuquerque Pereira, W.C.: Parametric simulated study of high intensity focused ultrasound (Hifu) for treatment of cancer. In: The Conference Proceedings of American Society of Thermal and Fluids Engineers, vol. 3, pp. 1131–1134 (2018). https://doi.org/10.1615/tfec2018.bio.021734

20. Majchrzak, E., Paruch, M., Dziewoński, M., Freus, S., Freus, K.: Sensitivity analysis of temperature field and parameter identification in burned and healthy skin tissue. In: Muñoz-Rojas, P.A.A. (ed.) Computational Modeling, Optimization and Manufacturing Simulation of Advanced Engineering Materials. ASM, vol. 49, pp. 89–112. Springer, Cham (2016). https://doi.org/10.1007/978-3-319-04265-7_5

21. Guittet, C., Ossant, F., Remenieras, J.P., Pourcelot, L., Berson, M.: High-frequency estimation of the ultrasonic attenuation coefficient slope obtained in human skin: Simulation and in vivo results. Ultrasound Med. Biol. **25**, 421–429 (1999)

22. Deeba, F., et al.: Attenuation coefficient estimation of normal placentas. Ultrasound Med. Biol. **45**, 1081–1093 (2019)

23. Lerch, T.P., Cepel, R., Neal, S.P.: Attenuation coefficient estimation using experimental diffraction corrections with multiple interface reflections. Ultrasonics **44**, 83–92 (2006)

24. Rouyer, J., Cueva, T., Portal, A., Yamamoto, T., Lavarello, R.: Attenuation coefficient estimation of the healthy human thyroid in vivo. Phys. Procedia **70**, 1139–1143 (2015)
25. Madsen, K., Nielsen, H.., Tingleff, O.: Method for Non-Linear Least Squares Problems. Informatics and Mathematical Modelling Technical University of Denmark, Denmark (2004)
26. Gill, P.E., Murray, W.: Algorithms for the solution of the nonlinear least-squares problem. SIAM J. Numer. Anal. **15**, 977–992 (1978)
27. Fu, Y.B., Chui, C.K., Teo, C.L., Kobayashi, E.: Elasticity imaging of biological soft tissue using a combined finite element and non-linear optimization method. Inverse Probl. Sci. Eng. **23**, 179–196 (2015). https://doi.org/10.1080/17415977.2014.880904
28. Tong, Q., Yuan, Z., Zheng, M., Liao, X., Zhu, W., Zhang, G.: A novel nonlinear parameter estimation method of soft tissues. Genomics, Proteomics Bioinforma. **15**, 371–380 (2017). https://doi.org/10.1016/j.gpb.2017.09.003
29. Pennes, H.H.: Analysis of tissue and arterial blood temperatures in the resting human forearm. J. Appl. Physiol. **1**, 93–124 (1948)
30. Comsol Multiphysiscs: Comsol Multiphysiscs Modelling Software. www.comsol.com. Accessed 08 Oct 2019
31. Pereyra, S., Lombera, G.A., Frontini, G., Urquiza, S.A.: Sensitivity analysis and parameter estimation of heat transfer and material flow models in friction stir welding. Mater. Res. **17**, 397–404 (2013). https://doi.org/10.1590/s1516-14392013005000184
32. Cortela, G.A., Pereira, W.C.A., Negreira, C.A.: Ex vivo determined experimental correction factor for the ultrasonic source term in the bioheat equation. Ultrasonics **82**, 72–78 (2018)
33. de Los Ríos Cárdenas, L., Bermeo Varón, L.A., de Albuquerque Pereira, W.C.: Sensitivity study in high intensity focused ultrasound therapy for cancer. In: Henriques, J., Neves, N., de Carvalho, P. (eds.) MEDICON 2019. IFMBE Proceedings, vol. 76, pp. 1337–1342. Springer, Cham (2020). https://doi.org/10.1007/978-3-030-31635-8_163

Machine Learning for Optimizing Technological Properties of Wood Composite Filament-Timberfill Fabricated by Fused Deposition Modeling

Germán O. Barrrionuevo[1(✉)] and Jorge A. Ramos-Grez[1,2]

[1] Department of Mechanical and Metallurgical Engineering,
School of Engineering, Pontificia Universidad Católica de Chile,
Av. Vicuña Mackenna, 4860 Macul, Santiago, Chile
gobarrionuevo@uc.cl
[2] Research Center for Nanotechnology and Advanced Materials (CIEN-UC),
Av. Vicuña Mackenna, 4860 Macul, Santiago, Chile

Abstract. This work evaluates the applicability of machine learning (ML) tools in additive manufacturing (AM) processes. One of the most employed AM techniques is fused deposition modeling (FDM), where a part is created from a computer-aided design (CAD) model using layer-by-layer deposition of a feed-stock plastic filament material extruded through a nozzle. Owing to the large number of parameters involved in the manufacturing process, it is necessary to identify printing parameters ranges to improve mechanical properties as yield and ultimate strength. In that sense, ML has proven to be a reliable tool in engineering and materials processing, where hybrid ML algorithms are the best alternative since one-part acts as a forecaster, and the other part acts as an optimizer. To evaluate the performance of wood composite filament fabricated by FDM a uniaxial tensile test was performed at room temperature. The experimental procedure was carried out with a design of experiments of four factors at three levels, where the statistical significance of layer thickness, fill density, printing speed and raster angle was obtained as well as their interactions. Furthermore, ML's algorithm accuracy was explored, where a neuro-fuzzy system (ANFIS) was trained and tested with the experimental data. Through the development of the present work, it is concluded that layer thickness and raster angle play a significant role in FDM of a wood composite filament where fibers presence increases the layer thickness accelerating the FDM process.

Keywords: Machine learning · ANFIS · Additive Manufacturing · Fused deposition modeling · Wood composite filament

1 Introduction

1.1 Additive Manufacturing: Fused Deposition Modelling

Additive Manufacturing (AM) also known as 3D printing technology, was developed as a rapid prototyping tool for visualization and testing new concept designs [1]. This

© Springer Nature Switzerland AG 2020
M. Botto-Tobar et al. (Eds.): ICAT 2019, CCIS 1194, pp. 119–132, 2020.
https://doi.org/10.1007/978-3-030-42520-3_10

technology expands design freedom and brings the possibility to process almost any material since metals, polymers, ceramics to composites [2]. Within polymers fabrication, fused deposition modeling (FDM) technology stands out, where an object is created from a computer-aided design (CAD) model through layer deposition and it is possible to build complex geometries, impossible to achieve by conventional processes [3]. Owing to its flexibility, FDM is currently the 3D printing technique that dominates the market [4]. In order to transform this technology from rapid prototyping to rapid manufacturing it is necessary to understand how the process parameters, material properties and environmental conditions affects the final product. Structural integrity is the key to the successful adoption of this technique towards mass production [5]. Mechanical properties of a part fabricated by FDM depends on several process parameters where build orientation, layer thickness, printing speed, extruder temperature, infill pattern are some of the most commonly explored parameters [6, 7, 12].

Another important aspect to promote the massification of FDM technology is the total cost involved in the final part's production, thus; engineers have to guarantee the suitable process planning to achieve desired mechanical properties at minimum manufacturing cost [8]. In addition, to establish this technology as a standardized manufacturing process, it is important to know how a sample manufactured through FDM compares to a conventionally processed one, such as injection molding (IM) or extrusion. There are some studies which compare mechanical behavior of parts manufactured by FDM and IM. Uddin et al. [9] evaluate acrylonitrile butadiene styrene (ABS) samples fabricated by FDM and compared it with their injected counterparts, where if the suitable process parameters are selected, it is possible to achieve mechanical properties as good as the injected samples. Dawoud et al. [10] study the mechanical behavior of ABS samples fabricated by FDM and IM techniques, the FDMed parts exhibit tensile strength around 80% of the injected specimens, as long as the processing parameters had been optimized [11]. Two of the most commonly materials employed in FDM are ABS and polylactide (PLA), PLA shows a stronger response with specific strength 1.2 to 1.5 times that of ABS [4, 12, 13], typical values of yield strength are around 39 and 29 MPa respectively [9, 14]. The development of the FDM technique and its better understanding has generated the study of more materials, such as ethylene vinyl acetate (EVA) [15], polyether-ether-ketone (PEEK) [2, 16], polycarbonate (PC) [4] and composites [17, 18].

Composites materials are made from two or more constituent materials with different physical or chemical properties that, when combined, produce a material with improved characteristics [19]. Roj et al. [20] categorize compound filaments in four categories: metal, carbon, wood, and stone, where carbon shows the highest strength and the lowest weight, and wood composites show attractive applications for environmental and eco-friendly reasons. Tao et al. [17] explore the performance of wood flour (WF) filled PLA composite filament for 3D printing, where the initial deformation resistance of the composite is enhanced after adding WF, compared to pure PLA. Kariz et al. [18] study the effect of the content of natural wood (beech) on the PLA matrix, the addition of wood particles increases the number of porous which decreasing overall mechanical strength. In contrast, Gao et al. [21] investigate the effect of carbon fibers and talc fillers in PLA matrix, where the interfacial diffusion is improved due to the decrease of melt viscosity which cause lower degree of mechanical anisotropy.

With the motivation of optimizing the mechanical response of wood composite filament fabricated by FDM, this work focuses on the effect of process parameters on the yield strength, where layer thickness, fill density, printing speed and raster angle are analyzed.

1.2 Computational Intelligence: Machine Learning

Computational intelligence or artificial intelligence (AI) refers to the ability of a computer to learn a specific task from data [22]. One of these AI techniques is machine learning (ML), ML algorithms use computational methods to learn information directly from data without relying on a predetermined equation or model [23]. Within ML algorithms, hybrid models appears as a good alternative, since one part acts as an estimator or predictor, and the other part acts as an optimizer [24]. Artificial neural networks (ANN) are self-learning, excel in areas where the solution is difficult to express in traditional algorithms [25]. On the other hand, fuzzy logic (FL) is a method of reasoning that resembles human reasoning, FL lets to rely on the experience of experts who already know the process, which is different to neural networks, which takes training data and generates close models [22, 25].

Neuro-fuzzy system is a hybrid intelligent system which integrates both ANN and FL principles and has the potential to capture the benefits of both in a single framework [25]. Adaptive neuro-fuzzy inference system (ANFIS) is a ML hybrid model that have been applied in different areas for optimization and prediction in engineering and materials processing [26, 27], disease detection [28], ship speed prediction [29], quality control [30], energy consuming [24], etc.

Owing to computational advance researchers have developed modelling techniques such as regression analysis and ANN for the modelling of AM processes; Garg et al. [31] compare a hybrid genetic algorithm to those of support vector regression (SVR) and (ANFIS) for ensuring greater trustworthiness of prediction ability in modelling of FDM process. Bayraktar et al. [32] apply ANN for predicting mechanical properties of PLA fabricated by FDM and results are contrasted with analysis of variance (ANOVA), where the raster orientation has the most statistically significant effect. Zhang et al. [33] develop a tensile strength prediction model based on deep learning (DL) in FDM, considering material properties, process parameters and sensor signals to feed the neural network, the obtained error was less than 7%. The applicability of DL requires a very large amount of data, high training process and expensive processing power; the aim of this work is to develop a flexible computing tool that allows predictions and determine suitable parameters for the FDM process, for this reason, ANFIS appears as a good solution.

Barzani et al. [34] apply FL to predict the roughness of Al–Si–Cu–Fe surface of a turned part and contrast the obtained values using ANOVA, the obtained results through FL reached an error of less than 5.4%. Sen et al. [35] applied ANFIS to predict the performance of a CNC milling machine on Inconel 690, evaluating the surface roughness, the cutting force and the cutting temperature, the predictions reached encouraging values with an error of less than 3.8%. Saw et al. [36] define optimal processing parameters in drilling process to reduce tool's wear, applying a hybrid approach between ANFIS and genetic algorithms, obtaining good predictions with a relative error less than 3%. Bagchi et al. [37] perform a numerical simulation to optimize the processing parameters in laser welding, comparing the results obtained by simulation with the results by ANN, also apply Taguchi's methodology where errors less than 8% were obtained. Garg et al. [38] apply response surface methodology (RSM) to feed an ANN of genetic algorithms and thus be able to predict the surface roughness and waviness of a piece manufactured by selective laser melting (SLM).

Regarding to the current work, two methodologies are applied to study the performance of a wood composite filament fabricated by FDM and select suitable processing parameters to improve mechanical response; first, an experimental design of four factors at three levels is employed to find a range of parameters to maximize elastic limit, where layer thickness (0.2 mm, 0.3 mm, 0.4 mm), fill density (50%, 75%, 100%), printing speed (30 mm/min, 40 mm/min, 50 mm/min) and raster angle (0°, 45°, 90°) are evaluated using an L27 Taguchi orthogonal array method, the statistical significance of each parameter is obtained applying ANOVA as well as their interactions; second, ANFIS tool is applied to obtain the combination of parameters that give the maximum yield strength.

2   Material and Method

2.1   Material and Sample Preparation

The material employed in this study is a filament made of PLA (70 wt%) with contain of fibers from beech wood (30 wt%), hereafter referred to as timberfill, commercialized by Fillamentum [39]. Timberfill filament with diameter of 1.75 mm was used, this material has an appearance like real wood. Table 1 shows typical values of mechanical properties of timberfill.

Table 1. Mechanical properties of timberfill [39].

Properties	Value	Test method	Test condition
Density	1.26 g/cm^3		
Melting temperature	145–160 °C		
Tensile strength	39 MPa	ISO 527	At break, 5 mm/min
Elongation at break	2%	ISO 527	5 mm/min
Tensile modulus	3200 MPa	ISO 527	1 mm/min
Charpy impact strength	22 kJ/m^2	ISO 179	23 °C
Hardness	77 Shore D	ISO 7619	

The geometry of the 3D printed samples was modelled using Autodesk Inventor software exported as an STL file and imported to the 3D printing software (MakerBot Print). The main dimensions of the specimens are shown in Fig. 1. Timberfill samples were manufactured using a commercial desktop 3D printer (Makerbot Replicator 2X), with a 0.5 mm nozzle size, extrusion and bed temperature of 150 and 50° respectively, the remaining processes parameters are given in Table 2. In this study, the ASTM

D638-14 standard was applied for testing tensile properties, the tensile test was performed by using universal testing machine (Instron, Zwick) with 2-kN capacity.

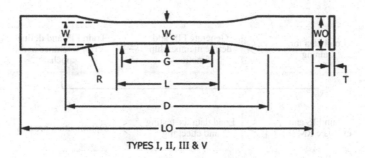

TYPES I, II, III & V

Fig. 1. ASTM D638-14 standard specimen for tensile testing [40].

2.2 Design of Experiments

A classical design of experiments (DOE) of four factors at three levels (3^4), corresponds to a total of 81 experimental runs. For design optimization an L27 Taguchi orthogonal array method was employed, providing a systematic and efficient method which reduces the experiments number to 27. Table 2 shows the factors and levels used in this work, these values were chosen through a thorough literature review [4, 12].

Table 2. Factors and levels used for the DOE.

Factor	Symbol	Levels		
Layer thickness (mm)	LT	0.2	0.3	0.4
Fill density (%)	FD	50	75	100
Printing speed (mm/min)	PS	30	40	50
Raster angle (°)	RA	0	45	90

All the results obtained from the experimental work were analyzed using Minitab 19® statistical software, where a Taguchi analysis was conducted to analyze the main effects as well as their interaction graphs.

2.3 Machine Learning Algorithm: ANFIS

A fuzzy inference system (FIS) was designed in Matlab where a Takagi-Sugeno type of ANFIS was used to train, test and check the results, the workflow is showed in Fig. 2 and ANFIS model structure are show in Fig. 3. Input parameters were the same as the experimental procedure: layer thickness, fill density, printing speed and raster angle;

the output parameter was yield strength (YS). A total of 21 datasets were used for the training process and 6 datasets for testing and checking. A complete explanation of the ANFIS architecture could be found in [36].

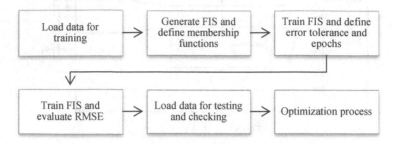

Fig. 2. Scheme of the ANFIS workflow.

The FIS was designed with 21 membership functions (MF) for every input, Gaussbell (gauss2mf) was selected as MF type, a total of 60 epochs was defined and error tolerance of 1×10^{-5} to train the system were stablished.

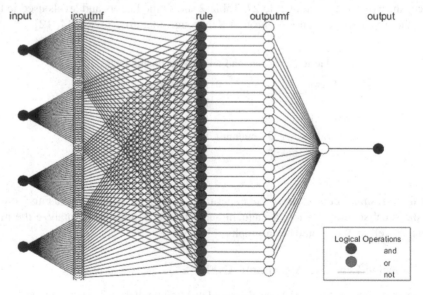

Fig. 3. ANFIS architecture

3 Results

3.1 Experimental Results

The yield strength of timberfill samples manufactured by FDM were tested following the ASTM D638-14 standard, Table 3 shows the experimental results. The highest values were obtained with a fill density of 100% and a raster angle of 90°. The statistical significance of each parameter is shown in Fig. 4 and their interaction in Fig. 5.

Table 3. Experimental data results.

ID	Layer thickness (mm)	Fill density (%)	Printing speed (mm/min)	Raster angle (°)	Yield stress [MPa]
1	0,2	50	30	0	19,66
2	0,2	50	40	45	19,89
3	0,2	50	50	90	24,63
4	0,2	75	30	45	22,14
5	0,2	75	40	90	27,88
6	0,2	75	50	0	21,92
7	0,2	100	30	90	33,04
8	0,2	100	40	0	28,09
9	0,2	100	50	45	27,87
10	0,3	50	30	45	19,56
11	0,3	50	40	90	27,36
12	0,3	50	50	0	21,69
13	0,3	75	30	90	32,07
14	0,3	75	40	0	24,52
15	0,3	75	50	45	22,58
16	0,3	100	30	0	24,83
17	0,3	100	40	45	25,25
18	0,3	100	50	90	39,89
19	0,4	50	30	90	26,6
20	0,4	50	40	0	25,22
21	0,4	50	50	45	21,04
22	0,4	75	30	0	27,93
23	0,4	75	40	45	24,25
24	0,4	75	50	90	36,31
25	0,4	100	30	45	28,67
26	0,4	100	40	90	39,65
27	0,4	100	50	0	35,28

The analysis of variance (ANOVA) shows that printing speed does not play a significant role in FDM of wood composite filaments (Table 4).

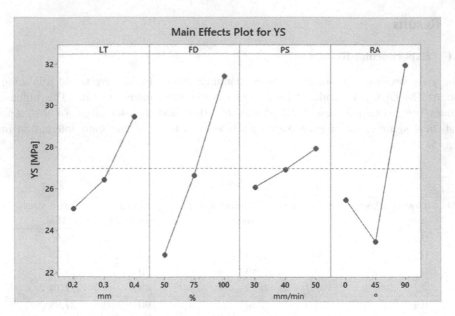

Fig. 4. Main effects of experimental factors on the yield strength

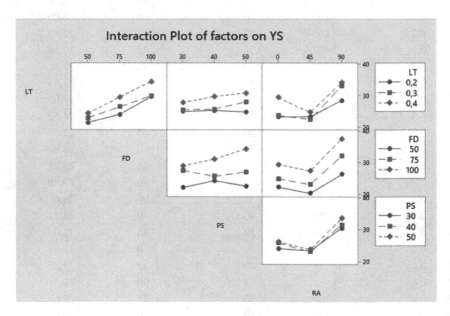

Fig. 5. Interaction plot of factors on the overall yield strength

Table 4. ANOVA results for yield strength

Source	DF	Seq SS	Adj SS	Adj MS	F	P
LT	2	92,07	92,07	46,033	16,24	0,004
FD	2	330,21	330,21	165,106	58,24	0,000
PS	2	15,55	15,55	7,777	2,74	0,143
RA	2	352,64	352,64	176,318	62,19	0,000
LT * FD	4	10,82	10,82	2,704	0,95	0,495
LT * PS	4	12,08	12,08	3,021	1,07	0,449
LT * RA	4	42,56	42,56	10,640	3,75	0,073
Residual error	6	17,01	17,01	2,835		
Total	26	872,94				

3.2   ANFIS Results

The relationship between yield strength and process parameters could be seen in Fig. 6, the training, testing and checking results are shown in Fig. 7.

4   Discussions

For the timberfill fabrication through FDM, the effect of layer thickness, fill density, printing speed and raster angle on yield strength are discussed below.

4.1   Experimental Results

Results shows that the highest yield strength (YS) is obtained by using a fill density (FD) of 100%, this is because the number of pores which act as stress concentrators, are reduced. These results are validated applying ANOVA, where FD appears as the main factor which affect the YS response.

The second effect statistically significant is raster angle (RA) showing a change in the slope, when the fibers are aligned with load direction (RA = 90°) the YS shows the highest value, fibers perpendicular (0°) to load direction show lower strength, but the lowest YS is obtained at 45°, these results are in agreement with Motaparti et al. [41].

The last effect statistically significant is layer thickness (LT) which shows a linear tendency, the highest YS is obtained at 0.4 mm and the lowest at 0.2 mm, these results are attributed to the presence of wood fibers which improve adhesion between layers [19]. The higher the LT, the lower manufacturing time [4].

Furthermore, the interaction between LT, FD, PS and RA is analyzed through ANOVA where just the interaction between LT and RA shows statistical significance with 95% confidence. The highest YS, 39.89 MPa is obtained with LT = 0.4 mm and RA = 90°. Finally, the effect of printing speed does not play a significant role in the yield strength response.

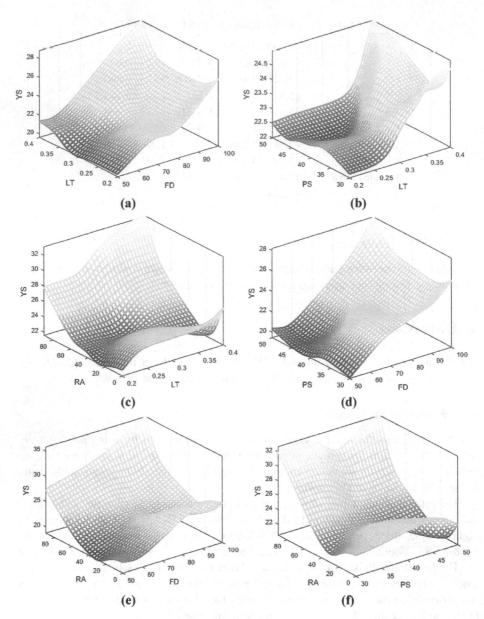

Fig. 6. Yield strength (YS) surface response, interactions between; (a) LT-FD, (b) PS-LT, (c) RA-LT, (d) PS-FD, (e) RA-FD, (f) RA-PS

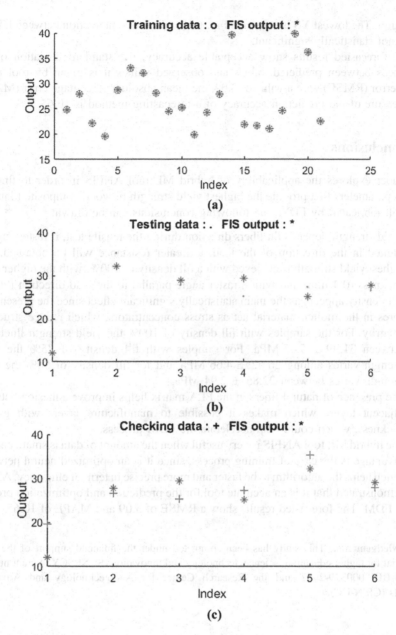

Fig. 7. ANFIS procedure; (a) training, (b) testing, (c) checking

4.2   ANFIS Results

The interaction of the surface response shows the highest yield strength in the Fig. 6(e) it is because of FD and RA are statistically significant effects, as well as their interaction. The lowest YS is shown in Fig. 6(b), where the interaction between LT and PS are not statistically significant.

The forecasted results show acceptable accuracy, the standard deviation of the differences between predicted values and observed values it is given by root mean square error (RMSE) with a value of 3.09, the mean absolute percentage error (MAPE) as a measure of the prediction accuracy of a forecasting method is 10.65.

5   Conclusions

This paper explores the applicability of hybrid ML tool ANFIS in order to find the process parameters that provide the highest yield strength in wood composite filament-timberfill fabricated by FDM, the following conclusions can be drawn:

(1)   Yield strength depends on fibers direction during the tensile test, if raster angle is aligned in the direction of the load, a greater resistance will be obtained. The highest yield strength is achieved with a fill density of 100%, with the higher layer thickness (0.4 mm) and with a raster angle parallel to the load direction (90°).

(2)   Fill density appears as the main statistically significant effect since the presence of pores in the molten material act as stress concentrators, which reduce structural integrity. For the samples with fill density of 100% the yield strength fluctuates between 31.99 ± 5.47 MPa. For samples with fill density of 75% the yield strength varies among 26.62 ± 4.66 MPa and for fill density of 50% the yield strength varies between 22.85 ± 2.94 MPa.

(3)   The presence of natural fibers in the PLA matrix helps improve adhesion between adjacent layers which makes it possible to manufacture layers with greater thickness, which could help accelerate the FDM process.

(4)   The hybrid ML tool ANFIS is very useful when the amount of data is small, another advantage is the ease of training process, since it is an optimized neural network, which helps the algorithm to be faster and more precise in term of efficiency. ANFIS demonstrated that it is an accurate tool for the prediction and optimization process in FDM. The forecasted results show a RMSE of 3.09 and MAPE of 10.65.

Acknowledgements. This study has been completed under the financial support of the State Secretariat for higher education, science, technology and innovation (SENESCYT) grant number ARSEQ-BEC-000329-2017 and the Research Center for Nanotechnology and Advanced Materials (CIEN-UC).

References

1. Alafaghani, A., Qattawi, A., Alrawi, B., Guzman, A.: Experimental optimization of fused deposition modelling processing parameters: a design-for-manufacturing approach. Procedia Manuf. **10**, 791–803 (2017)
2. Rinaldi, M., Ghidini, T., Cecchini, F., Brandao, A., Nanni, F.: Additive layer manufacturing of poly (ether ether ketone) via FDM. Compos. Part B Eng. **145**, 162–172 (2018)

3. Kozior, T., Kundera, C.: Evaluation of the influence of parameters of FDM technology on the selected mechanical properties of models. Procedia Eng. **192**, 463–468 (2017)
4. Gordelier, T.J., Thies, P.R., Turner, L., Johanning, L.: Optimising the FDM additive manufacturing process to achieve maximum tensile strength: a state-of-the-art review. Rapid Prototyp. J. **25**(6), 953–971 (2019)
5. Keles, O., Wayne, C., Keith, B.: Effect of build orientation on the mechanical reliability of 3D printed ABS. Rapid Prototyp. J. **23**(2), 320–328 (2017)
6. Chacón, J.M., Caminero, M.A., García-Plaza, E., Núñez, P.J.: Additive manufacturing of PLA structures using fused deposition modelling: effect of process parameters on mechanical properties and their optimal selection. Mater. Des. **124**, 143–157 (2017)
7. Durgun, I., Ertan, R.: Experimental investigation of FDM process for improvement of mechanical properties and production cost. Rapid Prototyp. J. **20**(3), 228–235 (2014)
8. Raut, S., Jatti, V.S., Khedkar, N.K., Singh, T.P.: Investigation of the effect of built orientation on mechanical properties and total cost of FDM parts. Procedia Mater. Sci. **6**, 1625–1630 (2014)
9. Uddin, M.S., Sidek, M.F.R., Faizal, M.A., Ghomashchi, R., Pramanik, A.: Evaluating mechanical properties and failure mechanisms of fused deposition modeling acrylonitrile butadiene styrene parts. J. Manuf. Sci. Eng. **139**(8), 081018 (2017)
10. Dawoud, M., Taha, I., Ebeid, S.J.: Mechanical behaviour of ABS: an experimental study using FDM and injection moulding techniques. J. Manuf. Process. **21**, 39–45 (2016)
11. Song, Y., Li, Y., Song, W., Yee, K., Lee, K.Y., Tagarielli, V.L.: Measurements of the mechanical response of unidirectional 3D-printed PLA. Mater. Des. **123**, 154–164 (2017)
12. Popescu, D., Zapciu, A., Amza, C., Baciu, F., Marinescu, R.: FDM process parameters influence over the mechanical properties of polymer specimens: a review. Polym. Test. **69**(April), 157–166 (2018)
13. Subramaniam, S.R., et al.: 3D printing: overview of PLA progress. In: AIP Conference Proceedings, vol. 2059, January 2019
14. Liu, Z., Wang, Y., Wu, B., Cui, C., Guo, Y., Yan, C.: A critical review of fused deposition modeling 3D printing technology in manufacturing polylactic acid parts. Int. J. Adv. Manuf. Technol. **102**(9–12), 2877–2889 (2019)
15. Kumar, N., Jain, P.K., Tandon, P., Pandey, P.M.: The effect of process parameters on tensile behavior of 3D printed flexible parts of ethylene vinyl acetate (EVA). J. Manuf. Process. **35**, 317–326 (2018)
16. Wu, W., Geng, P., Li, G., Zhao, D., Zhang, H., Zhao, J.: Influence of layer thickness and raster angle on the mechanical properties of 3D-printed PEEK and a comparative mechanical study between PEEK and ABS. Materials **8**(9), 5834–5846 (2015)
17. Tao, Y., Wang, H., Li, Z., Li, P., Shi, S.Q.: Development and application of wood flour-filled polylactic acid composite filament for 3D printing. Materials **10**(4), 1–6 (2017)
18. Kariz, M., Sernek, M., Obućina, M., Kuzman, M.K.: Effect of wood content in FDM filament on properties of 3D printed parts. Mater. Today Commun. **14**, 135–140 (2018)
19. Fazeli, M., Florez, J.P., Simão, R.A.: Improvement in adhesion of cellulose fibers to the thermoplastic starch matrix by plasma treatment modification. Compos. Part B Eng. **163**, 207–216 (2019)
20. Roj, R., Theiß, R., Dültgen, P.: Mechanical properties of 16 different FDM-plastic types. Mater. Test. **61**(10), 999–1006 (2019)
21. Gao, X., Zhang, D., Qi, S., Wen, X., Su, Y.: Mechanical properties of 3D parts fabricated by fused deposition modeling: effect of various fillers in polylactide. J. Appl. Polym. Sci. **136**(31), 1–10 (2019)
22. Siddique, N., Adeli, H.: Computational Intelligence: Synergies of Fuzzy Logic (2013)

23. Chelly, S.M., Denis, C.: Introducing machine learning. Med. Sci. Sports Exerc. **33**(2), 326–333 (2001)
24. Mosavi, A., Salimi, M., Ardabili, S.F., Rabczuk, T., Shamshirband, S., Varkonyi-Koczy, A. R.: State of the art of machine learning models in energy systems, a systematic review. Energies **12**(7), 1301 (2019)
25. Casalino, G.: Computational intelligence for smart laser materials processing. Opt. Laser Technol. **100**, 165–175 (2018)
26. Mathew, J., Griffin, J., Alamaniotis, M., Kanarachos, S., Fitzpatrick, M.E.: Prediction of welding residual stresses using machine learning: comparison between neural networks and neuro-fuzzy systems. Appl. Soft Comput. J. **70**, 131–146 (2018)
27. Zaharuddin, M.F.A., Kim, D., Rhee, S.: An ANFIS based approach for predicting the weld strength of resistance spot welding in artificial intelligence development. J. Mech. Sci. Technol. **31**(11), 5467–5476 (2017)
28. Huang, M.L., Chen, H.Y., Huang, J.J.: Glaucoma detection using adaptive neuro-fuzzy inference system. Expert Syst. Appl. **32**(2), 458–468 (2007)
29. Valčić, M., Antonić, R., Tomas, V.: ANFIS based model for ship speed prediction. Brodogradnja **62**(4), 373–382 (2011)
30. Huang, C.-W., Baron, L., Balazinski, M., Achiche, S.: Comprehensive model optimization in pulp quality prediction: a machine learning approach. PeerJ PrePrints **5**, 1–18 (2017)
31. Garg, A., Tai, K., Lee, C.H., Savalani, M.M.: A hybrid M5'-genetic programming approach for ensuring greater trustworthiness of prediction ability in modelling of FDM process. J. Intell. Manuf. **25**(6), 1349–1365 (2014)
32. Bayraktar, Ö., Uzun, G., Çakiroğlu, R., Guldas, A.: Experimental study on the 3D-printed plastic parts and predicting the mechanical properties using artificial neural networks. Polym. Adv. Technol. **28**(8), 1044–1051 (2017)
33. Zhang, J., Wang, P., Gao, R.X.: Deep learning-based tensile strength prediction in fused deposition modeling. Comput. Ind. **107**, 11–21 (2019)
34. Barzani, M.M., Zalnezhad, E., Sarhan, A.A.D., Farahany, S., Ramesh, S.: Fuzzy logic based model for predicting surface roughness of machined Al-Si-Cu-Fe die casting alloy using different additives-turning. Meas. J. Int. Meas. Confed. **61**, 150–161 (2015)
35. Sen, B., Mandal, U.K., Mondal, S.P.: Advancement of an intelligent system based on ANFIS for predicting machining performance parameters of Inconel 690 – A perspective of metaheuristic approach. Meas. J. Int. Meas. Confed. **109**, 9–17 (2017)
36. Saw, L.H., et al.: Sensitivity analysis of drill wear and optimization using Adaptive Neuro fuzzy –genetic algorithm technique toward sustainable machining. J. Clean. Prod. **172**, 3289–3298 (2018)
37. Bagchi, A., Saravanan, S., Kumar, G.S., Murugan, G., Raghukandan, K.: Numerical simulation and optimization in pulsed Nd: YAG laser welding of Hastelloy C-276 through Taguchi method and artificial neural network. Optik **146**, 80–89 (2017)
38. Garg, A., Lam, J.S.L., Savalani, M.M.: Laser power based surface characteristics models for 3-D printing process. J. Intell. Manuf. **29**(6), 1191–1202 (2018)
39. Dillard, D.A.: Physical properties. In: Handbook of Adhesion Technology, 2nd edn, vol. 1–2, pp. 433–457 (2018)
40. American Society for Testing and Materials. D638 – 14 Standard Test Method for Tensile Properties of Plastics. Standard Test Method for Tensile Properties of Plastics, p. 17 (2014)
41. Motaparti, K.P., Leu, M.C., Chandrashekhara, A.K., Dharani, L.R.: Effect of build parameters on mechanical properties of Ultem 9085 parts by fused deposition modeling. In: Solid Freeform Fabrication. 2016 Proceedings of 26th Annual International Solid Freeform Fabrication. 2016 Proceedings of 27th Annual International Solid Freeform Fabrication Symposium – An Additive Manufacturing Conference, pp. 964–977 (2016)

Automated Systems for Detecting Volcano-Seismic Events Using Different Labeling Techniques

Enrique V. Carrera[✉], Alexandra Pérez, and Román Lara-Cueva

Departamento de Eléctrica y Electrónica,
Universidad de las Fuerzas Armadas ESPE, Sangolquí 171103, Ecuador
{evcarrera,asperez1,ralara}@espe.edu.ec

Abstract. Several systems have been developed in the last years to automatically detect volcanic events based on their seismic signals. Many of those systems use supervised machine learning algorithms in order to create the detection models. However, the supervised training of these machine learning techniques requires labeled-signal catalogs (*i.e.*, training, validation and test data-sets) that in many cases are difficult to obtain. In fact, existing labeling schemes can consume a lot of time and resources without guarantying that the final detection model is accurate enough. Moreover, every labeling technique can produce a different set of events, without being defined so far which technique is the best for volcanic-event detection. Hence, this work proves that the labeling scheme used to create training sets definitely impacts the performance of seismic-event detectors. This is demonstrated by comparing two techniques for labeling seismic signals before to train a system for automated detection of volcanic events. The first technique is automatic and computationally efficient, while the second one is a handmade and time-consuming process carried out by expert analysts. Results show that none of the labeling techniques is completely trustworthy. As a matter of fact, our main result reveals that an improved detection accuracy is obtained when machine learning classifiers are trained with the conjunction of diverse labeling techniques.

Keywords: Volcanic seismic events · Labeled data-sets · Machine learning algorithms · Automated detection systems

1 Introduction

Volcanic hazards are part of the daily life of many communities around the world. In order to monitor, prevent and mitigate the risks related to these volcanic hazards, technological tools are increasingly needed. Thus, a wide variety of systems has been developed in the last decades to automatically detect volcanic events based on their seismic signals [1,8–11]. Many of the proposed systems use machine learning techniques based on supervised training to create compelling

© Springer Nature Switzerland AG 2020
M. Botto-Tobar et al. (Eds.): ICAT 2019, CCIS 1194, pp. 133–144, 2020.
https://doi.org/10.1007/978-3-030-42520-3_11

detection models [12,14]. Machine learning approaches allow to automate the analysis of recorded or real-time signals in order to monitor events at a larger scale. However, supervised training of machine learning algorithms require existing labeled data-sets.

Fortunately, over the past years, the amount of available volcanic-earthquake data has grown dramatically. But, the creation of seismic-signal catalogs with identification labels for each event has been a tedious and delayed task [14]. In fact, depending on the technique used to label each seismic event, we can finish with a different set of reported events [17]. Approaches combining automatic detection schemes with revisions by analysts are commonly used to confirm the occurrence or not of these seismic events. However, due to a lack of human resources or excessive seismic activity, organizations have been forced to only review events within smaller regions or magnitude ranges. Hence, the confidence in the actual detection of a volcanic earthquake is not high enough.

Since seismic-event catalogs with labels for each event are needed by supervised training schemes (*i.e.*, training, validation, and test data-sets), this work studies the dependency of machine learning techniques on the type of approach used to label volcanic-seismic signals. In particular, we have used seismic signals from the Cotopaxi volcano, located in Ecuador, which have labels provided by analysts of the Ecuadorian Geophysical Institute (IGEPN) that show the beginning and the end of each volcanic earthquake [8,9]. In addition, we have implemented a traditional detector knows as STA/LTA (*short-term average/long-term average*) to automate the process of labeling [7]. The labeled seismic catalogs are then used to train two machine learning algorithms: artificial neural networks and support vector machines. The training of these algorithms is based on 22 features extracted from the time and frequency domains of the signals. The results of such an evaluation show up to 97% of accuracy for detecting volcanic events when the data-set includes only events detected simultaneously by the IGEPN analysts and the STA/LTA algorithm. When the classifiers are trained with data-sets generated by isolated labeling methods, the overall accuracy is reduced.

The rest of this paper is organized as follows. Section 2 introduces the creation of labeled seismic-signal catalogs. Section 3 presents the approach designed to evaluate the efficiency of supervised machine learning techniques with different signal catalogs. The main results and their implications are analyzed in Sect. 4. Finally, Sect. 5 concludes this paper.

2   Seismic-Signal Catalogs

Seismic-signal catalogs have traditionally been created using tedious and delayed processes through expert analysts. However, some automatic and computationally efficient techniques like STA/LTA have also been proposed [7,15].

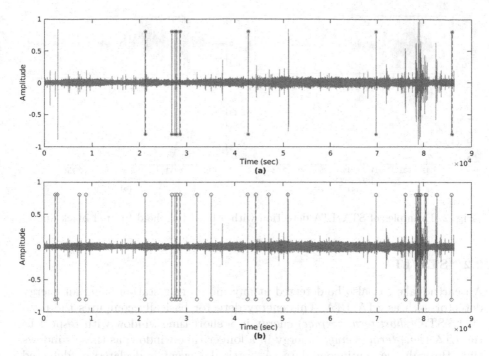

Fig. 1. Processed seismic signals with (*a*) IGEPN (red), and (*b*) STA/LTA (green) labels. (Color figure online)

2.1 Original Data-Set

The raw seismic signals from the Cotopaxi volcano, located in Ecuador, were provided by the IGEPN in several files formatted according to the standard SAC. Each file contains 20 min of the seismic signal sampled to 100 Hz and belonging to the vertical-axis of station VC2 during the year 2012 [8,9]. The station VC2 is a broadband seismograph that is part of the monitoring network installed by the IGEPN, presenting a good frequency response between 0.01 and 50 Hz. In order to restore the seismic signal that belongs to a particular day, 72 SAC files must be read, concatenated, and processed.

Besides the raw seismic signals, the IGEPN also provides the identification of 241 seismic events in the year 2012. The events identified by the IGEPN are 23 volcano-tectonic, 208 long duration, 7 hybrids and 3 tremors [8,9]. The labels of each event were determined by analysts of the IGEPN showing the beginning and the end of every volcanic earthquake. Figure 1(a) presents a day-long window of the VC2 seismic signal with all the events (vertical red lines) identified by the IGEPN in that specific day.

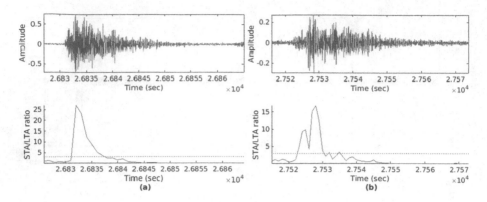

Fig. 2. Examples of STA/LTA detection with a fixed threshold (dotted black line).

2.2 STA/LTA

An earthquake can also be detected at any monitoring station using an energy detector such as STA/LTA. This energy detector basically computes the ratio of the STA (*short-term average*) energy in a short time window with respect to the LTA (*long-term average*) energy in a longer time window, as these windows slide through the continuous data. A particular event is declared as detected when the STA/LTA ratio exceeds certain threshold [15]. Due to its simplicity, the STA/LTA method is the most popular one among the automatic detection processes. In fact, STA/LTA is widely used in operational context and published works [4,7]. However, this method has also some disadvantages [16]; for instance, it requires a careful setting of some parameters including the trigger threshold level and two window lengths (*i.e.*, the short-term and long-term windows). Figure 2 shows two examples of identified seismic events where it is clear that a single trigger threshold is not always an optimal solution.

In general, STA/LTA identifies successfully earthquakes with impulsive, high signal-to-noise ratio (SNR) arrivals. However, STA/LTA fails to detect earthquakes in more challenging situations such as low SNR, waveforms with emergent arrivals, or overlapping events [4,16]. Although these problems, STA/LTA is able to detect a wide variety of earthquakes without prior knowledge of the event waveform or source information. Thus, Fig. 1(b) presents a day-long window of the VC2 seismic signal with all the events (vertical green lines) identified by the STA/LTA algorithm in that day.

3 Seismic-Event Detector Based on Machine Learning

As explained before, conventional detection techniques applied to seismic events are time-consuming processes and/or their results are not completely trustworthy. This is why machine learning techniques have been proposed, but labeled signal catalogs are still required by the supervised learning algorithms. Hence,

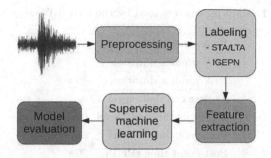

Fig. 3. Processing system for detecting volcano-seismic events using different labeling techniques.

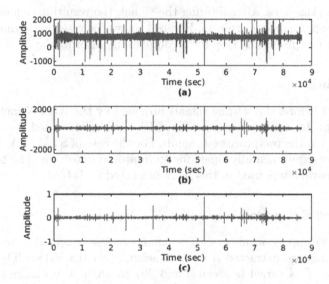

Fig. 4. Pre-processing of the volcano-seismic signals: (a) raw signal, (b) filtered signal, and (c) normalized signal.

the system presented in Fig. 3 was developed to evaluate the influence of labeling techniques in the performance of detectors based on machine learning algorithms. The whole system was implemented in Matlab (release 2015a), starting from the raw signals provided by the IGEPN. The main components of the system presented in Fig. 3 are described in the following subsections.

3.1   Pre-processing

Since raw signals are not completely useful for seismic-event detection, the system starts filtering the original signal using a 128-order band-pass filter with a Hamming window and cutoff frequencies of 0.5 Hz and f_{high}. The cutoff frequency f_{high} is varied between 10 and 25 Hz in order to maximize the perfor-

Table 1. Main parameters used by the STA/LTA algorithm.

Parameter	Value
STA window duration	1 s
LTA window duration	30 s
Trigger threshold level	8
Detrigger threshold level	0.5
Pre-event time (PEM)	6 s
Post-event time (PET)	20 s

mance of the classifiers. After filtering the signal, the resulting sequence is normalized through a *min-max* rule [6]. Figure 4 shows the resulting signal obtained through each pre-processing step for a complete monitoring day.

3.2   Labeling

As mentioned before, the seismic signals provided by the IGEPN include labels for 241 events. In similar way, the STA/LTA algorithm is used for identifying seismic events in the pre-processed signals. For the case of STA/LTA, its parameters [15] were experimentally tuned for an improved detection. The final values of the main parameters used in this work are listed in Table 1.

3.3   Feature Extraction

Using non-overlapping windows of T seconds, 22 features in the time and frequency domains are extracted from each seismic-signal window. The duration of the window T is varied between 5 and 20 s in order to optimize the performance of the classifiers. Moreover, only windows containing events from start to end (positive case), or containing just noise (negative case), are used for feature extraction.

The features extracted in the time domain are: mean value, maximum value, minimum value, standard deviation, absolute mean deviation, standardized magnitude area, energy, inter-quartile range, entropy, variance, and auto-regressive model of order 5 (*i.e.*, four coefficients). In addition, the features in the frequency domain are: maximum power, frequency with maximum power, weighted average frequency, skewness, kurtosis, and 3 band energies (corresponding to 3 equal-size bands between 0 and f_{high}). It is also important to mention that the feature matrix is normalized using a z-score rule [6] before presenting these data to the classifiers.

A feature selection scheme based on information gain [2] is also applied to the 22 basic features. The feature selection scheme seeks to improve the performance of the machine learning classifiers reducing the numbers of actual inputs.

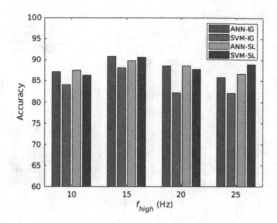

Fig. 5. Accuracy of ANN and SVM detectors in function of the higher cutoff frequency.

3.4 Machine Learning Classifiers

The classifiers used in this study are artificial neural networks (ANN) and support vector machines (SVM) [3,13]. Both models are based on supervised training and require the typical training and test data-sets. Both algorithms are implemented using the default Matlab functions in a Windows computer. In the case of ANN, its implementation includes a hidden layer containing between 5 and 40 processing units.

The evaluation of the trained models is made using a k-fold cross-validation technique with $k = 10$, and the metrics assessed for each model are accuracy, recall, specificity, and precision [5].

4 Results

Using the events identified by the IGEPN analysts and the STA/LTA algorithm, ANN and SVM models have been trained to automatically detect these seismic events. Note that in a 5-day period, 109 12-s windows were identified as belonging to a seismic event for the IGEPN. On the other hand, STA/LTA identified 252 12-s windows belonging to an event. Both labeling techniques coincided in 65 of these 12-s positive windows. Since there are some parameters that are important to understand the obtained results, a parameter space study is presented first.

4.1 Parameter Space Study

Three main parameters of our proposal have been varied in order to find their best combination that maximizes the accuracy of the trained detectors.

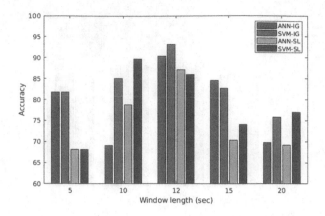

Fig. 6. Accuracy of ANN and SVM detectors in function of the window size.

High Cutoff Frequency. As described in Sect. 3, the band-pass filter is a key component of the pre-processing stage. Although the low cutoff frequency is well understood, there is no agreement on which high cutoff frequency (f_{high}) is the best. Thus, it has been varied f_{high} between 10 and 25 Hz, keeping the low cutoff frequency in 0.5 Hz. ANN and SVM algorithms were trained with the events labeled by the IGEPN analysts and by the STA/LTA algorithm using a 15-s window and a data set belonging to 5 consecutive days.

Figure 5 summarizes the accuracy of the detectors based on ANN and SVM using the labels provided by the IGEPN (IG) and STA/LTA (SL). Accuracy results shows that a band-pass filter with cutoff frequencies of 0.5 and 15 Hz produces a consistent good performance for all the studied detectors.

Windows Size. The size of the window analyzed to extract the 22 features of each seismic event is also evaluated. That size is varied from 5 to 20 s, keeping constant the cutoff frequencies of the band-pass filter in 0.5 and 15 Hz. In this case, ANN and SVM algorithms were trained with the events labeled by the IGEPN analysts and by the STA/LTA algorithm using a data-set belonging to 5 consecutive days.

Figure 6 shows that the accuracy results are maximized for a window size of 12 s in average. Although the SVM model trained with labels generated by the STA/LTA algorithm has a better performance for a window size of 10 s, the other 3 detectors achieve the best performance with a window size of 12 s.

Number of Days. The size of the data-set is also evaluated to see any dependency between the achieved accuracy and the number of days used to create the training and test data-sets. In this case, the number of days is varied between 1 and 20. In Fig. 7, it is possible to see that the results obtained for ANN and SVM algorithms do not present any significant improvement after a 5-day period.

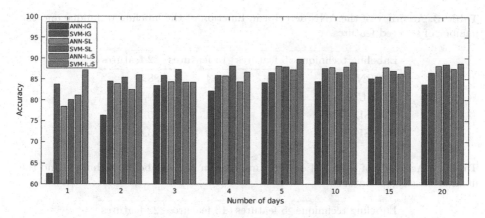

Fig. 7. Accuracy of ANN and SVM detectors in function of the data-set size in days.

Table 2. Accuracy of the ANN detector in function of the labeling technique.

Labeling technique	Hidden layer size	Accuracy (%)
IGEPN	20	86.7
STA/LTA	30	92.1
IG ∪ SL	20	91.2
IG ∩ SL	35	95.4

Note additionally that the evaluation of this last parameter includes a new data-set labeled as 'I ∪ S', containing the union of the two previously mentioned sets, besides the data-sets labeled by the IGEPN (IG) and the STA/LTA algorithm (SL).

4.2   Artificial Neural Networks

In the case of the ANN detector, a free parameter is the size of the hidden layer. Thus, this detector was trained with the events labeled by the IGEPN analysts, by the STA/LTA algorithm, by the union of both labeled data-sets (IG ∪ SL), and by the intersection of the same two sets (IG ∩ SL). The data-sets belong to 5 consecutive days, pre-processed with a band-pass filter using cutoff frequencies of 0.5 and 15 Hz, and a window size of 12 s.

Table 2 shows the accuracy results for the best ANN when the number of neurons in the hidden layer is varied between 5 and 40. We can see that the optimal number of neurons is around 26 in average. Moreover, the STA/LTA algorithm presents a better performance that the IGEPN labeling. However, when the events included in the intersection (*i.e.*, coincidences) of the two labeling techniques are exclusively used, we obtain the best performance: an accuracy of 95.4%.

Table 3. Accuracy of the ANN detector in function of the labeling technique and the number of selected features.

Labeling technique	5 features	13 features	22 features
IGEPN	89.0	89.4	88.5
STA/LTA	87.5	91.1	89.7
IG∪SL	90.8	91.4	85.7
IG∩SL	93.1	96.9	93.8

Table 4. Accuracy of the SVM detector in function of the labeling technique and the number of selected features.

Labeling technique	5 features	13 features	22 features
IGEPN	89.0	91.8	93.2
STA/LTA	86.9	92.7	89.3
IG∪SL	93.1	89.5	87.4
IG∩SL	83.7	95.4	93.0

Table 5. Overall performance of the ANN and SVM detectors in function of the labeling technique for a 13-feature set.

ANN				
Labeling technique	Accuracy	Recall	Specificity	Precision
IGEPN	89.4	82.6	96.3	95.7
STA/LTA	91.1	84.6	97.6	91.2
IG∪SL	91.4	83.6	99.2	99.1
IG∩SL	96.9	93.8	100	100
SVM				
Labeling technique	Accuracy	Recall	Specificity	Precision
IGEPN	91.8	83.9	97.6	96.3
STA/LTA	92.7	87.0	97.7	97.1
IG∪SL	89.5	79.8	98.9	98.5
IG∩SL	95.3	91.7	100	100

In order to optimize the ANN model, a feature selection stage has also been implemented to reduce the size of the input to the network. Using *information gain* as the metric for feature selection, subsets of 5 and 13 features were created according to the methodology proposed in [2]. Table 3 shows the accuracy results for these new feature sets. As we can see, the 13-feature set presents the best results for the neural network model. In this case, the intersection of the two labeling techniques achieves a 96.9% accuracy.

4.3   Support Vector Machines

In addition to the ANN model, a SVM algorithm was used to train a new detector. The SVM model was trained with the three sets of selected features (*i.e.*, 5, 13, and 22 features) and with the 4 types of data-sets (*i.e.*, IGEPN, STA/LTA, IG ∪ SL, and IG ∩ SL). The accuracy results for these 12 combinations are presented in Table 4. Although there are some small changes, the results and their trends are very similar to the obtained with the ANN model.

4.4   Overall Performance

Finally, in order to present a more complete evaluation of the overall behavior of the implemented detectors, Table 5 shows results for accuracy, recall, specificity, and precision when a set of 13 features is extracted from each 12-s window. It is important to note that for the intersection of the two labeling techniques (*i.e.*, IG ∩ SL), the specificity and precision are 100%. Furthermore, this conjunction of data-sets presents the best accuracy and recall for both machine learning algorithms: ANN and SVM.

In addition, although the IGEPN labeling was done by expert analysts, the STA/LTA labeling presents better results that the former labeling technique. This demonstrates that even time-consuming processes do not always guarantee to finish with better models.

5   Conclusions

Machine learning algorithms are widely used today for detecting and classifying volcano-seismic events. However, these algorithms require preexisting labeled catalogs that are hard to obtain or generate. In fact, the existing labeled data-sets must be validated before using them for training machine learning algorithms, since results can vary significantly for the different labeling techniques. This work showed that the labeling techniques used in such data-sets must be understood and contrasted with other labeling schemes in order to obtain useful models and results.

Furthermore, a handmade or time-consuming labeling technique does not always guarantee better results. In other words, besides the specific machine learning algorithm and the tuning of its main parameters, the labeling technique used to create the training and test data-sets is an important characteristic to be considered in any automated detection system.

Acknowledgments. This work was partially supported by the Universidad de las Fuerzas Armadas ESPE under research grants 2015-PIC-004 and 2016-EXT-038. We would also like to thank to the Ecuadorian Geophysical Institute (IGEPN) for providing the labeled data-sets used in this study.

References

1. Bhatti, S.M., et al.: Automatic detection of volcano-seismic events by modeling state and event duration in hidden Markov models. J. Volcanol. Geoth. Res. **324**, 134–143 (2016)
2. Carrera, E.V., Rodriguez, J.: Feature selection for continuous attributes in physical-activity classification tasks. RECI-Revista Iberoamericana de las Ciencias Computacionales e Informática **5**(10), 79–100 (2016)
3. Chen, C.H.: Handbook of pattern recognition and computer vision. World Scientific (2015)
4. Ebrahimi, M., Moradi, A., Bejvani, M., Tafreshi, M.D.: Application of STA/LTA based on cross-correlation to passive seismic data. In: Sixth EAGE Workshop on Passive Seismic (2016)
5. Hossin, M., Sulaiman, M.: A review on evaluation metrics for data classification evaluations. Int. J. Data Min. Knowl. Manag. Process **5**(2), 1 (2015)
6. Jayalakshmi, T., Santhakumaran, A.: Statistical normalization and back propagation for classification. Int. J. Comput. Theory Eng. **3**(1), 1793–8201 (2011)
7. Jones, J.P., van der Baan, M.: Adaptive STA–LTA with outlier statistics. Bull. Seismol. Soc. Am. **105**(3), 1606–1618 (2015)
8. Lara-Cueva, R., Benitez, D., Carrera, E.V., Ruiz, M., Rojo-Álvarez, J.: Feature selection of seismic waveforms for long period event detection at cotopaxi volcano. J. Volcanol. Geoth. Res. **316**, 34–49 (2016)
9. Lara-Cueva, R., Benitez, D., Carrera, E.V., Ruiz, M., Rojo-Álvarez, J.L.: Automatic recognition of long period events from volcano tectonic earthquakes at cotopaxi volcano. IEEE Trans. Geosci. Remote Sens. **54**(9), 5247–5257 (2016)
10. Lara-Cueva, R., Carrera, E.V., Morejon, J.F., Benitez, D.: Comparative analysis of automated classifiers applied to volcano event identification. In: 2016 IEEE Colombian Conference on Communications and Computing (COLCOM), pp. 1–6. IEEE (2016)
11. Maggi, A., Ferrazzini, V., Hibert, C., Beauducel, F., Boissier, P., Amemoutou, A.: Implementation of a multistation approach for automated event classification at Piton de la Fournaise volcano. Seismol. Res. Lett. **88**(3), 878–891 (2017)
12. Malfante, M., Dalla Mura, M., Métaxian, J.P., Mars, J.I., Macedo, O., Inza, A.: Machine learning for volcano-seismic signals: challenges and perspectives. IEEE Signal Process. Mag. **35**(2), 20–30 (2018)
13. Marsland, S.: Machine Learning: An Algorithmic Perspective. CRC Press, Boca Raton (2015)
14. Reynen, A., Audet, P.: Supervised machine learning on a network scale: application to seismic event classification and detection. Geophys. J. Int. **210**(3), 1394–1409 (2017)
15. Trnkoczy, A.: Topic understanding and parameter setting of STA/LTA trigger algorithm. New Manual Seismol. Observatory Pract. **2**, 1–20 (1999)
16. Vaezi, Y., Van der Baan, M.: Comparison of the STA/LTA and power spectral density methods for microseismic event detection. Geophys. Suppl. Monthly Not. R. Astron. Soc. **203**(3), 1896–1908 (2015)
17. Yoon, C.E., O'Reilly, O., Bergen, K.J., Beroza, G.C.: Earthquake detection through computationally efficient similarity search. Sci. Adv. **1**(11), 1–13 (2015)

Fraud Prediction in Smart Supply Chains Using Machine Learning Techniques

Fabián-Vinicio Constante-Nicolalde[1,2] , Paulo Guerra-Terán[1] ,
and Jorge-Luis Pérez-Medina[1(✉)]

[1] Intelligent and Interactive Systems Lab (SI² Lab),
Universidad de Las Américas (UDLA), Quito, Ecuador
{fabian.constante,paulo.guerra.teran,jorge.perez.medina}@udla.edu.ec
[2] School of Technology and Management, Polytechnic Institute of Leiria,
Leiria, Portugal
2162316@my.ipleiria.pt

Abstract. In the domain of Big Data, the company's supply chain has
a very high-risk exposure and this must be observed from a preventive perspective, that is, act before such situations occur. As a company
grows and diversifies the number of suppliers, customers and therefore
increases its number of daily transactions and associated risks. Despite
the innovation and improvements that have been incorporated into financial management, credit and debit cards are the main means of exchanging cash online, with the expansion of e-commerce, online shopping has
also increased number of extortion cases that have been identified and
that continues to expand greatly. It takes a lot of time, effort and investment to restore the impact of these damages. In this paper, we work
with machine learning techniques, used in predicting smart supply chain
fraud, are valuable for estimating, classifying whether a transaction is
normal or fraudulent, and mitigating future dangers.

Keywords: Big Data Analysis · Classification approaches · Fraud
prediction

1 Introduction

Organizations seek to boost innovation considering the vision of customers.
Today, Predictive Modeling Analysis is carried out, for this purpose each application requires a large database size, so that a large amount of data can be stored in
them [14]. Data mining involves an automated approach identifying interesting
patterns from large databases helping to generate descriptive, understandable
and predictive models [20].

The use of credit and debit cards in any transaction today is common; so
that, it is important to perfect a model for the detection of fraud with credit
or debit cards in all areas. Card fraud is done for any unqualified activity of a
person [14]. It occurs when the user and the issuer are not aware that another
user tries to handle the card. There-fore, this is known as fraud.

© Springer Nature Switzerland AG 2020
M. Botto-Tobar et al. (Eds.): ICAT 2019, CCIS 1194, pp. 145–159, 2020.
https://doi.org/10.1007/978-3-030-42520-3_12

A critical factor in the Supply Chain of companies is to detect anomalies in a timely manner, illegal transactions on the cards since it has become an important activity for payment processors. The combinations of knowledge-based rules are compatible with fraud detection systems; since they allow to verify specific things discovered by the researchers. A fraud scenario arises when "the cardholder can avoid negotiating in a given country, between the next few weeks, he can negotiate for a specific amount in another country". If the described case is detected between transactions, then the anomaly detection system will issue an alert. Machine learning algorithm rules learn patterns and categorize new transactions [22].

The main objective is to show a diversity of Semi-supervised Machine Learning techniques to apply them and compare their results to a fraud detection problem. The explanation of the use of statistically supervised semi-supervised models considers fraud as a special case of outliers, that is, of points in the data set that differ significantly from the analysis data set. [17]. Supervised models are inadequate because they face a conflict in the misclassification when it relates to fraud detection. The use of a supervised model has as a limitation the incorporation of biases to the confusion matrix. This leads to the detection of serious biases in false negatives. As a consequence fraudulent cases can be predicted incorrectly, that is, not fraudulent [17].

This work discusses different Machine Learning techniques that are used to detect fraud in different transactions, the implementation of predictive analysis using these open source techniques and technologies to process data in a scalable, efficient and fault-tolerant manner, applied to an Architecture for IoT Big Data Analysis in Smart Supply Chain Fields. The remain of this paper is structured as follows: Next section discusses some works related to fraud prediction and detection. Section 3 shows an overview of the Machine Learning algorithms for the treatment of Big Data in the context of the fraud prediction. In Sect. 4, the Proposed Approach of work is described. Section 5 shows experimental results of implementation of Machine Learning algorithms. Section 6 presents a discussion based on the experimental results obtained. Finally, the article concludes with some research perspectives and future work in Sect. 7.

2   Related Work

Big Data Analysis (BDA) [1] is classified as real-time analysis, (commonly used in e-commerce and finance cases [12]) and offline analysis [4]. The main architectures for real-time analysis incorporate two main aspects: (a) parallel processing based on relational and traditional data; (b) computer platforms based on memory calculations [11]. Offline analysis considers those applications that have a high response time. Examples of these applications include recommendation algorithms [25], statistical analysis [12] and Machine Learning [19]. Below, we present some works related to the detection and prediction of fraud in different fields.

The Naïve Bayes classifier [15] is used in [14], the researchers used algorithm, for the categorization of normal and fraudulent transactions. The categorization is based on a set of data, of a private nature, used to produce predicted results in terms of accuracy and recovery [14]. In their research, the authors did not use other algorithms with which they can compare and conclude that the Naïve Bayes classifier is the best in fraud prediction.

In [17], the authors address a specific fraud detection problem focused on property insurance claims. The study makes use of a set of highly unbalanced public data [21]. Within the results, the authors propose an innovative methodology based on semi-supervised routes and incorporate a new classification, the Cluster-Score defined by Sebastian Palacio [17], a metric for the measurement of abnormal similarity, allows the transmutation of models not supervised to supervised models and defines the objective limits between the groups and evaluates the uniformity of the anomalies in the elaboration of the group. The proposed objectives of their work focus on reducing the proportion of claims investigated as non-fraudulent and detecting the largest number of fraudulent claims [17]. Although the described methodology has proven to be able to solve the problem with great success, it presents a disadvantage in terms of implementation due to its complexity in the detection of outliers.

A field study of fraud prediction, when the user clicks was presented in [3]. Basically, the study is focused on learning automatically labeled data. This technique offers a new perspective to produce click profiles for publishers of online advertisements. Supervised learning is integrated from these data labeled with an appreciation of the conflicting predictions between the fundamental model and the new classifier. In the study, the authors selected the Random Forest model [23] based on the average accuracy [3]. We found that in this work no comparisons were made with other algorithms that allow determining the efficiency of Random Forest as the most optimal.

The authors in [13], presents data mining approaches to face fraud detection, through a set of rules written by human experts, where the rules are specific to the banking context. These rules are difficult to maintain as they grow in size and complexity. With this issue in mind, as a solution the authors propose a system that determines the best adaptation to the existing rules to capture all fraudulent transactions. In this work, the rules have the property of being explain-able unlike most data mining models that do not own this property. Additionally, the study presents a survey conducted on the most applied Machine Learning methods, for this purpose in July 2018 the researchers explored the Web of Science (WoS) data-bases and focused on peer-reviewed articles in journals and publications of conferences published in the last 10 years (between 2009 and 2018) corresponding to the category "Artificial intelligence of informatics", among which are highlighted: Neural Networks [10], Random Forest [23], Logistic Regression [8], Vector Support Machines (SVM) [2], Decision Trees [9] and Naïve Bayes [15]; while some novel methods, such as deep learning, appear only in more recent publications [13]. The limitation of this work is that it is not indicated based on what parameters the researchers identified the most used algorithms, but determine it based on a survey. Nor do they mention the advantages of each in terms of usability.

Based on these works, there are some deficiencies that are not addressed when choosing between a supervised learning model or another for fraud prediction. The researchers only show the benefits of using a specific Machine Learning algorithm or in turn mention a set of most used algorithms and do not take into account aspects such as the management of memory resources used at the time of their operation, nor the time of computational processing in its execution, considering its importance in the work environment, for which they were implemented. It is for this reason that the following sections present the implementation of some Semi-supervised Machine Learning algorithms, used by researchers and others that have not been mentioned or implemented in any of the related works. In addition, through the use of open source tools such as R, perform data processing that not only comes from credit cards but also from transactions made in the Smart Supply Chains flow, with the purpose of predicting fraud more accurately, Choose the most optimal model based on parameters such as: number of false positives, computational processing time and less memory resource usage.

3   Overview of Machine Learning Algorithms for Fraud Prediction

This section describes an overview of Machine Learning techniques of fraud prediction, used in Smart Supply Chains.

3.1   Rpart Method

The R package partykit offers a flexible toolkit for summarizing, representing, learning and showing a wide variety of tree-structured regression and classification models. The fundamental infrastructure is encompassed by the functionality for interpreting trees deduced by any algorithm so that usable print/plot/predict methods are available [7]. Rpart is a dedicated method for trees with constant fits in the leaves (or terminal nodes). It builds classification or regression models. The resulting models can be represented as binary trees [7].

3.2   C5.0 Algorithm

The C5 Algorithm based on Decision Trees gives recognition of noise, lack of data and can anticipate which attributes are relevant and which are not in the classification. Classify the result set with low memory usage and high precision. Compared to other techniques, it generates fewer rules with fast results [18]. The classifier is initially evidenced to catalog unknown data and for this purpose, it uses the resulting decision tree. To bypass overfitting in the decision tree, use two different settings. The first adjustment is not to originate the tree before it reaches the point where the training data is correctly classified. The second approach is to prune the tree when the tree exceeds the data [18]. For dimensionality reduction, we use the feature selection technique. We choose a small subset of the relevant characteristics of the real set that generally provides better learning performance and a better interpretation of the model [18].

3.3   Random Forest Model

Random Forest is a well known assembly Machine Learning algorithm in which the decision tree functions as basic classifiers, creating different decision trees and then merging the result produced by each of them, used to solve the classification problem, in which the dependent variable is categorical [23].

Decision trees are informal and largely moderate models to know; although their predictive ability is lower, which is why they are called weak students [23]. The operation of Random Forest is like a set of random trees. Syndicates the performance of multiple random trees and then originates with its own output. Random Forest is similar to the Tress Decision, however it will not choose all the data points and variables in each of the trees. Subjectively groups data points and variables in each of the trees it generates and then unifies to reach the exit at the end, joins to reach the exit at the end, eliminating the prejudices found in a decision tree model and improving predictive power [23].

3.4   Support Vector Machine (SVM)

In classification and regression problems, SVMs are used, which are Supervised Machine Learning algorithms. In the case of a classification problem, a support vector machine will determine the best method to classify the data [2]. Once the training data is plotted on an n-dimensional plane, where n is the number of factors analyzed, the SVM generates equations for multiple hyperplanes that can linearly separate the data points by category [2].

A hyperplane exists as a line, a plane or a hyper-plane if two, three or more than three factors are analyzed, respectively. The letters A and B in the Fig. 1.

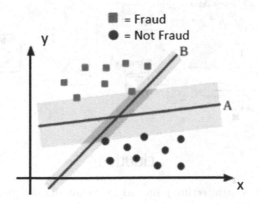

Fig. 1. Support Vector Machine separation [2].

Support Vector Machine Separation represent a sample of generated hyperplanes. The data points on the right of hyperplanes are classified as non-fraudulent, while the other points fall into the fraudulent category [2]. The optimal hyperplane is selected in interpretation of the distance from the line to the point closest to each side. This distance is understood as margin, and the points that define the margin are referred to as support vectors. To make predictions in the SVM Algorithm the hyperplane with greater margin is chosen [2].

4   Proposed Approach

Based on the type of result variable, four different Machine Learning models are handled: Rpart Model, Model C5.0, Random Forest Model and SVM. To determine which one best fits the test set based on the ROC curve (Receiver Operating Characteristic) and accuracy. Each model is tested against 100,000 random transactions categorized as "No Fraud", to estimate the false positive rate. Processing time is also captured by estimating the computational intensity of each model.

4.1   Work Environment

Machine Learning techniques applied to fraud prediction are evaluated in the environment of a Big Data for Smart Supply Chains Analysis Architecture [6]. The scope of the architecture used is limited to generation, extraction, ingestion in HDFS [16], visualization and analysis of data corresponding to clickstream and transactions, facilitating the development of Smart Supply Chains, specifically focused on handling corresponding to payment methods [6]. The R language is used to execute Algorithms of automatic learning that help to make predictions fraud prediction. The Fig. 2 shows the evaluation architecture.

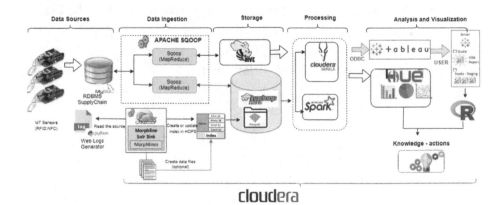

Fig. 2. BDA architecture proposed for Smart Supply Chains [6].

4.2   Data Source

For fraud analysis and prediction, the variables corresponding to the transactions included in the DataCoSupplyChainDataset.csv file as shown in Table 1. For more information about the description of each of the variables that make up the used dataset, check the DescriptionDataCoSupplyChain.csv file. Both files are available at [5].

Table 1. Selection of fraud predictor variables.

Variable	Type	Description
Hour-month	integer	Indicates the time of the month in which this data was captured, this variable is calculated based on the variable "order-date"
Type	char	Indicates the type of transaction performed, it is a categorical variable calculated based on the variable "order-status"
Sales-per-customer	double	Shows amount of money invested in each transaction
Customer-State	char	Categorical variable indicating the place of origin of the transaction
Order-State	char	Categorical variable that shows the destination
isFraud	integer	Calculated binary categorical variable that indicates 0: No Fraud and 1: Fraud

4.3   Processing of Data Set

The "type" and "isFraud" columns are categorical and changed by factors. The "isFraud" attribute, which contains the values 0 and 1, are re-coded: 0 = "No", 1 = "Yes". All the transactions that present fraud are obtained and it is observed in Fig. 3, that the type of transaction "PAYMENT" does not present fraud.

```
 isFraud            type          hour_month          Sales_per_customer
 No :134726   CASH    :20189   Min.   :-1.720470   Min.   :-1.4555
 Yes:  4068   DEBIT   :69295   1st Qu.:-0.867463   1st Qu.:-0.6527
              PAYMENT:     0   Median : 0.004395   Median :-0.1589
              TRANSFER:49310   Mean   : 0.000000   Mean   : 0.0000
                               3rd Qu.: 0.862114   3rd Qu.: 0.5330
                               Max.   : 1.781100   Max.   :14.5550
```

Fig. 3. Review of fraudulent transactions.

4.4 Dimensionality Reduction

Insignificant variables are removed and filtered by the words "CASH", "DEBIT" and "TRANSFERS". There are no transactions per product higher than 1939.99, so it is also possible to filter by this amount. The "hour month" variable indicates the time of the month in which this data was captured, so they are considered time series, as shown in Fig. 4. There is a positive correlation between "hour-month" and "Sales-per-costumer" variables of 0.0272. To determine the existence of some pattern of fraud, then the scatter plot is presented in Fig. 5, with a confidence level of 90%.

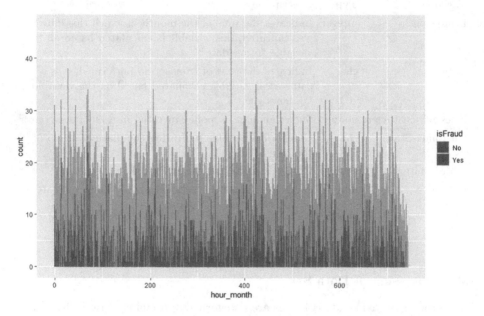

Fig. 4. Time series of fraudulent and non-fraudulent transactions.

The existence of linear relationships between predictors was not observed and a sample containing 134726 non-fraudulent transactions and 4068 fraudulent transactions are included for training. A control is created for all models to be evaluated using three 10-fold cross-validation iterations for each model to compare their performance and computational response time.

4.5 Machine Learning Algorithms Application

The Dataset was splitted into 2 parts: for training, 70% of random data was chosen and test set 30%. The models presented in Table 2 follow a consistent pattern, which is:

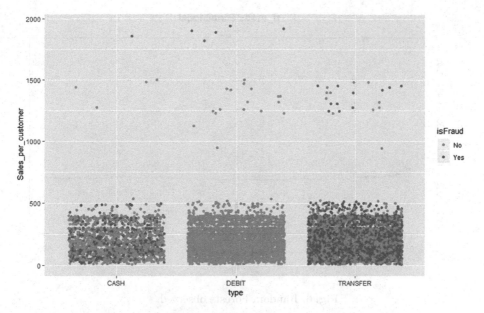

Fig. 5. Search for fraud patterns by transaction amount.

Table 2. Prediction in fraud detection with Machine Learning.

Prediction models	Training data	Test data	Large dataset without fraud	AUC	Training time	Execution time in large datasets
Rpart	0.7571	0.748	0.761	0.8255	6.699935 seg	5.048236 seg
C5	0.7162	0.7453	0.6706	0.814	2.532301 seg	4.110189 seg
Random Forest	0.9971	0.7332	0.8155	0.8197	3.375788 min	4.616288 seg
SVM	0.7204	0.7417	0.7022	0.8129	19.4246 min	38.09374 seg

- Fit the model to the data
- Model compared to the data in which it was trained
- Model compared to the set of test data that was unknown for the construction of the model
- Comparison of the model with a sample of 100,000 cases of No Fraud to determine expected false positives

With Rpart the resulting models are represented as binary trees and their training time was 6.699935 s. The C5 Algorithm anticipates which attributes are relevant and which are not in the classification, with a training time of 2.532301 s. Using the Random Forests Algorithm in the classification, its training time was 3.375788 min and it was observed that the prediction error is leveled at around 100 trees without any significant negative impact on performance, as shown in Fig. 6.

rf_model$finalModel

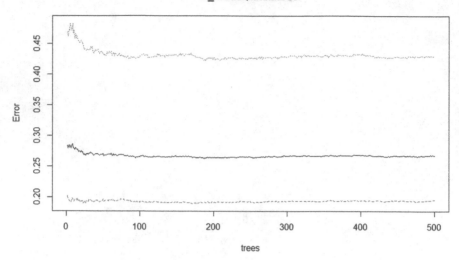

Fig. 6. Random Forests observed.

The variable "type DEBIT", contains the most significant information, "type TRANSFER" and "hour-month", are also influential variables, as shown in Fig. 7. In the analysis with SVM, for the generation of non-linear decision limits in the separation of non-linear data the Radial Kernel support vector classifier was used, the model training time was 19.4246 min.

5   Experimental Results

In this section, we detail the experimental results obtained in the comparison of the ROC (Receiver Operating Characteristic) curves for the choice of the best model.

ROC diagnostic performance curves are used, since it is a global and independent measure of the cut-off point. The choice is made by comparing the Area Under the Curve (AUC), corresponding to each of the models in use [24]. Figure 8 shows the outline of each ROC curve. Area below the ROC curves for each model used is shown in Table 3. The results indicate that the Random Forest and Rpart models have a fairly high AUC, for the choice of the best model to use, an evaluation is carried out with the results indicated in Table 4, corresponding to the metrics used in large datasets without fraud.

Random Forest has a lower number of false positives and less execution time compared to RPart, which means that there is an 81.97% probability that the prediction made to a fraudulent transaction is more correct than that of a normal transaction chosen at random, chance, fate.

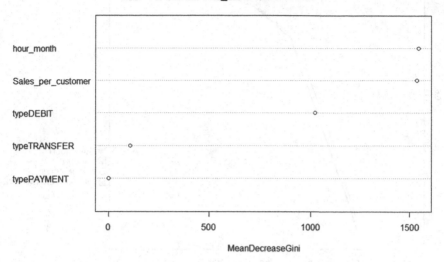

Fig. 7. Evaluation of significant variables with Random Forest.

Table 3. Area below the ROC curves obtained.

SVM	C5	RF	Rpart
0.8128819	0.8139833	0.8196976	0.8255087

Table 4. Model choice metrics.

Model Evaluation	Random Forest	Rpart
Accuracy	0.8155	0.761
False-positives	18446	23900
Run-time	4.616 seg	5.048 seg

6 Discussion

The experimental results evaluated the performance of the models: Rpart, C5, Random Forest and SVM. However, several classification approaches were used for prediction analysis and choice of the best technique, based on a public data set.

The Random Forest model proved to be the most robust in terms of precision and number of false positives found, when working with categorical variables in the management of outliers within the classification of fraudulent transactions.

A deficiency found in the use of this algorithm lies in the training period since it is greater compared to Rpart and C5, because it generates many trees and makes the decision to combine their results, causing greater consumption of computational resources and much longer runtime.

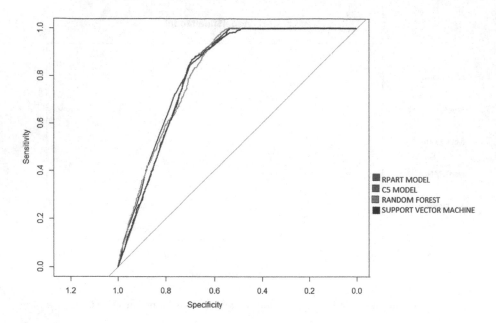

Fig. 8. ROC curve comparison.

The false positive rate, thus minimal, is an inconvenience, not only in Random Forest, but in any Machine Learning model used, because it causes small classification errors, since it is the handling of thousands of transactions can cause large economic losses and thus reduce significantly the number of customers. This scenario can be generated in the way in which the dataset to be analyzed is presented, that is, in how it shows its data, considering previous aspects before the analysis such as: the user's own errors at the time of entering the data, transactions made at unusual times, the way in which the categorization of the variables is carried out prior to its analysis.

However, the analysis strategies presented in this work showed some benefits such as ease of implementation with R, fault tolerance, scalability, speed in training and response time.

It is clear that there are also many more approaches to address this problem that are unfortunately difficult to compare, but it is also notable that some appear more than others such as Random Forest. Fraud detection is also complicated, from a technological point of view, since transactions are a flow of data that must be processed almost in real time. This area is constantly changing and is unlikely to be "resolved" in the near future.

7   Conclusions and Perspectives

The risk of exposure is high in the supply chains of companies, it is not necessary to wait for incidents to occur to seek information and make decisions that add value to the business and its ecosystem.

In this work, we have implemented and evaluated fraud prediction models in Smart Supply Chains stage. One of the most important applications for BDA was fraud detection. Based on the type of outcome variable, four different models were used to determine which one best fit the set of tests based on the ROC curve and accuracy. Each model was tested against 100,000 random transactions, capturing processing time and estimating the false positive rate, resulting in the best model to use Random Forest.

The main contribution was to propose the use of some semi-supervised Machine Learning algorithms, which have not been used as C5 and Rpart and others used as Random Forests, SVM, to show their benefits and utility through the use of open source tools such as R, in the processing of data that not only comes from credit cards but also from transactions made in the flow of Supply Chains with the aim of predicting fraud with greater precision, less use of memory resources and low computational processing time.

As future work, we will conduct a more thorough study of methods to control the probability of detection of false positives. Also, we will present some experiments in an architecture for IoT Big Data Analysis in Smart Supply Chain Fields, which will be carried out through the application of Machine Learning algorithms, to perform clickstream analysis, in order to study and know which they are the most common behavior patterns of visitors who visit the web, when making their transactions within the supply flow and market basket analysis.

Acknowledgements. This work was made possible thanks to the financial support of "Universidad de Las Américas" from Ecuador and thanks to the participation of Polytechnic Institute of Leiria from Portugal.

References

1. Acharjya, D.P., Ahmed, K.: A survey on big data analytics: challenges, open research issues and tools. Int. J. Adv. Comput. Sci. Appl. **7**(2), 511–518 (2016). https://doi.org/10.14569/IJACSA.2016.070267
2. Banerjee, R., Bourla, G., Chen, S., Kashyap, M., Purohit, S., Battipaglia, J.: Comparative analysis of machine learning algorithms through credit card fraud detection. New Jersey's Governor's School of Engineering and Technology (2018)
3. Berrar, D.: Learning from automatically labeled data: case study on click fraud prediction. Knowl. Inf. Syst. **46**(2), 477–490 (2016). https://doi.org/10.1007/s10115-015-0827-6
4. Chen, M., Mao, S., Zhang, Y., Leung, V.C., et al.: Big data: related technologies, challenges and future prospects (2014)
5. Constante, F., Silva, F., Pereira, A.: DataCo smart supply chain for big data analysis, pp. 1–13. Mendeley Data, v5 (2019). https://doi.org/10.17632/8gx2fvg2k6.5

6. Constante-Nicolalde, F.-V., Pérez-Medina, J.-L., Guerra-Terán, P.: A proposed architecture for IoT big data analysis in smart supply chain fields. In: Botto-Tobar, M., León-Acurio, J., Díaz Cadena, A., Montiel Díaz, P. (eds.) ICAETT 2019. AISC, vol. 1066, pp. 361–374. Springer, Cham (2020). https://doi.org/10.1007/978-3-030-32022-5_34

7. Hothorn, T., Zeileis, A.: partykit: a modular toolkit for recursive partytioning in R. J. Mach. Learn. Res. **16**(1), 3905–3909 (2015)

8. Huang, R., Xu, W.: Performance evaluation of enabling logistic regression for big data with R. In: 2015 IEEE International Conference on Big Data (Big Data), pp. 2517–2524, October 2015. https://doi.org/10.1109/BigData.2015.7364048

9. Jadhav, S.D., Channe, H.: Comparative study of K-NN, Naive Bayes and decision tree classification techniques. Int. J. Sci. Res. (IJSR) **5**(1), 1842–1845 (2016)

10. Lata, L.N., Koushika, I.A., Hasan, S.S.: A comprehensive survey of fraud detection techniques. Int. J. Appl. Inf. Syst. **10**(2), 26–32 (2015). https://doi.org/10.5120/ijais2015451471

11. Li, G.: Big data related technologies, challenges and future prospects. Inf. Technol. Tourism **15**(3), 283–285 (2015). https://doi.org/10.1007/s40558-015-0027-y

12. Marjani, M., et al.: Big IoT data analytics: architecture, opportunities, and open research challenges. IEEE Access **5**, 5247–5261 (2017). https://doi.org/10.1109/ACCESS.2017.2689040

13. Mekterović, I., Brkić, L., Baranović, M.: A systematic review of data mining approaches to credit card fraud detection. WSEAS Trans. Bus. Econ. **15**, 437 (2018)

14. Monika, E., Kaur, E.A.: Fraud prediction for credit card using classification method. Int. J. Eng. Technol. **7**(3), 1083–1086 (2018). https://doi.org/10.14419/ijet.v7i3.12577. https://www.sciencepubco.com/index.php/ijet/article/view/12577

15. Mukherjee, S., Sharma, N.: Intrusion detection using Naive Bayes classifier with feature reduction. Procedia Technol. **4**, 119–128 (2012). https://doi.org/10.1016/j.protcy.2012.05.017. http://www.sciencedirect.com/science/article/pii/S2212017312002964. 2nd International Conference on Computer, Communication, Control and Information Technology (C3IT-2012) on 25–26 February 2012

16. Nagdive, A.S., Tugnayat, R.M.: A review of hadoop ecosystem for bigdata. Int. J. Comput. Appl. **180**, 35–40 (2018)

17. Palacio, S.M.: Detecting outliers with semi-supervised machine learning: a fraud prediction application. XREAP WP 2018-02 **2** (2018). https://doi.org/10.2139/ssrn.3165318. https://ssrn.com/abstract=3165318

18. Pandya, R., Pandya, J.: C5.0 algorithm to improved decision tree with feature selection and reduced error pruning. Int. J. Comput. Appl. **117**(16), 18–21 (2015)

19. Patil, K.: A survey on machine learning techniques for insurance fraud prediction. HELIX **8**, 4358–4363 (2018). https://doi.org/10.29042/2018-4358-4363

20. Prasanna, P.L., Rao, D.R., Meghana, Y., Maithri, K., Dhinesh, T.: Analysis of supervised classification techniques. Int. J. Eng. Technol. **7**(1.1), 283–285 (2017). https://doi.org/10.14419/ijet.v7i1.1.9486. https://www.sciencepubco.com/index.php/ijet/article/view/9486

21. Qian, J., Saligrama, V.: Spectral clustering with unbalanced data. arXiv preprint arXiv:1302.5134 (2013)

22. Shukur, H.A., Kurnaz, S.: Credit card fraud detection using machine learning methodology (2019)

23. Vadakara, J.M., Kumar, D.V.: Aggrandized random forest to detect the credit card frauds. Adv. Sci. Technol. Eng. Syst. J. 4(4), 121–127 (2019). https://doi.org/10.25046/aj040414
24. Vogel, R., Bellet, A., Clémençon, S.: A probabilistic theory of supervised similarity learning for pointwise ROC curve optimization. arXiv preprint arXiv:1807.06981 (2018)
25. Zarzour, H., Maazouzi, F., Soltani, M., Chemam, C.: An improved collaborative filtering recommendation algorithm for big data. In: Amine, A., Mouhoub, M., Ait Mohamed, O., Djebbar, B. (eds.) CIIA 2018. IAICT, vol. 522, pp. 660–668. Springer, Cham (2018). https://doi.org/10.1007/978-3-319-89743-1_56

Selection of Mental Tasks for Brain-Computer Interfaces Using NASA-TLX Index

Jhon Freddy Moofarry⑩, Kevin Andrés Suaza Cano⑩,
Diego Fernando Saavedra Lozano⑩,
and Javier Ferney Castillo García$^{(\boxtimes)}$⑩

Grupo de Investigación en Electrónica Industrial y Ambiental – GIEIAM,
Universidad Santiago de Cali, Cali, Colombia
Javier.castillo00@usc.edu.co

Abstract. The brain-computer interfaces - BCIs allow people with disabilities to interact with the outside world using different communication channels than conventional ones. This article deals with the selection of tasks in the protocols for the development of BCI based on the paradigm of mental tasks. It is proposed to use the NASA-TLX index to evaluate the effect of the mental load of each of the tasks and contrast the performance of the interface task by task. In the implementation of BCI, the OPENBCI hardware was used for signal acquisition and the MATLAB software for processing. Five mental tasks were defined that activated different regions of the cerebral cortex. The acquisition protocol consisted of defining the rest time, execution and recovery for the tasks. The extraction methods used temporal, frequency and time-frequency combination characteristics. The classifiers used were neural networks, nearby neighbors and support vector machines. The evaluation of the TLX index seeks to quantify the appreciation of the effort, frustration and complexity of the task, therefore after the acquisition of signals for each task, the participant proceeded to evaluate the mental overload using the NASA-TLX index. The results obtained show that those tasks that require greater complexity to be performed presented a greater repeatability and higher success rate.

Keywords: Task mental · BCI · NASA-TLX index · Workload

1 Introduction

Brain-computer interface (BCI) is a communication system that is not dependent on the normal input of the brain but on the exit pathways of the peripheral nerves and muscles [1, 2]. It is foreseen that the more direct the connection between a computer and the neuronal commands, it should give the BCIs a possible advantage over the other methods of alternative communication. This technology holds the promise of providing functionality and independence for people with motor disabilities by allowing them to directly control assistive devices such as computers, wheelchairs, or prosthetics.

M. Botto-Tobar et al. (Eds.): ICAT 2019, CCIS 1194, pp. 160–172, 2020.
https://doi.org/10.1007/978-3-030-42520-3_13

Previous BCI studies have shown that humans can move a computer cursor toward a target using surface electrode signals (i.e., electroencephalogram or EEG) [1, 3, 4], subdural electrodes (electrocorticogram or ECoG) [5, 6], or electrodes implanted within the brain (local field potentials and action potentials) [7, 8]. Although methods vary, cursor control is achieved by training the participant to use motor images to modulate neuronal signals [2, 4–6].

Although BCI research is advancing rapidly, there are still limitations to overcome before BCIs become widely available assistive technology. The problem lies in success rates, this is because you don't have a methodology for selecting mental tasks. Consequently, it is not well understood how people perceive the difficulty of performing BCI tasks, what specific aspects of the workload contribute most, and whether there is a difference in perceived workload between users who are healthy and disabled. Similarly, it is important to understand how to make the selection of the best mental tasks to reduce mental workload and prevent fatigue, two factors that are likely to limit current technology [9–11].

This study evaluated mental overload, using NASA's Task Load Index (NASA-TLX) for the acquisition of EEG-based BCI targets. The experience of the target acquisition task involved teaching participants the established protocol, in which mental tasks are performed to discriminate between them with a NASA-TLX coefficient, which was directly related to the user's mental workload.

In previous works [29], present a new chain for the classification of two levels of mental workload fatigue, which appear after having passed by time considered, from the EEG signal is filtered initially in a certain frequency band and 15 electrodes of 32 are selected using a method based on the geometry of Riemann, then a step of spatial filtering is made with 6 filters of common spatial pattern (CSP), finally, a binary classification is made by linear discriminant analysis of Fisher (FLDA). The problem is that there is no way to measure mental overload and this is what is proposed in this research.

1.1 Mental Workload

The mental workload must be considered when designing tasks or procedures that require the user's attention. Research on mental workload began in the 1960s, when researchers evaluated human-machine systems such as ground transportation, air traffic control, and process control. Many studies have evaluated strategies to reduce mental effort when performing computer tasks [12], they tend to require high visual and cognitive demands, which leads to a significant increase in the mental workload for the use of the BCI [13].

BCI research is typically conducted in a controlled environment in an isolated, distraction-free laboratory, where participants are instructed to sit still and concentrate on the task. There are different types of mental tasks such as visual ones, which refers to images and involves the occipital, parietal and frontal regions; auditory, which consists of imagining sounds that incite the temporal and frontal region; kinesthetic, which uses elements of the somatosensory region of the brain. We focused our research on the mental motor tasks, the latter include kinesthetic, which consists of imagining movements of the limbs. It will be the case of people who could use the BCI as their main mode of communication.

1.2 Mental Workload Index

NASA-TLX was developed by the Human Performance Group at NASA's Ames Research Center [14]. It is a multidimensional classification procedure with six subscales: Mental demands, physical demands, temporal demands, performance, effort and frustration. The first three subscales refer to the demands of the task and the last three are related to the participant's interaction with the task. Numerical classifications for each are obtained by having the participant mark a line divided into 20 equal intervals, which becomes one on a scale of 0–100. The line is anchored to each side with bipolar descriptors (e.g., low and high). The gross workload score is calculated by adding the score given to each of the six subscales and dividing it by six [15].

The most commonly used overall workload grade is calculated based on a weighted average of the subscales. The weights are determined by the participant's assessment of how much each factor contributed to the workload of the task. After completing the initial scores, 15 comparisons of the 6 subscales in pairs are presented individually, and the participant chooses the subscale that contributes most to the workload of the task. The weight of each subscale is the total number of times it has been selected divided by 15. The overall workload score is calculated by multiplying each rating by the weight given to that subscale and can range from 0–100 [14]. As a multidimensional scale, NASA-TLX is useful for obtaining more detailed data and diagnostics than a one-dimensional scale [16]. NASA-TLX has been used in many human-computer interaction investigations as a method for value mental workload [10, 17, 18].

Figure 1 shows the questions related to six fundamental aspects for the execution of a task, such as: mental demand, physical demand, temporal demand, performance, effort and frustration. Each aspect is related to a subjective evaluation by the user, which was contrasted by an assessment that allowed quantifying the overload perceived by the user.

It is also presented in Fig. 2, one of the 15 combinations of characteristics to quantify the user's perception of overload. The evaluation of the most relevant aspects for the execution of a task.

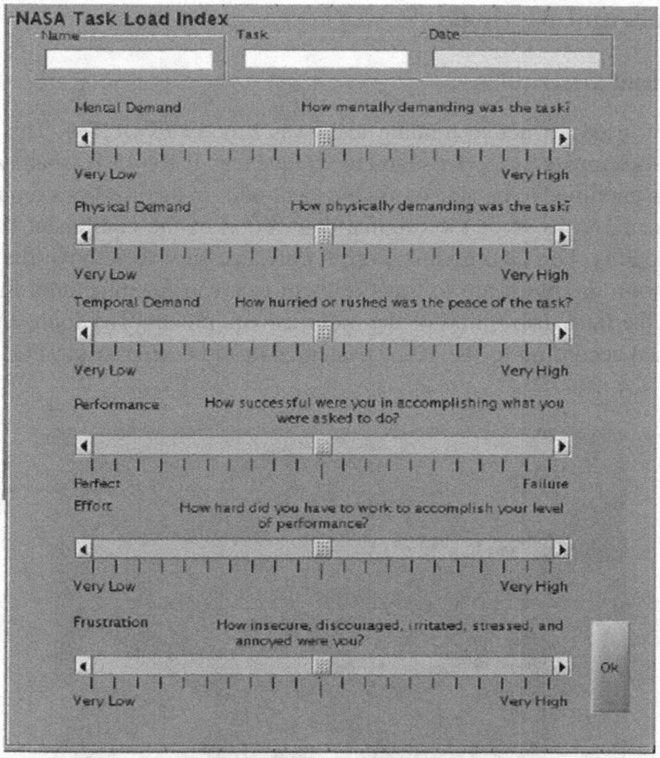

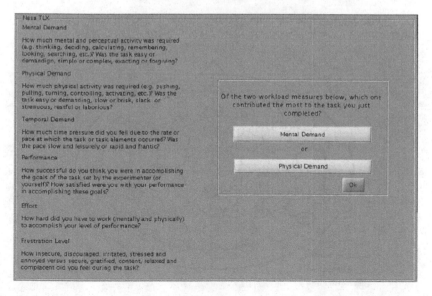

Fig. 1. NASA TLX overload index.

Fig. 2. Evaluation of a pair of characteristics to quantify the NASA TLX index.

2 Materials and Methods

2.1 Equipment Description

The equipment used for active training tests was Emotiv EPOCTM. This equipment consists of 14-channels (128 Hz sampling frequency; 0.16–45 Hz bandwidth). The equipment was modified to be able to use the electrodes in the positions required for the analysis of mental tasks since the original disposition of the equipment these studies related to emotions. For this modification it was necessary to remove the electrodes, and together with the hardware to transfer them to a cap manufactured in Neoprene, Fig. 3 shows the final disposition of the modified equipment. The fourteen electrodes were positioned according to the 10/20 system, as can be seen in the Fig. 4.

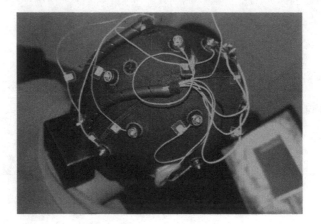

Fig. 3. Neoprene cap with modified Emotiv EPOCTM hardware.

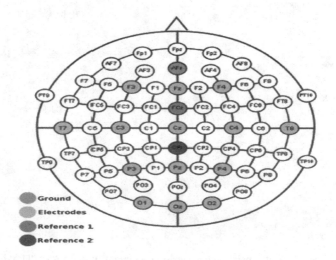

Fig. 4. Electrode positioning.

2.2 Description of Users

The users who participated in the different tests were university students, who did not present or recorded any type of disability. The ages ranged from 24 to 37 years of age. Users agreed to participate in the tests by signing the terms of consent that were presented in a timely manner. Any questions from the participants were clarified.

2.3 Pre-processing

This stage has the purpose of improving the signal/noise ratio, for this purpose CAR (common average reference) was implemented, it has been suggested as a solution to the problem of the reference electrode [19]. The CAR supposes a bipolar register from several electrodes, all referred to a single site. In order to obtain the car, the mean is calculated to all the EEG signal, averaging through the electrodes, and the punctual result is subtracted from each electrode.

Feature Extraction: In order to select the most suitable classifier for a BCI system, it is essential to clearly understand what characteristics are used, what their properties are and how they are used. This section describes the features, their properties and how they are used in this research.

Transform of Wavelets: The Fourier power spectrum analysis of a symphony will of course reveal the tones and their harmonics, as well as other frequencies are repeated in modulations and vibrations. If we play the parts in another order, the power spectrum does not change at all, but for the listening it will be a completely different piece, and even more so if we exchange parts within the parts, at a finer scale. On the contrary, the analysis of Wavelets not only gives us the main frequencies, but also tells us when they occur and how long they last. The Wavelets transform was originally designed to study non-stationary signals, it is a time-frequency analysis. It can reveal aspects of the data such as trends, breakpoints, discontinuities in derivatives, and self-similarity. The Wavelet transform is more effective in characterizing EEG signals than traditional techniques [20, 21]. This is calculated in chunks of a time function $\psi(t)$, described in Eq. 1

$$\psi(t) = \frac{1}{\sqrt{a}}\psi\left(\frac{t-b}{a}\right) \tag{1}$$

Where a, is the scale parameter for the frequency represented by the Wavelet, and b, a displacement factor in time, the central point of the Wavelet. When arbitrary scales are used between the sampling intervals contained in the time series, reference is made to the transformed continuous Wavelet (CWT) of a time function, $f(t)$, is expressed as Eq. 2.

$$W(f) = \int_{-\infty}^{+\infty} f(t)\psi(f) \tag{2}$$

Classification: Understand, the classification process as one that allows mapping a set of input data to a space for the labels of the keys to be learned. The classification has as function to decide which is the class to which a data belongs, without label, depending on its characteristics. To carry out this decision, a classifier that incorporates a priori knowledge needs tagged examples. The acquisition of this knowledge is called the machine learning phase or the classifier training phase [22, 23].

K-Nearest Neighbors
The K-Nearest Neighbors (KNN) method is a supervised classification method used to estimate the $F(x/C_j)$, density function of the x predictors for each C_j class. This is a non-parametric classification method, which estimates the value of the probability density function or directly a posteriori probability that an element x, belongs to class C_j, from the information provided by the set of prototypes. In the learning process no assumption is made about the distribution of predictor variables. In the topic of pattern recognition, the K-NN algorithm is used as a method of classifying objects (elements) based on a training through close examples in the space of the elements.

The K-NN is a type of "Lazy Learning", where the function approaches only locally and all the calculation is deferred to the classification. Training examples are vectors in a characteristic multidimensional space, each example is described in terms of p attributes considering q classes for classification. The values of the attributes of the i-th, example where $1 \leq i \leq n$, is represented by the vector p-dimensional by Eq. 3.

$$x_i = (x_{1i}, x_{2i}, \ldots, x_{pi}) \in X \tag{3}$$

The space is partitioned into regions by locations and training example labels. A point in the space is assigned to class $C_{j,}$, if this is the most frequent class among the closest training examples. The Euclidean distance, given by Eq. 4, is generally used:

$$d(x_i, x_j) = \sqrt{\sum_{r=1}^{p} (x_{ri} - x_{rj})^2} \tag{4}$$

The training phase of the algorithm consists of storing the characteristic vectors and labels of the training examples classes. In the classification phase, the evaluation of the example (of which its class is not known) is represented by a vector in the characteristic space. The distance between the stored vectors and the new vector is calculated, and the nearest k examples are selected. The new example is classified with the class that most repeats itself in the selected vectors.

This method assumes that the nearest neighbors give the best classification, and this is done using all the attributes; the problem with this assumption is that it is possible to have many irrelevant attributes that undermine the classification: two relevant attributes would lose weight among twenty irrelevant ones. To correct the possible bias, a weight can be assigned to the distances of each attribute, thus giving greater importance to the most relevant attributes. Another possibility is to try to determine or adjust the weights with known training examples. Finally, before assigning weights, it is advisable to identify and remove the attributes that are considered irrelevant.

2.4 Selection Based on Silhoutt's Distance

The Silhouette width -SW allows to relate the cohesion and the separation for cluster. The SW refers to a method for cluster interpretation and validation [24]. This index reflects cohesion; intracluster distance and separation measurement; intercluster distance measurement, can be applied to various metrics. In Fig. 5 a representation of the measure of cohesion and separation for a cluster can be seen from the perspective of the data. For each trial i, its SW $s(i)$ is defined in Eq. 5:

$$s(i) = \begin{cases} 1 - \frac{a(i)}{b(i)}, & if\ a(i) < b(i) \\ 0, & if\ a(i) = b(i) \\ \frac{b(i)}{a(i)} - 1, & if\ a(i), \end{cases} \tag{5}$$

Where,

(i), is the average distance of trial i, where $a(i)$, is the average distance of trial i, for the other trials in the same cluster, $b(i)$, is the average distance of trial i, to the trials of the neighbouring clusters. The average of $s(i)$, across all trials reflects the quality of the resulting cluster. The SW as a performance index can allow to evaluate which characteristics provide the best performance (accuracy) in a mental task for be implemented in a BCI [25]. The other trials in the same cluster, $b(i)$, is the average distance from the trial i, to the trials of the neighboring clusters. The average of $s(i)$, across all trials reflects the quality of the resulting cluster. The SW as a performance index can allow to evaluate which characteristics provide the best performance (accuracy) in a mental task to be implemented in a BCI [25].

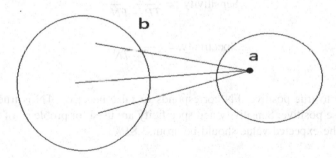

Fig. 5. Representation SW.

2.5 Performance Metrics

For the evaluation of the system, the proposed methodology [26] will be used, in order to compare the results of the experiments, so that their differences can be quantified and to say objectively which of the results is better, for the classification of mental tasks. According to this methodology, cross validation with different groups of trainings, cross validation (k-folds) and performance measures will be used.

Cross Validation: To validate the system, the so-called "cross validation" strategy is described in more detail in the book [27]. This strategy consists of dividing the data set into two randomly disjointed subsets, using 80% for training and 20% for evaluation. At the end of the separation process, all available data are used to train and validate the model. The estimation of the generalization of the model is the average of the classification rates obtained with each of the test subsets, which allows to calculate confidence intervals for the presentation of results, giving greater robustness to the system. The results are in terms of accuracy, sensitivity, specificity and Kappa.

Accuracy: It's related to the success rate of the classifier.

Kappa: Is a parameter representing the concordance between the target value and the predicted value. In this sense, the index used here is the one proposed by Cohen [22]. The Kappa coefficient is defined in Eq. 6.

$$Kappa = \frac{\sum_{i=1}^{a} p_{ii} - \sum_{i=1}^{q} p_i - p - i}{1 - \sum_{i=1}^{a} p_i - p - i} \tag{6}$$

where $\sum_{i=1}^{a} pii$, is the accuracy and $\sum_{i=1}^{q} p_i - p - i$, is the percentage due to chance. Kappa > 0.61, indicates a good level of concordance [28]. Then, for five classes the accuracy should be $>0.69\%$.

Sensitivity and Specificity: Those are measures that provide information about the ability to detect a class (true positive or negative condition). Sensitivity and specificity are calculated according to Eqs. 7 and 8 respectively.

$$\text{Sensitivity} = \frac{TP}{TP - FN} \tag{7}$$

$$\text{Specificity} = \frac{TN}{FP - TN} \tag{8}$$

Where,
TP refers to true positive; FN corresponds to false negative; TN is true negative, and FP is false positive. Sensitivity and specificity are used for problems of two kinds; in this case the expected value should be around 81%.

2.6 Description of Mental Tasks

Mental tasks have their foundation in the principles that govern the functionalities of the different regions of the brain. In order to explore the range of possibilities it offers; the following mental tasks were established:

Visual Mental Tasks: Under this term were selected, those that use the channel of representation of reality based on images and that correspond to the use of the occipital, parietal and frontal regions.

Auditory Mental Tasks: These consisted of imagining sounds to use the auditory representation channel using the temporal and frontal regions specifically.

Kinesthetic Mental Tasks: These were intended for the user to use proprioceptive elements located in the somatosensory region of the brain.

Motor Mental Tasks: These are part of the kinesthetic mental tasks, only they are widely referred to in literature and it was decided to give a specific space to imagine movement of the limbs, this is where we focus our work.

2.7 Active Learning

Active training is a strategy proposed for the learning process of an BCI, since it allows to obtain a better quality of the data. Traditional training is implemented to obtain a mode what best represents the data set of the BCI system. Each experience requires a fixed number of training samples which are not fed back to the user. The active training system allows the user to observe the output of the system and use those samples that improve the performance of the system.

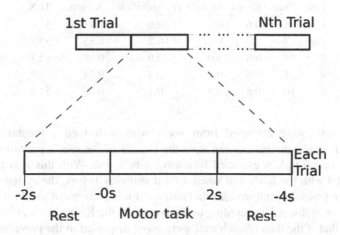

Fig. 6. Experimental protocol for active training for the BCI.

The experimental protocol is presented in Fig. 6. The tasks defined for this type of training were 5 tasks of motor imagination:

1. Movement of the right hand
2. Movement of the left hand
3. Movement of both hands
4. Manipulation of a bucket using both hands
5. Movement of both legs

To establish when the system considers a dataset as valid, each entry is classified and if the kappa coefficient is >0.6 (good concordance) the system completes the training process for this task. If the number of trials is >10 the first trial is discarded, and the system performs a re-learning (active learning with forgetting). After the system has been trained, it goes into online operation mode to validate the test and store the training trials. Each mental task is trained with an idle state relaxation phase. To select the best mental task, the overload index (NASA TLX) was implemented.

3   Results

Table 1 presents the results of each task performed, the number of trials used, the Kappa coefficient value, the accuracy and the calculation of the TLX value for each task. The task that had the highest NASA TLX index was the one that required the greatest number of repetitions and the highest success rate, which allows inferring that a mental task presents a greater cognitive load when the user requires a higher level of concentration in the execution of the same compared to other tasks. This allows validating the approach proposed when using the NASA TLX index.

Table 1. Performance for each mental task.

Task	Trials	Kappa	Sensitivity	Specificity	Accuracy	TLX
1	5	0.6	0.8	0.9	0.91	54.6
2	7	1.0	0.8	0.2	0.55	53.3
3	24	0.8	1.0	0.9	0.98	62.1
4	5	0.6	0.9	0.4	0.74	52.9
5	10	0.6	0.9	0.1	0.56	51.9

These results were extracted from users who performed a mental task, then remained at rest and the system can save the signals of the user's previous state, i.e. rest, then a second task is executed following a new rest. With this it is possible to create a model with the KNN and when a third stimulus arrives, the system calculates the SW of the possible combinations and defines if it is convenient to add it to create a better model or to discard it according to the criteria of the KNN. As an exit criterion, it was defined that if the data of each trial were more dispersed to the previous ones, the session was terminated. This to create a more robust model when performing the evaluation of the NASA-TLX index. It is obtained that task 3 generates more difficulty to perform it, but when compared with the success rate is observed to be the highest. What was evaluated is whether the tasks more difficult than implications the classification of that task and the state of rest.

Note that unlike Task 1, not so many trials were performed because the success rate was already at an optimal value with the model generated and has a relationship with the NASA-TLX index.

4   Conclusions and Discussions

The process of selecting tasks according to the registered in the literature has basically arisen from the need to use the channels of representation of reality easily used by users, this means that the use of tasks based on motor imagination are the most widely used for the ease they represent for the user. Currently, there is no mechanism to evaluate and rate which would be the best tasks to be used in an interface, what exists is to choose those that have the best possible success rate. The proposed active training scheme allows to choose the number of trials necessary for a user to perform in the best possible way a mental task, regardless of the type of representation channel used (visual, kinesthetic and others). It is not possible to assert that those tasks with a higher value in terms of the NASA TLX index, would be the best, but there is something clear in this aspect, related to its repeatability. Because the fact of presenting a greater degree of difficulty for the user, this will make a greater effort to execute the task. The disadvantage of this selection is related to the cognitive load in the use of this type of task. This is an aspect that affects the comfort and usability of the interface.

The success rate is the main measure of a BCI. However, sensitivity and specificity provide information about the ability to recognize classes. The Kappa coefficient provides information on whether the accuracy is close to the value of chance. Performance measures and statistical significance have been analyzed to evaluate the proposed method.

About the method of selecting characteristics using the SW, it can be inferred that it is very useful, mainly because it reduces the computational cost that would be involved in carrying out the whole process of cross validation to select the characteristics that could best model a BCI. The SW by relating the cohesion and separation of trials in relation to the data clustering process.

A methodology is established for the selection of mental tasks for their posterior classification in a BCI system, in this way high success rates are generated.

The limitations of NASA-TLX is that the selection of mental tasks depends on the cognitive abilities of the user and it is difficult to generalize these. The best mental tasks are related to the ability of the user to reproduce the same brain patterns in the execution of a given mental task and are not related to the cognitive load of the task, for example, if tasks are selected that require a high cognitive load facilitate the user's concentration, but brings with it a higher level of user fatigue that would not allow its use continuously.

References

1. Wolpaw, J., et al.: Brain-computer interface technology: a review of the first international meeting (2000)
2. Wolpaw, J., Birbaumer, N., Mcfarland, D.J., Pfurtscheller, G., Vaughan, T.M.: Brain–computer interfaces for communication and control. Clin. Neurophysiol. 113, 767–791 (2002)
3. Kübler, A., et al.: Patients with ALS can use sensorimotor rhythms to operate a brain-computer interface. Neurology 64(10), 1775–1777 (2005)
4. Wolpaw, J., McFarland, D.J.: Control of a two-dimensional movement signal by a noninvasive brain-computer interface in humans. Proc. Natl. Acad. Sci. U.S.A. 101(51), 17849–17854 (2004)
5. Leuthardt, E., Schalk, G., Wolpaw, J.R., Ojemann, J.G., Moran, D.W.: A brain-computer interface using electrocorticographic signals in humans. J. Neural Eng. 1(2), 63–71 (2004)
6. Felton, E., Wilson, J.A., Williams, J.C., Garell, P.C.: Electrocorticographically controlled brain-computer interfaces using motor and sensory imagery in patients with temporary subdural electrode implants: report of four cases. J. Neurosurg. 106(3), 495–500 (2007)

7. Hochberg, L., et al.: Neuronal ensemble control of prosthetic devices by a human with tetraplegia. Nature 442(7099), 164–171 (2006)
8. Kennedy, P., Bakay, R.A.E., Moore, M.M., Adams, K., Goldwaithe, J.: Direct control of a computer from the human central nervous system. IEEE Trans. Rehabil. Eng. 8(2), 198–202 (2000)
9. Curran, E., Stokes, M.J.: Learning to control brain activity: a review of the production and control of EEG components for driving brain-computer interface (BCI) systems. Brain Cogn. 51(3), 326–336 (2003)
10. Riccio, A., et al.: Workload measurement in a communication application operated through a P300-based brain-computer interface. J. Neural Eng. 8(2), 1–6 (2011)
11. Parasuraman, R., Hancock, P.A.: Adaptive control of mental workload. In: Stress, Workload, and Fatigue, pp. 305–320. Lawrence Erlbaum Associates Publishers, Mahwah (2001)
12. Wickens, C.D., Hollands, J.G., Banbury, S., Parasuraman, R.: Engineering Psychology and Human Performance, vol. 13, no. 1 (1987)
13. Hjortskov, N., Rissén, D., Blangsted, A.K., Fallentin, N., Lundberg, U., Søgaard, K.: The effect of mental stress on heart rate variability and blood pressure during computer work. Eur. J. Appl. Physiol. 92(1–2), 84–89 (2004)
14. Hart, S., Staveland, L.E.: Development of NASA-TLX (task load index): results of empirical and theoretical research. Adv. Psychol 52, 139–183 (1988)
15. Byers, J.A., Högberg, H.E., Unelius, C.R., Birgersson, G., Löfqvist, J.: Structure-activity studies on aggregation pheromone components of Pityogenes chalcographu. J. Chem. Ecol. 15, 685–695 (1989)
16. Hill, C., Jones, T.M.: Stakeholder-agency theory. J. Manag. Stud. 29(2), 131–154 (1992)
17. Vitense, H., Jacko, J., Emery, V.: Multimodal feedback: establishing a performance baseline for improved access by individuals with visual impairments, pp. 49–56 (2002)
18. Castillo-Garcia, J.: Interfaz cerebro computador adaptativa basada en agentes software para la discriminacion de cuatro tareas mentales. Doctoral theses, Universidad del valle (2015)
19. Bertrand, P.: A theoretical justification of the average reference in topographic evoked potential studies. Justification théorique de la référence moyenne dans les études topographiques de potentiels évoqués 62, 462–464 (1985)
20. Daubechies, I.: Ten Lectures of Wavelets, p. 357. Springer, Heidelberg (1992)
21. Trejo, L., Shensa, M.J.: Feature extraction of event-related potentials using wavelets: an application to human performance monitoring. Brain Lang. 66(1), 89–107 (1999)
22. Jordan, M., Kleinberg, J., Schölkopf, B.: Pattern Recognition and Machine Learning (2017)
23. Bishop, C.: Pattern Recognition and Machine Learning (Information Science and Statistics). Springer, New York (2006)
24. Rousseeuw, P.: Silhouettes: a graphical aid to the interpretation and validation of cluster analysis. J. Comput. Appl. Math 20, 53–65 (1987)
25. Castillo-Garcia, J., Hortal, E., Caicedo Bravo, E., Bastos, T., Azorin, J.: Feature selection based on Silhouette's width for spontaneous brain computer interface. In: Conference: Proceedings of the 1st International Workshop on Assistive Technology, Vitoria - Brasil (2015)
26. Dubitzky, W., Granzow, M., Berrar, D.P.: Fundamentals of Data Mining in Genomics and Proteomics. Springer, Heidelberg (2007)
27. Duda, R.O., Stork, D.G., Hart, P.E.: Pattern Classification, p. 738 (2000)
28. Cohen, J.: A coefficient of agreement for nominal scales. Educ. Psychol. Meas. 20(1), 37–46 (1960)
29. Roy, R.N., Charbonnier, S., Bonnet, S.: Detection of mental fatigue using an active BCI inspired signal processing chain. IFAC Proc. 19, 2963–2968 (2014)

Algorithms for the Management of Electrical Demand Using a Domotic System with Classification of Electrical Charges

Kevin Andrés Suaza Cano and Javier Ferney Castillo Garcia[(✉)]

Universidad Santiago de Cali, Calle 5 no. 62-00, Cali, Colombia
javier.castillo00@usc.edu.co

Abstract. Electricity demand management is the process of making appropriate use of energy resources. This process is carried out with the aim of achieving a reduction in electricity consumption. The electrical demand management algorithms are implemented in a domotic system that has the capacity to identify electrical loads using artificial neural networks. An analysis was carried out on the most important physical variables in the home, which have a direct relationship with energy consumption, and strategies were proposed on how to carry out a correct control over these, in search of generating energy savings without affecting comfort levels in the home. It was obtained, as a result that it is possible to generate an energy saving of 63% in comparison to a traditional house, this without affecting to a great extent the comfort of the user and allowing a great level of automation in the home.

Keywords: Domotic system · Neural network · Demand management

1 Introduction

Currently the energy consumption in home is high, this consequence of a very low energy efficiency, which translates into a high impact on the environment, this makes necessary to have a greater degree of control over the electrical devices used in the home. In the beginning, domotic was restricted to be a tool at the hand of the user, providing the ability to make some adjustments on the house like turning lights on or off, or setting timers, however, the technology has advanced very quickly and domotic has been assigned more complex tasks over the years.

In the current market you can find different advances in terms of safety, energy and comfort, like intelligent cameras that brings the opportunity of access their images in real time from anywhere in the world, another of the great aspects in which domotic has evolved in terms of security, is being able to control the doors, windows and blinds of the home through an Internet connection [1, 2]. One of the most important features of home automation and one of its main characteristics is the ability to control different appliances in the home that directly influence energy consumption, in areas such as air conditioning and lighting, all this in order to generate energy savings and have a better use of the resources [3, 4]. One of the major trends in terms of home automation is the use of personal assistants such as Google Home, or Amazon Alexa, this type of devices

M. Botto-Tobar et al. (Eds.): ICAT 2019, CCIS 1194, pp. 173–183, 2020.
https://doi.org/10.1007/978-3-030-42520-3_14

have the ability to recognize voice commands from the user and receive direct commands from it, although one of its main features is the ability to control different devices on home [5, 6].

The objective of this paper is present the implementation of an artificial neural network in an embedded system for the characterization of electrical loads for the management of electrical demand in a domotic system and present the algorithms necessary to carry out the control of a house for a traditional family.

2 State of the Art

Studies related to domotic systems based on demand management or load connection/disconnection control systems are presented. Widely used systems share similar characteristics such as hardware definition for communication protocols, load control system (contactors or solid state), temporization for the connection/disconnection of the different electrical devices.

In the first instance [7] addresses the problem of excessive consumption of electrical resources in the home and raises the need to implement automation systems on a massive scale.

The most basic aspect of home automation is the ability to control one or more aspects of the home, such as turning on/off lights or devices connected to the electrical network, [8] and to be able to provide information on basic aspects such as temperature, humidity [9] and in some cases more detailed information such as light intensity, motion detection and magnetic fields to detect the opening of doors or windows [10]. In the research work of [11] the implementation of a traditional domotic system using KNX technology is carried out in a 67 m^2 house. This system, although reliable, becomes obsolete when compared to technological proposals such as the one carried out by [12]. In its research work, the implementation of a decentralized on/off system is carried out, using ESP8266 modules connected through a wireless network, which allows a greater level of flexibility and ease of implementation, all this together with a lower total cost of the system.

One of the areas where domotic has greater capacity for evolution, is in automatic decision making, providing the domotic the ability to decide which electrical devices are connected to the electrical grid, allows a more precise control over the electricity consumption of the home. Different approaches to this idea have been made in the literature and proposals have been found that have the capacity to generate electricity savings, [13] in this work, he carries out a system of scheduling of electrical charges, with which he manages to make the appropriate use of the photovoltaic resource that a home has, making use to a lesser extent of the home electrical network, however, a variety of works have been found in which the objective is to carry out a system that can classify electrical charges on the basis of their voltage and current signals, [14] makes an approach to this idea, making use of Artificial Neural Networks achieves the identification of fundamental characteristics in a radio signal, for its part, [15] develop a system with the ability to correctly identify a cooler using an ANN trained solely from the power data of the device, [16] makes a similar system, however in its development are used the general consumption data of the house and with the use of the transformed

Hilbert-Huang achieves a correct characterization of elements with high energy consumption such as the washing machine, clothes dryer and refrigerator.

The proposed home automation system will characterize and identify electrical loads, which can identify an appliance and its mode of operation (full load or low consumption). The characterization uses electrical parameters that model the different loads from their characteristics in power, harmonic distortion among others. The demand management uses the information of the appliances connected to the network and establishes their mode of operation, those in low consumption are disconnected after their identification. The ecosystem is made up of outlet modules, switch modules and a master module where the actions that depend on the identified appliance are defined. The activation of the loads will be carried out with solid state power electronic devices.

3 Materials and Methods/Methodology

3.1 Embedded System

An embedded system was development, which allowed the acquisition of voltage and current data of each one of the most common household appliances, and then these data were used to extract the electrical parameters that allowed differences between the several appliances and their states of consumption.

Hardware. The selection of the platform carries out to meet the needs of the development phase and that could be used as a final product. The requirements were raised, an embedded system that can have readings on voltage and alternating current (AC), that counts on different means of wireless communication, for its later implementation in a domotic system, that count with great capacities of memory and a high power of compute for the implementation of the system of characterization. Having as main objective a low-cost platform, easy in the implementation of a wireless communication network with a high level of processing and enough memory capacity. The development platform selected was the ESP32, given that its low price, has a powerful dual core processor Xtensa LX6 32 Bits at 240 MHz, 4 MB flash memory for the program and 540 KB of integrated RAM, WI-FI b/g/n and Bluetooth 4.2 Low Energy integrated and the ability to perform analog voltage reading with a resolution of 12 Bits. Other platforms with similar characteristics have the disadvantage of being more expensive, making it impossible to develop a low-cost system.

In the data acquisition phase, voltage and electrical current samples were taken from the most common household appliances in the home. To perform this task, circuits were implemented to couple these physical magnitudes to the working ranges of the ESP32 platform, and a storage system was implemented through an external SD memory connected to one of the SPI modules available on the development platform.

The last requirement of the embedded system was to provide it with the ability to control on and off on the load that was connected to it, for this was designed and implemented a control system, using a triac BTA20 semiconductor device.

Software. The first stage of the project consisted of collecting voltage and current data from the most commonly used appliances in the home, to fulfill this purpose the necessary hardware was developed and an algorithm was developed, which was intended to perform the storage of this information in an SD memory using parameters previously specified, these parameters are, have the ability to generate a text file for each appliance analyzed, the data was taken at a fixed frequency of 1 kHz.

Data Acquisition. The acquisition of voltage and current data of the most commonly used household appliances and their characteristics as model, manufacturers and type are presented in the Table 4.

3.2   Electrical Characteristics

Taking as information the voltage and current data collected from household electrical devices, a correct differentiation must be made between the different types of devices recorded, but just analyzing this information is not possible to give a correct analysis, it is necessary to have electrical features from these signals that provide information, allowing the classification of different household devices. The following features were used: Instantaneous Power (P), Maximum Power (P_{Max}), Distorted Power (D), Power Factor (Fp), Reactive Power (Q), Apparent Power (S), Offset Angle (ϕ), Total Harmonic Distortion (TDH), Current Variance (σ_n^2), Root Mean Square (I_{RMS}) [17].

3.3   Learning Machines

The learning machines are computational techniques used to develop prediction algorithms based on data samples, there are different classification techniques which are divided into two classes, unsupervised learning machines and supervised learning machines [18].

   In the training process of all classification methods used, (Cross-Validation) was performed, defined by [19] As the process of making a random separation of the input data and their corresponding outputs to make use of part of the data for system training and with the remaining data to perform its validation. In the implementation three types of learning machines were analyzed, which will present ease in their development in an embedded system, the types of machines used were: Neural networks, decision trees and K nearest neighbors.

   An artificial Neural Network (ANN) is a mathematical abstraction of the process of communication and information generation of physical neural networks. One type of implementation of artificial neural networks widely used are the multilayer perceptron MLP (Multi-Layer Perceptron), this are an evolution of the single-layer neural network, its operation is based on the use of a series of hidden layers, which is ideal for solving non-linear problems [20] a MLP is formed by at least 3 layers, the first is the input layer, this is where the data is received to be used by the ANN, the second layer is part of the group of hidden layers being necessary to have at least one layer of that type and finally has the output layer which delivers the data generated by the ANN [21].

Equation 1 is a mathematical representation of a neural network, where x are the inputs to the neuron, w is the matrix of synaptic weights and b is the value of the tracks [22].

$$y = f(x \cdot w + b) \tag{1}$$

3.4 Implementation of the Load Characterization System on the Embedded Platform

The implementation of the load characterization system consists of providing the embedded system with the ability to take voltage and current data to perform the electrical feature extraction, making use of these electrical features in the process of generate a result using the ANN previously trained, the training is done on a desktop computer. When the training process ends, the information of the synaptic weight matrix, bias value, network topology and characteristic normalization values are transferred.

To execute the neural network, a series of for cycles are used to realize the calculation of the output of each one of the neurons, in total 3 for cycles was used, the first one to go through all the layers, the second one used to go through all the neurons on a layer and the last one to do the calculations for all the inputs on a neuron. The output of each one of the neurons is evaluated on the activation function.

3.5 Implementation of Control Algorithms

Considering the capabilities of the hardware used, algorithms are implemented that can control the lights and outlets of the house.

To make the control of the lights it is used as base the information of a sensor of presence by radio frequency and the hour of the day, if the hour is between the 6 PM and the 6 AM and the sensor detects movement it makes the ignition of the light by a short period of time. On the other hand, if an acoustic signature (applause) is detected and the light conditions warrant it, the light is switched on for a slightly longer period. Finally, if the touch sensor is used and the light conditions warrant it, it is switched on for a period of 30 min, in this way the light can be precisely controlled according to the user's needs and allowing the maximum possible savings to be generated.

The algorithm to control the lights is quite simple, because having the ability to know precisely which device is connected to each socket gives the freedom to make a very fine control without the need for a complicated algorithm. If the connected device charges a battery (cell phone, laptop computer), the system can recognize when the device is fully charged and can disconnect the device to not generate consumption per standby.

Because the system can recognize when a device is in standby mode, it can disconnect those elements that are not being used but generate a parasite charge, such as computers and televisions, which generally remain connected 24 h a day and generate a high electrical consumption.

Finally, the system can recognize certain heavy loads that should not be on for long periods of time such as blenders and irons and this in the ability to generate an alarm to the user or disconnect the load in order to keep the consumption low.

4 Results and Discussion

The Table 1 shows the average values of each of the features of all household appliances, the value in brackets corresponds to the standard deviation of the value of each feature.

Table 1. Average and standard deviation of the electrical features used.

Feature/device	D (W)	Fp	P (W)	Phi (°)	P_{Max} (W)	Q (VAR)	S (VA)	TDH	σ_n^2 (A²)	I_{RMS} (A)
Cell	0,62 (0.05)	0,60 (0.05)	8,17 (3.38)	51,49 (4.33)	73,02 (21.9)	10,16 (3.43)	13,20 (4.76)	0,20 (0.12)	0,01 (0.01)	0,11 (0.04)
Iron	0,49 (0.44)	0,48 (0.45)	346,9 (362)	50,97 (37.7)	822,3 (748)	91,10 (118)	377,9 (362)	0,41 (0.49)	26,83 (28.1)	3,73 (3.59)
Computer	0,85 (0.05)	0,85 (0.02)	26,58 (2.80)	30,76 (4.77)	90,48 (14.0)	15,81 (2.16)	31,14 (2.51)	0,05 (0.13)	0,06 (0.01)	0,25 (0.02)
Blender	0,42 (0.04)	0,41 (0.05)	61,36 (10.9)	64,94 (3.03)	410,8 (150)	132,1 (6.03)	146,2 (9.39)	0,15 (0.11)	1,95 (0.23)	1,39 (0.08)
Laptop	0,63 (0.05)	0,61 (0.05)	30,59 (4.38)	50,81 (3.99)	261,2 (20.9)	37,94 (1.34)	49,22 (3.10)	0,18 (0.07)	0,19 (0.05)	0,44 (0.05)
TV	0,76 (0.20)	0,75 (0.21)	39,26 (20.4)	34,28 (21.7)	171,5 (18.4)	21,07 (7.72)	47,97 (13.3)	0,12 (0.14)	0,20 (0.10)	0,43 (0.12)
Low consumption	0,19 (0.10)	0,16 (0.09)	3,90 (12.5)	78,88 (6.99)	54,71 (62.8)	10,17 (21.7)	11,12 (25.1)	0,53 (0.67)	0,06 (0.25)	0,10 (0.23)

A Decision Tree classifier was trained because this algorithm has a low computational cost. It was implemented using conditionals and basic mathematical operations. Cross validation results in an error rate of 23.4%. The decision tree is implemented in the embedded system and its operation is verified, obtaining as a result, that the system does not have the capacity to carry out the correct characterization of the electrical appliances that present strong variations in their electrical features, due to the fact that this classifier uses the definition of thresholds, therefore it does not have the capacity to identify correctly elements that produce high levels of noise and cannot use features that have a high level of variation such as reactive power and current variance.

A k-Neighbor closer KNN classifier was trained. This classifier is a much more robust system than the decision tree and had an average error rate of 1.1538% and a standard deviation of 0.7278%, however, its implementation requires 38 KB of volatile memory and very long execution times which makes it impossible to implement in an embedded system.

The ANN training process was carried out and a list of values of synaptic weights, biases, values for data normalization and activation functions was generated. Different network topologies were used, which implied different capacities in the size of bytes to use number of neurons to use, the Table 2 shows the results of the different trained topologies. The topology with two hidden layers, each with 10 neurons, provides an average error level of 2.051% and a standard deviation of 1.2385%, without greatly compromising system memory and execution times. An adjustment was made to the KNN classifier, approximating the value of volatile memory that consumes to the values of use of the artificial neural network, this in order to appreciate under equal conditions which classifier presents a higher performance, it was found that the KNN presents error levels higher than the artificial neural network, this information is shown in the Table 3.

Table 2. Tests performed on different neural network topologies. Source Authors.

Topology	Neurons	Average error	Size (bytes)	Standard deviation	Time (μs)
10 4	14	3,71	836	3,27	215
10 8	18	7,05	1044	7,93	277
10 10	20	2,05	1148	1,23	310
10 4 4	18	13,20	936	28,53	266
10 4 8	22	7,17	1048	7,38	326

Table 3. Comparison between different classifiers.

Classifier	Average error	Standard deviation	Size (bytes)
Decision tree	23.4%	1.32	100
KNN	1.15%	0.72	34320
KNN (Adjusted)	83.8%	18.65	1232
Neuronal network	2.0%	1.23	1148

The artificial neural network implemented in the embedded system made it possible to evaluate the signal establishment times of the connected appliances, the average times was 4.5 ± 2 s.

The demand management proposed for use in an intelligent home automation system, with the ability to recognize that appliances are connected to the system and to exercise control actions that ensure low electricity consumption, without making major effects on the user, requires a correct characterization and identification of the electrical loads connected to the system. To evaluate the proposed system, a simulation is carried out based on the time that the devices remain connected to the electrical network and what would be the consumption without load management actions what we call normal consumption, a load management based on traditional domotic systems and load management, where the characterization of the household appliances is used as a basis for intelligent demand management. The results of this simulation and the corresponding reduction in terms of carbon dioxide production are presented. Simulated systems include the electrical consumption of demand monitoring and management devices.

The Fig. 1 shows the result of simulating a home using a home automation system with load characterization, a traditional home automation system with finely adjusted timers and a home without any type of home automation control. From the figure it can be extracted that by means of the use of a domotic system with characterization of loads an average saving of 34% can be achieved compared to a house without any type of control and 12% compared to a traditional domotic system with a complex configuration of timers.

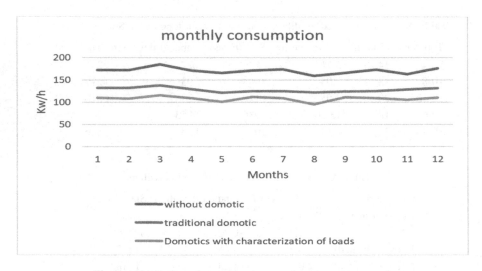

Fig. 1. Simulation of monthly consumption per system used.

Considering the consumption obtained with each of the simulations, it is possible to reflect this electricity consumption in an annual cost, which corresponds to the value in USD to be paid with each of the proposed systems. In the Fig. 2 you can see the graphical representation of the costs, for this simulation we took as a base the cost of a Kw/h in New York City, which corresponds to $0.18 USD. From this information it can be concluded that the use of a domotic system with load characterization can generate annual savings of approximately $137 USD.

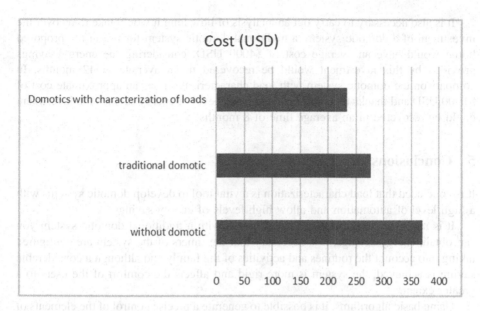

Fig. 2. Simulation of the cost of the electric bill in one year.

Based on the results of the energy consumption it was possible to make an analysis on the generation of Co2 based on the domotic system used, it can be seen in the Fig. 3 that using the domotic system with load characterization could generate savings of approximately 280 kg of Co2 per year.

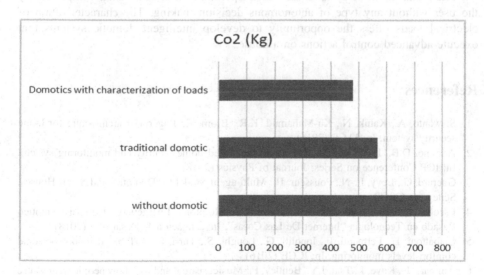

Fig. 3. Simulation of the amount of Co2 generated in a year

It is also necessary to carry out an analysis of how long it would take to recover the investment of a domotic system, a traditional domotic system for use in the proposed home would have an average cost of $4000 USD, considering the energy savings provided by this investment would be recovered in an average of 42 months. In comparison the domotic system with load characterization has an approximate cost of $1000 USD and thanks to its low cost and better level of energy savings the investment could be recovered in an average time of 8 months.

5   Conclusions

It is concluded that load characterization is a vital tool to develop domotic systems with a high level of automation and allow high levels of energy savings.

It is necessary to emphasize that, although with a traditional domotic system you can obtain energy savings, it is necessary that the timers of the system are configured taking into account the routines and activities of the family and although a considerable saving is achieved, the system is more rigid and affects the comfort of the users to a greater extent.

Using basic algorithms, it is possible to generate a precise control of the elements of a house, allowing to generate energy saving without affecting the level of comfort of the user and keeping the costs low.

The implementation of mathematical and statistical tools, such as learning machines for the process of characterization of electrical loads require an adequate definition and selection of technical and economic specifications for implementation in embedded systems or low-cost development systems.

Traditional domotic systems base their operation on actions taken completely by the user without any type of autonomous decision making. The characterization of electrical loads offers the opportunity to develop intelligent domotic systems that execute advanced control actions on a house.

References

1. Siswanto, A., Katuk, N., Ku-Mahamud, K.R.: Biometric fingerprint architecture for home security system. In: IACE (2016)
2. Adriano, D.B., Budi, W.A.C.: IoT-based integrated home security and monitoring system. In: IOP Conference on Series: Journal of Physics (2018)
3. Guerard, G., Levy, L.-N., Pousseur, H.: Multi-agent Model for Domotics and Smart Houses. Scitepress (2018)
4. Escobar Gallardo, E., Villazón, A.: Sistema de Monitoreo Energético y Control Domótico Basado en Tecnología "Internet De Las Cosas". Investigación & Desarrollo (2018)
5. Ciabattoni, L., Ferracuti, F., Ippoliti, G., Longhi, S., Turri, G.: IoT based indoor personal comfort levels monitoring. In: ICCE (2016)
6. Ammari, T., Kaye, J., Tsai, J.Y., Bentley, F.: Music, search, and IoT: how people (really) use voice assistants. ACM Trans. Comput.-Hum. **26**, 1–28 (2019)
7. Hassanpour, V., Rajabi, S., Shayan, Z., Hafezi, Z., Arefi, M.M.: Low-cost home automation using arduino and modbus protocol. In: ICCIA (2017)

8. Ovalles, F.O., Bolivar, A.E., Rodriguez, A.J.: Use of an embedded system with WiFi technology for domotic. IOP Publishing (2018)
9. Dobrescu, L.: Domotic embedded system. In: ECAI (2014)
10. Banu, B., Merlin, S., Suhitha, S.: Domotics using labview. Int. J. Res. **5**, 1–7 (2018)
11. Morón, C., Payán, A., García, A., Bosquet, F.: Domotics project housing block. In: MDPI (2016)
12. García, V.H., Vega, N.: Low power sensor node applied to domotic using IoT. In: Mata-Rivera, M.F., Zagal-Flores, R. (eds.) WITCOM 2018. CCIS, vol. 944, pp. 56–69. Springer, Cham (2018). https://doi.org/10.1007/978-3-030-03763-5_6
13. Moya, F., Da Silva, L., Lopez, J.: A mathematical model for the optimal scheduling of smart home electrical loads. WSEAS Trans. Power Syst. **13**, 300–310 (2018)
14. Toro Betancur, V.: Algoritmo de Estimación de Parámetros y Modulación de Una Señal Recibida Por Un SDR. EAFIT (2017)
15. Cherraqi, E.B., Maach, A.: Load signatures identification based on real power fluctuations. In: Noreddine, G., Kacprzyk, J. (eds.) ITCS 2017. AISC, vol. 640, pp. 143–152. Springer, Cham (2018). https://doi.org/10.1007/978-3-319-64719-7_13
16. García Ortiz, M.: Adaptación y Aplicación de la Transformada de Hilbert-Huang a Sistemas Eléctricos: Aplicaciones en el Estudio de la Gestión de la Demanda y Caracterización de Transitorios. Springer, Cham (2016)
17. Bollen, M.H.J., Gu, I.Y.H.: Signal Processing of Power Quality Disturbances. IEEE Press Series on Power Engineering, Danvers (2006)
18. Witten, I.H., Eibe, F., Hall, M.A., Pal, C.J.: Data Mining: Practical Machine Learning Tools and Techniques. Elsevier, Cambridge (2017)
19. Zhang, Y., Yang, Y.: Cross-validation for selecting a model selection procedure. J. Econom. **187**, 95–112 (2015)
20. Chen, X., Liu, G., Shi, J., Xu, J., Xu, B.: Distilled binary neural network for monaural speech separation. In: IJCNN (2018)
21. Zhang, H., Wang, Z., Liu, D.: A comprehensive review of stability analysis of continuous-time recurrent neural networks. IEEE Trans. Neural Netw. Learn. Syst. **25**, 1229–1262 (2014)
22. Khosrojerdi, S., Vakili, M., Kalhor, K.: Thermal conductivity modeling of graphene nanoplatelets/deionized water nanofluid by MLP neural network and theoretical modeling using experimental results. Int. Commun. Heat Mass Transf. **74**, 11–17 (2016)

Cluster Analysis for Abstemious Characterization Based on Psycho-Social Information

Pablo Torres-Carrión[1]([⊠]) [iD], Ruth Reátegui[1], Byron Bustamante[1],
Jorge Gordón[2], María José Boada[3], and Pablo Ruisoto[4]

[1] Universidad Técnica Particular de Loja, San Cayetano Alto S/N, Loja, Ecuador
{pvtorres, rmreategui, bfbustamante}@utpl.edu.ec
[2] Universidad Técnica del Norte, Ibarra, Ecuador
jegordon@utn.edu.ec
[3] Universidad Politécnica Salesiana, Quito, Ecuador
mboada@ups.edu.ec
[4] Universidad de Salamanca, Salamanca, Spain
ruisoto@usal.es

Abstract. The consumption of alcohol, tobacco and other drugs are a health and social problems worldwide. According to several studies, abstemious people are a minority population among university students. The objective of this research is to identify clusters of abstemious students and their psycho-socials patterns. Based on information obtained through five psychological questionnaires (Patient health questionnaire PHQ-9, avoidance and action questionnaire AAQ 7, loneliness scale UCLA-R, the satisfaction with life scale SLQ, the Barratt impulsiveness scale BISS-1, and perceived stress scale PSS-10) a cluster analysis was conduct using Sparse K-means algorithm. The sample comprised 510 abstemious college students from three Ecuadorian universities. Two clusters were obtained: satisfaction with life, loneliness, and avoidance and action are the most representative variables contributing to the cluster distribution.

Keywords: Cluster analysis · Abstemious · University students · Ecuador

1 Introduction

The consumption of alcohol, tobacco and other drugs is a health and social worldwide problem, around 5.5% of the global population aged 15–64, had used drugs in 2018, there is an increase of 30% in the last ten years (210 to 271 millions of users), while 35 million people are estimated to be suffering from drug use disorder [1].

In Ecuador and another Andean countries, the highest prevalence and greatest alcohol and drugs consumption rates occurs in young people aged 18 to 25 years [2]. In the Ecuadorian university context, the alcohol and drugs consume are the main problem in Ecuadorian college students [3]. According [2] students who don't consume alcohol in the last year is around one third, and this proportion is lower in particular universities, around 8% in men and 17% in women [3]. On the other hand, around half of man and a quarter of women, in one particular Ecuadorian college, have a problem with

M. Botto-Tobar et al. (Eds.): ICAT 2019, CCIS 1194, pp. 184–193, 2020.
https://doi.org/10.1007/978-3-030-42520-3_15

alcohol consumption [3]. In a national research with eleven universities, the study has reported similar results [4]. There are few studies in the consumption of other drugs; the most important reports that in Ecuadorian college students, the actual tobacco consumers are 17.2%. 12.5% are electronic cigarettes users, 20% have consumed two or more psychoactive substances, 12.8% have used one illicit substance. 11.7% are recent cannabis users and 1.45% are recent cocaine users [2].

Knowing the psychosocial abstemious profile is important because this knowledge can guide prevention and promotion health actions; therefore, some variables obtained from psychological tests are necessary to know the user profile and predict the problematic use of alcohol, tobacco, and other drugs. In our study applying cluster techniques, the following indicators are relevant: stress, depression, life satisfaction, loneliness, psychological inflexibility, personality traits, and cognitive impulsivity. To relate this knowledge, a systematic search has been carried out to support previous studies in relation to these indicators. So, stress, alcohol, and drug use is considered a daily stress avoidance mechanism [5]. High stress level is associated with alcohol problems use and abuse [6]. In a Ecuadorian college study, Ruisoto [3] reports that the problematic use of alcohol is associated with increase of perceived stress level in men and women. Considering depression, there are a lot evidence about bidirectional relation among depression, alcohol, and drugs use [7–9]. Likewise, the low-life satisfaction is associated with alcohol, tobacco, and other substances that teens and young people use [10–12]. The problem with alcohol use is associated with dissatisfaction life. The level of life satisfaction can predict the alcohol abuse in men and women [13]. In [14] reported that, the low-life satisfaction join adverse life circumstances in teens and youth can increase the intensity and frequency of alcohol, cannabis, and tobacco use.

The loneliness in combination with depression are associated with relapse in drug dependents [15]. There is a lot of evidence that loneliness is associated with drug dependency because it can cause a maintenance factor or relapse trigger factor [16, 17]. About psychological inflexibility, some empirical studies have verified the association between substance consume and psychological (in) flexibility [3, 18]. Reporting that college students with high psychological inflexibility have more problems with alcohol use. In Personality traits, some studies report that extraversion, affability, and conscientiousness are associated with alcohol use in teens [19, 20] while in youth the neuroticisms joins low affability and conscientiousness can be an explanation of why the alcohol use is more [21].

About the psychosocial abstemious profile, some researchers report a strong relation with high cognitive impulsivity and problematic consume of drugs [22, 23], however is not clear the direction of this relation. To summarize, stress, depression, life satisfaction, loneliness, psychological inflexibility, cognitive impulsivity, and some personality traits can explain the alcohol and other drugs use. For us, it is necessary to use these variables to explain the abstemious behavior and explore the diversity in this rare group in college students. Maybe in abstemious characteristics we can find answers to promote and prevent use, consume, abuse, and dependence of substances.

Artificial Intelligence (AI) is a field of cross-sectional research into several areas of knowledge. Drug use is studied as a health problem, and the state of art focuses on these lines of research. Drug-related AI applications have been an educational tool using a Bayesian network that provides legal advice to users, adapting the recommendation to their profile [24]; and through classification models based on vector machines to establish possible drug distribution routes, supply sources and clandestine laboratories [25]. In related health fields, for example, in [26] classifiers are constructed to predict the mental health of an individual, applying ML models (SVM, decision three, naive Bayes classifier, K-nearest neighbor classifier and logistic regression), and in [27] ontologies are applied to determine patient health risk factors, and recommendation of prescriptions in diabetes.

These computer solutions that include AI are focused on helping people who want to stop drugs. So, several investigations have been found. In [28] the m-health architecture is proposed, as a solution based on mobile platforms, which identifies the needs of the smoker and medical staff, and includes a decision-making system based on data mining techniques. [29] describes a system to quit smoking with the help of an Android app and user movement patterns, which from a previously configured custom plan, monitor user activity, add a trained LSTM algorithm from the movements made at the time of smoking or not, and apply them as a warning. A comparative study [30] uses text mining techniques applied to tweets, to establish medical aspects considered by US and UK professionals regarding the use of electronic cigarettes. In [31], ML techniques are applied to the data obtained from an environmental carbon monoxide gas sensor emulated with a mobile app, to determine remotely from a gamified space if the user has smoked. A classification model of the smoke sample characteristics is used to determine the state of the smoker.

Obtaining data to work with groups of abstemious is a great challenge due to the limited existing data and the difficulty in obtaining them [32]. The main solutions for the classification of abstainers and consumers have focused on tobacco. [33] details the classification of smokers from user-generated content in an online community to quit smoking, and knowledge of experts in the domain labeling data. Applying ML will establish patterns with important user characteristics to determine their intervention. Related study [34], uses 2578 publications of online communities to quit smoking and also applies keywords and the analysis of the keyword network to identify types of social support and analyze their similarities and differences at different stages to give up smoking. With a variation in the type of data, and focus groups of abstemious, in [35] algorithms are applied to classify (alcoholic or abstemious) from EEG signals, obtaining relevant results from the algorithms: Quadratic SVM 95%, Cubic SVM 92%, fine KNN 91% and ensemble bagged tree 90%; the data have been obtained from EEG dataset of the University of California (UCAKDD), in two groups: (a) alcoholic and (b) abstemious control group. In all cases, although varying the type of data, ML methods have been adapted to classify abstemious and consumers with acceptable percentages of acceptance.

Most studies that analyze specific characteristics of abstemious have interest in tobacco, as a so-called social vice. In [36] they make a comparison of four ML techniques: logistic regression (LR), support vector machine (SVM), random forest (RF) and Naïve Bayes (NB) to predict whether or not to quit smoking; They analyze 60 characteristics associated with smoking cessation, among these sociodemographic, age at which smoking begins, duration and others, through the use of a filter-based feature selection method, identifying eight factors associated with quitting smoke. In [37] its objective is the social commitment, communication attributes, and emotional environment of those who have successfully quit smoking; It takes as a base the interactions between peers of an online health community to quit smoking and develops a model based on ML, adaptable to real-time interventions with emotional and communicational focus. In [38] having magnetic resonance imaging (fMRI) images of patients before and after quitting smoking (240,000 inputs) as a data base, with an accuracy of 74%, classifies people who undergo smoking cessation treatment, compared to those who receive a placebo. Of the ML algorithms used with the PCA and the genetic programming classifier the best accuracy is obtained.

In recent research, applying a systematic review methodology in the Scopus and WoS databases [39], no research has been found with a wealth of abstemious data that has psychological profiles to explicitly describe the resulting groups. This makes the present research more important, and gives the scientific community a significant contribution in terms of the behavior of the abstemious population. Another fact to consider is that the studies mainly focus on tobacco and alcohol; the database of our study has information related to poly-consumption. This is a first approach to the analysis of these data, and as a research group we will be sharing subsequent analyzes that allow us to deepen each of the behavioral aspects of the population of abstainers.

The following section details the method used to perform the cluster analysis, including a detail of the database with psychological tests related to mental health and psychosocial characteristics. The results section details the characteristics of clusters obtained. Including the values of the six psycho-logical tests (PHQ-9, AAQ-7, UCLA-R, SLQ, BISS-11, PSS-10, TIPI-SPA, and socio-demographic). In the final sections the discussion of results and the main conclusions of the study are carried out.

2 Materials and Methods

2.1 Dataset

From a sample of 3981 college students from Ecuador who validly answered a psychosocial questionnaire, 510 students, who have not used alcohol, tobacco, or any other drug, were selected. The data are obtained as part of the "CEPRA XII-2018-05 Prediction of drug use" project funded by CEDIA (Ecuadorian Corporation for Research Development and the Academy, by Spanish acronym), and in which researchers from three Ecuadorian and two Spanish universities: Universidad Técnica Particular de Loja, Universidad Técnica del Norte, Universidad Politécnica Salesiana, Universidad de Salamanca and Universidad Loyola Andalucía. The data correspond to research conducted with the entire student population of the three Ecuadorian universities, applying a test from an online platform. The data has gone through a filtering and cleaning, separating all those that presented anomalies. From the questionnaire that consist of 10 test, 11 variables detailed below in Table 1 were selected:

Table 1. List of questionnaires and variables

Test/Questionnaires	Description	Variables
Patient health questionnaire (PHQ-9) [40]	This test allows to screen the presence and severity of depression	1 variable that resume this questionnaire
Avoidance and action questionnaire (AAQ 7) [41]	This test permits to assess the psychological inflexibility	1 variable that resume this questionnaire
Loneliness scale (UCLA-R) [42]	This test measures one's subjective feeling of loneliness and social isolation	1 variable that resume this questionnaire
The satisfaction with life scale (SLQ) [43]	This scale measures the life satisfaction component of subjective well-being	1 variable that resume this questionnaire
The Barratt impulsiveness scale (BISS-11) [44]	This test measures the impulsiveness	1 variable that resume this questionnaire
Perceived stress scale (PSS-10) [45]	This test measures the perception of stress	1 variable that resume this questionnaire
Personality inventory questionarie (TIPI-SPA) [46]	This test measures personality traits	5 variables: extroversion, affability, responsibility, emotional stability, opening to experience
Total variables	11	

2.2 Cluster Analysis

Sparse K-means algorithm was used to conduct a cluster analysis. This algorithm simultaneously finds clusters and a subset of cluster variables. Sparse K-means assigns weight to the variables with the most representative ones having the highest weight. Also, this algorithm is useful when the dataset has noise variables. Before use sparse K-means, we applied gap statistics to estimate the number of clusters. Gap statistics is a standard method for detecting the number of clusters [47].

Before cluster analysis all the variables were normalized. Then the Sparse K-means [48] algorithm was used to identify clusters. This algorithm simultaneously finds groups of objects and a subset of cluster variables. Sparse K-means assigns weight to the variables with the most representative ones having the highest weight. Furthermore, this algorithm is useful when the dataset has noise variables. Also, gap statistics was used in order to estimate the number of clusters. Gap statistics is a standard method for detecting the number of clusters [47]. We used R Studio[1] that have implemented gap statistics and Sparse K-means.

---

[1] https://rstudio.com/.

3 Results

Cluster analysis allows to identify two clusters of abstemious students, CLA1 and CLA2. Table 1 shows the distribution of the clusters to the tests PHQ-9, AAQ 7, UCLA-R, SLQ, BISS-11, and PSS-10. Table 2 shows details of the clusters distribution to 5 personality variables. The values of the questionnaires are represented by quartiles and the quantity of students in each range are represented by percentage according the number in each cluster.

Table 2. Clusters distribution according PHQ-9, AAQ 7, UCLA-R, SLQ, BISS-11 and PSS-10 tests

Patient health questionnaire (PHQ-9)		Avoidance and action questionnaire (AAQ 7)		UCLA loneliness scale (UCLA-R)		Satisfaction with life scale (SLQ)		Barratt impulsiveness scale BISS-11		Perceived stress scale (PSS-10)	
W = 0.004		W = 0.254		W = 0.397		W = 0.849		W = 0.182		W = 0.04	
Values	%	Values	%	Values	%	Values	%	Values	%	Values	%
CLA1 (187)											
Depression treatment with antidepressants is justified	16	7–11	6	3–4	0	They love their lives and feel that things are going very well	13	3–11	18	0–12	6
The doctor should use his judgment about the treatment	64	12–17	10	5–6	16	They love their lives, despite this they have identified areas of dissatisfaction	21	12–14	30	13–16	17
Probably, the patient does not need treatment for depression	20	18–25	32	7	22	They consider that there are areas of their life that need improvement	27	15–16	22	17–20	34
		26–49	51	8–12	61	They have small but significant problems in various areas of their life	26	17–26	29	21–39	43
						They are significantly dissatisfied with their lives	9				
						They are extremely unhappy with their current life	3				
CLA2 (323)											
Depression treatment with antidepressants is justified	2	7–11	37	3–4	54	They love their lives and feel that things are going very well	45	3–11	36	0–12	37
The doctor should use his judgment about the treatment	22	12–17	33	5–6	42	They love their lives, despite this they have identified areas of dissatisfaction	33	12–14	30	13–16	36
Probably, the patient does not need treatment for depression	76	18–25	21	7	3	They consider that there are areas of their life that need improvement	14	15–16	19	17–20	20
		26–49	10	8–12	0	They have small but significant problems in various areas of their life	5	17–26	15	21–39	7
						They are significantly dissatisfied with their lives	2				
						They are extremely unhappy with their current life	2				

As we can see, the variables SQL, UCLA-R and AAQ 7 are the most representative variables with the highest weights 0.849, 0.397 and 0.254 respectively. Cluster CLA1 has 187 students, 21% are students that love their lives, despite this they have identified areas of dissatisfaction, 27% consider that there are areas of their life that need improvement, and 26% have small but significant problems in various areas of their life. In this cluster, most of the students (61%) are in the range of 8–12 of loneliness scale. Also, according to the AAQ 7 test, 32% are in the ranges 18–25, and 51% are in the ranges 26–49.

Cluster CLA2 has 323 students, 45% love their lives and feel that things are going very well, 33% love their lives, despite this they have identified areas of dissatisfaction. Considering the loneliness scale, most of the students (54%) are in the range 3–4, and 42% are in the range of 5–6. Taking account, the AAQ 7 questionnaire, 37% of the students are in the ranges 7–11, and 33% in the range 12–17 (Table 3).

Table 3. Clusters distribution according personality variables.

Ten-item personality inventory (TIPI-SPA)									
Extraversion		Affability		Responsibility		Emotional stability		Opening to experience	
W = 0.037		W = 0.088		W = 0.014		W = 0.029		W = 0.108	
Values	%	Values	%	Values	%	Values	%	Values	%
CLA1 (187)									
2–7	56	4–9	40	2–8	34	2–8	50	2–8	41
8	21	10	27	9–10	36	9–10	36	9–10	43
9	13	11–12	21	11–12	20	11	6	11	7
10–14	11	13–14	12	13–14	10	12–14	7	12–14	9
CLA2 (323)									
2–7	29	4–9	32	2–8	24	2–8	25	2–8	27
8	21	10	20	9–10	29	9–10	35	9–10	35
9	16	11–12	30	11–12	29	11	10	11	15
10–14	33	13–14	19	13–14	19	12–14	30	12–14	24

4 Discussion

Our aim was identify the characteristics of abstemious Ecuadorian college; in our study has been identified two types of abstemious, and the principal variables to discriminate are life satisfaction, loneliness and psychological inflexibility, consistency with other studies that report the importance of life satisfaction [10–14], loneliness [15–17] and psychological inflexibility [3, 18] to predict the substance consume.

Two types of abstemious were identified: (1) First cluster (CLA1) characterized by low life satisfaction, high loneliness and psychological inflexibility. (2) Second cluster (CLA2) characterized by high life satisfaction, good social support and low psychological inflexibility. The second cluster is theoretical explain by the scientific evidence, however the first cluster maybe is a risk group, and others individuals or family variables can explain its abstention from substance use.

5 Conclusions

Considering five psychological questionnaires and a sample of 510 abstemious college students from three Ecuadorian universities, two clusters were obtained. The variables: satisfaction with life, loneliness, and avoidance and action permit to describe the groups. Clusters CLA1 with 187 students is a group that has feelings of loneliness, they are impulsive and have the feeling that their life needs to improve. CLA1 is a group that could need attention to avoid future consumption of drugs and alcohol. The biggest cluster CLA2 with 323 students represent a group that do not feel loneliness, they are less impulsive and satisfied with their lives.

Acknowledgement. This research was carried out with the funds of the project "CEPRA XII-2018-05 Prediction of Drug Consumption", winner of the CEPRA contest of CEDIA-Ecuador. The researchers thank CEDIA for their contribution in the development management of the project.

References

1. Oficina de las Naciones Unidas contra la Droga y el Delito (UNODC): World Drug Report 2019. UNODC Research, Lima - Perú (2019)
2. Oficina de las Naciones Unidas contra la Droga y el Delito (UNODC): III Estudio epidemiológico andino sobre consumo de drogas en la población universitaria, Informe Regional, 2016. Oficina de las Naciones Unidas Contra la Droga y el Delito (UNODC), Lima - Perú (2017)
3. Ruisoto, P., Cacho, R., López-Goñi, J.J., Vaca, S., Jiménez, M.: Prevalence and profile of alcohol consumption among university students in Ecuador. Gac. Sanit. **30**, 370–374 (2016)
4. López, V., Paladines, B., Vaca, S., Cacho, R., Fernández-Montalvo, J., Ruisoto, P.: Psychometric properties and factor structure of an Ecuadorian version of the Alcohol Use Disorders Identification Test (AUDIT) in college students. PLoS ONE **14**, e0219618 (2019)
5. Armendáriz García, N.A., Villar Luis, M.A., Alonso Castillo, M.M., Alonso Castillo, B.A., Oliva Rodríguez, N.N.: Eventos estresantes y su relación con el consumo de alcohol en estudiantes universitarios. Investig. en enfermería Imagen y Desarro. **14**, 97–112 (2012)
6. Velten, J., et al.: Lifestyle choices and mental health: a representative population survey. BMC Psychol. **2**, 58 (2014)
7. Paljärvi, T., Koskenvuo, M., Poikolainen, K., Kauhanen, J., Sillanmäki, L., Mäkelä, P.: Binge drinking and depressive symptoms: a 5-year population-based cohort study. Addiction **104**, 1168–1178 (2009)
8. Churchill, S.A., Farrell, L.: Alcohol and depression: evidence from the 2014 health survey for England. Drug Alcohol Depend. **180**, 86–92 (2017)
9. Boden, J.M., Fergusson, D.M.: Alcohol and depression. Addiction **106**, 906–914 (2011)
10. Diulio, A.R., Cero, I., Witte, T.K., Correia, C.J.: Alcohol-related problems and life satisfaction predict motivation to change among mandated college students. Addict. Behav. **39**, 811–817 (2014)

11. Lew, D., Xian, H., Qian, Z., Vaughn, M.G.: Examining the relationships between life satisfaction and alcohol, tobacco and marijuana use among school-aged children. J. Public Health (Bangkok) (2018)
12. Tartaglia, S., Gattino, S., Fedi, A.: Life satisfaction and alcohol consumption among young adults at social gatherings. J. Happiness Stud. **19**, 2023–2034 (2018)
13. Koivumaa-Honkanen, H., Kaprio, J., Korhonen, T., Honkanen, R.J., Heikkilä, K., Koskenvuo, M.: Self-reported life satisfaction and alcohol use: a 15-year follow-up of healthy adult twins. Alcohol Alcohol. **47**, 160–168 (2012)
14. Masferrer, L., Font-Mayolas, S., Pérez, M.E.G.: Satisfacción con la vida y consumo de sustancias psicoactivas en la adolescencia. Cuad. Med. psicosomática y Psiquiatr. enlace. **6** (2012)
15. Yang, Y.-J., Xu, Y.-M., Chen, W.-C., Zhu, J.-H., Lu, J., Zhong, B.-L.: Loneliness and its impact on quality of life in Chinese heroin-dependent patients receiving methadone maintenance treatment. Oncotarget **8**, 79803 (2017)
16. Mannes, Z.L., Burrell, L.E., Bryant, V.E., Dunne, E.M., Hearn, L.E., Whitehead, N.E.: Loneliness and substance use: the influence of gender among HIV+ Black/African American adults 50+. AIDS Care **28**, 598–602 (2016)
17. Segrin, C., Pavlich, C.A., McNelis, M.: Transitional instability predicts polymorphous distress in emerging adults. J. Psychol. **151**, 496–506 (2017)
18. Levin, M.E., Lillis, J., Seeley, J., Hayes, S.C., Pistorello, J., Biglan, A.: Exploring the relationship between experiential avoidance, alcohol use disorders, and alcohol-related problems among first-year college students. J. Am. Coll. Heal. **60**, 443–448 (2012)
19. Gallego, S., Mezquita, L., Moya-Higueras, J., Ortet, G., Ibáñez, M.I.: Contribution of the five factors of personality and peers on adolescent alcohol use: a cross-national study. Span. J. Psychol. **21** (2018)
20. Ibáñez, M.I., Camacho, L., Mezquita, L., Villa, H., Moya-Higueras, J., Ortet, G.: Alcohol expectancies mediate and moderate the associations between Big Five personality traits and adolescent alcohol consumption and alcohol-related problems. Front. Psychol. **6**, 1838 (2015)
21. Coëffec, A.: Les apports du modèle des cinq grands facteurs dans le domaine de l'alcoolodépendance. Encephale **37**, 75–82 (2011)
22. Verdejo, A., López-Torrecillas, F., Orozco, C., Perez, M.: Impact of the neuropsychological impairments associated to substance abuse on clinical practice with drug addicts. Adicciones **14** (2002)
23. Rogers, R.D., Robbins, T.W.: Investigating the neurocognitive deficits associated with chronic drug misuse. Curr. Opin. Neurobiol. **11**, 250–257 (2001)
24. Kurniawan, R., Jamal, K., Nur, A., Nazri, M.Z.A., Kholilah, D.: Advise-giving expert systems based on Islamic jurisprudence for treating drugs and substance abuse. J. Theor. Appl. Inf. Technol. **96**, 4941–4952 (2018)
25. Maione, C., et al.: Establishing chemical profiling for ecstasy tablets based on trace element levels and support vector machine. Neural Comput. Appl. **30**, 947–955 (2018)
26. Srividya, M., Mohanavalli, S., Bhalaji, N.: Behavioral modeling for mental health using machine learning algorithms. J. Med. Syst. **42** (2018)
27. Ali, F., et al.: Type-2 fuzzy ontology–aided recommendation systems for IoT–based healthcare. Comput. Commun. **119**, 138–155 (2018)
28. Alsharif, A.H., Philip, N.Y.: A framework for smoking cessation in the Kingdom of Saudi Arabia using smart mobile phone technologies (Smoke Mind) (2015)
29. Chen, T., et al.: Are you smoking? Automatic alert system helping people keep away from cigarettes. Smart Heal. **9–10**, 158–169 (2018)

30. Glowacki, E.M., Lazard, A.J., Wilcox, G.B.: E-cigarette topics shared by medical professionals: a comparison of tweets from the United States and United Kingdom. Cyberpsychol. Behav. Soc. Netw. **20**, 133–137 (2017)
31. Valencia, S., Smith, M.V., Atyabi, A., Shic, F.: Mobile ascertainment of smoking status through breath: a machine learning approach (2016)
32. Li, W., Cui, X., Amaral, K.M., Sadasivam, R., Chen, P.: Smoking cessation recruitment analysis: a case study (2017)
33. Wang, X., et al.: Mining user-generated content in an online smoking cessation community to identify smoking status: a machine learning approach. Decis. Support Syst. **116**, 26–34 (2019)
34. Qian, Y., Li, B., Yao, Z., Lv, H., Che, M., Cheng, Z.: Mining user-generated content to identify social support in Chinese online smoking cessation community (2019)
35. Jiajie, L., Narasimhan, K., Elamaran, V., Arunkumar, N., Solarte, M., Ramirez-Gonzalez, G.: Clinical decision support system for alcoholism detection using the analysis of EEG signals. IEEE Access **6**, 61457–61461 (2018)
36. Davagdorj, K., Yu, S.H., Kim, S.Y., Van Huy, P., Park, J.H., Ryu, K.H.: Prediction of 6 months smoking cessation program among women in Korea. Int. J. Mach. Learn. Comput. **9**, 83–90 (2019)
37. Sridharan, V., Cohen, T., Cobb, N., Myneni, S.: The portrayal of quit emotions: content-sensitive analysis of peer interactions in an online community for smoking cessation (2018)
38. Tahmassebi, A., et al.: An evolutionary approach for fMRI big data classification (2017)
39. Torres-Carrión, P.V., Reátigui, R., Valdiviezo, P., Bustamante-Granda, B., Vaca-Gallegos, S.: Application of techniques based on Artificial Intelligence for predicting the consumption of drugs and substances. A Systematic Mapping Review. In: International Conference on Applied Technologies ICAT 2019, Quito, Ecuador (2019)
40. Kroenke, K., Spitzer, R.L., Williams, J.B.W.: The PHQ-9: validity of a brief depression severity measure. J. Gen. Intern. Med. **16**, 606–613 (2001)
41. Ruiz, F.J., Luciano, C., Cangas, A.J., Beltrán, I.: Measuring experiential avoidance and psychological inflexibility: the Spanish version of the Acceptance and Action Questionnaire-II. Psicothema **25**, 123–129 (2013)
42. Hughes, M.E., Waite, L.J., Hawkley, L.C., Cacioppo, J.T.: A short scale for measuring loneliness in large surveys: results from two population-based studies. Res. Aging **26**, 655–672 (2004)
43. Moyano, N.C., Tais, M.M., Muñoz, M.P.: Propiedades psicométricas de la Escala de Satisfacción con la Vida de Diener. Rev. Argentina Clínica Psicológica **22**, 161–168 (2013)
44. Steinberg, L., Sharp, C., Stanford, M.S., Tharp, A.T.: New tricks for an old measure: the development of the Barratt Impulsiveness Scale-Brief (BIS-Brief). Psychol. Assess. **25**, 216 (2013)
45. Cohen, S., Kamarck, T., Mermelstein, R.: A global measure of perceived stress. J. Health Soc. Behav. 385–396 (1983)
46. Renau, V., Oberst, U., Gosling, S., Rusiñol, J., Chamarro, A.: Translation and validation of the ten-item-personality inventory into Spanish and Catalan. Aloma Rev. Psicol. Ciències l'Educació i l'Esport. **31** (2013)
47. Tibshirani, R., Walther, G., Hastie, T.: Estimating the number of clusters in a data set via the gap statistic. J. R. Stat. Soc. Ser. B (Stat. Methodol.) **63**, 411–423 (2001)
48. Witten, D.M., Tibshirani, R.: A framework for feature selection in clustering. J. Am. Stat. Assoc. **105**, 713–726 (2010)

Toward Automatic and Remote Monitoring of the Pain Experience: An Internet of Things (IoT) Approach

Juan José Rodríguez Rodríguez⬡, Javier Ferney Castillo García⬡, and Erick Javier Argüello Prada(✉)⬡

Facultad de Ingeniería, Universidad Santiago de Cali (USC), Cali, Colombia
erick.arguello00@usc.edu.co

Abstract. Automatic and remote monitoring of patients with pain may decrease treatment costs and improve the quality of care. In this sense, the Internet of Things (IoT) emerges as a suitable candidate for developing solutions enabling continuous and remote assessment of pain experience. However, only a few efforts have been devoted to adopting IoT-based solutions for pain assessment. In the present work, an IoT-based system for pain monitoring is proposed on the basis of a performance assessment of several communication protocols for IoT: TCP/IPv4, TCP/IPv6, UDP, MQTT, and HTTP. The peripheral blood flow and the skin's ability to conduct electricity were chosen as the physiological parameters through which it is possible to measure pain. The capabilities of the aforementioned IoT communication protocols for transmitting the physiological data stream to a cloud server were evaluated by implementing each of those protocols and using the Wireshark protocol analyzer to compute the mean byte rate, the mean packet rate, the mean error value, and the network reliability for 1 h. Results show that the TCP/IPv4 and TCP/IPv6 protocols showed the highest packet transmission rate as well as the highest network reliability. Moreover, given the characteristics of the chosen physiological parameters, the proposed solution does not require a high transmission data rate, so there would be no limitation regarding the wireless communication protocol that could be used for implementing it. Nevertheless, a wider range of parameters needs to be considered in order to carry out a more rigorous performance assessment.

Keywords: Internet of Things (IoT) · Pain assessment · Cloud server · Photoplethysmography (PPG) · m-health

1 Introduction

Pain is one of the most (if not the most) common reasons for seeking medical attention and it continues to be a clinical, economic and social problem [1]. Most of the pain assessment tools used in clinical practice mainly focus on the patient's self-report. However, all these methods rely on the patient's ability to inform the magnitude of perceived pain, which could not be possible (or reliable) under certain circumstances [2]. Furthermore, pain is usually evaluated at a specific time of the day (i.e., at the medical appointment) [3], which means that the information about how the patient feels

© Springer Nature Switzerland AG 2020
M. Botto-Tobar et al. (Eds.): ICAT 2019, CCIS 1194, pp. 194–206, 2020.
https://doi.org/10.1007/978-3-030-42520-3_16

during his/her daily routine or how other factors may affect pain intensity might be ignored. Delayed access to medical care and prolonged waiting time may also aggravate patients' suffering as well as increase treatment costs.

The continuous and remote monitoring of patients as well as communication between professionals, relatives, and patients will especially benefit from the use of mobile devices. This leads to the appearance of a new concept that is redefining healthcare services: mobile health (m-health). Likewise, recent advances in sensing and processing technologies have led to the emergence of new technological paradigms, such as the Internet of Things (IoT). The IoT consists in to allow multiple devices with unique identities to exchange information in order to provide customized services for process automation and remote monitoring [4]. Based on the advantages that IoT is able to provide to healthcare, some efforts have been devoted to adopting IoT-based solutions for automatic, real-time and remote monitoring of pain experience. In [5], for instance, an IoT-based system to evaluate the pain experienced by infants through their facial expressions is presented. The proposed solution automatically captures video of facial expressions of the infants and streams the video to the remote pain center, where healthcare professionals perform the pain assessment. More recently, a wearable sensing device capable of simultaneously collecting surface electromyographic signals (sEMG) from several facial muscles was integrated into an IoT system for remote pain monitoring [6]. The system adopts a sensor-gateway-cloud architecture, thereby enabling scalability and real-time visualization. Nevertheless, the authors do not explain the rationale for the use of one or another communication protocol, and while others [7–10] have conducted performance evaluations to compare between different protocols, none of them used their results to suggest which protocol is the most suitable for implementing a pain assessment system based on IoT. Therefore, an IoT-based system for pain monitoring is proposed herein on the basis of a performance assessment of several communication protocols for IoT: TCP/IPv4, TCP/IPv6, UDP, MQTT and HTTP. The main contributions of this work are summarized as follows: (i) developing a low-power and miniaturized circuit for capturing patients' peripheral blood flow and skin conductance; (ii) evaluating the capabilities of several IoT communication protocols for transmitting the physiological data stream to a cloud server; and (iii) proposing an IoT-based solution for pain monitoring that involves a cloud-based platform supporting patients who have access to Wi-Fi connection at home and healthcare professionals using a computer or a smart device with web browsing feature.

2 Materials and Methods

2.1 Physiological Parameters

Given the effects of pain on the ANS, as well as the increasing number of devices for monitoring physiological variables that are available in healthcare facilities, the observation of vital signs can be considered as a quick, simple and objective method for pain assessment [11, 12]. The aforementioned IoT-based solutions for pain assessment [5, 6] use the subject's facial expression as a pain indicator, which might be cumbersome and computationally resource-consuming. As an alternative, this study proposes to measure both the changes in peripheral blood flow and the skin's ability to conduct electricity for pain assessment.

PPG Amplitude (PPGA). Photoplethysmography (PPG) has been widely used to assess HR and vasomotor tone, which have been found to be affected by pain experience. Therefore, PPG and some parameters derived from PPG signals (e.g., the PPG signal's amplitude - PPGA) could potentially be used for pain assessment. For instance, several studies have found that experimentally induced pain [13], as well as the nociceptive response during general anesthesia [14], produces changes in PPGA.

Heart Rate (HR). Several studies have reported an increase in heart rate (HR) as a response to pain [15, 16]. Likewise, HR has been found to decrease after relieving pain in adult patients receiving assisted ventilation [17]. In summary, HR promises to be an important tool in revealing crucial information for pain assessment, provided that it is used in conjunction with other physiological variables [18].

Pulse Rate Variability-High Frequency (PRV-HF) Power. The term heart rate variability (HRV) is used to describe fluctuations in the inter-beat intervals [19] also known as RR intervals. These oscillations reflect how the ANS modulates cardiac activity and it can provide relevant information about the patient's condition. It has been shown that the PRV can be used as a surrogate measurement of HRV, not only during stationary conditions [20, 21] but also under non-stationary conditions [22], although further research is needed to extend its use in clinical practice [23]. Regarding the effect of pain on the HRV, several authors [18, 24] have pointed out that experimentally induced pain in healthy adults produces an increase in the low frequency (LF: 0.04–0.15 Hz) spectral component of the HRV and a decrease in the HF component.

Skin Conductance Response (SCR). Electrodermal activity refers to all those electrical phenomena occurring at the skin level, including the alterations in its ability to conduct electricity [25]. As reported by some studies [26, 27], the electrical conductivity of the skin may increase in response to pain evoked by thermal or mechanical stimulation, a phenomenon that is also referred to as Skin Conductance Response (SCR). Other authors [28] point out that, in comparison with HR, SCR has a greater correlation with the reports provided by the patient in relation to the intensity of pain he/she experiences. Moreover, since SCR is not affected by respiratory rate, changes in blood volume or the action of numerous medications, this physiological variable acts as a much more robust pain indicator than those derived from cardiac activity.

2.2   Network Protocols

IEEE 802.11 Standards. The set of standards IEEE 802.11 enables wireless local area network (WLAN) computer communication in the 2.4, 3.6 and 5 GHz frequency bands [29]. Since 802.11b and 802.11g use the 2.4 GHz industrial, scientific, and medical radio band, such systems may occasionally suffer intervention from cordless telephony, microwave ovens, and Bluetooth devices. Nowadays, by using direct-sequence spread spectrum (DSSS) and orthogonal frequency-division multiplexing (OFDM) [30] signaling methods, 802.11b and 802.11g control, respectively, its vulnerability and interference.

Transmission Control Protocol (TCP). Amongst the wide range of Internet protocols, the Transmission Control Protocol (TCP) is the one used by major Internet applications such as World Wide Web, remote administration, and file transfer [31]. Due to its connection-oriented nature, TCP enables reliable and ordered delivery of a byte stream from one computer to another.

User Datagram Protocol (UDP). The User Datagram Protocol (UDP) is one of the foundation protocol members of the Transmission Control Protocol & Internet Protocol (TCP/IP) Suite. Unlike TCP, UDP is not a connection-oriented protocol. UDP sends Datagrams (i.e., Packets) to other hosts on an Internet Protocol (IP) network without the need for earlier communications to set up special transmission channels or data paths. Thus, UDP provides an unreliable service and datagrams may arrive out of order, appear duplicated, or go missing without notice [32].

Hyper-Text Transfer Protocol (HTTP). The Hyper-Text Transfer Protocol (HTTP) protocol is a light-weight text-based standard consisting of requests and responses, developed to allow users to interact with documents described by a URI. Its initial version only included the GET method [33], and described HTTP requests as idempotent (i.e., they can be called many times without different outcomes). The current version of HTTP states that the methods GET, HEAD, PUT, and DELETE should be idempotent, and have no side-effects.

Message Queuing Telemetry Transport (MQTT). The Message Queuing Telemetry Transport (MQTT) protocol is an open, and easy to implement messaging protocol and it works on default TCP/IP with port 1883 [34]. On the other hand, MQTT can only be used for lightweight implementations because it cannot maintain active sessions and the amount of information per message is very limited.

2.3 Data Acquisition and Preprocessing

The electric diagram of the circuit used for this study is shown in Fig. 1. Whereas the SCR was acquired through two reusable Ag-AgCl electrodes, peripheral blood volume changes were collected through a NellcorTM adult finger clip (model DS-100A). A 2N3904 NPN transistor supplies the infrared light-emitting diode (IR-LED) driving current, which is controlled by the potentiometer R2. The signal from the phototransistor is filtered and amplified by a single pole, band-pass active filter (0.7–2.34 Hz) with a gain of 100. The output of the first stage is regulated by the potentiometer R7 and additionally amplified by 10. Both signals are sampled at 100 Hz and digitalized by the 8-channel analogue-to-digital converter (12-bit of resolution) of an ESP32 microcontroller (GPIO, pin33), where PPG systolic peaks and pulse onsets (troughs) are detected by the peak detection algorithm peak detection method proposed in [35].

The whole circuit was powered with a rechargeable Ni-MH battery (3.6 V/1000 mAh, Huawei) and integrated into a single printed circuit board (PCB) using high-quality components, in order to reduce the signal contamination as much as possible.

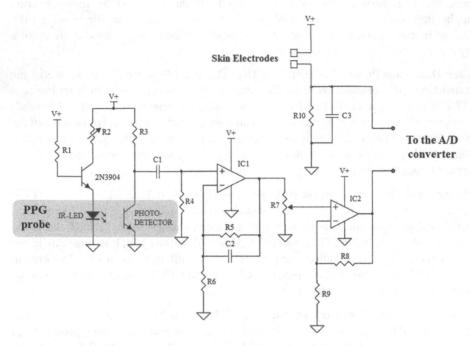

Fig. 1. Diagram of the circuit for acquisition and conditioning of PPG and SCR signals.

2.4 Data Transmission and Protocol Implementations

The subsequent paragraphs are intended to briefly outline the technical specifications for the implementation of the TCP/IPv4, TCP/IPv6, UDP, MQTT, and HTTP protocols from both the client- and the server-side.

TCP/IPv4, TCP/IPv6, and UDP Protocols. For these protocols, the client was implemented in Arduino environment and installed on the ESP32 microcontroller. The WiFi and WiFiUdp libraries provided a client connection which was established through the host port 56200. Peak and trough values of the PPG signal, as well as their temporal locations, and the SCR signal were wirelessly transmitted every 10 s. The data frame is depicted in Eq. (1) as follows:

$$ID0;\ value0, \ldots, valueN;\ ID1;\ value0, \ldots, valueN; \ldots; IDN;\ value0, \ldots, valueN \quad (1)$$

Data indexing was performed using the capabilities of these protocols to maintain an active session. Data transmission from the client-side is depicted in Fig. 2.

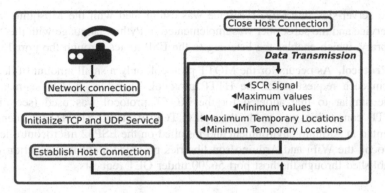

Fig. 2. Data transmission diagram for the TCP/IPv4, TCP/IPv6 and UDP protocols implemented in the ESP32 microcontroller.

At the server-side, the TCP/IPv4, TCP/IPv6, and UDP services were implemented in Python programming language (v3.2.3). The low-level network interface library provided the tools to establish a socket service under the TCP and UDP protocols. The listening port was established as 56200.

MQTT Protocol. As done for TCP/IPv4, TCP/IPv6, and UDP protocols, the client was implemented in Arduino environment and installed on the ESP32 microcontroller. The WiFi, PubSubClient and ArduinoJson libraries provided a client connection which was established through the port 1883. Peak and trough values of the PPG signal, as well as their temporal locations, and the SCR signal were wirelessly transmitted every 10 s. However, since the MQTT protocol limits severely the transmitted information volume, it was necessary to subdivide the data frame by using a data exchange format (JavaScript Object Notation - Json). Figure 3 shows the transmission scheme for the MQTT protocol as well as the structure of the Json data format.

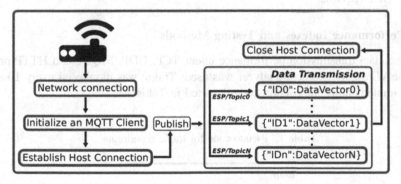

Fig. 3. Data transmission diagram for the MQTT protocol implemented in the ESP32 microcontroller.

At the server-side, the MQTT service was established with the Mosquito MQTT broker service and the subscriber was implemented in Python language with the "paho-mqtt" library, which enables the listening to the ESP topics through the port 1883.

HTTP Protocol. As occurs for the MQTT protocol, only a small amount of data can be sent in each request under the HTTP protocol. Therefore, a data segmentation technique similar to that adopted for the MQTT protocol was used (see Fig. 4). The HTTP connection was made through a TCP/IP client. The HTTP client was implemented in Arduino environment and installed on the ESP32 microcontroller. For this protocol, the WiFi and ArduinoJson libraries provided a client connection which was established through the host port 56200 under GET requests.

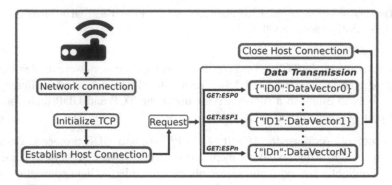

Fig. 4. Data transmission diagram for the HTTP protocol implemented in the ESP32 microcontroller.

At the server-side, the HTTP service was established in Python programming language with the Flask framework, which allows users to create Web Server Gateway Interface (WSGI) applications. The service was configured to listen on port 56200 and retrieve GET requests.

2.5 Performance Indexes and Testing Methods

To assess data transmission performance under TCP, UDP, MQTT and HTTP protocols, the Wireshark protocol analyzer was used. Traffic was monitored every 10 s for 1 h by implementing the pseudo code depicted in Table 1.

Table 1. Pseudo code for traffic monitoring.

loop
readLine[i].txt
AllPacket/seg > 0
Data[i]=save.readLine[i].txt
end loop

The performance indexes used in this study were the mean byte rate, the mean packet rate, the mean error value, and the network reliability, defined as follows in Eqs. (2) through (5):

$$meanBytes_{seg} = mean.(bytes[Data.Bytes[i], \ldots, Data.Bytes[i]]) \qquad (2)$$

$$meanAllPacket_{seg} = mean.(AllPacket[Data.AllPacket[i], \ldots, Data.AllPacket[i]]) \quad (3)$$

$$meanErrors_{seg} = mean.(Errors[Data.Errors[i], \ldots, Data.Errors[i]]) \qquad (4)$$

$$NetworkReliability = 100*(1 - (meanErrors/meanAllPackets)) \qquad (5)$$

The analysis was run on an Intel(R) Core(TM) i5-4210U CPU @ 1.70 GHz, con OS Linux 4.15.0-65-generic. Traffic was captured using the dumpcap application of the Wireshark package (Git v2.6.10 packaged as 2.6.10-1~ubuntu16.04.0). The WiFi controller was set at a CPU rate of 240 MHz. Each transaction was composed of 20 attributes which corresponded to the peak and trough values of the PPG signal, as well as their temporal locations, and the SCR signal. Tests were performed between 12:00 a. m. and 1:00 a.m from a local home network.

3 Results and Discussion

One of the purposes of the present work is to propose an IoT-based system for automatic and remote monitoring of the pain experience. To this end, it is crucial to ensure that pain-related data is correctly transmitted. Each IoT communication protocol has particular characteristics regarding data representation and indexing, transmission speed, and power consumption, so it is important to evaluate the capabilities of the proposed system for data transmission when different communication protocols are implemented.

Table 2 summarizes the results achieved by the proposed system when five different IoT communication protocols (TCP/IPv4, TCP/IPv6, UDP, MQTT, and HTTP) were used for data transmission. The TCP/IPv4 and TCP/IPv6 protocols showed the highest packet transmission rate as well as the highest network reliability. On the other hand, although the UDP protocol showed the lowest packet transmission rate, it is able to maintain the connection open during the transmission of the whole data frame. Unlike UDP and HTTP protocols, the MQTT protocol is unable to do that, so it was necessary to subdivide the data frame by using a data exchange format. This issue was also observed for the HTTP protocol, mainly due to the limitations of the buffer capacity. In the absence of mechanisms for detecting frame errors, this can potentially cause data corruption which in turn might cause a serious impact on the analysis and treatment of painful conditions.

Table 2. Results achieved by the proposed system when five different IoT communication protocols (TCP/IPv4, TCP/IPv6, UDP, MQTT, and HTTP) were implemented.

	TCP/IPv4	TCP/IPv6	UDP	MQTT	HTTP
Mean bytes/s	22105.92	23471.77	334.21	214.76	281.89
Mean all packet/s	44.79	47.73	1.13	2.63	3.88
Mean errors/s	0.24	0.14	0.10	0.02	1.88
Network reliability	99.46	99.70	91.18	94.79	51.49

The selection of the most appropriate communication protocol for implementing a pain assessment system based on IoT depends on several parameters, among which the mean packet transmission rate and the percentage of error in data transmitting play a crucial role [36]. The relation between the mean packet transmission rate and the mean error rate is represented in Fig. 5. All the aforementioned protocols are characterized by an average loss of information less than 4% of the total amount of transmitted information. Figure 6 shows the values of the mean packet transmission rate and the mean error rate achieved by the proposed system during 24 h of a typical work day implemented on a home network that supports the TCP/IPv6 protocol. As can be seen, even in the range of hours within which users are continually transmitting large volumes of data (e.g., 4:00 p.m. and 9:00 p.m.), both the packet transmission rate and the percentage of error in data transmitting are kept within ranges that allow the transactions between the data acquisition module and the data transmission system of the proposed IoT-based solution.

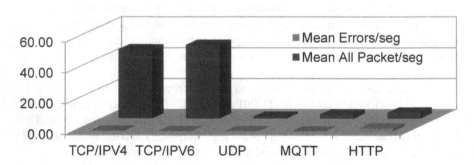

Fig. 5. The comparative between the mean packet transmission rate and the mean error rate for five different IoT communication protocols (TCP/IPv4, TCP/IPv6, UDP, MQTT, and HTTP).

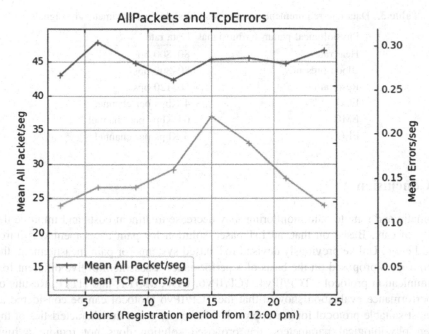

Fig. 6. Packets and error rates for a period of 24 h under the TCP/IPV6 transport layer protocol.

It is worth mentioning that the results of the performance assessment conducted in this study are significant only within the context of the physiological parameters that have been chosen for pain measurement. Previously proposed IoT-based systems for pain assessment [5, 6] use the patient's facial expression as a pain indicator, which involves a high data transmission rate and large storage capacity, as shown in Table 3 [37]. Instead, the present work proposes to measure both the peripheral blood flow and the skin's ability to conduct electricity, two physiological parameters whose frequency content does not extend beyond some tens of Hz unlike, for instance, electromyographic (EMG) signals [38]. Furthermore, it has been demonstrated that analysis of variability in the time and frequency domain from pulse rate obtained through PPG signals may be as reliable as that derived from the analysis of the electrocardiogram (ECG), provided that a sampling rate greater than 25 Hz is used [39]. All this suggests that the proposed system does not require high data transmission rates and, therefore, there is virtually no limitation regarding the wireless communication protocol that could be used for implementing the proposed system.

Table 3. Data rate requirements for various physiological parameters/biosignals.

Physiological parameter/biosignal	Data rate
Hear rate	80–800 bps
Blood pressure	80–800 bps
Respiration	50–120 bps
ECG	4 Kbps per channel
EMG	64 Kbps per channel
EEG	3 Kbps per channel

4 Conclusion

Automatic and remote pain monitoring may decrease treatment costs and improve the quality of care. Based on that, an IoT-based solution for pain assessment was introduced herein. Unlike previously devised IoT-based systems for pain measurement, the current one is proposed on the basis of a performance assessment of five different IoT communication protocols: TCP/IPv4, TCP/IPv6, UDP, MQTT, and HTTP. Results of the performance evaluation showed that the TCP/IPv6 protocol can be considered as the most suitable protocol for our approach. Moreover, given the characteristics of the chosen physiological parameters, the proposed solution does not require a high transmission data rate, so there would be no limitation regarding the wireless communication protocol that could be used for implementing it. Further work should aim to include a performance assessment in terms of a wider range of parameters (e.g., the network's latency), in order to conduct a more rigorous validation.

References

1. Duke, G., Haas, B.K., Yarbrough, S., Northam, S.: Pain management knowledge and attitudes of baccalaureate nursing students and faculty. Pain Manag. Nurs. **14**(1), 11–19 (2013)
2. Barr, J., et al.: Clinical practice guidelines for the management of pain, agitation, and delirium in adult patients in the in-tensive care unit. Crit. Care Med. **41**(1), 263–306 (2013)
3. Chhikara, A., Rice, A.S.C., McGregor, A.H., Bello, F.: In-house monitoring of low back pain related disability (impaired). In: 2008 30th Annual International Conference of the IEEE Engineering in Medicine and Biology Society, pp. 4507–4510. IEEE (2008)
4. Silva, B.N., Khan, M., Han, K.: Internet of Things: a comprehensive review of enabling technologies, architecture, and challenges. IETE Tech. Rev. **35**(2), 205–220 (2018)
5. Zhong, Y., Liu, L.: Remote neonatal pain assessment system based on Internet of Things. In: 2011 International Conference on Internet of Things and 4th International Conference on Cyber, Physical and Social Computing, pp. 629–633. IEEE, Dalian (2011)
6. Yang, G., et al.: IoT-based remote pain monitoring system: from device to cloud platform. IEEE J. Biomed. Health Inf. **22**(6), 1711–1719 (2017)
7. Gündoğan, C., Kietzmann, P., Lenders, M., Petersen, H., Schmidt, T.C., Wählisch, M.: NDN, CoAP, and MQTT: a comparative measurement study in the IoT. In: Proceedings of ACM ICN. ACM (2018)

8. Yokotani, T., Sasaki, Y.: Comparison with HTTP and MQTT on required network resources for IoT. In: 2016 International Conference on Control, Electronics, Renewable Energy and Communications (ICCEREC), pp. 1–6. IEEE (2016)
9. Yokotani, T., Sasaki, Y.: Transfer protocols of tiny data blocks in IoT and their performance evaluation. In: 2016 IEEE 3rd World Forum on Internet of Things (WF-IoT), pp. 54–57. IEEE (2016)
10. Gosai, A.M., Goswami, B.H.: Experimental performance testing of TCP and UDP protocol over WLAN standards, 802.11b and 802.11g. Karpagam J. Comput. Sci. 7(3), 168–183 (2013)
11. Jaenig, W., Baron, R.: Sympathetic nervous system and pain. In: Schmidth, R.F., Gebhart, G.F. (eds.) Encyclopedia of Pain, 2nd edn, pp. 3763–3779. Springer, Berlin (2013). https://doi.org/10.1007/978-3-642-28753-4_4327
12. Schlereth, T., Birklein, F.: The sympathetic nervous system and pain. NeuroMol. Med. 10(3), 141–147 (2008)
13. Shelley, K.H.: Photoplethysmography: beyond the calculation of arterial oxygen saturation and heart rate. Anesth. Analg. 105(6), S31–S36 (2007)
14. Magerl, W., Geldner, G., Handwerker, H.O.: Pain and vascular reflexes in man elicited by prolonged noxious mechano-stimulation. Pain 43(2), 219–225 (1990)
15. Gélinas, C., Johnston, C.: Pain assessment in the critically ill ventilated adult: validation of the Critical-Care Pain Observation Tool and physiologic indicators. Clin. J. Pain 23(6), 497–505 (2007)
16. Terkelsen, A.J., Mølgaard, H., Hansen, J., Andersen, O.K., Jensen, T.S.: Acute pain increases heart rate: differential mechanisms during rest and mental stress. Auton. Neurosci. 121(1–2), 101–109 (2005)
17. Gélinas, C., Arbour, C.: Behavioral and physiologic indicators during a nociceptive procedure in conscious and unconscious mechanically ventilated adults: similar or different? J. Crit. Care 24(4), 628-e7 (2009)
18. Treister, R., Kliger, M., Zuckerman, G., Aryeh, I.G., Eisenberg, E.: Differentiating between heat pain intensities: the combined effect of multiple autonomic parameters. Pain 153(9), 1807–1814 (2012)
19. Kleiger, R.E., Stein, P.K., Bigger Jr., J.T.: Heart rate variability: measurement and clinical utility. Ann. Noninvasive Electrocardiol. 10(1), 88–101 (2005)
20. Lu, S., et al.: Can photoplethysmography variability serve as an alternative approach to obtain heart rate variability information? J. Clin. Monit. Comput. 22(1), 23–29 (2008)
21. Lu, G., Yang, F., Taylor, J.A., Stein, J.F.: A comparison of photoplethysmography and ECG recording to analyse heart rate variability in healthy subjects. J. Med. Eng. Technol. 33(8), 634–641 (2009)
22. Gil, E., Orini, M., Bailon, R., Vergara, J.M., Mainardi, L., Laguna, P.: Photoplethysmography pulse rate variability as a surrogate measurement of heart rate variability during nonstationary conditions. Physiol. Measur. 31(9), 1271 (2010)
23. Georgiou, K., Larentzakis, A.V., Khamis, N.N., Alsuhaibani, G.I., Alaska, Y.A., Giallafos, E.J.: Can wearable devices accurately measure heart rate variability? A systematic review. Folia Med. 60(1), 7–20 (2018)
24. Koenig, J., Jarczok, M.N., Ellis, R.J., Hillecke, T.K., Thayer, J.F.: Heart rate variability and experimentally induced pain in healthy adults: a systematic review. Eur. J. Pain 18(3), 301–314 (2014)
25. Boucsein, W.: Electrodermal Activity. Springer, Boston (2012). https://doi.org/10.1007/978-1-4614-1126-0

26. Loggia, M.L., Juneau, M., Bushnell, M.C.: Autonomic responses to heat pain: heart rate, skin conductance, and their relation to verbal ratings and stimulus intensity. Pain **152**(3), 592–598 (2011)
27. Breimhorst, M., Sandrock, S., Fechir, M., Hausenblas, N., Geber, C., Birklein, F.: Do intensity ratings and skin conductance responses reliably discriminate between different stimulus intensities in experimentally induced pain? J. Pain **12**(1), 61–70 (2011)
28. Storm, H.: The capability of skin conductance to monitor pain compared to other physiological pain assessment tools in children and neonates. Pediatr. Therapeutic **3**, 168 (2013)
29. Marks, R.B., Gifford, I.C., O'Hara, B.: Standards in IEEE 802 unleash the wireless Internet. IEEE Microw. Mag. **2**(2), 46–56 (2001)
30. Prasad, R.: OFDM for wireless communications systems. Universal Personal Communication (2004)
31. Vinton, G.C., Robert, E.K.: A protocol for packet network intercommunication. IEEE Trans. Commun. **22**(5), 637–648 (1974)
32. Postel, J.: User Datagram Protocol RFC 768. DDN Protocol Handbook. ISI, 2.175–2.177 (1982)
33. Berners-Lee, T.: Hyper text transfer protocol. Technical report, CERN (1992). http://www.w3.org/History/19921103-hypertext/hypertext/WWW/Protocols/HTTP.html
34. Nastase, L.: Security in the Internet of Things: a survey on application layer protocols. In: Proceedings of the 2017 21st International Conference on Control Systems and Computer Science, Bucharest, Romania, pp. 29–31 (2017)
35. Argüello-Prada, E.J.: The mountaineer's method for peak detection in photoplethysmographic signals. Revista Facultad de Ingeniería **90**, 9–17 (2019)
36. Jain, R., Durresi, A., Babic, G.: Throughput fairness index: an explanation. In: ATM Forum contribution, vol. 99, no. 45 (1999)
37. Touati, F., Tabish, R.: U-healthcare system: state-of-the-art review and challenges. J. Med. Syst. **37**(3), 9949 (2013)
38. Nagel, J.H.: Biopotential amplifiers. In: Bronzino, J.D. (ed.) The Biomedical Engineering Handbook, p. 1300. CRC Press LLC, Boca Raton (2000)
39. Choi, A., Shin, H.: Photoplethysmography sampling frequency: pilot assessment of how low can we go to analyze pulse rate variability with reliability? Physiol. Measur. **38**(3), 586 (2017)

Enterprise Architecture an Approach to the Development of IoT Services Oriented to Building Management

Maria Camargo-Vila, German Osma-Pinto$^{(\boxtimes)}$, and Homero Ortega-Boada

Universidad Industrial de Santander, Bucaramanga, Colombia
alejandravila_2232@hotmail.com, gealosma@uis.edu.co,
homero.ortega@radiogis.uis.edu.co

Abstract. Currently, various sectors including industry, health, academia and others, are facing huge challenges regarding the fast I.T. development, and more broadly, the IoT (Internet of things). The renewable energy sector is not exempt from this new paradigm: for example, an application for monitoring the photovoltaic panel system integrated with a green roof will offer the opportunity to improve resources and energy efficiency. This requires a structured approach to integrate all the elements involved in this context; and although the established technology exists, there is not a methodology to build platforms that integrate hardware and software and can interoperate effectively. Therefore, this document presents a methodology under an enterprise architecture approach, to design and implement an IoT-based service platform for monitoring a photovoltaic system integrated with vegetation. A case study is presented where the developed concept is applied.

Keywords: IoT · Building management · Photovoltaic system · Green roof · Enterprise architecture

1 Introduction

The internet of things (IoT) and its constant evolution has transformed the world in different fields such as health, industry, academia, business, science, home, among others; turning them into intelligent environments by integrating heterogeneous objects connected to the internet, with the potential to capture, analyze and share information with any other individual, be it a machine or a human. The leading network technology company CISCO estimates that by 2020 there will be 50 million connected devices [1].

A sector involved in the world of IoT is the energy with emphasis on buildings, for this study, commonly known as intelligent energy in buildings, an important research area that studies the energy efficiency of buildings, the impact on the environment and global sustainability [2]. For Colombia in 2017, the main source of energy consumption was the residential sector with 37.91% of energy consumption, accompanied by the commercial and public sector with 26.93%

© Springer Nature Switzerland AG 2020
M. Botto-Tobar et al. (Eds.): ICAT 2019, CCIS 1194, pp. 207–221, 2020.
https://doi.org/10.1007/978-3-030-42520-3_17

and the industrial sector with 21.46% [3]. For years, 80% of energy consumption has been concentrated in sectors characterized by medium and large buildings.

Consequently, the search for new sources of clean energy and energy efficiency improvement for buildings has become a vital task to counteract the negative impacts on the environment generated by humans and reach an ecological balance for the entire planet. From this perspective, the IoT combined with construction applications promote intelligent energy savings through technological tools and sensor networks for online monitoring, remote control and/or building process automation. Through using the sensing and processing of information from different systems integrated into the construction as well as the surrounding environment. The IoT in this sector is becoming stronger, thanks to the significant decrease in the cost of its sensors, its computational potential and the ease of storing large volumes of data [4].

In this scenario there are multiple systems and subsystems responsible for the supervision, control and management of the energy efficiency of the building. In most cases, these cannot be well integrated with each other, due to restrictions from the manufacturer or the technology they implement: this hinders the development of a system capable of managing them in a collaborative platform that allows their entire administration. Therefore, there are several building management systems (BSM), whose main objective is to carry out this demanding task.

On the other hand, as part of this energy saving initiative, the GRIPV (Green Roof Integrated Photovoltaics) systems, are one of the green technologies that seek to harmonize the use of green roofs in buildings with the use of photovoltaic panels (PV). These two technologies are recognized in the American LEED (Leadership in Energy & Environmental Design) building sustainability certification system. In Colombia to date, the Colombian Council for Sustainable Construction (CCCS), in partnership with the Green Business Certification Inc. (GBCI) for the LEED Program, has managed to certify 151 projects and 223 more are in this process [5].

With this work, we face the initial challenges that arise in GRIPV systems from the point of view of Information and communications technologies (ICT). The development of different research works around this type of systems requires the monitoring of electrical and environmental variables, which describe the behavior of the PV system, the green roof, the irrigation system and the building, for which it is counted with different measuring equipment, such as energy meters, dataloggers, development boards, sensors and actuators [6,7]. This diversity of equipment used for the supervision and control of the systems makes the research work complicated, since they use different software for the acquisition and storage of data.

Therefore, in this research work advances in ICT are used to support future Smart-GRIVP systems within an IoT platform that provides solutions to the need for current integration of all sensing and remote control techniques and, that serves as seed for the future ecosystem that will be required to satisfy the requirements of the academy, and the business sector, and the government, etc.

The design has been carried out with a business architecture (EA) approach with the purpose of addressing the development of IoT services from another point of view that allows guiding the transformation of systems into intelligent environments, aligning the interests of interested parties with the new IoT paradigm, in an architecture that responds to the needs of agility and, flexibility that the digital world demands today.

As is well known, the EA is a methodology that allows aligning processes, data, applications and technological infrastructure with the strategic objectives of the business. This understanding of the organization as a complex whole [8] translates into facilities for making timely decisions and the possibility of creating a transferable abstraction of the system description [9]. In this project, ToGAF was used as a reference, but in the EA, there are other frames of reference to follow as DoDAF which is used by the United States Department of Defense and the Zharman.

The platform is mainly composed of the technological and physical infrastructure layer, the application layer and the business layer. In the first layer, it is possible to solve the connectivity problems between the heterogeneous technologies available; in the application layer, storage and data processing capabilities were implemented; in the business layer applications were implemented for the final users.

The rest of the document is structured as follows: In Sect. 2 we present the literature related to this document. In Sect. 3, we describe the proposed methodology for the design and implementation of the IoT service platform. Section 4 presents the study case where the methodology will be applied, and Sect. 5 describes its implementation. Finally, the conclusions in Sect. 6 are presented and future work is outlined.

2 Related Literature

There are several approaches to the development of BSM, studies such as [10] specialized in these open type systems, mentioning existing architectures that facilitate hardware abstraction– however they do not contemplate the new IoT paradigm. Therefore they join their efforts in an architecture whose main objective is to take advantage of all the concepts of the IoT, focusing on the development of functional blocks that facilitate its implementation in three levels of the IoT systems: edge level, platform level and business level. In [11] they develop a fog computing platform under the separate design of hardware, software and communication architectures that are eventually integrated under certain conditions; but the connections between them and their components are not specified.

The platform proposed by [12] bases its architecture on a component called Enterprise Network Integrated Building System (ENIBS), consisting of reconfigurable intelligent resources (resources, sensors and actuators) that communicate via cable or wirelessly with each other, with system control and with the information management layer. They propose the development of a graphical human process interface that can be accessed from a PC, smartphones and tablets; and the creation of a virtual space for the intelligent building industry (IB), called the IB cloud, where it is possible to acquire, sell, and develop resources and knowledge of both hardware and software. This architecture is applied to a simple system, which leaves many gaps to be used in a robust one, as it does not provide details about how to develop each of the components of the architecture.

On the other hand, in [13] they present the implementation of a wireless sensor network system (WSN) in a large hotel in Mexico, which monitors construction energy consumption from different points. The sensors are integrated into the nCore platform and use the management software it provides, Polaris. This platform specializes in wireless sensors. As did the previous work, [14] presents the implementation of a wireless sensor network for parallel measurement of the internal temperature and energy consumption of a building through the Zig-Bee communication protocol. Its architecture is not completely described, but it bases its structure on the sensing and acting components and on the control and management unit.

The BBData project [15] provides a full-featured data processing platform for the Big data of buildings; with an architecture that supports sensor decoupling, data storage and application levels. They use the concept of virtual objects as a solution measure for the wide variety of sensors from different manufacturers; despite this, they do not specify how they achieve the interconnection of all these. In the case of [16] a model is presented for the control of the temperature and brightness of conferences rooms, through a master actuator system (MAS) that receives orders from the SCADA platform connected to the Matlab software through the OPC protocol. The operations are performed by distributed PLCs and the communication flow between them, and the SCADA is performed on an Ethernet bus. This system follows a vertical approach integrated with the use of routers; but even so it does not present compatibility information with other devices that can be added to the system.

The platform developed by [17] presents a layered architecture, technology layer, middleware layer, management layer and service layer. In each of them components are specified but the relations between them are not clear. In [18] they present the SOrBet (Smart Objects for Intelligent Building Management) architecture, which is made up of 6 basic layers: the physical, virtualization, automation, application, trust and configuration. Its methodology consists of the design of the functional system architecture, which consists of components responsible for the connection of the system equipment to the interfaces of the service administrator.

Finally, works [19,20] show the development of a platform with a layered architecture based on the main component, which is responsible for collecting information about different devices. These devices must meet pre-established conditions. From this component, the rest of the service is deployed. Despite the development in software and hardware matter, these works do not present a conceptual architecture that is independent of the technology to use and, that also allows the views to be homologated so anyone interested can understand it.

Each of the previously related works presents models and architectures for the design and implementation of intelligent building management systems, from which it is possible to rescue approaches and functionalities to rebuild and make improvements. Therefore, one of the shortcomings that we want to address in this work is the description of a methodology that allows the development of these types of systems under design and implementation guidelines based on the IoT reference model [21] and with a focus on enterprise architecture. This yields a graphic representation of the architecture which has one sole interpretation and serves as a central axis for decision making throughout the design, execution and continuous improvement process; exposing the relationships between and within the layers of each of its structural components. With this, it is possible to establish clear semantics among all interested parties and give true value to the proposed architecture for informed decision making.

3   Methodology of System Design

The monitoring of integrated photovoltaic systems with vegetation involves different types of electrical and climatic variables, so it is necessary to use different sensing, measurement, control and communication technologies, each of which offers various alternatives. To carry out the design and implementation of a monitoring system with a high degree of complexity, as in this case, we propose systematizing these processes under a methodology that combines the enterprise architecture approach and the IoT reference model.

Figure 1 shows an overview of the proposed methodology, where three main stages are identified: Definition of the problem context, Architecture design and Architecture implementation. These stages feed each other, in order to constantly evaluate the results of each process and to answer them with the requirements set by the interested parties, future users, designers and developers.

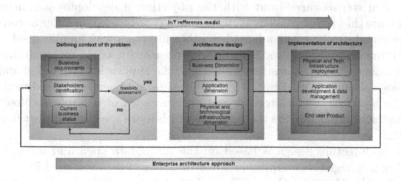

Fig. 1. Overview of the proposed methodology stages

The approach undertaken for the construction of the proposed methodology is part of the approach to enterprise architecture and each of its phases is based on the framework TOGAF, one of the most recognized and used in the world. TOGAF is utilized in this work because it focuses its efforts on activities, allowing easy understanding and a common language that can be understood by all persons associated with the business. Its ADM method allows the generic development of architectures, which can be modified and adapted to meet the needs of all stakeholders. Next, the three stages of the proposed methodology are analyzed in detail.

3.1 Defining Context of the Problem

This stage of defining the context of the problem is the fundamental basis for the design of the IoT service platform. It consists of four steps: current state of the business, where the concept to be developed and the general definition of the problem are presented based on a general description of the business, history, background, current status and future vision; stakeholders identification, that is the recognition of interested parties, future users, designers and developers in order to compromise their participation in the specification of business requirements and thereby ensure every important aspect of each of these sectors; formalized business requirements, an iterative process that refines them and represents UML use case diagrams without specifying implementation details; and finally, a feasibility assessment of the requirements formulated to determine whether the requirements, before anything else, cover the needs of stakeholders and end users. On the other hand, to assess the viability of the proposed solutions, in case there is any impediment in executing any of the requirements (whether due to the availability of resources or limitations due to external factors) it would be necessary to reformulate them to comply with the stipulations.

3.2 Architecture Design

The second step is carried out with the objective of developing a conceptual architecture that is independent of the technology to be used, in order to consider it generic and can be easily adapted to other use cases with similar needs. The architecture design is designed from an enterprise architecture approach through its three dimensions: the business dimension, the application dimension and the technological and physical infrastructure dimension. Each of these dimensions form the layers of the architecture and are composed of reusable architectural building blocks that provide the basis on which more specific architectures can be built.

The architecture design is based on the ArchiMate open and independent modeling language for enterprise architectures, one of the most used in this field and defined by The Open Group. The technical document [22] provides a general description of the language specification. Next, each of the dimensions (layers) of the architecture are detailed.

Business Dimension. This dimension is perhaps the cornerstone of architecture. It is based on the definition of requirements made in the previous stage because it contains the current vision of the business and the situation it is desired to reach. It defines the product(s) that the end user will obtain with the relevant characteristics and aspects that align with the business objective but that are simultaneously independent of the systems and technologies to be used, working in a field purely conceptual.

Application Dimension. This dimension focuses on the components that will support the product that will be delivered to the end user in relation to the definition of the data required by the business and the applications for its processing; providing an interface between the applications used by the end customer and the underlying network in which the information is transmitted. This layer consists of two main entities from which different services are deployed: the database server and the application server. These two types of servers work together to perform a collective application behavior, through information transactions and are supported within an operating system mounted on the virtual machine.

Dimension of Technological and Physical Infrastructure. The technological and physical infrastructure layer includes all the necessary software and hardware capabilities to support the implementation of application and business services. This layer includes the base that supports the entire system and has both physical and virtual resources that are responsible for managing the flow, storage, processing and analysis of data. In this stage the designers evaluate the technologies that the business has and propose solutions to align them with the desired functionalities and restrictions or, instead of that, include new ones that fulfill this purpose.

The main components of this layer are defined as nodes of servers, development teams and IoT terminals, all connected to a local network as well as to the internet. From these large entities, the rest of the elements are displayed; but special emphasis is placed on the IoT terminal, which is described below.

The IoT terminal is one of the base components of the architecture, as it is responsible for data collection (sensor) and/or execution of specific orders or actions (actuator) and for sending and receiving information; therefore it is important to identify each of the equipment that conforms the system and perform a classification to determine which are IoT terminals and which are not.

The simplest definition describes an IoT terminal as any object that has internet connectivity; but within this work a more complex model is described, due to research efforts done by the RadioGIS group. RadioGIS defines an IoT terminal as a system consisting of four layers: sensor and actuator layer, administrable system layer, application layer premises and communication interface layer, as shown in Fig. 2. This model represents an IoT terminal with a high intelligence capacity; but along with the initial definition, which highlights the most important feature, the internet connection is fundamental.

The equipment of the system that complies with the model is called IoT terminals and is ready to be integrated into the platform. Conversely, in the case of devices that do not have these characteristics, they must be equipped with them through other devices or technology, for example, development boards such as Intel Galileo, Arduino, Raspberry, among others, that allow communication with the network (internet).

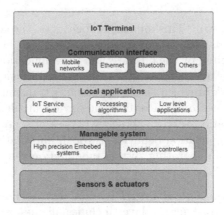

Fig. 2. IoT terminal model with a high intelligence capacity, which consists of four layers.

3.3   Architecture Implementation

The last stage of the methodology consists of the implementation of the architecture designed modeling and creating prototypes of the system, which must be approved by all the stakeholders of the business. This becomes an iterative process that gradually refines the architecture and evaluates its performance, to meet all the needs posed.

This process is divided into three steps that can be executed in parallel, since once the architecture is defined, experts in each area can work simultaneously in the development of the solutions of each layer. The first of these is the deployment of physical and technological infrastructure, where experts in this area dedicate their efforts in aligning the components of this layer with the guidelines of the architecture and implementing new solutions that contribute to the fulfillment of the objectives. These people are responsible for supporting all the devices that collect and send business data.

The second step is the development of applications and data management by the developers, who set up the data model with the creation of the databases as with the applications that will support the processing of that information, establishing the bridge between the end user and the physical and technological infrastructure layer. Finally, the last step is the product for the end user, where it is finished and they can make use of it, validating that it meets all the established requirements. In the case of not satisfying these needs or needing some improvement, the designers must correct these errors or complement any voids, verifying each one of the stages developed until all the stakeholders agree that the system is ready.

4   Case of Study

At the Universidad Industrial de Santander, the smart building line has been developed, led by the GISEL Electric Energy Systems Research Group through different research projects that support this initiative. For this, the conditions of experimentation have been created in the School of Electrical, Electronic and Telecommunications Engineering E³T building, at UIS.

Within this context of energy sustainability in buildings, this construction has a hybrid system of lighting and natural overhead lighting; a hybrid air conditioning system; forced natural ventilation; and plant covers on the terraces accompanied by a photovoltaic system connected to the power grid, (known as the GRIPV system and shown in Fig. 3) which seeks to reduce the environmental impact on your operation.

Fig. 3. GRIPV system of E³T building

In the case of the GRIPV system, the execution of projects that involve this type of system demands the monitoring of the electrical and climatic variables that describe the behavior of the PV system, the green roof, the intelligent irrigation system for the green roof and the building. For this, there is various measurement equipment, such as power meters, dataloggers, development boards and sensors. For this system, the proposed methodology is applied, the result of which is a prototype of an IoT service platform for GRIPV systems, as shown in the following section.

5   Implementation of Proposed Methodology

This section details the implementation of the proposed methodology, describing each of the three phases that compose it. The result is the prototype of an IoT-based service platform for monitoring the GRIPV system of the E³T building of the Universidad Industrial de Santander.

5.1 Definition of Problem Context

Once the business context is explored, the functional and non-functional requirements that the platform prototype must meet are formally established, so that they can meet the needs detected and through these efficiently manage the monitoring of the GRIPV system. The functional requirements are the monitoring of the GRIPV E^3T system, data processing, centralized storage of information and its visualization; and non-functional ones are associated with availability, security and scalability.

5.2 Architecture Design

In accordance with the guidelines established in the proposed methodology, the architecture design was carried out under three layers with the help of the free and open source tool, of visual design and modeling, Archi.

The modeling of each of the dimensions of the architecture results in the architectural model of the prototype of the platform, represented in Fig. 4. The general view of all the layers highlights each one of the components of the architecture and how they are related in and out of each layer.

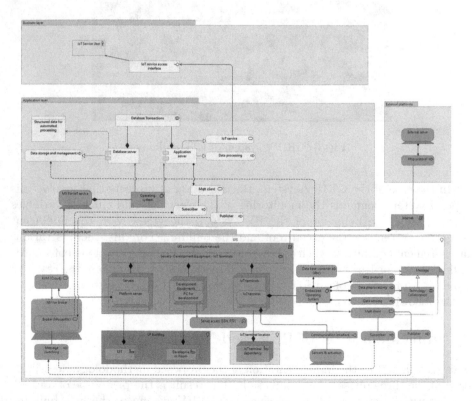

Fig. 4. Architectural model of the service platform prototype based on IoT

Each of the layers of the architecture are detailed below.

Business Dimension. Based on the requirements posed in the previous phase, the business layer is modeled representing the final product that is delivered to the customer, a web page for access and administration of the entire system and its information.

Application Dimension. This dimension is responsible for supporting the applications with which the end user interacts, establishing a bridge between them and the underlying layer from which the information is transmitted. The model of this layer consists of the two main elements, mentioned in the previous phase: the database server and the application server. These are called application components within the ArchiMate modeling language, and include the functionalities of the application, and is behavior and data, exposing the services and making them accessible through interfaces.

Technological and Physical Infrastructure Dimension. The technological and physical infrastructure layer consists of three main blocks: the IoT terminal node, the server node and the development team node. Nodes are both computational and physical resources, in which other resources of this type are housed, manipulated and/or interact. This infrastructure is modeled under the ArchiMate language, along with the connection of external platforms through the network.

5.3   Architecture Implementation

In the previous section the service platform prototype was modeled, therefore, in the following section it is implemented. The steps that are included this phase can be performed in parallel, so we present in detail the execution process starting with the layer of technological and physical infrastructure until reaching the end user product.

Technological and Physical Infrastructure Layer. This layer begins with the development and structuring of the physical and technological infrastructure, that will support the processes is the application layer. The first components identified were the sensing devices: once inspected, those that did not meet the characteristics of an IoT terminal were equipped with them through the Raspberry Pi development board.

This is the case of advanced energy meters, in which the Modbus RTU protocol transmits the requested data to the development, which in turn sends them to the database through the database connector and an Ethernet connection.

On the other hand, the sensors and actuators of the intelligent irrigation system were combined with Arduino NANO boards (secondary modules) and an Arduino development board (central module), the latter being in charge of receiving the information collected by the secondary modules through the protocol ZigBee and sending it to the database through a DBC and an Ethernet connection.

An inspection of the IoT terminals, an inspection of the information transfer process was carried out by the IoT terminals. The manufacturer of the microinverters provides a JSON-based API, through which connecting the platform and sharing information stored on its servers is possible through the HTTP protocol.

Application Layer. The storage and processing of information is supported by the system server that is in the facilities of the E^3T building. It has a Linux Mint operating system and a distribution of community GNU/Linux. On this server, the database server and the application server are installed, each with tasks defined as explained in previous sections; for this reason, they are detailed separately below.

The database server corresponds to MySQL, which allows the administration of relational databases. This server can only be accessed by a console, and for this reason Workbench and SQLyog managers are needed. This server supports the database that serves the intelligent irrigation system and the energy meters of the GRIPV system.

On the other hand, the application server belongs to an Apache Tomcat, an HTTP server specialized in Java applications. This server is responsible for the IoT service; therefore, it receives the client's requests and performs the respective transactions with the database server to deliver the response to its request to the user, under the client server model.

Business Layer. The final product for the customer is the development of a Java application (website), whose functionalities were established at the beginning. The website design was based on Material Design, a CSS framework that allows you to design websites and applications through predefined components and types that facilitate its creation. This website runs within the application server and can be accessed from any web browser even, including a mobile device. Figure 5 is a copy of the home page of the web application.

Through the platform it is possible to consult a list of system indicators that were predefined in the design stage, through which users can verify the system status in the selected date range, having access to the information of installed photovoltaic system, and electrical measurements, and building in general, and intelligent irrigation system. Figure 6 is a sample of the data that can be accessed and downloaded.

Figure 7 represents the information associated with the intelligent irrigation system, where users can to observe the temperature, volumetric content of the zones and the condition of the solenoid valves.

Finally, Fig. 8 represents the information associated with the energy produced by the installed photovoltaic system.

Fig. 5. Home page of the web application

Fig. 6. Monitoring of electrical measurements of the building

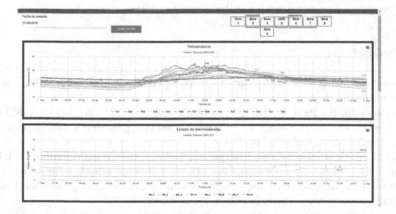

Fig. 7. Information associated with the intelligent irrigation system

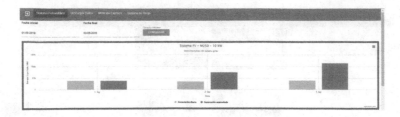

Fig. 8. Photovoltaic system power generation

6   Conclusions

The IoT impulses systems towards a more technological future, turning each scenario into an intelligent environment, capable of being conscious of what is happening around them and making decisions accordingly. In this sense, the GRIPV systems do not escape the IoT and thanks to the interconnected network of objects its monitoring and operation are possible remotely, boosting its capabilities. With this, we can state that IoT more than a great ally of green technologies is a fundamental part of them.

The business architecture approach has been used for a complete IoT design that brought advantages such as the alignment of IoT with business, under a standard language that allows the understanding of the system by each of the stakeholders; a global vision of the business for future decision making in response to changing and growing system needs; and a contribution to the growth of GRIPV systems from the perspective of ICT in combination with EAs, supporting research work at UIS.

A prototype platform was designed and implemented under the methodology proposed in this document, proposed as a flexible scenario that analyzes the main challenges in the adoption of green technologies and the concepts of energy efficiency in buildings; as well as supporting the development of new research in related areas and systems with similar characteristics.

References

1. Cisco Networking Academy. Cisco Networking Academy Builds IT skills & education for future careers
2. Pan, J., Jain, R., Paul, S., Tam, V., Saifullah, A., Sha, M.: An Internet of Things framework for smart energy in buildings: designs, prototype, and experiments. IEEE IoT J. **2**(6), 527–537 (2015)
3. Unidad de Planeación Minero Energética. Balance energético colombiano- BECO
4. Minoli, D., Sohraby, K., Occhiogrosso, B.: IoT considerations, requirements, and architectures for smart buildings-energy optimization and next-generation building management systems. IEEE IoT J. **4**(1), 269–283 (2017)
5. Consejo Colombiano de Construcción Sostenible. Programa LEED® en Colombia - Consejo Colombiano de Construcción Sostenible - CCCS

6. GISEL. Validación experimental del beneficio energético de la implementación de sistemas fotovoltaicos (FV) integrados con techos verdes para entornos tropicales con clima cálido, pp. 1–40 (2014)
7. GISEL, GIEMA, and GIEFIVET. Viabilidad técnica de la implementación de sistemas fotovoltaicos (FV) integrados con vegetación como estrategia de generación distribuida y horticultura en entornos urbanos de clima cálido tropical, pp. 1–46 (2016)
8. Castillo Santos, R., Castillo García, H.: Arquitectura Empresarial y las Organizaciones Estatales, vol. 1, no. 1, pp. 1–9 (2014)
9. Sarasty España, H.F.: Documentación y análisis de los principales frameworks de arquitectura de software en aplicaciones empresariales. Ph.D. thesis (2015)
10. McGibney, A., Rea, S., Ploennigs, J.: Open BMS - IoT driven architecture for the internet of buildings. In: IECON Proceedings (Industrial Electronics Conference), pp. 7071–7076 (2016)
11. Faruque, M.A.A., Vatanparvar, K.: Energy management-as-a-service over fog computing platform. IEEE IoT J. 3(2), 161–169 (2015)
12. Brad, B.S., Murar, M.M.: Smart buildings using IoT technologies. Constr. Unique Buildings Struct. 5(20), 15–27 (2014)
13. Garcia, O., Alonso, R.S., Tapia, D.I., Corchado, J.M.: Electrical power consumption monitoring in hotels using the n-Core Platform. In: Clemson University Power Systems Conference, PSC 2016 (2016)
14. Nguyen, N.H., Tran, Q.T., Leger, J.M., Vuong, T.P.: A real-time control using wireless sensor network for intelligent energy management system in buildings. In: IEEE Worskshop on Environmental, Energy, and Structural Monitoring Systems, Proceedings EESMS 2010–2010, pp. 87–92 (2010)
15. Linder, L., Vionnet, D., Bacher, J.P., Hennebert, J.: Big building data-a big data platform for smart buildings. Energy Procedia 122, 589–594 (2017)
16. Figueiredo, J., Costa, J.S.D.: A SCADA system for energy management in intelligent buildings. Energy Build. 49, 85–98 (2012)
17. Moreno, M.V., Zamora, M.A., Skarmeta, A.F.: User-centric smart buildings for energy sustainable smart cities. Trans. Emerg. Telecommun. Technol. 25, 41–55 (2014)
18. Tragos, E.Z., et al.: An IoT based intelligent building management system for ambient assisted living. In: 2015 IEEE International Conference on Communication Workshop, ICCW 2015, pp. 246–252 (2015)
19. Jamborsalamati, P., Fernandez, E., Hossain, M.J., Rafi, F.H.M.: Design and implementation of a cloud-based IoT platform for data acquisition and device supply management in smart buildings. In: 2017 Australasian Universities Power Engineering Conference, AUPEC 2017, 1–6 November 2017 (2018)
20. Cheng, B., Longo, S., Cirillo, F., Bauer, M., Kovacs, E.: Building a big data platform for smart cities: experience and lessons from santander. In: Proceedings of the 2015 IEEE International Congress on Big Data, BigData Congress 2015, pp. 592–599 (2015)
21. UIT. Visión general de la Internet de las cosas (ITU-T Y.4000/Y.2060 (06/2012)), p. 20 (2012)
22. Josey, A., Lankhorst, M., Band, I., Jonkers, H., Quartel, D.: An introduction to the archimate® 3.0 specification, June 2016

Characterization of Functions Using Artificial Intelligence to Reproduce Complex Systems Behavior

Takagi Sugeno Kang Order 2 to Reproduce Cardiac PQRST Complex

Jesús Rodríguez-Flores and Víctor Herrera-Pérez[(✉)]

Facultad de Informática y Electrónica,
Escuela Superior Politécnica de Chimborazo, Riobamba, Ecuador
{jesus.rodriguez,isaac.herrera}@espoch.edu.ec

Abstract. In the field of signal processing, for forecasting purposes, the characterization of functions is a key factor to be faced. In most of the cases, the characterization can be achieved by applying least square estimation (LSE) to polynomial functions; however, it is not fully in all cases. To contribute in this field, this article proposes a variant of artificial intelligence based on fuzzy characterization patterns initialized by Lagrange interpolators and trained with neuro-adaptive system. The aim is to minimize a cost function based on the absolute value between samples and their prediction. The proposal is applied to the characterization of cardiac PQRST complex as case study. The results show a satisfactory performance providing an error of around 1.42% compared to the normalized PQRST complex signal.

Keywords: Characterization of functions · Fuzzy system · Cost function · Cardiac PQRST complex · Neuro-adaptive system · Lagrange interpolator

1 Introduction

In this article, a method to make prediction based on mathematical analysis based on a justification supported by a statistical is presented. This method is applied to reproduce the cardiac PQRST complex, allowing in principle an application on a standard cardiac signal.

It is noteworthy that, based on regression analysis strategies, such as linear regression applied to polynomial functions, used as a statistical indicator or cost function lead to closed mathematical expressions. However cases such as the proposed by Xiloyannis et al. [1] defines the problem when the mathematical model of the system is unknown or when nonlinearities or uncertainties are present in this. Under these conditions, one of the possible solutions is to incorporate computational methods capable of processing nonlinear cost functions and operate on systems with characteristics of nonlinearity and uncertainty.

M. Botto-Tobar et al. (Eds.): ICAT 2019, CCIS 1194, pp. 222–234, 2020.
https://doi.org/10.1007/978-3-030-42520-3_18

The theory of error for the treatment of data, taking into account intervals, has allowed to extend the applicability of the regression methods in a satisfactory manner as demonstrated by Jun-peng et al. [2]. Additionally, the issue of uncertain or vague information, as well as the numerical indeterminacies presented by the data, can be ignored by means of the application of intelligent techniques such as fuzzy theory. Fuzzy tools allows to define the universe of possible operative solutions by means of linguistic variables which can be defined by arithmetic intervals with probable characteristics, such defined by Sun et al. [3] in their article entitled Diffuse extended TS model based on interval arithmetic and its application to nonlinear regression analysis by intervals. Similar context is proposed by Araujo et al. [4] in the study of a particular case of nonlinear oscillator.

Interest about reproducing the behavior of the heart, either through simulation or emulation is a topic cover for several studies in the literature. However, due to the new computational and mathematical tools, the study and optimization about reproducing the heart behavior keep the subject in effect, as evidenced by Bhowmick et al. [5] in their work entitled "Synthesis of ECG waveform using Simulink model" which, using classical mathematical tools and signal processing, introduce the modeling of heart waves using a graphical language.

The purpose of this article is to show the potential of mathematical tools, applying artificial intelligence, to reproduce the PQRST complex heart curve. In this case, the contribution is focused on the application of the concept of fuzzy logic, by means of Takagi-Sugeno-Kang (TSK) method, with a variant of order 2 [6, 7]. For this application, the neuro-fuzzy structure plays an important role in the characterization process being the process of training and learning strategy that introduces a filtering effect after the process of minimizing a cost function. For this study, the cost function chosen is the absolute value of error. It is important to highlight that; the proposal of this article can be applied under other data type to characterize their function.

To develop a characterization, for the PQRST cardiac complex, by applying TSK method allow to obtain an improved setting of data with a spline cubic. The advantage of this analysis is to obtain a synthesized mathematical expression, which can be used for any kind of studies related with medical or health topics where the cardiac behavior needs to be modeled. Besides, the characterization method proposed in this paper can be applied to reproduce other functions with the advantage of low computational cost compared to similar method applying TSK with sigmoidal functions.

2 Methodology for Characterizing Functions of Order 2 Using TSK to Reproduce Cardiac PQRST Complex

2.1 Theoretical Framework

The study of both physiological behavior as well as electrochemical heart, under different considerations, have led to many studies focused on modeling their behavior. For instance, Abramovich-Sivan et al. [8] conducted their to reproduce a response curve of modulation heartbeat to the interaction of a cardiac pacemaker. However, for many of the studies, it is necessary to reproduce the behavior under optimal conditions of health or under typified fault conditions, as well as arrhythmias.

On the other hand, a study without the knowledge of a standard curve, which can be considered as optimal, or the emulation for the calibration of equipment for registration might become in a challenge for the characterization study. In words of McSharry et al. [9], the PQRST complex is a signal which varies over time due to ionic current flow which makes cardiac fibers to contract and relax, being possible to observe it from a register called electro-cardiogram (ECG). Therefore, it is possible to record the potential difference between two electrodes placed on the skin surface. Besides, it is usually to place three electrodes in order to record a bio-potential difference capable of suppressing as review in [10].

Based on aforementioned reasons, in this work a study is presented where both statistical applicability in the treatment of the signal as the signal itself, are used to characterize the system. Part of the study lies in the results presented by Medina et al. [11], Bhowmick et al. [5], and McSharry et al. [9]. Under the considerations raised from these studies, the waveform of the PQRST complex is as shown in Fig. 1, which has been conceptually generated with a sampling of 500 Hz, with a duration of interval PR wave of 50 ms, 190 ms of QRS and 220 ms of ST. The voltage register is in the range of −0.4 to 1.2 mV, synthesized from registration made by MacSharry et al. [9].

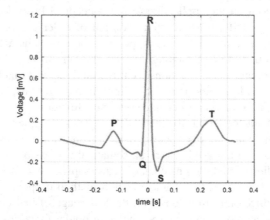

Fig. 1. Waveform synthesized from a record of a normal human PQRST complex.

The development of computing and digital signal processing have led to multiple mathematical applications reserved only to the conceptual level. However, as review Seising [12], electrical engineering makes one of his great contributions in the mathematical field, being led this contribution by Professor Lotfi Zadeh who introduced the concept of fuzzy set theory with linguistic treatment to treat vagueness of human language and use to take decisions about the operation of a system.

The application of fuzzy inference has evolved since its genesis, and it is constantly rediscovering and rethinking to apply the concept on different applications. For instance, Dutu et al. [13], proposed, under Mamdani's considerations, the classical concepts of discourse universe, such as the variable that constitutes the domain of a function, and which is characterized by a noun which refers to the variable under study, which in linguistic way are characterized by the use of adjectives and adverbs. In the mentioned study, the authors raise, the issue of generating rules that allow defining an inference that eventually leads to an aggregation of rules to obtain an output, using weighted average methods or methods such as the centroid.

The Mamdani method, to make fuzzy inferences, presents a very high computational cost, therefore, a simplified manner, both for creating membership functions, automatic generation of rules, and weighting values of degrees of truth in the form of behavior was already raised by Takagi, Sugeno and Kang [14, 15]. This method was recently presented as a review of the subject by Bacha et al. [16].

It can be highlighted that, the reduction of computational cost is a key factor to solve the overlapping between membership functions as demonstrated in the article presented by Zhang et al. [17] and Yu et al. [18]. Similarly, in the process of initialization of parameters of single values or singletons, that, for reasons of numerical simplicity, the Lagrange polynomial interpolator is chosen as solving method, as cited by Tjahjono et al. [19].

2.2 Standardization of Data

The first phase of the study leads to the normalization of the data. The purpose is to have control over the simulated heart rate depending in any case a control signal for future use in studies involving emulation. For normalization, the peak amplitude of R of the cardiac complex is considered as unit amplitude, and the length of the signal as unitary default period, centered on a zero-value unit. Figure 2 shows the effect of normalization on the data.

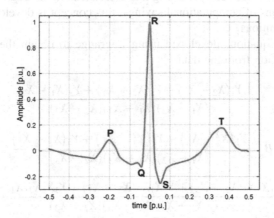

Fig. 2. Waveform normalized in amplitude and time PQRST complex.

2.3   Structuring Fuzzy Inference System

Due to the studied signal is a time series, the function domain corresponds to the standard time, which is divided into thirty-five subsets, which define the membership functions of triangular type (Fig. 3).

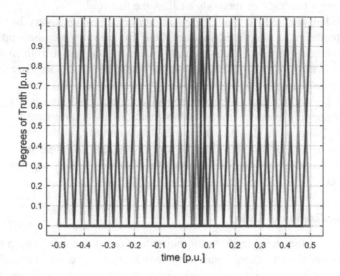

Fig. 3. Membership functions in the time domain of the function.

Initializing Singletons or unique values for the TSK structure of order 2 can be done by using the recorded sample taking into consideration the Lagrange interpolators [14, 15]. For reasons of computational savings, this concept is developed as shown in the following definitions.

Definition 1: Equations developed using Lagrange to obtain the quadratic polynomial approximation from the data.

$$A = \frac{Y_1(X_2 - X_3) + Y_2(X_3 - X_1) + Y_3(X_1 - X_2)}{(X_1 - X_2)(X_1 - X_3)(X_2 - X_3)} \tag{1}$$

$$B = \frac{Y_1(X_2^2 - X_3^2) + Y_2(X_3^2 - X_1^2) + Y_3(X_1^2 - X_2^2)}{(X_1 - X_2)(X_1 - X_3)(X_2 - X_3)} \tag{2}$$

$$C = \frac{Y_1 X_2 X_3(X_2 - X_3) + Y_2 X_1 X_3(X_3 - X_1) + Y_3 X_1 X_2(X_1 - X_2)}{(X_1 - X_2)(X_1 - X_3)(X_2 - X_3)} \tag{3}$$

Definition 2: This equation is used as a Lagrange interpolator for second order polynomial expression associated with each membership function.

$$\widehat{Y} = AX^2 + BX + C \tag{4}$$

The structure of neuro-adaptive system includes a synaptic structure in which neurons show activation functions of linear nature [20], linear saturated between 0 and 1 and production [21]. Figure 4 illustrates the neural network for the implementation of neuro-fuzzy inference system.

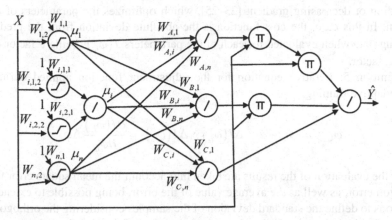

Fig. 4. Neuro-fuzzy network for characterizing cardiac PQRST complex using TSK order 2.

The intermediate layer with weights presented in Eqs. (1)–(3), are initialized using the Lagrange interpolator, while in the input layer, their weights depend on whether they are associated with intermediate membership functions, or if instead they are layers of membership functions associated with the ends of function domain. The following definition allows parameterizing these weights.

Definition 3: These equations are proposed to perform initialization of the weights of the input layers of the neuro-fuzzy system.

$$W_{1,1} = \frac{X_2}{X_2 - X_1} \bigwedge W_{1,2} = \frac{-1}{X_2 - X_1} \tag{5}$$

$$W_{n,1} = \frac{-X_{n-1}}{X_n - X_{n-1}} \bigwedge W_{n,2} = \frac{1}{X_n - X_{n-1}} \tag{6}$$

$$W_{i,1,1} = \frac{-X_{i-1}}{X_i - X_{i-1}} \bigwedge W_{i,1,2} = \frac{1}{X_i - X_{i-1}} \tag{7}$$

$$W_{i,2,1} = \frac{X_i}{X_{i+1} - X_i} \bigwedge W_{i,2,2} = \frac{-1}{X_{i+1} - X_i} \tag{8}$$

Once established the structure of fuzzy inference system, and properly initialized, it must be trained such that the weights associated to the interpolator reduce the cost function [22]. The cost function defined for this purpose is defined in the following definition.

Definition 4: Equation of the cost function of the absolute value of error.

$$J(\varphi) = \sum_{i=1}^{N} \left| Y_i - \widehat{Y}(\varphi)_i \right| \therefore \varphi = [W_1 \cdots W_n] \tag{9}$$

The function to adjust the parameters or weights is known as the method of descending or decreasing gradient [23–25], which optimizes the parameters of a cost function. In this case, the cost function is the absolute deviation from the prediction regarding data when evaluated for each of the parameters $J(\varphi_{i_n})$, being α factor called learning factor.

Definition 5: Iterative equation for the adjustment function applied during the neuro-adaptive training.

$$\varphi_{i_{n+1}} = \varphi_{i_n} - \alpha \times \Delta J(\varphi_{i_n}) \therefore \Delta J(\varphi_{i_n}) = \frac{\partial J(\varphi_{i_n})}{\partial \varphi_{i_n}} \Delta \varphi_{i_n} \tag{10}$$

For the evaluation of the results are taken into account the mean square error for the prediction error, as well as the average value of the error, being possible to extend both definitions to define the standard deviation of the samples considering the orthogonality between the deviation standard and the average value of the signal.

Definition 6: Average prediction error (rms value).

$$\varepsilon_{RMS}^2 = \frac{1}{N} \sum_{i=1}^{N} \left(Y_i - \widehat{Y}_i \right)^2 \tag{11}$$

Definition 7: Mean prediction error (rms value).

$$\varepsilon_{DC}^2 = \left[\frac{1}{N} \sum_{i=1}^{N} \left(Y_i - \widehat{Y}_i \right) \right]^2 \tag{12}$$

Definition 8: Standard deviation of prediction.

$$\sigma = \sqrt{\varepsilon_{RMS}^2 - \varepsilon_{DC}^2} \tag{13}$$

The multivariable correlation factor is defined as follows to ensure their value between 0 and 1.

Definition 9: Multivariable correlation factor.

$$r_y^2 = 1 - \left| 1 - \frac{\sum_{i=1}^{N} \left(\widehat{Y}_i - \overline{\widehat{Y}} \right)^2}{\sum_{i=1}^{N} \left(Y_i - \overline{Y} \right)^2} \right| \tag{14}$$

3   Results and Discussions

The characterization system optimally initialized is able to reproduce the system behavior. At the end of training, the characterization of the neuro-fuzzy TSK order 2 system led to a standard deviation of 0.0130 and a quadratic correlation multivariable 0.9970, showing a noticeable improvement and how relevant is TSK characterization method of order 2 for reproduction of trends and/or curves.

Figure 5 illustrates the behavior of the cardiac signal while applying the neuro-fuzzy system proposed. Figure 6 shows the evolution of the cost function during the training process.

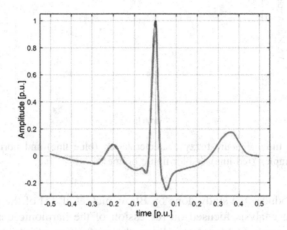

Fig. 5. Neuro-fuzzy characterization using TSK order 2 (blue line) and normalized signal of cardiac PQRST complex (red line). (Color figure online)

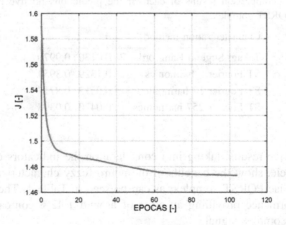

Fig. 6. Evolution of the cost function during the training of neuro-fuzzy system.

In Fig. 7, it is evident that the neuro-fuzzy TSK order 2 characterization is derivable in the whole function domain, which is presented as a kindness compared to the characterization method TSK of order 0.

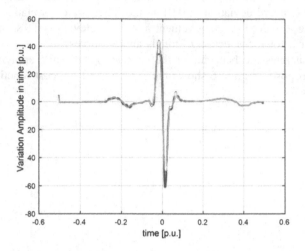

Fig. 7. Change in time of neuro-fuzzy characterization (blue line) and normalized signal of cardiac PQRST complex (red line). (Color figure online)

Figure 8 reproduces the study done by Bhowmick et al. [5] of the cardiac PQRST complex, used an analysis focused on regression of the harmonic components characterization using curve-fit Matlab, to obtain, the coefficients of a trigonometric series of Fourier considering only four harmonic study for 5, 17, 257 harmonics.

Table 1. Statistical comparison results of each of the predictions relative to the normalized signal of cardiac PQRST complex.

Characterization method	σ	r_y^2
Takagi Sugeno Kang order 2	0.0130	0.9970
ST Fourier 5 harmonics	0.1329	0.3954
ST Fourier 17 harmonics	0.0612	0.8717
ST Fourier 257 harmonics	0.0420	0.9398

The comparative results, taking into consideration the indicators of cost function raised in this article, show the benefits of the neuro-fuzzy characterization applied to reproduction cardiac PQRST complex, as can be seen in Table 1. The results show a satisfactory performance providing an error of around 1.42% compared to the normalized PQRST complex signal.

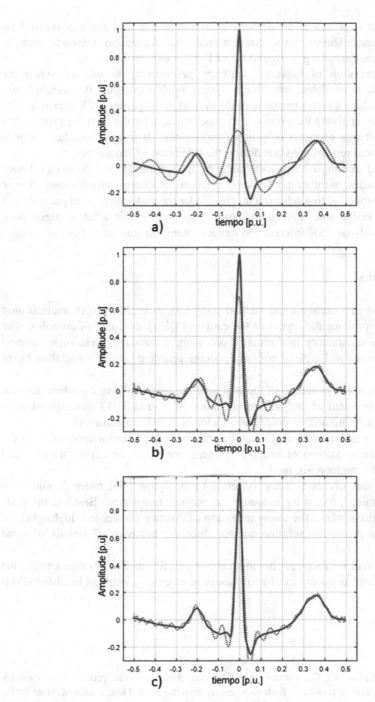

Fig. 8. Result of reproducing the study done by Bhowmick et al. [5] for 5 (a), 17 (b) and 257 (c) harmonics characterization (blue dashed line) compared to the normalized signal of cardian PQRST complex (red line). (Color figure online)

The comparative results allow demonstrating the potential of the method of TSK neuro-fuzzy characterization, over conventional characterization methods such as approximations method using a trigonometric Fourier expression.

The characterization of functions applying neuro-fuzzy techniques, when the experimental data is available, the ANFIS system is able to adjust the weights (singletons) until reaching a better response of the system in the process of forecasting. The main advantage of applying the proposed method is that, after the learning process, the characterization model becomes only in a fuzzy system. It means that, the functions weights (singletons) remain constant during the operation of the system.

The presented technique has a wide range of applications where the most relevant are: geodetic studies, weather predictions, behavior of charge profiles and electric consumption, instrumentation adjustment, etc. Under the wide range of applications, it is necessary to verify and validate the implementation of this kind of algorithms. Therefore, the study of TSK inference systems presents an important field of study.

4 Conclusions

This article proposed a mathematical method using TSK order 2 for the characterization of functions to reproduce the signal of the cardiac PQRST complex of complex. The results showed a satisfactory performance providing a standard deviation of around 0.0142 corresponding to 1.42% of error and a multivariable quadratic correlation factor of 0.96.

After 104 periods of evolution of the cost function, and applying a gradient descent method, an improvement of the standard deviation of around 0.013 (corresponding to 1.3% error) and a multivariable correlation factor of 0.997 are achieved.

Based on the simulation results, it can be concluded the relevant application of the TSK characterization method of order 2 to reproduce trends or curves with several applications in the engineering field.

The applications of neuro-fuzzy systems, by applying TSK order 2, allows to improve the learning process by reducing the computational cost. Besides, the wide range of applications where the characterization of functions is needed, highlights the TSK methods as promising solution to reproduce the behavior of signals or trend predictions.

Future work will be focus of the applications of TSK method to characterize the prediction of energy consumption by end-users in electricity markets in distributions grids.

References

1. Xiloyannis, M., Gavriel, C., Thomik, A.A.C., Faisa, A.A.: Gaussian process regression for accurate prediction of prosthetic limb movements from the natural kinematics of intact limbs. In: International IEEE/EMBS Conference on Neural Engineering NER, vol. 2015-July, pp. 659–662 (2015)

2. Jun-peng, G., Wen-hua, L.: Regression analysis of interval data based on error theory. In: Proceedings of the 2008 IEEE International Conference on Networking, Sensing and Control ICNSC, pp. 552–555 (2008)
3. Sun C., Xu, Z.: Extended T-S fuzzy model based on interval arithmetic and its application to interval nonlinear regression analysis. In: IEEE International Conference on Fuzzy Systems, pp. 1773–1778 (2009)
4. Araujo, E., Dos Santos Coelho, L.: Fuzzy model and particle swarm optimization for nonlinear identification of a Chua's oscillator. In: IEEE International Conference on Fuzzy System (2007)
5. Bhowmick, S., Kundu, P.K., Sarkar, G.: Synthesis of ECG waveform using simulink model. In: 2016 International Conference on Intelligent Control Power and Instrumentation, ICICPI 2016, pp. 61–64 (2017)
6. Kukolj, D., Levi, E.: Identification of complex systems based on neural and Takagi-Sugeno fuzzy model. IEEE Trans. Syst. Man Cybern. Part B Cybern. 34(1), 272–282 (2004)
7. Méndez, G.M., De Los Angeles Hernández, M., González, D.S., López-Juarez, I.: Interval singleton type-2 TSK fuzzy logic systems using orthogonal least-squares and backpropagation methods as hybrid learning mechanism. In: Proceedings of the 2011 11th International Conference on Hybrid Intelligent Systems HIS 2011, pp. 417–423 (2011)
8. Abramovich-Sivan, S., Akselrod, S.: The generation and modulation of the heart beat: a phase response curve based model of interacting pacemakers cells, pp. 589–592 (2002)
9. McSharry, P.E., Clifford, G.D., Tarassenko, L., Smith, L.A.: A dynamical model for generating synthetic electrocardiogram signals. IEEE Trans. Biomed. Eng. 50(3), 289–294 (2003)
10. Dijk, J., Van Loon, B.: Scanning our past from The Netherlands: the electrocardiogram Centennial: Willem Einthoven (1860–1927). Proc. IEEE 94(12), 2182–2185 (2006)
11. Medina, V.U., González-Camarena, R., Echeverría, J.C.: Effect of noise sources on the averaged PQRST morphology. Comput. Cardiol. 32, 743–746 (2005)
12. Seising, R.: When computer science emerged and fuzzy sets appeared: the contributions of Lotfi A. Zadeh and Other Pioneers. IEEE Syst. Man Cybern. Mag. 1(3), 36–53 (2016)
13. Dutu, L.C., Mauris, G., Bolon, P.: A fast and accurate rule-base generation method for mamdani fuzzy systems. IEEE Trans. Fuzzy Syst. 26(2), 715–733 (2018)
14. Sugeno, M., Kang, G.T.: Structure identification of fuzzy model. Fuzzy Sets Syst. 28(1), 15–33 (1988)
15. Takagi, T., Sugeno, M.: Fuzzy identification of systems and its applications to modeling and control. IEEE Trans. Syst. Man Cybern. SMC-15(1), 116–132 (1985)
16. Bacha, S.B., Bede, B.: On Takagi Sugeno approximations of mamdani fuzzy systems. In: Annual Conference of the North American Fuzzy Information Processing Society - NAFIPS (2017)
17. Zhang, J.D., Zhang, S.T.: Controller design of T-S fuzzy systems with standard fuzzy partition inputs. In: 2009 4th IEEE Conference on Industrial Electronics and Applications ICIEA 2009, pp. 3101–3106 (2009)
18. Yu, J.X., Ren, G., Zhang, S.T.: Stability analysis of discrete T-S fuzzy systems with standard fuzzy partition inputs. In: Proceedings of the 2006 International Conference on Machine Learning Cybernetics, vol. 2006, pp. 438–442 (2006)
19. Tjahjono, A., Sudiharto, I., Suryono, Anggriawan, D.O.: Modelling non-standard over current relay characteristic curves using combined lagrange polynomial interpolation and curve fitting. In: Proceeding of the 2016 International Seminar Intelligent Technology and Its Applications ISITIA 2016, pp. 589–594 (2017). Recent Trends Intell. Comput. Technol. Sustain. Energy

20. Baldi, P.F., Hornik, K.: Learning in linear neural networks: a survey. IEEE Trans. Neural Netw. **6**(4), 837–858 (1995)
21. Matich, D.J.: Redes Neuronales: Conceptos Básicos y Aplicaciones. Universidad Tecnológica Nacional - Facultad Regional Rosario (2001)
22. Salas, R.: Redes Neuronales Artificiales-Rodrigo Salas, p. 7 (2005)
23. Zhang, N., Zeng, S.: A gradient descending solution to the LASSO criteria. In: Proceedings International Joint Conference on Neural Networks, vol. 5, pp. 2942–2947 (2005)
24. Cruz, A., Acevedo, A.: Reconocimiento de voz usando redes neuronales artificiales backpropagation y coeficientes lpc. Cómputo en June, pp. 89–99 (2008)
25. Caicedo, B., Lopez, J.: Redes Neuronales Artificiales. Charlas Fis, p. 276 (2010)

Snake Hunting System Supplied with Solar Energy Based on Cages Installed in the Jungle for Strictly Curative Purposes, Promoting Ancestral Knowledge, Natural Medicine and Indigenous Cultural Products from Rural Areas

Daniel Icaza$^{(\boxtimes)}$ (ID), Carlos Flores-Vázquez (ID),
and Santiago Pulla Galindo (ID)

GIRVyP Grupo de Radiación Visible y Prototipado,
Universidad Católica de Cuenca, Cuenca, Ecuador
{dicazaa, cfloresv, gpullag}@ucacue.edu.ec

Abstract. In the present article a basic design for the snake hunt in the Ecuadorian Jungle at a domestic scale is presented, minimizing the risk of bites and poisoning to the people who carry out these activities in these zones. The mechanism considers a source of autonomous solar energy supply to transform it into electrical energy. In the present work of investigation also the simulations in Matlab and their design characteristics are presented. The research process consisted of collecting data on solar radiation, temperature, through a meteorological station located in the Amazon rainforest.

The exposed system does not have the purpose of exploitation on an industrial scale, which is intended to reduce the risk in the snake hunt and promote natural medicine, ancestral knowledge and maintain the cultural production of rural areas, peoples and nationalities that Throughout the times they have been developing in Ecuador.

In the end, we present the results of the simulations and general design of the system for the snake hunt with the use of materials from the area and other materials that contribute to the conservation of the environment.

Keywords: Renewable energy · Modeling · Solar panel · Snakes · Rural development · Ancestral medicine · Cultural products

1 Introduction

The systematic use of plants and animals with medicinal attributes goes back hundreds of years where the primitives healed the wounds especially of the skin. The first prosimians, probable ancestral of the man, looked for in the immense jungles the palliatives for eventual organic disturbances. As evolution rehearsed and sculpted

© Springer Nature Switzerland AG 2020
M. Botto-Tobar et al. (Eds.): ICAT 2019, CCIS 1194, pp. 235–244, 2020.
https://doi.org/10.1007/978-3-030-42520-3_19

intelligence in the first hominids, instinct gave way to reason, and the search for flora, now empirical, came to have a strong ally: the investigative spirit. Plant and animal life, despite having disappeared in a proportion, have always been useful for human beings. Errors and successes forged the knowledge in the evolution of time to the point that nowadays research has dabbled with great interest in native species and native plants as is the case in the Ecuadorian and Peruvian Jungle.

The present ancestral medicine continues to play an important role in the lives of people in Andean countries in South America, such as Peru, Ecuador, Colombia, Bolivia. The native people of places of high mountain and particularly of the Jungle Amazonian Equatorial keep rooted the native customs and the very particular and interesting healing forms.

It also highlights that the flora and fauna that dazzles in the indigenous communities on the border between Ecuador and Peru is quite interesting, especially in the towns near Cueva de los Tayos formerly explored in 1976 by the Former NASA Astronaut Neil Armstrong [2]. In this Andean mountain range there are special attributes of very particular native plants and species that are very useful for the rural population.

Traditional medicine, its techniques, know-how and its knowledge are explained through the mentality and the culture of the social groups that they generate, reproduce and practice, because it is part of their life, of their daily life, finally of sacred, historical and real time [1, 10].

The indigenous peoples of the Ecuadorian and Peruvian Amazon find healing aspects in plants and animals of their jungle, but above all they identify healing aspects in the venom of snake, its fat and the skin. However, the risks are imminent for people who perform these activities of being bitten by snakes. In Fig. 1 you can see the way in which snakes and the objects they commonly use hunt [2, 8].

Fig. 1. Traditional system for snake hunts.

With regard to the preparation and use of these poisons they are surrounded by a series of ritual prescriptions and prohibitions that signal their connection with female body secretions, especially menstrual blood. The poison enters the same category of other bitter, hot and rotten substances that cause disease and death. [3] and [4] disclose a wide literature of the consequences generated by snake bites and also the benefits they have for curing very unique diseases of people.

Fig. 2. Originals exposing themselves to being bitten by snakes.

Taking into account these aspects, we proceeded to design the demolished but sufficient equipment for the snake hunt in the Amazon on the border between Ecuador and Peru as it is the equipment shown below in Fig. 3, it forms a single body where internally they are found the snakes that entered were screwed in, thus avoiding exposing themselves to being bitten by a snake as shown in Fig. 2.

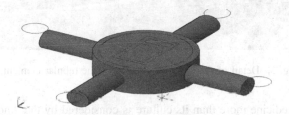

Fig. 3. Unique snake hunting equipment.

This equipment is designed so that by means of its tubes they enter during the day or night to the central part of the snake collector where once entered they will roll up and the tube that served as an entrance will automatically close once the tail of the snake has passed. snake, this process will be repeated in all the team's income. The sensors will be on the edge of the pipe as shown in Figs. 4 and 5.

Allow energy from sensors and sensors to realize the median system autonomously of the energy, specifying the median energy used by the panel to produce the energy required for sensing the battery. The energy required for a 12 V power supply.

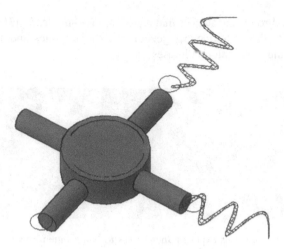

Fig. 4. Location of sensors at the tubular ends.

Fig. 5. Detail of the entry of a snake by the tubular element.

Traditional medicine more than its culture is considered by the Andean peoples of South America fundamentally sacred. It is surrounded by a religious mentality, a mythical worldview of the universe and its practice crystallizes with a harmoniously mythical ritual. It not only deals with the sick body, but goes beyond; travels towards a world full of mysticism. He takes possession, travels through it, makes it his own and is interested in restoring cosmic disorder. Address the rupture between body and the universe. Heal the spirit and man.

2   Location of Research

The project that refers to the snake hunt for strictly healing purposes is part of an investigation to benefit the eastern areas of Ecuador and the areas of influence with Peru, since they are sites that have a similar culture and rely on medicine ancestral as one of the ways to live in peace with nature and with their peers. See Fig. 6.

Fig. 6. Location of the research.

This investigation and development of this mechanism for snake hunting are of strict healing purpose and at no time is it sought to exploit on a large scale since it would undermine the preservation of the species and therefore their natural habitat.

Then, in Fig. 7 we can identify the general scheme where the contribution of photovoltaic energy, which is the purpose of the investigation, which is subject to simulations in Matlab to see its behavior according to the different operating conditions. It is clear that in our case study we contemplate it without connection to the public distribution network [6, 9].

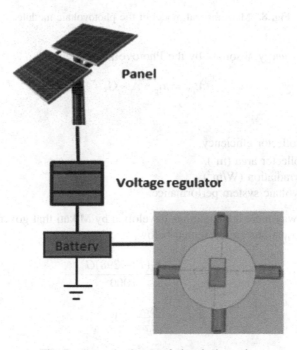

Fig. 7. General scheme of electrical supply.

3 Mathematical Modeling

In the mathematical model indicated below, the most representative parameters that will influence the final objective, which is the production of electrical energy from the sun and supply the sensing system and closure of the tubular bodies before the entry of snakes. This process involves finding the best option for the design of the system provided by a solar panel, which leads to even better developments based on mathematical models [5, 10, 11].

3.1 Photovoltaic System

The photovoltaic panel is formed by the union of several cells or solar cells, in the same way by means of an equivalent circuit the solar cells can be represented as can be seen in Fig. 8.

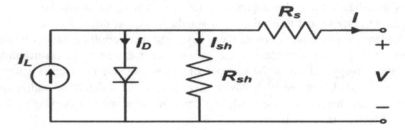

Fig. 8. Mathematical model of the photovoltaic module.

The thermal energy absorbed by the Photovoltaic system is;

$$P_{pv} = \eta_p * A * G_1 \tag{1}$$

Where;

η_p = Solar collector efficiency.
A = Solar collector area (m^2).
G_1 = Solar irradiation (W/m^2).
η_{pv} = Photovoltaic system performance.

In the same way there are equations developed by Mikati that govern the behavior of cells. These equations are presented below:

$$I = I_{Scr} + \frac{K_i(T - 298)G}{1000} \tag{2}$$

$$I_{rs} = \frac{I_{Scr}}{e^{\frac{qV_{oc}}{N_s kAT}} - 1} \tag{3}$$

$$I_D = I_{Scr} \left(\frac{T}{T_r}\right)^3 * e^{\frac{qV_{oc}}{Bk}*\left(\frac{1}{T_r}-\frac{1}{T}\right)} \tag{4}$$

$$I = N_p I_L - N_p I_D * \left(e^{\frac{q(V_{PV}+I_{PV}*R_s)}{N_s A k T}} - 1\right) \tag{5}$$

Where:

N_s, N_p: Number of solar cells in series and parallel.

K: Boltzman constant.

qe: Charge of the electron.

T: Working temperature of the solar panel in °C.

R_s: Series resistor.

R_p: Parallel resistance.

I_L (G1, T1): Short-circuit current Isc.

I_D: Inverse saturation current of the diode.

Voc: Open circuit voltage.

4 Blocks Diagram

Figure 9 shows the mathematical model developed in Matlab, the simulation is developed according to the characteristic parameters of the photovoltaic system as the area of the panels, number of panels, usual radiation.

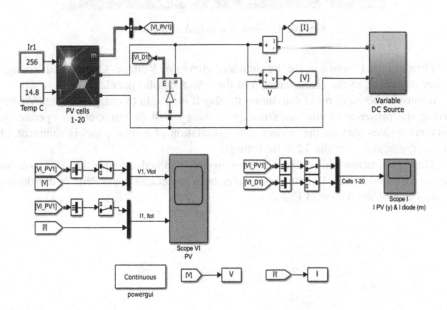

Fig. 9. Block diagram in Matlab/Simulink

5   Results and Discussions

To solve Eqs. (1) to (5) of according to the previously defined literature related to the solar generation system for the automation of snake caser mechanism in the Amazon Rainforest of the Ecuador and Peru, this will be the only one source of energy that will provide to supply the load that in this case is the system of sensors installed in the tubular elements that is in an energetically isolated area and there is the possibility of hunting snakes. It is taken into account that there is a battery bank for energy storage and to provide energy simultaneously and sufficient for when the snake enters the equipment and can be closed when the tail is pierced, see Fig. 10. The fundamental objective is the validation of this model with the use of dynamic software such as Matlab/Simulink, which we have used in several similar situations and presented interesting results and then have been compared with measurements made in the field. In addition, to validate and adjust the simulated results, the most coherent data became reality and, according to the measurements made, they were designed as shown in Fig. 11.

Fig. 10. Equipment installed in the field.

Then in Fig. 11 you can see the simulated curves of Voltage- Current and Voltage-Power according to the characteristics of the selected solar panel.

It must be borne in mind that during the day it is sought to charge the battery when having the presence of the sun's rays, the energy will be sufficient to operate the installed devices such as the sensors. The installation of a solar panel is sufficient, all the energy is stored in the 12 V dc batteries.

The figure below shows the radiation profile obtained through the ACT weather station indicated in Fig. 12 in the border area between Ecuador and Peru, in the vicinity of the Cave of the Tayos in Ecuador.

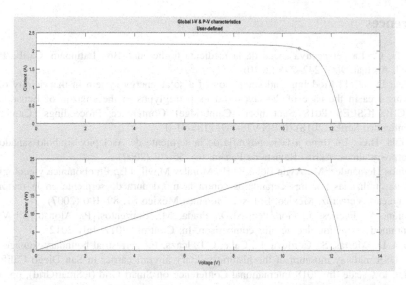

Fig. 11. Simulation in Matlab regarding the characteristics of the solar panel.

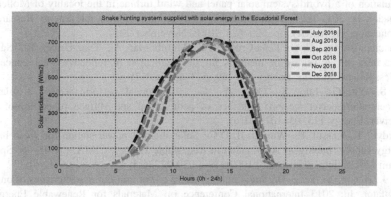

Fig. 12. Solar irradiances (W/m²) profile July 2018–Dec 2018.

Figure 12 shows the solar radiation profile that was measured in the vicinity where the location of this airplane type hotel is located in the Valle. It should be noted that the measurements with the weather station were not made on the site since security could not be installed in this secluded site. The station was located in the Quingeo Parish and its measurements correspond to several months of the year 2018 at different times of the day.

References

1. Chifa, C.: La perspectiva social de la medicina tradicional. Bol. Latinoam. Caribe Plant. Med. Aromát. **9**(4), 242–245 (2010)
2. Icaza, D., et al.: Modeling and simulation of a solar energy system in the vicinity of the hiking area in the cave of the tayos and its petroglyphs in the amazon of Ecuador. In: AICIBS-ICSEES 2018 September (Cambridge) Conference Proceedings. Cambridge Conference Series (2018). ISBN 978-1-911185-71-0
3. Zymla, H.G.: En torno a la iconografía de la serpiente de Asclepio: símbolo sanador de cuerpos y almas. Akros: Revista de Patrim. (6), 55–72 (2007)
4. Ramos Hernández, M., Ávila Bello, C.H., Morales Mávil, J.E.: Etnobotánica y ecología de plantas utilizadas por tres curanderos contra la mordedura de serpiente en la región de Acayucan, Veracruz, México. Bol. soc. Botánica México **81**, 89–100 (2007)
5. Morán, A., Fuertes, J., Gonzalez, M.D., Prada, M., Barrientos, P., Alonso, S.: Visual continuous maps for electric bills comparison. In: Controlo 2012, July 2012
6. Icaza, D., Mestas, S., Córdova, F., Calle, C.J., Icaza, F.: Terrestrial boundary signage with the USS midway museum of the historic military aircraft carrier in San Diego California using low scale. In: 2018 International Conference on Smart Grid (icSmartGrid), pp. 142–147. IEEE, December 2018
7. Alvarez, D.I., Castro, C.J.C., Gonzalez, F.C., Uguña, A.L., Toledo, J.F.T.: Modeling and simulation of a hybrid system solar panel and wind turbine in the locality of Molleturo in Ecuador. In: 2017 IEEE 6th International Conference on Renewable Energy Research and Applications (ICRERA), pp. 620–625. IEEE, November 2017
8. Arancibia-Bulnes, C.A., Peña-Cruz, M.I., Mutuberría, A., Díaz-Uribe, R., Sánchez-González, M.: A survey of methods for the evaluation of reflective solar concentrator optics. Renew. Sustain. Energy Rev. **69**, 673–684 (2017)
9. Li, S., Xu, G., Luo, X., Quan, Y., Ge, Y.: Optical performance of a solar dish concentrator/receiver system: influence of geometrical and surface properties of cavity receiver. Energy **113**, 95–107 (2016)
10. Baydyk, T., Kussul, E., Wunsch II, D.C.: Solar thermal power station for green building energy supply. Intelligent Automation in Renewable Energy. CIMA, pp. 45–75. Springer, Cham (2019). https://doi.org/10.1007/978-3-030-02236-5_4
11. Du, J., Huang, X., Tang, D.: Study on the radiative heat transfer characteristics of one solar simulator. In: 2013 International Conference on Materials for Renewable Energy and Environment, vol. 1, pp. 12–16. IEEE, August 2014

SMCS: Automatic Real-Time Classification of Ambient Sounds, Based on a Deep Neural Network and Mel Frequency Cepstral Coefficients

María José Mora-Regalado⬤, Omar Ruiz-Vivanco$^{(\boxtimes)}$⬤,
Alexandra González-Eras⬤, and Pablo Torres-Carrión⬤

Universidad Técnica Particular de Loja, Loja, Ecuador
oaruiz@utpl.edu.ec

Abstract. This paper presents a model to classify ambient sounds in an automatic and real-time way using the sound dataset provided in the Kaggle free sounds competition. For this, two data preprocessing techniques are performed, the first, length normalization that unifies the audio inputs to a single time interval and the second, property normalization that standardizes the sampling frequency and bit depth; This also includes a DNN (Deep Neural Network) capable of classifying common environmental sounds, the input for the network is formed by MFCC (Mel Frequency Cepstral Coefficients) vectors, which reduces the processing time improving the response capacity of the model for detect sounds, especially those that are considered warning signs about environmental threats, facilitating the mobility of people with hearing impairment.

Keywords: MFCC · DNN · Sounds classification · FSDkaggle2018

1 Introduction

In the field of sound classification several architectures are proposed, most of which relate to DNN, and some with emphasis on RNN. Our proposal combines vectors resulting from MFCC extraction and data normalization techniques, for the preparation of common environmental sounds that are classified by a DNN without the need to transform the input signal to spectra. In this way, the model performs the automatic classification of environmental sounds, especially those that identify threats to the integrity of people with partial deficits in the perception of sounds, and in this way, it constitutes an additional support to people with hearing impairment.

In this work we classify the sounds of the dataset of Kraggle competition, using a model based on Mel Frequency Cepstral Coefficients vectors input as networks and deep learning neural model. Then, we compare the results against a learning model based on spectrograms input and RNN and CNN neural networks. In this way, our hypothesis is to establish which of the two approaches obtains greater precision and achieves a better performance in the classification of sounds.

M. Botto-Tobar et al. (Eds.): ICAT 2019, CCIS 1194, pp. 245–253, 2020.
https://doi.org/10.1007/978-3-030-42520-3_20

1.1 Preliminary

Evaluate models for a cross-sectional environment such as environmental sounds, has been considered a research area for many equipment and laboratories worldwide. So, in general, [22] applying deep learning contribute from the study of the temporal relationships between the characteristics of audio and video; and [1] develop a neural network architecture, with a variant of automatic encoders, which combines audio and video, obtaining a signal with reduced interference effect. In the medical area, sound engineering has a wide scope of application, as evidenced by it [23] with its "deep learning-based cardiac sound recognition framework", and SRMT (spectral restoration and training of various styles) method, achieving relevant recognition results in clean noise environments.

Adapting methods form other areas like as Psychology [25] proposed a method trains the long short-Term memory (LSTM) using parameters of the RNN model with the standard intrusive intelligibility estimation method, applied in reference speech signal. Researchers and their teams are very creative for the design of new architectures and training models and evaluation of algorithms in this area.

Besides, there are several investigations using RNN models in speech classification such as: a "novel speaker-dependent speech separation framework for the challenging CHiME-5 acoustic environments" [21]; "NELE (near-end listening enhancement) System by RNN-based noise cancellation and speech modification (RNC-SM)", with a noise-cancelling function, adapted to the modification of speech [12]; others using variable network configuration options, such as the number of microphones and network topology [6]; and an architecture for acoustic classification, applying hybrid feature extraction methods (DNN and CNN) [2].

Otherwise, there are works using deep learning models and multi-speaker DOA (direction of arrival) estimation, to solve multi-class label classification problem [4], this architecture combines in lineal form several deep learning sub-models (views of the same objective concept) [11]. We cannot find one scientific article with Scopus (year >= 2017) with a mobile architecture for the automatic classification of environmental sounds oriented to cloud.

The most used architecture according to the studies found in the systematic search are based on deep neural networks (DNN). So, Cohen [9] propose a "deep encoder-decoder-based neural network architecture", that separates speech from non-speech frames, obtaining a diffusion nets architecture joining the decoder with the encoder. Mirsamadi and Hansen [17] detail a "multi-domain adversarial training", who from an AMI meeting corpus assesses end-to-end speech recognition.

In this sense, several original architectures and frameworks are adapted and proposed: a system inspired by the "Modular Selection for Identification and Control (MOSAIC) framework" [3]. A "blind Room Parameter Estimator (ROPE)" which it uses as an RNN training input the "ranges of the reverberation time (RT)" and "he early-to-late reverberation ratio (ELR)" from a single-microphone speech signals [24]. A method for "automatic context window composition based on a gradient analysis", obtaining a more effective training DNN in reverberant scenarios [20].

As a DNN, but with an emphasis on a recurrent neural network (RNN), Li [12] shares an DNN architecture, to predict noise signals that allow to eliminate noise without the feedback microphone. A DNN architecture that improves speech sounds by reducing input noise, adding a feature to the RNN (MRACC multiple resolution auditory cepstral coefficient) to capture the spectrum-temporal context and then an adaptive mask to track the noise change [13]. From the long short-Term memory (LSTM) structure design an architecture to "non-intrusive objective speech intelligibility estimation method" [24]. Finally, Crocco et al. [5] calculate a "classification map (CM) based on spectral signatures to cope noisy environments characterized by persistent and high energy interfering sources". Several architectures have been shared, most of which relate to DNN, and a few with an emphasis on RNN.

There are some techniques applied to solve this problem or similar. The most common are RNN algorithms used by [12, 13] to reduce the noise, and [25] training the LSTM-RNN model parameters, where input and output are the "MFCC vector" and the "frame-wise STOI value", respectively; Mirsamadi and Hansen [17] expose a strategy for training NNA (neural network acoustic) models, using a adversarial training with environment labels and video as inputs; in related study, Tao and Busso [22] propose a "bimodal recurrent neural network (BRNN) framework", where audio and visual features are learned from the raw data. Applying a convolutional neural network (CNN), Chakrabarty and Habets [4] explain a method for estimating the "direction of arrival (DOA) of multiple speakers" from synthesized noise signals as input to train method, and Crocco et al. [5] detail their classification map (CM) based on spectral signatures.

Using DNN, Ravanelli and Omologo [20] expose a method for automatic context window composition based on a gradient analysis, aiming for distant voice recognition; Koh and Woo [11] obtain better consensus between the "number of DL sub-models", applying the "cost function of the Laplacian eigenmap" and the "weights of the linear combination". Finally, Sun et al. [21] separate "target speech from multi-talker mixed speech collected with multiple microphone arrays" using both DL based and conventional preprocessing techniques, improving the signal in the reference matrix, managing to avoid the problem of speaker permutation. These methods show the application of innovative solutions with ML and DL algorithms, which are facilitating the discrimination of sounds in noisy environments.

2 Material and Method

The data set used is obtained from the Kaggle competition: Free-sound General-Purpose Audio Tagging Challenge, which provides a set of Free-sound audios, known as Free-sound Dataset Kaggle 2018 [10], this dataset has 9473 audios, with a duration between 300 ms (milliseconds) and 30 s (seconds). To improve the training, the fragments without audio sound are eliminated [7] and the length of the audios is standardized in 5 s (seconds), in case of the audios with shorter length, the length is completed by doubling or tripling the sound [18].

2.1 Deep Neural Network (DNN)

According to the architecture proposed in [13], DNN's strong non-linear mapping capability provides a better classification of sounds when there is background noise.

248   M. J. Mora-Regalado et al.

Deep learning follow these steps: first training and testing, after validation and finally inference phase [8]. DNN is a collection of neurons organized in a sequence of multiple hidden layers, strengthens classification and regression. The input layer contains the MFCC vectors, the hidden layer contains several layers, and the output consists of by a 64-dimensional vector. The activation function of the output layer is a sigmoid function. Figure 1 shows the structure of the DNN network architecture of the experimentation, which is configured as follows: MFCC vector input layer; hidden layers 1024-1024-1024-1024; output layer 64.

The classification success rate is calculated on the comparison of the results obtained in the tests and the sound label of each input of the supervised learning in a range of 0 to 100%. The training is carried out with three numbers of times, 1000, 5000 and 10000, according to the references found in previous experiments.

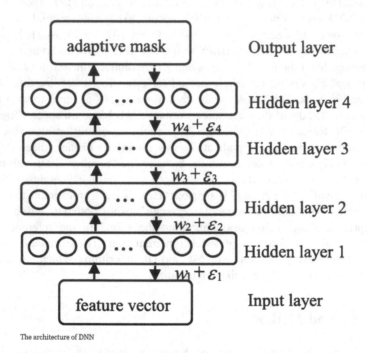

The architecture of DNN

Fig. 1. DNN architecture model used in [13]

2.2 Data Normalization

In order to obtain better results from the neuronal model, two datasets are generated: the first one where sounds are normalized in length, so that sounds between 300 ms and 30 s are normalized to a length of 5 s; and in the second the sounds are normalized according to the properties of the sampling frequency and the depth of bits. To obtain these properties, the load () function of the Python Librosa package [16] is used, which by default converts the sampling frequency to 22.05 kHz and the bit depth values to oscillate between −1 and 1. Figure 2 shows an example of the application of the technique, wherein the audio signal is normalized according to frequency and bit properties.

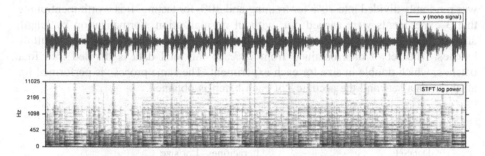

Fig. 2. Example of the application librasa.load() according to [16]

2.3 Extraction of MFCC

The MFCC (Cepstral Coefficients of Mel - Mel Frequency Cepstral Coefficients) show the local characteristics of the audio signal, commonly used in the classification of acoustic events [23]. MFCC vectors are used as inputs for each frame that enters the neural network [25]. By representing a broad spectrum of sounds in a compact manner, MFCCs have become the most used feature extraction technique in speech recognition [14].

Figure 3 represents the steps of the calculation of MFCCs summarized as follows:

a. Apply a pre-emphasis to the signal.
b. Hamming window.
c. Fast Fourier Transformation (FFT).
d. Calculate the values of the filter bank distributed in frequency, according to the Mel scale.
e. Logarithm of the transformed signal.
f. Discrete cosine transformation (DCT).
g. MFCC vector.

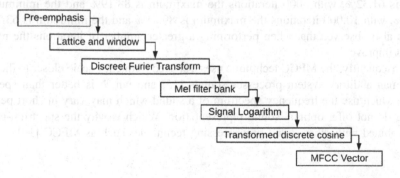

Fig. 3. Process for obtaining the MFCC coefficients according to [15]

3 Results

Table 1 shows the results of the sound classification model. As can be seen, the data set was randomly divided into 70% for training and 30% for testing. It is worth mentioning that 3 experiments were carried out: without normalization, normalization by length and normalization by audio properties. The MFCC is extracted from the three resulting batches. The set of data obtained from 70% of the total data is divided into four batches, which double in size for each experiment. The same process is performed for all three data sets and the accuracy of each batch is obtained.

Table 1. Results of enhanced experimentation.

DataSet	Iteraciones	Lot size			
		1657	3314	4971	6628
Without normalization	1000	61,32	62,76	61,32	63,35
	5000	69,45	64,22	69,45	63,74
	10000	63,76	68,93	63,76	68,76
Normalization by length	1000	76,32	83,23	79,97	80,08
	5000	80,63	83,97	83,35	83,69
	10000	83,75	81,76	89,41	86,46
Normalization by audio properties	1000	71,27	80,62	88,34	83,29
	5000	79,95	82,94	86,82	88,19
	10000	82,46	83,68	89,06	88,86

As observed in the results obtained by the sound classification model, without the normalization of the data, the maximum precision reached is 69.45% and the minimum is 61.32%. On the other hand, with the normalization of length a maximum is reached is 89.41% and the minimum is 76.32% in precision, and, with the normalization of properties the maximum is 89.06% and the minimum is 71.27%.

Then we can affirm that the tests performed with the unstandardized dataset obtain the worst results, and consequently, applying a normalization generates better results. Since with 1000 iterations a maximum precision of 88.34% is obtained and the minimum is 61.32%, with 5000 iterations the maximum is 88.19% and the minimum is 63.74%, with 10,000 iterations the maximum is 89.41% and the minimum is 63.76%; so it is also observed that when performing a greater number of iterations the results tend to improve.

Consequently, the MFCC technique uses a spaced frequency scale closer to the way the human auditory system processes sounds [15] and which is better than spectrograms, which use the frequency spectrum of a sound which may vary in short periods of time do not offer optimal sound representation. Which is why the spectrum-grams are combined with other sound preprocessing techniques such as MFCC [19].

4   Discussions

As mentioned in [18], the objective of the SMCS system is to propose the best model for automatic classification of environmental sounds, so it becomes essential to compare this work with the mentioned one, the most relevant points being the following:

The dataset used in both works is Free-sound Dataset Kaggle 2018 obtained in [10], composed of 9473 audios, it is highlighted in our proposal the batch separation of the original dataset [25] being incremental according to the following: 1st lot from 0 to 1657, 2nd lot from 0 to 3314, 3rd lot from 0 to 4971 and 4th lot from 0 to 6628, action that offers diversity in the generation of results.

In addition to the preprocessing performed in [18] on the audio segments that are normalized to a level of −0.1 dB and the elimination of silence spaces, in this work a tool [16] is used to balance the frequency of the audios.

Besides, an alternative to the aforementioned process is the use of a DNN using the vectors obtained from the MFCC as inputs, thus avoiding the transformation of these coefficients to a graphic representation. In [18] an experiment is presented for classification of sounds with the dataset of [10], using as a CNN classification model based on AlexNet with the preprocessing technique of graphic representation of audio spectrograms, compared with the experimentation of this work with the mentioned techniques there is an increase of 10 percentage points in the successful classification of audios (Fig. 4).

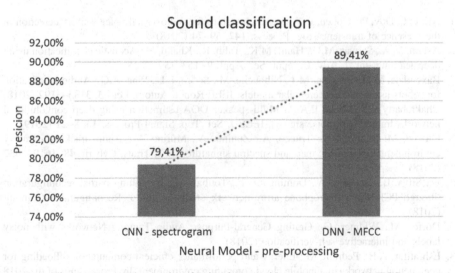

Fig. 4.   Present the results of using inputs with spectrogram and MFCC.

The sound classification model applied in this DNN work in comparison to that used in [18] CNN, represents the most marked difference that both works have, directly related to the input source for the neural network, while in [18] spectrograms of the sound frames are obtained, here MFCC vectors proposed in [17] are implemented as an improvement of the input technique to be used by DNN. In this way, the capacities of the classification model are maximized, establishing a 10% increase in the effectiveness of classification of environmental sounds in favor of this work.

5 Conclusions

This paper proposes a classification model of environmental sounds that result in danger alerts for people with hearing impairment, using audio preprocessing techniques, such as MFCC and normalization of the length of and properties of sounds as input for the DNN training. According to the results obtained, the classification model presents better precision results when it comes to audio signals, than by graphic representation of spectrograms, data that are evidenced with a prediction rate of 89.41% accuracy, obtained with the normalization of length and with 10,000 iterations, surpassing the experimentation carried out in [18] where the best result was 79.41%.

The following work steps are oriented towards the integration of this model into the SMCS system, increasing its capacity in the detection and classification of environmental sounds and the development of the evaluation of the SMCS system in real contexts.

References

1. Ariav, I., Dov, D., Cohen, I.: A deep architecture for audio-visual voice activity detection in the presence of transients. Sig. Process. **142**, 69–74 (2018)
2. Aslam, M.A., Sarwar, M.U., Hanif, M.K., Talib, R., Khalid, U.: Acoustic classification using deep learning. Int. J. Adv. Comput. Sci. Appl. **9**, 153–159 (2018)
3. Baxendale, M.D., Pearson, M.J., Nibouche, M., Secco, E.L., Pipe, A.G.: Audio localization for robots using parallel cerebellar models. IEEE Robot. Autom. Lett. **3**, 3185–3192 (2018)
4. Chakrabarty, S., Habets, E.A.P.: Multi-speaker DOA estimation using deep convolutional networks trained with noise signals. IEEE J. Sel. Top. Signal Process. **13**, 8–21 (2019)
5. Crocco, M., Martelli, S., Trucco, A., Zunino, A., Murino, V.: Audio tracking in noisy environments by acoustic map and spectral signature. IEEE Trans. Cybern. **48**, 1619–1632 (2018)
6. Da Silva, B., Braeken, A., Domínguez, F., Touhafi, A.: Exploiting partial reconfiguration through PCIe for a microphone array network emulator. Int. J. Reconfigurable Comput. (2018)
7. Dorfer, M., Widmer, G.: Grating General-Purpose Audio Tagging Networks with noisy labels and interactive self-verification (2018)
8. Eshratifar, A.E., Pedram, M.: Energy and performance efficient computation offloading for deep neural networks in a mobile cloud computing environment. In: Proceedings of the 2018 on Great Lakes Symposium on VLSI, pp. 111–116. ACM, New York (2018)
9. Ivry, A., Berdugo, B., Cohen, I.: Voice activity detection for transient noisy environment based on diffusion nets. IEEE J. Sel. Top. Signal Process. **13**, 254–264 (2019)
10. Kaggle: Freesound General-Purpose Audio Tagging Challenge (2018). https://www.kaggle.com/c/freesound-audio-tagging
11. Koh, B.H.D., Woo, W.L.: Multi-view temporal ensemble for classification of non-stationary signals. IEEE Access. **7**, 32482–32491 (2019)

12. Li, G., Hu, R., Wang, X., Zhang, R.: A near-end listening enhancement system by RNN-based noise cancellation and speech modification. Multimed. Tools Appl. **78**, 15483–15505 (2019)
13. Li, R., Sun, X., Liu, Y., Yang, D., Dong, L.: Multi-resolution auditory cepstral coefficient and adaptive mask for speech enhancement with deep neural network. EURASIP J. Adv. Signal Process. (2019)
14. Logan, B.: Mel frequency cepstral coefficients for music modeling. In: International Symposium on Music Information Retrieval (2000)
15. Martínez Mascorro, G.A., Aguilar Torres, G.: Reconocimiento de voz basado en MFCC, SBC y Espectrogramas. INGENIUS Rev. Cienc. Tecnol. **10**, 12–20 (2013)
16. McFee, B., et al.: Librosa: audio and music signal analysis in python. In: Proceedings of the 14th Python in Science Conference (2015)
17. Mirsamadi, S., Hansen, J.H.L.: Multi-domain adversarial training of neural network acoustic models for distant speech recognition. Speech Commun. **106**, 21–30 (2019)
18. Mora-Regalado, M.J., Ruiz-Vivanco, O., Gonzalez-Eras, A.: SMCS: mobile model oriented to cloud for the automatic classification of environmental sounds. In: Botto-Tobar, M., León-Acurio, J., Díaz Cadena, A., Montiel Díaz, P. (eds.) ICAETT 2019. AISC, vol. 1066, pp. 464–472. Springer, Cham (2020). https://doi.org/10.1007/978-3-030-32022-5_43
19. Ozer, I., Ozer, Z., Findik, O.: Noise robust sound event classification with convolutional neural network. Neurocomputing **272**, 505–512 (2018). https://doi.org/10.1016/j.neucom.2017.07.021
20. Ravanelli, M., Omologo, M.: Automatic context window composition for distant speech recognition. Speech Commun. **101**, 34–44 (2018)
21. Sun, L., Du, J., Gao, T., Fang, Y., Ma, F., Lee, C.-H.: A speaker-dependent approach to separation of far-field multi-talker microphone array speech for front-end processing in the CHiME-5 challenge. IEEE J. Sel. Top. Signal Process. **13**, 827–840 (2019)
22. Tao, F., Busso, C.: End-to-end audiovisual speech activity detection with bimodal recurrent neural models. Speech Commun. **113**, 25–35 (2019)
23. Tsao, Y., Lin, T.-H., Chen, F., Chang, Y.-F., Cheng, C.-H., Tsai, K.-H.: Robust S1 and S2 heart sound recognition based on spectral restoration and multi-style training. Biomed. Signal Process. Control **49**, 173–180 (2019)
24. Xiong, F., Goetze, S., Kollmeier, B., Meyer, B.T.: Exploring auditory-inspired acoustic features for room acoustic parameter estimation from monaural speech. IEEE/ACM Trans. Audio Speech Lang. Process. **26**, 1809–1820 (2018)
25. Yun, D., Lee, H., Choi, S.H.: A deep learning-based approach to non-intrusive objective speech intelligibility estimation. IEICE Trans. Inf. Syst. **E101D**, 1207–1208 (2018)

Artificial Intelligence Techniques to Detect and Prevent Corruption in Procurement: A Systematic Literature Review

Yeferson Torres Berru[1,2](✉) (iD), Vivian Félix López Batista[1](✉),
Pablo Torres-Carrión[3](✉), and Maria Gabriela Jimenez[2]

[1] University of Salamanca, Plaza de la Merced, s/n, 37008 Salamanca, Spain
{ymtorresb, vivian}@usal.es
[2] Instituto Superior Tecnológico Loja,
Av. Granada y Turunuma, 1101608 Loja, Ecuador
[3] Universidad Técnica Particular de Loja,
San Cayetano Alto S/N, 1101608 Loja, Ecuador
pvtorres@utpl.edu.ec

Abstract. Transparency International estimates that the costs of corruption in public procurement reach between 20 and 25% of the contract value, sometimes reaching 40–50%. In this study, we analyzed differentness kinds of corruption like (bribery, collusion embezzlement, misappropriation, fraud, abuse of discretion, favoritism, nepotism), and six types of Artificial Intelligence techniques (classification, regression, clustering, prediction, outlier detection, and visualization). The methodology proposed by Torres-Carrion was used, and four research questions were raised, which allow knowing the types of research carried out, the characteristics of the organizations in which the investigations are carried out, the technological tools, and data mining methodologies and techniques. The search was done in the Scopus and Web of Science databases, getting 102 articles published between 2015 and 2019. The primary data mining techniques used are logistic models, neural networks, Bayesian networks, supported vector machines, and decision trees.

Keywords: Artificial Intelligence · Corruption · Data mining · Procurement · Systematic literature review

1 Introduction

Corruption expenditures are equivalent to 5% of global GDP, according to the G-20 [1], being the third most lucrative "industry" of all those in the world. Transparency International estimates that the costs of corruption in public contracts average 20–25% of the contract value, and can reach 40–50% in some cases [2]; public procurement accounted for 32.5% of government expenditure. The highest risk of corruption in hiring occurs during the planning stage, potential frauds in the procurement system take very diverse forms starting from bribery, collusion embezzlement, misappropriation, fraud, abuse of discretion, favoritism, nepotism [3].

© Springer Nature Switzerland AG 2020
M. Botto-Tobar et al. (Eds.): ICAT 2019, CCIS 1194, pp. 254–268, 2020.
https://doi.org/10.1007/978-3-030-42520-3_21

Data mining combines Artificial Intelligence techniques (classification, regression, clustering, prediction, outlier detection, and visualization) with statistical analysis techniques (clustering, dimensional analysis, etc.), which allow analyzing information such as data, text, images, audio. In automatic learning, algorithms can be classified as supervised, unsupervised, and reinforcement [4]. These fields of knowledge, discussed for this systematic review of the literature (SLR), are detailed in Fig. 1, which allows establishing the theoretical knowledge base on which the review has its basis to contribute to the knowledge about data mining research and artificial intelligence for the detection and prevention of anomalies in contracts.

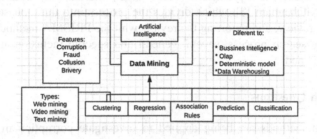

Fig. 1. Theoretical representation of data mining in contracts

1.1 Related Systematic Review of Literature

After the SLR, we proceed to look for previous literature review works related to the theoretical constructs: AI, data mining, and corruption. In the results obtained (see Table 1), the studies on the different types of fraud, at the private and governmental levels, are discussed, the different techniques of automatic learning and data mining are also considered in the process of fraud detection and prevention. All research is more than five years old and does not answer the research questions proposed in this study.

Table 1. Related reviews

Article	Analysis	#Papers
Fraud detection system: a survey	Survey about fraud prevention systems five areas of electronic frauds (e-fraud): credit card, telecommunication, health care insurance, automobile insurance and online auction in papers between 1994 and 2014	70
A survey of machine-learning and nature-inspired based credit card fraud detection techniques [5]	The survey revealed that various machine learning-based and nature-inspired algorithms had been used to handle credit card fraud detection between 2010 and 2014	47
The application of data mining techniques in financial fraud detection: a classification framework and an academic review of the literature [6]	Review data mining techniques in financial fraud, the articles were published between 1997 and 2008 analyzing fraud in bank, insurance and securities categories, the six data mining techniques (classification, regression, clustering, prediction, outlier detection, and visualization) and the main artificial intelligence methods logistic models, neural networks, the Bayesian belief network, and decision trees	49

Section two details the applied methodology, research questions, inclusion criteria, and semantic structure of the search.; in section three, research questions are answered based on the 102 articles found; finally, in section four, conclusions and future work are presented.

2 Method

2.1 Method for SLR

We used the method for a systematic review of the literature by Torres-Carrión [7] adapted from Kitchenham [8], which divides the process into three phases: planning, conducting the review, and reporting the review whit the PRISMA method [9]. As part of planning the search process, several general and specific inclusion and exclusion criteria were defined.

2.2 Research Questions

RQ1: What methods are being applied to investigate corruption in public procurement contracts?
RQ2: What are the characteristics of the organizations in which the research has been carried out?
RQ3: What technological tools are being used to investigate the detection and prevention of corruption?
RQ4: What algorithms, methodologies, and data analysis tools are used to detect corruption?

2.3 Quality Inclusion and Exclusion Protocols

As inclusion criteria, scientific articles published in the Web of Science (WOS) and Scopus databases during the years 2015–2019 are considered to be the type of document (article, review, editorial or conference proceedings), research area (Computer Science, Social Sciences and Decision Sciences); it excludes repeated documents, short papers, posters, and book chapters. As a quality criterion, a detailed review of the articles is carried out, filtering the studies that do not analyze corruption and its forms, or that do not apply data mining and artificial intelligence.

2.4 Semantic Search Structure

From the theoretical constructs (see Fig. 1), synonyms are sought in the scientific thesaurus; operators are applied (OR, AND, W/) to optimize the search, although Table 2 only shows the search query for Scopus, WOS search replaces the operators with the correspondents to its database. The procedure is detailed in levels with the resulting number of articles, as the inclusion, exclusion, and quality criteria are applied. One hundred two articles are obtained, identified as valid and explicitly related to the problem raised, and with which this SLR is working.

Table 2. Semantic search structure

Level	Thesaurus	SCOPUS script	Scopus	WOS
L1	Data mining			
	Mining W/4 (data or video or text or web)) Classificat* Cluster* Regression Association rules Detection Prediction Sequential Patterns	(Mining W/4 (data OR video OR text OR web)) OR classificat* OR cluster* OR regression OR (association W/2 rules) OR detection OR prediction OR (sequential W/2 patterns) OR (learning W/4 (machine OR deep OR reinforced))	6,962,346	8,072,286
L2	Corruption	Corruption OR	9,653	35,781
	Bribery, collusion, embezzlement, fraud, abuse of discretion, favoritism, nepotism	Bribery OR collusion OR (embezzlement OR misappropriation) OR fraud OR (abuse W/0 of W/0 discretion) OR favoritism OR nepotism)		
L3	Contracts	(Contract OR	691	1.811
	Contract, purchase, investment, procurement, acquisition, acquirement, tendering	Purchase OR investment OR procurement OR acquisition OR acquirement OR tendering)		
L4	Review protocol			
	(Last 5 years) from 2015		351	796
	(Research Areas) Computer Science, Social Sciences, Decision Sciences		222	275
	Article or review or editorial or conference proceedings		210	255
	Quality criteria		58	54
L5 Combination of results in Scopus and WOS (repeated = 5)			102	

3   Results

The results are presented in conformity with the research questions established in the methodology, and their corresponding variables and indicators.

RQ1: What methods are being applied to investigate corruption in public procurement contracts?

Knowing the methodology used by other researchers gives light to the planning of new studies. 87% of investigations work with a previously structured database; it is also observed that techniques such as web Scraping (3%) are rarely used for data collection (see Table 3). Quantitative research is the most used, as well as the statistical method of correlation. As for the time of validity of the data for the study, most of them are 1–3 years, and most of the research is an experimental type, with contributions to computer science. In this sense, the research highlights [10], which performs research with multivariate analysis and correlation, using web scraping and 4-year data.

Table 3. Research question 1.

Data collection instrument [11, 12]		f
Survey	[13–18]	6
Web Scraping	[13, 19]	2
Database	[10, 20–76]	57
Type of research [12]		
Qualitative	[77–79]	3
Quantitative	[10, 11, 13–102]	90
Mixt	[5, 103–110]	9
Statistics to evaluate results [12]		
Univariate	[14, 24, 26, 34, 64, 66, 80–90]	17
Multi-varied	[70, 91, 93]	6
Correlation	[10, 13, 15–23, 25, 27–33, 35–39, 41–48, 50–61, 65, 67–69, 71–77, 81, 89, 92, 95, 97–100, 111]	79
Data period		
<1 year	[26, 28, 29, 32, 40, 81]	6
1> & ≤ 3	[41, 43, 46, 47, 52, 53, 59, 61, 64–67, 71, 72, 75, 76, 95, 101, 102]	23
>3& ≤ 5	[10, 11, 13, 21, 27, 39, 49, 86, 97]	9
>5& ≤ 10	[92, 99]	2
>10	[20, 46, 52, 60, 82]	5
Research design [12]		
Experimental	[10, 11, 13, 16, 17, 19, 20, 22–26, 29–35, 37–76, 80, 86, 88, 91, 92, 95–97, 99, 100, 111]	70
No experimental	[5, 14, 23, 24, 36, 77–79, 82–85, 87–89, 94, 95, 98, 101, 104, 110]	23
Quasi experimental	[15, 18, 21, 81]	4
Field of science		
Computer Science	[11, 13, 17–26, 29–34, 36–40, 45–59, 62–76, 80, 81, 86–88, 90, 92, 94, 95, 99, 103, 104, 110, 111]	70
Economy	[27, 28, 35, 77–79, 82, 84, 85, 100]	10
Mathematic	[14, 34, 83, 89]	4
Statistics	[10, 16, 39, 61, 80, 91, 96]	6

RQ2: What are the characteristics of the organizations in which the research has been carried out?

50% of the works focus on the private sector, with a significant presence of experimental type works that relate the public and private sectors (17%). The main commercial activity of the organizations is the provision of services, having the banking sector as the most studied (64%) (see Table 4). In the government sector, tax fraud is prominent (33%), and public purchases account for only 6% of total investigations.

Table 4. Research question 2.

Activity sector		f
Public	[10, 16, 18, 24, 34, 35, 39, 42, 45, 47, 48, 50, 54, 57, 65, 66, 68, 71, 76, 77, 81, 84–87, 89, 91, 92, 97, 100]	31
Private	[11, 13, 15, 16, 18, 19, 25–29, 31–33, 36–38, 43, 46, 51–53, 58–64, 70, 72, 75, 88, 90, 91, 101, 102, 104–107, 111]	49
Mixt	[10, 11, 17, 30, 44, 49, 78–80, 83, 92, 93]	17
Commercial activity		
Services	[5, 13, 18, 21–23, 25–29, 33, 36–38, 47, 50–53, 55–57, 59, 62, 63, 70–72, 76, 81, 88, 90, 92, 104, 111]	36
Commercial	[19, 31, 32, 49, 58, 64, 65, 94]	8
Government	[10, 11, 16, 24, 35, 39, 41–43, 45, 46, 48, 54, 61, 67, 68, 86, 87, 89, 91, 100]	22
Organization		
Bank	[5, 21, 22, 25, 26, 29, 33, 36, 37, 47, 51, 55, 59, 62, 63, 70–72, 76, 79, 88, 90, 96, 104, 111]	25
Buys online	[32, 65]	2
Public buys	[10, 16, 35, 42, 43, 54, 69]	7
Taxes	[41, 67, 87, 91, 100]	5

In the mixed activity sector, the research conducted by Dhurandhar [30] proposes a bigdata-based solution for risk analysis in public and private sector procurement; focused on the public sector, in [46] a framework for the detection of crimes in contracts is presented, supported by information provided by the World Bank; in the private sector in [90] an experimental analysis of data mining tools in the banking sector is conducted.

RQ3: What technological tools are being used to investigate the detection and prevention of corruption?

Twenty articles (20/112) are selected that in their methodology propose the use or development of a technological tool to evaluate corruption (see Table 5); 89% are desktop tools; in terms of web platforms, these analyze corruption in contracts in public procedures [30, 69] and in the process of buying medicines [13].

The degree of reliability in the detection or prevention of corruption, as the case may be, has as its greatest interval between the 80%-90%; Baader et al. [17] obtain the lowest detection rate with 48.6%, applying the red flag approach with mining process to reduce the number of false positives in fraud analysis, whereas Darwish [23] obtains the best detection rate with a 98,5% in the analysis of credit card transactions in the banking environment; in the government environment, [43] work detects fake suppliers from the analysis of satellite images of the locations of companies, obtaining the best index of detection (97%). The pattern of software engineering and the computer security standard used in the tools were also analyzed, but no coincidences were found in the analysis of the work, excepting Carminati [71], which, in addition to the analysis tool, presents a mechanism for preventing the Mimicry Attack.

Table 5. Research question 3.

Platform		f
Desktop	[19, 23, 24, 31, 38–40, 46, 48, 52, 56, 58, 61, 64, 86, 92, 95]	17
Web	[13, 30, 69]	3
Percent of detection/prevention		
<70	[17, 24]	2
70–80	[46, 56]	2
80–90	[10, 48, 52–54, 60, 63, 64, 68]	9
91–95	[13, 32, 38, 57]	4
96–99	[23, 25, 29, 43, 65, 67]	6

RQ4: What algorithms, methodologies, and data analysis tools are used to detect corruption?

In this question (see Table 6), we evaluated factors like the data source, type, kind of mining, preprocessing methods, outliers values processing, evaluation metrics, artificial intelligence techniques, types and learning techniques, and technological tools.

Table 6. Research question 4

Data source		f
Re. Public	[18–24, 26, 30, 41, 43, 45, 46, 54, 61, 80, 86]	17
Re. Private	[11, 13, 25, 26, 38, 46–49, 55–57, 59, 63, 65, 67, 72, 92, 95, 111]	30
Type of data		
Data	[10, 13, 16, 17, 20–23, 25–27, 30–33, 38, 40–46, 48, 50, 53, 55, 57–61, 65, 66, 68, 70, 72, 74, 75, 78–82, 87, 90–92, 95–100, 102, 103, 105, 106, 109]	59
Text	[31, 49, 62, 64]	4
Audio	[54]	1

(*continued*)

Table 6. (*continued*)

Type of minning		
Predictive	[13, 21, 25, 46, 55–57, 59–61, 80, 88, 97]	13
Descriptive	[10, 11, 13, 16, 18–20, 22, 23, 25, 26, 29–32, 37–45, 47–49, 51–54, 63–65, 67–70, 72–75, 80, 86, 91, 92, 95, 96, 111]	51
Preprocessing of data		
Empirical assessment	[19, 25, 39, 41, 45–47, 55, 63, 75, 86]	13
Other	Cascade generalization [70] CKIP [48] Monroe and The [60]	3
Outlier values		
HBOS	[29, 45, 48, 57, 71, 80]	6
PSO	[48, 57, 59, 75]	4
Metric evaluation		
Confusion matrix	[13, 17, 18, 20, 23–25, 29, 32, 43, 49, 53, 60, 68, 80, 86, 111]	17
Curve ROC	[22, 26, 47, 54, 67, 71, 75]	7
Fraud Score	[53, 63]	3
K-fold	[57, 95]	2
Other	Matthews Correlation Coefficient [71]	
Artificial Intelligence technics		
Bayes based	[31, 38, 44, 45, 49, 52–54, 57, 60, 73, 99]	12
Neural network	[18, 19, 22, 37, 41, 47, 52, 68, 88]	9
SVM	[41, 44, 45, 48, 49, 52, 55, 87, 109]	9
Decision tree	[26, 29, 32, 51, 52, 55, 67]	7
Random Forest	[24, 25, 51, 55, 63, 67, 70]	7
Logistic- Linear regression	[14, 34, 41, 44, 45, 49, 51, 52, 67, 83, 87, 97]	12
Other	[11, 20, 36, 42, 44, 48–52, 56, 58, 59, 64, 70, 72, 86, 88]	17
Learning techniques		
Machine learning	[13, 18, 19, 23–26, 29–32, 37, 42, 46, 48–51, 53–57, 59, 64, 65, 67–70, 72–74, 76, 80, 81, 88, 91, 95, 111]	41
Deep learning	[38, 43, 47, 49]	4
Minning techniques		
Clustering	[18, 29, 30, 50, 59, 68, 70, 72, 73, 75, 76, 103]	41
Classification	[13, 18, 24–26, 29, 31, 32, 37, 38, 41, 43, 45–49, 52–54, 75, 76, 81, 95]	51
Regression	[41, 76]	2
Types of learning		
Supervised	[18–20, 25, 26, 31, 32, 37, 38, 42, 54, 55, 57, 68, 91]	15
No supervised	[18, 24, 30, 56, 59, 73]	6
Semi supervised	[29, 70, 71]	3

(*continued*)

Table 6. (*continued*)

Technological tools		
Java	[19, 64, 65, 95]	4
MatLab	[23, 41, 59, 64]	4
Python	[24, 31, 46, 65, 69, 80]	6
Weka	[44, 45, 52]	3
Others	[13, 38, 42, 50, 53, 57, 59, 61, 62, 70, 86, 92, 95]	13

Private datasets predominate in data collection sources 63% (see Table 6); this is because, as mentioned above, most work focuses on the private sector; the type of data used to generate and validate detection models in a large percentage is data (database or dataset). It should be noted that only 6% of jobs use text (documents), and only one job [54], the audio of telephone conversations are used to find fraud in public purchases.

The review focuses on the prevention and detection of corruption, with 79% of investigations of a descriptive type (detection) and 21% of a predictive type (prevention); in the techniques for the pre-processing of data, empirical assessment predominates with 81%, and amongst the methods are Cascade generalization [70] Chinese Knowledge Information Processing Group (CKIP) [48] Monroe and The [60]. In order to detect outliers in the data, the following method is used in similar percentages Particle Swarm Optimization (PSO) and the histogram-based outlier score (HBOS).

It is observed that the main techniques of artificial intelligence are those based on the theorem of Bayes, neural networks, Support Vector Machines (SVM), decision trees, Random Forest, and logistic and linear regression, reaching 76% among all of them; to a lesser extent technique such as: convolutional networks, tough set theory, graphs, natural language processing, kmeans, AdaBoost, genetic algorithms, bagging and logitBoos. The investigation with the highest number of applied techniques is [49], evaluating eleven artificial intelligence techniques to detect price manipulation in purchases.

The main evaluation metrics are: accuracy, efficiency, recalling using methods such as confusion matrix and ROC curves; in assessing of corruption, the significant contribution of the fraud score used as an evaluation indicator in some works, the confusion matrix is combined with the fraud score [53], and machine learning evaluation methods such as the Matthews Correlation Coefficient [71].

Concerning computer tools for programming, processing, and data storage Java, MatLab, Weka y Python have the highest percentage, with other tools such as R, RapidMiner, Hadoop, Spark, Neo4j, Casandra, Kafta, Visual Studio.

4   Conclusions and Future Work

Experimental quantitative research is the most widely used, as well as the statistical method of correlation. Most of the study is conducted with data over 1–3 years, with a significant contribution to computer science. The main commercial activity of the organizations is the provision of services, specifically the banking sector. In the government sphere, tax fraud is noteworthy, with a lesser presence of public procurement processes.

Web Scraping is a rarely used technique for obtaining data on corruption studies in contracts and can be used as a basis for future work. The few jobs related to contract analysis in public procurement use datasets and are not considered documents as the initial basis for data analysis. It is also evident that the computer tools created to carry out corruption analysis in contracts in both the public and private sectors are not considered computer security standards, and the percentage of tools in the web environment is very low.

The main artificial intelligence techniques found are logistic models, neural networks, bayesian networks, and supported vector machines. As future work, they would be enhanced with mixed learning methods. Fraud Score is proposed as a specific metric to assess the risk of corruption without leaving aside the metrics used to evaluate from the confusion matrix and ROC curves, with machine learning and supervised learning as the main types of technique.

References

1. Báez Gómcz, J.E.: Relación entre el Índice de Control de la Corrupción y algunas variables sociales, económicas e institucionales. Nómadas. Rev. Crítica Ciencias Soc. y Jurídicas **38** (2013)
2. Volosin, N.A.: Datos abiertos, corrupción y compras públicas (2015)
3. Padhi, S.S., Mohapatra, P.K.J.: Detection of collusion in government procurement auctions. J. Purch. Supply Manag. **17**, 207–221 (2011)
4. Martin, R.: A review of the literature of the followership since 2008: the importance of relationships and emotional intelligence. SAGE Open **5**, 2158244015608421 (2015)
5. Adewumi, A.O., Akinyelu, A.A.: A survey of machine-learning and nature-inspired based credit card fraud detection techniques. Int. J. Syst. Assur. Eng. Manag. **8**, 937–953 (2017)
6. Ngai, E.W.T., Hu, Y., Wong, Y.H., Chen, Y., Sun, X.: The application of data mining techniques in financial fraud detection: a classification framework and an academic review of literature. Decis. Support Syst. **50**, 559–569 (2011)
7. Torres-Carrion, P.V., Gonzalez-Gonzalez, C.S., Aciar, S., Rodriguez-Morales, G.: Methodology for systematic literature review applied to engineering and education. IEEE Global Engineering Education Conference EDUCON, pp. 1364–1373, April 2018
8. Kitchenham, B., Pearl Brereton, O., Budgen, D., Turner, M., Bailey, J., Linkman, S.: Systematic literature reviews in software engineering – a systematic literature review. Inf. Softw. Technol. **51**, 7–15 (2009)
9. Moher, D., et al.: PRISMA-P: preferred reporting items for systematic review and meta-analysis protocols (PRISMA-P). Syst. Rev. 1–9 (2015)
10. Auriol, E., Straub, S., Flochel, T.: Public procurement and rent-seeking: the case of paraguay. World Dev. **77**, 395–407 (2016)
11. Lei, M., Yin, Z., Li, S., Li, H.: Detecting the collusive bidding behavior in below average bid auction. In: ICNC-FSKD 2017 - 13th International Conference on Natural Computation, Fuzzy Systems and Knowledge Discovery, pp. 1720–1727 (2018)
12. Hernández Sampieri, R., Fernández Collado, C., Baptista Lucio, M.: Metodología de la Investigación (2010)

13. Kose, I., Gokturk, M., Kilic, K.: An interactive machine-learning-based electronic fraud and abuse detection system in healthcare insurance. Appl. Soft Comput. J. **36**, 283–299 (2015)
14. Charles Andoh, E., Ofosu-Hene, D.: Causes, effects and deterrence of insurance fraud: evidence from Ghana. J. Financ. Crime Iss. **5**, 39–44 (2016)
15. Huang, S.Y., Lin, C.C., Chiu, A.A., Yen, D.C.: Fraud detection using fraud triangle risk factors. Inf. Syst. Front. **19**, 1343–1356 (2017)
16. Seck, A.: Heterogeneous bribe payments and firms' performance in developing countries. J. African Bus. **21**, 1–20 (2019)
17. Baader, G., Krcmar, H.: Reducing false positives in fraud detection: combining the red flag approach with process mining. Int. J. Account. Inf. Syst. **31**, 1–16 (2018)
18. Choi, D., Lee, K.: An artificial intelligence approach to financial fraud detection under IoT environment: a survey and implementation. Secur. Commun. Netw. **2018** (2018)
19. Sadaoui, S., Wang, X.: A dynamic stage-based fraud monitoring framework of multiple live auctions. Appl. Intell. **46**, 197–213 (2017)
20. Yeh, C.C., Chi, D.J., Lin, T.Y., Chiu, S.H.: A hybrid detecting fraudulent financial statements model using rough set theory and support vector machines. Cybern. Syst. **47**, 261–276 (2016)
21. Ouenniche, J., Uvalle Perez, O.J., Ettouhami, A.: A new EDAS-based in-sample-out-of-sample classifier for risk-class prediction. Manag. Decis. **57**, 314–323 (2019)
22. Zakaryazad, A., Duman, E.: A profit-driven Artificial Neural Network (ANN) with applications to fraud detection and direct marketing. Neurocomputing **175**, 121–131 (2014)
23. Darwish, S.M.: An intelligent credit card fraud detection approach based on semantic fusion of two classifiers. Soft. Comput. **24**, 1243–1253 (2019)
24. Kehler, E., Paciello, J., Pane, J.: Anomaly detection in public procurements using the open contracting data standard (2019)
25. Van Vlasselaer, V., et al.: APATE: a novel approach for automated credit card transaction fraud detection using network-based extensions. Decis. Support Syst. **75**, 38–48 (2015)
26. Zareapoor, M., Shamsolmoali, P.: Application of credit card fraud detection: based on bagging ensemble classifier. Procedia Comput. Sci. **48**, 679–685 (2015)
27. Ngoc, B.H., Hai, D.B., Chinh, T.H.: Assessment of the should be effects of corruption perception index on foreign direct investment in ASEAN countries by spatial regression method. In: Anh, Ly H., Dong, L.S., Kreinovich, V., Thach, N.N. (eds.) ECONVN 2018. SCI, vol. 760, pp. 421–429. Springer, Cham (2018). https://doi.org/10.1007/978-3-319-73150-6_33
28. Burböck, B., Macek, A., Podhovnik, E., Zirgoi, C.: Asymmetric influence of corruption distance on FDI. J. Financ. Crime (2018)
29. Carminati, M., Caron, R., Maggi, F., Epifani, I., Zanero, S.: BankSealer: a decision support system for online banking fraud analysis and investigation. Comput. Secur. **53**, 175–186 (2015)
30. Dhurandhar, A., Graves, B., Ravi, R., Maniachari, G., Ettl, M.: Big data system for analyzing risky procurement entities. In: Proceedings of the 21st ACM SIGKDD International Conference on Knowledge Discovery and Data Mining, pp. 1741–1750 August 2015
31. Hooi, B., et al.: BIRDNEST: Bayesian inference for ratings-fraud detection. In: 16th SIAM International Conference on Data Mining 2016, SDM 2016, pp. 495–503 (2016)
32. Snyder, P., Kanich, C.: Characterizing fraud and its ramifications in affiliate marketing networks. J. Cybersecur. **2**, 71–81 (2016)

33. Moalosi, M., Hlomani, H., Phefo, O.S.D.: Combating credit card fraud with online behavioural targeting and device fingerprinting. Int. J. Electron. Secur. Digit, Forensics (2019)
34. Anh, N.N., Minh, N.N., Tran-Nam, B.: Corruption and economic growth, with a focus on Vietnam. Crime, Law Soc. Change **45**, 307–324 (2016)
35. Ferwerda, J., Deleanu, I., Unger, B.: Corruption in public procurement: finding the right indicators. Eur. J. Crim. Policy Res. **23**, 245–267 (2017)
36. Amanze, B.C., Onukwugha, C.G.: Credit card fraud detection system in nigeria banks using adaptive data mining and intelligent agents: A review. Int. J. Sci. Technol. Res. **7**, 175–184 (2018)
37. Zanin, M., Romance, M., Moral, S., Criado, R.: Credit card fraud detection through parenclitic network analysis. Complexity **2018** (2018)
38. Randhawa, K., Loo, C.K., Seera, M., Lim, C.P., Nandi, A.K.: Credit card fraud detection using AdaBoost and majority voting. IEEE Access **6**, 14277–14284 (2018)
39. Ausloos, M., Cerqueti, R., Mir, T.A.: Data science for assessing possible tax income manipulation: the case of Italy. Chaos, Solitons Fractals **104**, 238–256 (2017)
40. Helmy, T.H., Zaki, M., Salah, T., Badran, K.: Design of a monitor for detecting money laundering and terrorist financing. J. Theor. Appl. Inf. Technol. **85**, 425–436 (2016)
41. Rahimikia, E., Mohammadi, S., Rahmani, T., Ghazanfari, M.: Detecting corporate tax evasion using a hybrid intelligent system: a case study of Iran. Int. J. Account. Inf. Syst. **25**, 1–17 (2017)
42. Van Erven, G.C.G., Carvalho, R.N., De Holanda, M.T., Ralha, C.: Graph database: a case study for detecting fraud in acquisition of Brazilian Government (Banco de Dados em Grafo: Um Estudo de Caso em Detecção de Fraudes no Governo Brasileiro). In: Iberian Conference on Information Systems and Technologies CISTI, pp. 1–6 (2017)
43. Wacker, J., Ferreira, R.P., Ladeira, M.: Detecting fake suppliers using deep image features (2018)
44. Kim, Y.J., Baik, B., Cho, S.: Detecting financial misstatements with fraud intention using multi-class cost-sensitive learning. Expert Syst. Appl. **62**, 32–43 (2016)
45. Dutta, I., Dutta, S., Raahemi, B.: Detecting financial restatements using data mining techniques. Expert Syst. Appl. **90**, 374–393 (2017)
46. Grace, E., Rai, A., Redmiles, E., Ghani, R.: Detecting fraud, corruption, and collusion in international development contracts: the design of a proof-of-concept automated system (2016)
47. Gómez, J.A., Arévalo, J., Paredes, R., Nin, J.: End-to-end neural network architecture for fraud scoring in card payments. Pattern Recognit. Lett. **105**, 175–181 (2018)
48. Chen, Y.J., Wu, C.H., Chen, Y.M., Li, H.Y., Chen, H.K.: Enhancement of fraud detection for narratives in annual reports. Int. J. Account. Inf. Syst. **26**, 32–45 (2017)
49. Wang, Q., Xu, W., Huang, X., Yang, K.: Enhancing intraday stock price manipulation detection by leveraging recurrent neural networks with ensemble learning. Neurocomputing **347**, 46–58 (2019)
50. Tan, M., Lee, W.-L.: Evaluation and improvement of procurement process with data analytics. Int. J. Adv. Comput. Sci. Appl. **6**, 70–80 (2015)
51. Correa Bahnsen, A., Aouada, D., Stojanovic, A., Ottersten, B.: Feature engineering strategies for credit card fraud detection. Expert Syst. Appl. **51**, 134–142 (2016)
52. Li, H., Wong, M.-L.: Financial fraud detection by using grammar-based multi-objective genetic programming with ensemble learning (2015)
53. Throckmorton, C.S., Mayew, W.J., Venkatachalam, M., Collins, L.M.: Financial fraud detection using vocal, linguistic and financial cues. Decis. Support Syst. **74**, 78–87 (2015)

54. Arief, H.A.A., Saptawati, G.A.P., Asnar, Y.D.W.: Fraud detection based-on data mining on Indonesian E-Procurement System (SPSE). In: Proceedings of 2016 International Conference on Data and Software Engineering, ICoDSE 2016 (2017)
55. Vimala Devi, J., Kavitha, K.S.: Fraud detection in credit card transactions by using classification algorithms. In: International Conference on Current Trends in Computer, Electrical, Electronics and Communication, CTCEEC 2017, pp. 125–131 (2018)
56. Zhou, H., Chai, H.F., Qiu, M.L.: Fraud detection within bankcard enrollment on mobile device based payment using machine learning. Front. Inf. Technol. Electron. Eng. **19**, 1537–1545 (2018)
57. Hooda, N., Bawa, S., Rana, P.S.: Fraudulent firm classification: a case study of an external audit. Appl. Artif. Intell. **32**, 48–64 (2018)
58. Fu, Y., Liu, G., Papadimitriou, S., Xiong, H., Li, X., Chen, G.: Fused latent models for assessing product return propensity in online commerce. Decis. Support Syst. **91**, 77–88 (2016)
59. Demiriz, A., Ekizoğlu, B.: Fuzzy rule-based analysis of spatio-temporal ATM usage data for fraud detection and prevention1. J. Intell. Fuzzy Syst. **31**, 805–813 (2016)
60. Chimonaki, C., Papadakis, S., Vergos, K., Shahgholian, A.: Identification of financial statement fraud in Greece by using computational intelligence techniques. In: Mehandjiev, N., Saadouni, B. (eds.) FinanceCom 2018. LNBIP, vol. 345, pp. 39–51. Springer, Cham (2019). https://doi.org/10.1007/978-3-030-19037-8_3
61. Correa, M.A.O.S., Galindo Leal, A.: Identification of overpricing in the purchase of medication by the Federal Government of Brazil, using text mining and clustering based on ontoiogy (2018)
62. Alzaidi, A.A.: Impact of use of big data in decision making in banking sector of Saudi Arabia. Int. J. Comput. Sci. Netw. Secur. **18**, 72–80 (2018)
63. Kasa, N., Dahbura, A., Ravoori, C., Adams, S.: Improving credit card fraud detection by profiling and clustering accounts (2019)
64. Chen, Y.-J., Wu, C.-H.: On big data-based fraud detection method for financial statements of business groups (2017)
65. Weng, H., et al.: Online e-commerce fraud: a large-scale detection and analysis (2018)
66. Torres, C.F., Schütte, J., State, R.: Osiris: hunting for integer bugs in ethereum smart contracts (2018)
67. Lismont, J.: Predicting tax avoidance by means of social network analytics. Decis. Support Syst. **108**, 13–24 (2018)
68. Zhang, H., Wang, L.: Prescription fraud detection through statistic modeling (2018)
69. Martínez-Plumed, F., Casamayor, J.C., Ferri, C., Gómez, J.A., Vendrell Vidal, E.: SALER: a data science solution to detect and prevent corruption in public administration. In: Alzate, C., et al. (eds.) ECML PKDD 2018. LNCS (LNAI), vol. 11329, pp. 103–117. Springer, Cham (2019). https://doi.org/10.1007/978-3-030-13453-2_9
70. Carcillo, F., Dal Pozzolo, A., Le Borgne, Y.A., Caelen, O., Mazzer, Y., Bontempi, G.: SCARFF: a scalable framework for streaming credit card fraud detection with spark. Inf. Fusion. **41**, 182–194 (2018)
71. Carminati, M., Polino, M., Continella, A., Lanzi, A., Maggi, F., Zanero, S.: Security evaluation of a banking fraud analysis system. ACM Trans. Priv. Secur. **21**, 1–31 (2018)
72. Robinson, W.N., Aria, A.: Sequential fraud detection for prepaid cards using hidden Markov model divergence. Expert Syst. Appl. **91**, 235–251 (2018)
73. Ekin, T., Ieva, F., Ruggeri, F., Soyer, R.: Statistical medical fraud assessment: exposition to an emerging field. Int. Stat. Rev. **86**, 379–402 (2018)
74. Fauzan, A.C., Sarno, R., Ariyani, N.F.: Structure-based ontology matching of business process model for fraud detection. In: ICTS 2017, pp. 221–225 (2018)

75. Saghehei, E., Memariani, A.: Suspicious behavior detection in debit card transactions using data mining: a comparative study using hybrid models. Inf. Resour. Manag. J. **28**, 1–14 (2015)
76. El-kaime, H., Hanoune, M., Eddaoui, A.: The data mining: a solution for credit card fraud detection in banking. In: Mizera-Pietraszko, J., Pichappan, P., Mohamed, L. (eds.) RTIS 2017. AISC, vol. 756, pp. 332–341. Springer, Cham (2019). https://doi.org/10.1007/978-3-319-91337-7_31
77. Schlenther, B.O.: Addressing illicit financial flows in Africa: how broad is the whole of government approach supposed to be? J. Financ. Crime (2016)
78. Sadaf, R., Oláh, J., Popp, J., Máté, D.: An investigation of the influence of theworldwide governance and competitiveness on accounting fraud cases: a cross-country perspective. Sustain **10**, 1–11 (2018)
79. Wang, H., Chen, H.M.: Deterring bidder collusion: auction design complements antitrust policy. J. Compet. Law Econ. **12**, 31–68 (2016)
80. Wahid, A., Rao, A.C.S.: A distance-based outlier detection using particle swarm optimization technique. In: Fong, S., Akashe, S., Mahalle, Parikshit N. (eds.) Information and Communication Technology for Competitive Strategies. LNNS, vol. 40, pp. 633–643. Springer, Singapore (2019). https://doi.org/10.1007/978-981-13-0586-3_62
81. Coma-Puig, B., Carmona, J.: A quality control method for fraud detection on utility customers without an active contract. In: Proceedings of the 33rd Annual ACM Symposium on Applied Computing, pp. 495–498 (2018)
82. Fazekas, M., Cingolani, L.: Breaking the cycle? How (not) to use political finance regulations to counter public procurement corruption. Slav. East Eur. Rev. **95**, 76–116 (2017)
83. Lehne, J., Shapiro, J.N., Vanden Eynde, O.: Building connections: political corruption and road construction in India. J. Dev. Econ. **131**, 62–78 (2018)
84. Cieślik, A., Goczek, Ł.: Control of corruption, international investment, and economic growth – Evidence from panel data. World Dev. **103**, 323–335 (2018)
85. Lourenço, I.C., Rathke, A., Santana, V., Branco, M.C.: Corruption and earnings management in developed and emerging countries. Corp. Gov. **18**, 35–51 (2018)
86. van Erven, G.C.G., Holanda, M., Carvalho, Rommel N.: Detecting evidence of fraud in the brazilian government using graph databases. In: Rocha, Á., Correia, A.M., Adeli, H., Reis, L.P., Costanzo, S. (eds.) WorldCIST 2017. AISC, vol. 570, pp. 464–473. Springer, Cham (2017). https://doi.org/10.1007/978-3-319-56538-5_47
87. Rad, M.S., Shahbahrami, A.: Detecting high risk taxpayers using data mining techniques. In: 2016 2nd International Conference of Signal Processing and Intelligent Systems ICSPIS 2016, pp. 14–15 (2017)
88. Monirzadeh, Z., Habibzadeh, M., Farajian, N.: Detection of violations in Credit Cards of Banks and financial institutions based on artificial neural network and Metaheuristic optimization algorithm. Int. J. Adv. Comput. Sci. Appl. **9**, 176–182 (2018)
89. Bramoullé, Y., Goyal, S.: Favoritism. J. Dev. Econ. **122**, 16–27 (2016)
90. Saxena, A., Sharma, N., Saxena, K., Parikh, Satyen M.: Financial data mining: appropriate selection of tools, techniques and algorithms. In: Deshpande, A.V., Unal, A., Passi, K., Singh, D., Nayak, M., Patel, B., Pathan, S. (eds.) SmartCom 2017. CCIS, vol. 876, pp. 244–251. Springer, Singapore (2018). https://doi.org/10.1007/978-981-13-1423-0_27
91. Bogdanov, D., Jõemets, M., Siim, S., Vaht, M.: How the Estonian tax and customs board evaluated a tax fraud detection system based on secure multi-party computation. In: Böhme, R., Okamoto, T. (eds.) FC 2015. LNCS, vol. 8975, pp. 227–234. Springer, Heidelberg (2015). https://doi.org/10.1007/978-3-662-47854-7_14

92. Kültür, Y., Çağlayan, M.U.: Hybrid approaches for detecting credit card fraud. Expert Syst. **34**, 1–13 (2017)

93. Indrajani, Prabowo, H., Meyliana: Learning fraud detection from big data in online banking transactions: a systematic literature review. J. Telecommun. Electron. Comput. Eng. **8**, 127–131 (2016)

94. Hutchings, A.: Leaving on a jet plane: the trade in fraudulently obtained airline tickets. Crime Law Soc. Change **70**, 461–487 (2018)

95. Saia, R., Boratto, L., Carta, S.: Multiple behavioral models: a divide and conquer strategy to fraud detection in financial data streams (2015)

96. Lee, P.S., Owda, M., Crockett, K.: Novel methods for resolving false positives during the detection of fraudulent activities on stock market financial discussion boards. Int. J. Adv. Comput. Sci. Appl. **9**, 1–10 (2018)

97. Fazekas, M.: Red tape, bribery and government favouritism: evidence from Europe. Crime Law Soc. Change **68**, 403–429 (2017)

98. Yaseen, M., et al.: Secure sensors data acquisition and communication protection in eHealthcare: review on the state of the art. Telemat. Inform. **35**, 702–726 (2018)

99. Jetter, M., Parmeter, C.F.: Sorting through global corruption determinants: institutions and education matter – Not culture. World Dev. **109**, 279–294 (2018)

100. Jagger, P., Shively, G.: Taxes and Bribes in Uganda. J. Dev. Stud. **51**, 66–79 (2015)

101. Williams, M.J.: The political economy of unfinished development projects: corruption, clientelism, or collective choice? Am. Polit. Sci. Rev. **114**, 705–723 (2017)

102. Kussainov, D.S.: The problems of qualification of illegal alienation of ownership of residential premises. Asian Soc. Sci. **11**, 188 (2015)

103. Ahmed, M., Mahmood, A.N., Islam, M.R.: A survey of anomaly detection techniques in financial domain. Future Gener. Comput. Syst. **55**, 278–288 (2016)

104. Moro, S., Cortez, P., Rita, P.: Business intelligence in banking: a literature analysis from 2002 to 2013 using text mining and latent Dirichlet allocation. Expert Syst. Appl. **42**, 1314–1324 (2015)

105. Dal Pozzolo, A., Boracchi, G., Caelen, O., Alippi, C., Bontempi, G.: Credit card fraud detection: a realistic modeling and a novel learning strategy. IEEE Trans. Neural Netw. Learn. Syst. **29**, 3784–3797 (2018)

106. Kumar, P., Iqbal, F.: Credit card fraud identification using machine learning approaches (2019)

107. Mahmoudi, N., Duman, E.: Detecting credit card fraud by modified fisher discriminant analysis. Expert Syst. Appl. **42**, 2510–2516 (2015)

108. Abdallah, A., Maarof, M.A., Zainal, A.: Fraud detection system: a survey. J. Netw. Comput. Appl. **68**, 90–113 (2016)

109. Rajak, I., Mathai, K.J.: Intelligent fraudulent detection system based SVM and optimized by danger theory. In: IEEE International Conference on Computer, Communication and Control IC4 2015, pp. 2–5 (2016)

110. Xu, J.J., Lu, Y., Chau, M.: P2P lending fraud detection: a big data approach. In: Chau, M., Wang, G.A., Chen, H. (eds.) PAISI 2015. LNCS, vol. 9074, pp. 71–81. Springer, Cham (2015). https://doi.org/10.1007/978-3-319-18455-5_5

111. Hajek, P., Henriques, R.: Mining corporate annual reports for intelligent detection of financial statement fraud – a comparative study of machine learning methods. Knowl.-Based Syst. **128**, 139–152 (2017)

Cluster Analysis Base on Psychosocial Information for Alcohol, Tobacco and Other Drugs Consumers

Ruth Reátegui[1], Pablo Torres-Carrión[1($\boxtimes$)] (iD), Víctor López[1],
Anabela Galárraga[2], Gino Grondona[3], and Carla López Nuñez[4]

[1] Universidad Técnica Particular de Loja, San Cayetano Alto S/N, Loja, Ecuador
{rmreategui,pvtorres,vmlopez5}@utpl.edu.ec
[2] Universidad Técnica del Norte, Ibarra, Ecuador
asgalarraga@utn.edu.ec
[3] Universidad Politécnica Salesiana, Quito, Ecuador
ggrondona@ups.edu.ec
[4] Universidad Loyola Andalucia, Seville, Spain
clopezn@uloyola.es

Abstract. The consumption of alcohol, tobacco and other drugs is considered a public health problem, being one of the main causes of academic failure of university students. The objective of this research was to identify psychosocial characteristics in clusters of alcohol, tobacco and other drug consumers. A sample of 3741 college students from Ecuador who complete a psychosocial questionnaire was used. Sparse K-means algorithm showed three clusters. Cluster CLNA1 represents students with low consume of tobacco and alcohol. Apparently, they do not have depression and are comfortable with their lives. CLNA2 presents low consume of tobacco and alcohol. This group shows signals of depression and they consider that there are aspects of their life to improve and small but significant problems of their life. CLNA3 presents the higher consume of tobacco and alcohol.

Keywords: Cluster analysis · Alcohol · Tobacco · Drugs · University · Ecuador

1 Introduction

The consumption of alcohol, tobacco, and other drugs is a health and social problem, for its consequences for the consumer and for those close to their environment, including health problems, violence and productivity decline [1]. According to United Nations Office on Drugs and Crime report [2] "In 2017, an estimated 271 million people (5.5 per cent of the global population) aged 15–64, had used drugs in the previous year, while 35 million people are estimated to be suffering from drug use disorders". Regarding the type of drugs, cannabis is the most common drug used, 1 in 10 has used opiates, similar to those injecting some kind of drug. The data is alarming in terms of health consequences, with more than 11 million people in the world who inject drugs (1.4 million with HIV, 5.6 million with hepatitis C, and with a number of

© Springer Nature Switzerland AG 2020
M. Botto-Tobar et al. (Eds.): ICAT 2019, CCIS 1194, pp. 269–283, 2020.
https://doi.org/10.1007/978-3-030-42520-3_22

dead consumers close to 600,000 in 2017. In Ecuador and the Andean region, the National Drug Observatories, the National Commission for Drug Development and Life (DEVIDA) and the Inter-American Drug Abuse Control Commission (CICAD) reaffirm the dimension of the problem in this region. So, we came to the conclusion that there is a high percentage of consumers who have serious physiological and psychosocial consequences. In Andean countries, the greatest alcohol and drugs consumption rates occurs in young people aged 18 to 25 years [3]. This age corresponds to the majority of university students (>80%) in Ecuador and Latin America, and that is the reason why they are the target population of this study.

Although it is estimated that there are more than 230 million drug users in the world. 90% of these consumers do not have problematic consumption [2]. Likewise, it is necessary to identify the psychosocial and health characteristics associated with controlled or regulated consumption. For example, many indigenous communities have a non-problematic consumption of hallucinogens even for therapeutic purposes and, although this phenomenon has been described from an anthropological point of view (description of the process of plant collection, product preparation, etc.), it is still interesting to identify psychosocial patterns of the individual and then focus on groups, to propose alternatives for intervention in cases in which this consumption tends to be considered a social problem [4–8]. Knowing these patterns is important, and since they are very relevant, four types of tests have been proposed: sociodemographic, consumption, health and psychosocial. With this input, it is expected to identify relevant patterns in consumer groups.

The classic proposals in the field of alcohol and drug problems have been conceptualized more as a medical problem than as a social problem [9]. Thus, the action strategies have focused on individualized interventions, mainly aimed at treating people with drug abuse or prevention programs aimed basically at promoting healthy lifestyles at the individual, family and institutional level, through training in social skills or access to information about the risks of illegal drug abuse [10]. The approach to the problem, of drug and substance use, should encompass all psychoactive drugs, regardless of their legal status, in this way high cases of deaths associated with medical prescription opioids can be avoided as in the US [11]. The investigation has considered a wide coverage of consumer options available in test ASSIST, AUDIT-C, TFDN (see Table 1), and from the patterns obtained it is possible to establish a dialogue between groups that use drugs considered legal or medicinal purposes.

Complementing the proposals that cover all drugs; prevention is the strategy suggested by international organizations and entities [1–3]; prevention requires acting on the social determinants of health, that is, living conditions or contexts associated with problematic use but also with controlled drug use [12–14]. Drug prevention proposals must be continuously connected to the social, political and economic conditions in which said consumption is inserted; its objective is to promote access to the resources and opportunities that are necessary to guarantee a gracious life, strengthening the individual and collective capacities of accompaniment and care, complying with continuous adaptation and contribution to the community (social environment), and promoting steadiness that transcends the temporality of the project [15–19]. This is how this proposal came to be, which based on AI methods and algorithms, proposes the identification of groups (cluster), with defined characteristics that allow the best practice in terms of decision-making for the application of prevention and intervention programs.

From related study in which ML techniques and methods are applied, the research sample is diverse: some use information obtained directly from people [20–22], others apply analysis of electrical wave signals [23, 24], some works use digital image files [25], chemical composed [26], tweets [27] and other type of document [28]. From studies in which people participate directly, in [20] 108 women participate (age 18–56, mean = 25, SD ± 9), 94.5% living in the urban area and 69.7% have a university degree; in [22] they use sample of young men (252 in the experimental group and 357 in the control group); in [21] the study was conducted with 241 participants (aged 16–25) in two Swiss cities (Zurich and Lausanne), and before the study began, participants were asked to install the two apps on their own smartphones and contribute data for 10 weekend nights. Studies in which the sample corresponds to health indicators, in [25] they uses YOLOv3 in cyclic voltammetry data for the detection of dopamine; in [23] they use data collected from 14 EEG channels located on an Emotiv Epoc BCI headset, sampled at 128 Hz; in [24] they classify (alcoholic or abstemious) from EEG signals; and [28] analyzes information from electronic health records (EHR), reduced the number of cases to 11.573, 873 types of lab tests and 51 types of vital signs, including lab tests, vital signs, medical procedures, prescriptions, and other data from millions of patients to predict opioid substance dependence. The effort of the global scientific community is broad to establish consumer patterns from the diversity of data types.

Regarding classification algorithms applied, the closest research is [20] that classifies and predicts women waterpipe smokers' level of nicotine dependence, considering 19 factors (WTSQ) and obtaining four cluster: CL1. Age, Residence, Level_Education, Work Status, Waterpipe-Income; CL2. Tobacco Past, Tobacco Age, Tobacco-Type, Number-of-Cigarette; CL3. Current-Waterpipe, Waterpipe-Week, Waterpipe-Month, Waterpipe-Need, Waterpipe Cigarette-Instead; and CL4. Waterpipe-Inhale, Waterpipe-Stop, Waterpipe-Alone, Waterpipe-Cigarette-Instead. Other researches use classification algorithms with different aims related to alcohol, tobacco and related health; So, in [28] they trained a machine learning model to classify patients, matched by age, gender, and status of HIV, hepatitis C, and sickle cell disease; Jiajie et al. [24] classifies (alcoholic or abstemious) from EEG signals, from Quadratic SVM 95%, Cubic SVM 92%, fine KNN 91% and ensemble bagged tree 90% algorithms; in [25] they improve the automatic dopamine identification using convolutional neural networks (CNN); in [23] apply BCI classifier with EEG signals, using logistic regression, support vector machines, decision trees, k-nearest neighbors and Naive Bayes; in [26] they do target classification using both docking scores and similarity scores; Santani et al. [21] obtains an automatically schema classifying the alcohol consumption status as a binary classification task, using Random Forests; and in [27] from tweets, they obtain behavior patterns in time series from people who consume tobacco and alcohol.

About approaches used when applying ML to determine patterns of consumer behavior, there is great diversity. Logistic regression is used in [23, 26, 27], Naive Bayes classifiers [20, 23], decision trees [20], CNN [25], SVM [20, 24, 26, 27], random forest [21, 26–28], Artificial Neural Networks [26], k-star and IBk [20], and KNN [24]. For models' assessment, accuracy is applied from the confusion matrix, complemented with analysis under the ROC curve and F1-score. The main tools and programming languages used for the construction of the models are Python [22, 26, 28], MathLab [24], SPSS [22], Weka [20] and C [25]. The models for this research were programmed using R and R Studio, as it is a solid and open platform, with the programming inputs (libraries, tools) that allow us to meet the objectives set.

Finally, analyzing previous studies conducted in the local context, preliminary study (2015–2017) allowed the simultaneous evaluation of alcohol and other drugs consumption in 13 universities throughout Ecuador through a digital platform, with more than 13,000 participants (students, professors and administration staff) and which highlighted the importance of handling large amounts of psychosocial information to detect key variables in alcohol consumption [29, 30]. There are few studies in the consumption of other drugs, the most important reports that in Ecuadorian college students, the actual tobacco consumers are 17.2%, 12.5% are electronic cigarettes users, 20% have consumed two or more psychoactive substances, 12.8% have used one illicit substance, 11.7% are recent cannabis users and 1.45% are recent cocaine users [3]. This new proposal is a joint effort of two research groups of the UTPL: Clinical and Health Psychology (KUUSA)[1], and inclusive Human Computer Interaction Research Group (iHCI)[2], with the purpose of advancing the use of artificial intelligence for the generation of predictive models that allow not only the detection of risk groups in problematic and controlled drug use [31, 32], but also the design of more effective prevention programs and decision-making on public policies aimed at the problem of evidence-based drugs.

In recent systematic literature review applied in the Scopus and WoS databases [33], as shown in previous paragraphs, no research results were found that provide psycho-social profiles of consumers. No research results were found that provide psycho-social profiles of consumers. Most studies focus on the consumption of tobacco and alcohol, and in some cases it applies to the consumption of one or two substances, but none have covered the ten types of drug and substance use studied in the ASSIST test [34]. Other interesting and original areas of research are the inclusion of tests concerning the mental health of the individual, such as Loneliness scale (UCLA-R) [35], the satisfaction with life scale (SLQ) [36], the Barratt impulsiveness scale (BISS-11) [37] y Perceived stress scale (PSS-10) [38] (see Table 1). This is the first approach to the analysis of these data, and as a research group we will be sharing subsequent analyzes that allow us to deepen each of the behavioral aspects of the population of consumers.

---

[1] https://investigacion.utpl.edu.ec/grupos/allikay.

[2] https://investigacion.utpl.edu.ec/grupos/ihci.

The following section details the method used to perform the cluster analysis, including a detail of the database with psychological tests related to mental health and psychosocial characteristics. The results section details the characteristics of clusters obtained, including the values of the eleven psychological tests (AUDIT-C, TFDN, ASSIST, PHQ-9, AAQ-7, UCLA-R, SLQ, BISS-11, PSS-10, TIPI-SPA and socio-demographic). In the final sections the discussion of results and the main conclusions of the study are carried out.

2 Materials and Methods

2.1 Dataset

This experiment worked with a sample of 3741 college students from Ecuador who validly answered a psychosocial questionnaire. The 23 variables detailed below were selected. The data are obtained as part of the "CEPRA XII-2018-05 Prediction of drug use" project funded by CEDIA (Ecuadorian Corporation for Research Development and the Academy, by Spanish acronym), and in which researchers from three Ecuadorian and two Spanish universities: Universidad Técnica Particular de Loja, Universidad Técnica del Norte, Universidad Politécnica Salesiana, Universidad de Salamanca and Universidad Loyola Andalucía. The collection of information is carried out through web-based questionnaires previously selected based on three criteria: brevity, psychometric properties, and open for research purposes. Both the questionnaires and the database with the answers were hosted on the web server of the observatory created for this purpose. For the purpose of the evaluation and commissioning of the prediction and classification models, the HPC server technology available at CEDIA is used for high-performance computing. The data has gone through a filtering and cleaning, separating all those that presented anomalies. From the questionnaire that consist of 10 test, 23 variables detailed in Table 1 were selected.

2.2 Cluster Analysis

Before cluster analysis all the variables were normalized. Then the Sparse K-means [45] algorithm was used to identify clusters. This algorithm simultaneously finds groups of objects and a subset of cluster variables. Sparse K-means assigns weight to the variables with the most representative ones having the highest weight. Furthermore, this algorithm is useful when the dataset has noise variables. Also, gap statistics was used in order to estimate the number of clusters. Gap statistics is a standard method for detecting the number of clusters [44]. We used R Studio[3] that have implemented gap statistics and Sparse K-means.

---

[3] https://rstudio.com/.

Table 1. List of questionnaires and variables

Test/Questionnaires	Description	Variables
The Alcohol Use Disorders Identification Test- Self-Report Version (AUDIT-C) [39]	The 10-item scale was designed to assess three conceptual domains: alcohol intake (1–3), dependence (4–6), and adverse consequences (7–10)	1 variable that resume this questionnaire
Test de Fagerström (TFDN) [40]	This test provide a short self-report measure of dependency on nicotine	1 variable that resume this questionnaire
The alcohol, smoking and substance involvement screening test (ASSIST) [34]	This test allows the detection of different levels of risk associated with the consumption of alcohol, tobacco and other substances	10 variables: tobacco, alcohol, cannabis, cocaine, amphetamines, inhalants, sedatives, hallucinogens, opiates and other drugs
Patient health questionnaire (PHQ-9) [41]	This test allows to screen the presence and severity of depression	1 variable that resume this questionnaire
Avoidance and action questionnaire (AAQ7) [42]	This test permits to assess the psychological inflexibility	1 variable that resume this questionnaire
Loneliness scale (UCLA-R) [35]	This test measures one's subjective feeling of loneliness and social isolation	1 variable that resume this questionnaire
The satisfaction with life scale (SLQ) [36]	This scale measures the life satisfaction component of subjective well-being	1 variable that resume this questionnaire
The Barratt impulsiveness scale (BISS-11) [37]	This test measures the impulsiveness	1 variable that resume this questionnaire
Perceived stress scale (PSS-10) [38]	This test measures the perception of stress	1 variable that resume this questionnaire
Personality inventory questionarie (TIPI-SPA) [43]	This test measures personality traits	5 variables: extroversion, affability, responsibility, emotional stability, opening to experience
Total of variables	23	

3   Results

Three clusters were obtained: CLNA1, CLNA2 and CLNA3 with 1898, 1084 and 489 students respectively. Table 2 shows the distribution of the clusters according PHQ-9 and SLQ tests. Table 3 shows the clusters distribution for AUDIT-C, AAQ 7, UCLA-R, TFDN, BISS-11, and PSS-10 tests. Table 4 shows the clusters distribution for the ASSIST test and Table 5 show details of the clusters distribution to 5 personality variables. Range values of some questionnaires are quartiles and the quantity of students are percentages according the number of students in each cluster.

The variables with the heights weight are AAQ 7 (0.458), UCLA-R (0.333), PHQ-9 (0.223), SLQ (0.123), PSS-10 (0.103) and from the ASSIST test the questions: "In the last 3 months, how often have you used tobacco (cigarettes, rolling tobacco, chewing tobacco, cigars, etc.)?" (0.75) and "In the last 3 months, how often have you consumed alcohol (beer, wine, spirits, cocktails, etc.)?" (0.123). Considering these variables, next an explanation of the cluster distribution is detailed.

Cluster CLNA1 has 1898 students, most of the students (73%) never have consumed tobacco, and 61% have consumed alcohol one or twice in the last three months. Also, most of the students (67%) do not need treatments for depression, most of them (80%) present low values (7 – 20) for the AAQ 7 test, 85% have low values (3–6) for the UCLA-R test, and 77% have low values (0-18) for the PSS-10 test. Moreover, 35% of them "love their lives and feel that things are going very well" and 34% "love their lives, despite this they have identified areas of dissatisfaction".

Cluster CLAN2 has 1084 students, most of the students (70%) never have consumed tobacco, and 53% have consumed alcohol one or twice in the last three months. Also, for most of the students (65%) the doctor should use his judgment about the depression treatment, for 24% of them, depression treatment with antidepressants is justified. Most of them (97%) present high values (21–49) for the AAQ 7 test, 80% have high values (7–12) for the UCLA-R test, and 77% have high values (19–40) for the PSS-10 test. Moreover, 31% of them "consider that there are areas of their life that need improvement" and 28% "have small but significant problems in various areas of their life".

Cluster CLAN3 has 489 students, in this cluster 29% of the people have consumed tobacco daily or almost daily, 34% of the people have consumed tobacco weekly, and 38% have consumed tobacco monthly. Also, 40% of the students have consumed alcohol weekly, and 33% have consumed alcohol monthly. Moreover, for the 47% of them the doctor should use his judgment about the depression treatment, ant for the 34% of them do not need treatment for depression. For the other variables above analyzed, the values are distributed in all the ranges.

Table 2. Clusters distribution according PHQ-9 and SLQ

Patient Health Questionnaire (PHQ-9)		Satisfaction with Life Scale (SLQ)	
W = 0.223		W = 0.123	
CLNA1 (1898)			
Values	%	Values	%
Depression treatment with antidepressants is justified	0	They love their lives and feel that things are going very well	**35**
The doctor should use his judgment about the treatment	33	They love their lives, despite this they have identified areas of dissatisfaction	**34**
Probably, the patient does not need treatment for depression	67	They consider that there are areas of their life that need improvement	23
		They have small but significant problems in various areas of their life	6
		They are significantly dissatisfied with their lives	1
		They are extremely unhappy with their current life	1
CLNA2 (1084)			
Values	%	Values	%
Depression treatment with antidepressants is justified	24	They love their lives and feel that things are going very well	7
The doctor should use his judgment about the treatment	**65**	They love their lives, despite this they have identified areas of dissatisfaction	20
Probably, the patient does not need treatment for depression	10	They consider that there are areas of their life that need improvement	**31**
		They have small but significant problems in various areas of their life	**28**
		They are significantly dissatisfied with their lives	11
		They are extremely unhappy with their current life	3
CLNA3 (489)			
Values	%	Values	%
Depression treatment with antidepressants is justified	19	They love their lives and feel that things are going very well	18
The doctor should use his judgment about the treatment	**47**	They love their lives, despite this they have identified areas of dissatisfaction	21

(*continued*)

Table 2. (*continued*)

Patient Health Questionnaire (PHQ-9)		Satisfaction with Life Scale (SLQ)	
W = 0.223		W = 0.123	
Probably, the patient does not need treatment for depression	**34**	They consider that there are areas of their life that need improvement	32
		They have small but significant problems in various areas of their life	20
		They are significantly dissatisfied with their lives	9
		They are extremely unhappy with their current life	2

Table 3. Clusters distribution according AUDIT-C, AAQ 7, UCLA-R, TFDN, BISS-11, and PSS-10

AUDIT-C		AAAQ 7		UCLA-R		TFDN		BISS-11		PSS-10	
W = 0.053		W = 0.458		W = 0.333		W = 0.015		W = 0.028		W = 0.103	
CLNA1 (1898)											
Values	%	Values	%	Values	%	Values	%	Values	%	Values	%
0–7	61	7–14	47	3–4	40	Does not consume	82	1–11	37	0–13	**39**
8–15	32	15–20	33	5–6	45	Low nicotine dependence	18	12–14	29	14–18	**38**
16–19	4	21–28	18	7–8	14	Moderate nicotine dependence	1	15–17	22	19–21	15
20–40	3	29–49	1	9–12	1		0	18–29	12	22–40	8
CLNA2 (1084)											
Values	%	Values	%	Values	%	Values	%	Values	%	Values	%
0–7	54	7–14	0	3–4	2	Does not consume	78	1–11	15	0–13	3
8–15	33	15–20	4	5–6	18	Low nicotine dependence	20	12–14	25	14–18	19
16–19	7	21–28	33	7–8	**34**	Moderate nicotine dependence	1	15–17	26	19–21	**25**
20–40	6	29–49	64	9–12	46	High nicotine dependence	1	18–29	33	22–40	**52**
CLNA3 (489)											
Values	%	Values	%	Values	%	Values	%	Values	%	Values	%
0–7	20	7–14	27	3–4	23	Does not consume	10	1–11	22	0–13	24
8–15	39	15–20	17	5–6	30	Low nicotine dependence	79	12–14	26	14–18	23
16–19	18	21–28	27	7–8	25	Moderate nicotine dependence	4	15–17	27	19–21	21
20–40	23	29–49	29	9–12	21	High nicotine dependence	6	18–29	25	22–40	32

Table 4. Clusters distribution according ASSIT test

The Alcohol, Smoking and Substance Involvement Screening Test (ASSIST)										
	In the last 3 months, how often have you used tobacco?	In the last 3 months, how often have you consumed alcohol?	In the last 3 months, how often have you used cannabis?	In the last 3 months, how often have you used cocaine?	In the last 3 months, how often have you used amphetamines?	In the last 3 months, how often have you used inhalants?	In the last 3 months, how often have you used sedatives or sleeping pills?	In the last 3 months, how often have you consumed hallucinogens?	In the last 3 months, how often have you used opiates?	In the last 3 months, how often have you used other psychoactive substances?
	W = 0.76	W = 0.123	W = 0.031	W = 0.004	W = 0.003	W = 0.002	W = 0.006	W = 0.003	W = 0.002	W = 0.006
CLNA1 (1898)										
Values	%	%	%	%	%	%	%	%	%	%
Daily or almost daily	0	0	1	0	0	0	0	0	0	0
Weekly	0	6	1	0	0	0	0	0	0	0
Monthly	0	17	1	1	0	0	0	0	0	0
Once or twice	27	**61**	7	1	1	1	3	1	1	2
Never	**73**	16	91	98	99	99	97	98	99	97
CLNA2 (1084)										
Values	%	%	%	%	%	%	%	%	%	%
Daily or almost daily	0	1	1	0	0	0	1	0	0	1
Weekly	0	8	1	0	0	0	1	0	0	0
Monthly	1	22	3	1	0	1	1	0	1	1

(continued)

Table 4. (*continued*)

The Alcohol, Smoking and Substance Involvement Screening Test (ASSIST)									
In the last 3 months, how have you used tobacco?	In the last 3 months, how have you consumed alcohol?	In the last 3 months, how often have you used cannabis?	In the last 3 months, how often have you used cocaine?	In the last 3 months, how often have you used amphetamines?	In the last 3 months, how often have you used inhalants?	In the last 3 months, how often have you used sedatives or sleeping pills?	In the last 3 months, how often have you consumed hallucinogens?	In the last 3 months, how often have you used opiates?	In the last 3 months, how often have you used other psychoactive substances?
$W = 0.76$	$W = 0.123$	$W = 0.031$	$W = 0.004$	$W = 0.003$	$W = 0.002$	$W = 0.006$	$W = 0.003$	$W = 0.002$	$W = 0.006$

CLNA1 (1898)

Values	%	%	%	%	%	%	%	%	%	%
Once or twice	29	53	9	2	1	2	7	3	1	4
Never	70	17	87	97	98	97	90	96	98	94

CLNA3 (489)

Values	%	%	%	%	%	%	%	%	%	%
Daily or almost daily	29	4	3		1	1	1	1	0	1
Weekly	34	40	7	2	2	1	1	2	1	3
Monthly	38	33	8	4	3	4	4	3	4	4
Once or twice	0	19	20	6	3	1	9	6	2	8
Never	0	4	62	88	91	92	85	89	92	84

Table 5. Clusters distribution according personality variables

Ten-Item Personality Inventory (TIPI-SPA)

Extraversion		Affability		Responsibility		Emotional stability		Opening to experience	
W = 0.02		W = 0.013		W = 0.02		W = 0.071		W = 0.0081	
CLNA1 (1898)									
Values	%	Values	%	Values	%	Values	%	Values	%
2–7	28	2–8	20	2–8	22	2–8	27	2–9	36
8	19	9–10	37	9–10	39	9	17	10	25
9–10	33	11	15	11	14	10	23	11	16
11–14	20	12–14	28	12–14	24	11-14	33	12–14	23
CLNA2 (1084)									
Values	%	Values	%	Values	%	Values	%	Values	%
2–7	43	2–8	31	2–8	40	2–8	63	2–9	50
8	22	9–10	40	9–10	36	9	15	10	21
9–10	22	11	14	11	10	10	11	11	13
11–14	12	12–14	15	12–14	14	11–14	10	12–14	16
CLNA3 (489)									
Values	%	Values	%	Values	%	Values	%	Values	%
2–7	34	2–8	34	2–8	44	2–8	**52**	2–9	44
8	19	9–10	37	9–10	33	9	14	10	19
9–10	24	11	13	11	10	10	15	11	13
11–14	23	12–14	16	12–14	14	11–14	18	12–14	24

4 Discussion

The analysis of sociodemographic, psychosocial and health variables associated with drug use through AI, has allowed the detection of patterns associated with the consumption of alcohol, drugs and substances, which will favor the design of action plans for other social problems directly related to drug abuse [46] such as violence, decrease in academic and work performance, address psychosocial problems of the university population, reduce academic drop-out rates, among others of great importance for Ecuadorian universities and of the world. The three groups obtained CLNA1 (1898), CLNA2 (1084) and CLNA3 (489), classify behaviors of consumption, psychosocial and repercussions in health indicators.

Unlike the classical statistical analyzes or models, the results obtained from the three clusters allow us to identify characteristics from consumer groups from psychosocial, health and socio-demographic variables. This relationship has not been found in previous studies analyzed in the literature review [47], the closest being the work of Kharabsheh et al. [20] that classifies and predict women waterpipe smokers' level of nicotine dependence from 19 factors (WTSQ), and generating four clusters that reference patterns of women who use tobacco. Also Jiajie et al. [24] classifies (alcoholic or abstemious) from EEG signals, from Quadratic SVM 95%, Cubic SVM 92%, fine KNN 91% and ensemble bagged tree 90% algorithms; and Ellis et al. [28] trained a machine learning model to classify patients, matched by age, gender, and status of HIV, hepatitis C, and sickle cell disease. There are no studies covering transcendental areas of consumption, health and psychosocial analysis, as proposed in this research.

As the UN report details [2], only 10% of consumers (>230 million) represent a problem group. This statement is only fulfilled in the first cluster (CLNA1), in which the AUDIT-C test applied to alcohol consumption reaches 7% with values (16–40) that represent problematic consumption; in CLNA2 it reaches 13% and in CLNA3 41%. This characteristic is maintained in the ASSIST test where the values are from a moderate consumer in CLNA1 and CLNA2 to a problematic consumer in CLNA3. In Ecuadorian college students, Ruisoto in [30] reports that the problematic use of alcohol is associated with increase of perceived stress level in men and women.

5 Conclusions

In this research, the Sparse K-means algorithm was applied in a sample of 3741 college students from Ecuador to identify psychosocial characteristic in tobacco, alcohol and drug consumers. Three groups were found with important difference about the tobacco and alcohol consumption. These differences are the psychological inflexibility, sales of loneliness, presence or severity of depression, satisfaction of life, and perception of stress. Cluster CLNA1 represents students with low consume of tobacco and alcohol. Apparently, they do not have depression and are comfortable with their lives. CLNA2 presents low consume of tobacco and alcohol. This group shows signals of depression and they consider that there are aspects of their life to improve and small but significant problems of their life. CLNA3 presents the higher consume of tobacco and alcohol. They present signs of depression.

Acknowledgement. This research was carried out with the funds of the project "CEPRA XII-2018-05 Prediction of Drug Consumption", winner of the CEPRA contest of CEDIA-Ecuador. The researchers thank CEDIA for their contribution in the development management of the project.

References

1. United Nations Office on Drugs and Crime: World Drug Report. www.unodc.org/wdr2017
2. United Nations Office on Drugs and Crime: World Drug Report. Austria, Vienna (2019)
3. Oficina de las Naciones Unidas contra la Droga y el Delito (UNODC): III Estudio epidemiológico andino sobre consumo de drogas en la población universitaria, Informe Regional, 2016. Oficina de las Naciones Unidas Contra la Droga y el Delito (UNODC), Lima, Perú (2017)
4. Werb, D., Rowell, G., Guyatt, G., Kerr, T., Montaner, J., Wood, E.: Effect of drug law enforcement on drug-related violence: evidence from a scientific review. International Centre for Science in Drug Policy, Vancouver (2010)
5. Rolles, S., Kushlick, D.: Prohibition is a key driver of the new psychoactive substances (NPS) phenomenon. Addiction **109**, 1589–1590 (2014)

6. Global Commission on Drug Policy: Taking control: pathways to drug policies that work. Global Commission on Drug Policy (2014)
7. Hughes, C.E., Stevens, A.: What can we learn from the Portuguese decriminalization of illicit drugs? Br. J. Criminol. **50**, 999–1022 (2010)
8. Rosmarin, A., Eastwood, N.: A quiet revolution: drug decriminalisation policies in practice across the globe. Drugs, Law Hum. Rights, Release, London 2012 (2012)
9. Buchanan, J.: Understanding problematic drug use: a medical matter or a social issue? (2006)
10. Carey, G., Malbon, E., Crammond, B., Pescud, M., Baker, P.: Can the sociology of social problems help us to understand and manage 'lifestyle drift'? Health Promot. Int. **32**, 755–761 (2016)
11. Vashishtha, D., Mittal, M.L., Werb, D.: The North American opioid epidemic: current challenges and a call for treatment as prevention. Harm Reduct. J. **14**, 7 (2017)
12. Braveman, P., Gottlieb, L.: The social determinants of health: it's time to consider the causes of the causes. Public Health Rep. **129**, 19–31 (2014)
13. Pega, F., Veale, J.F.: The case for the World Health Organization's Commission on Social Determinants of Health to address gender identity. Am. J. Public Health **105**, e58–e62 (2015)
14. Braveman, P., Egerter, S., Williams, D.R.: The social determinants of health: coming of age. Annu. Rev. Public Health **32**, 381–398 (2011)
15. Marmot, M., Allen, J.J.: Social determinants of health equity (2014)
16. Marmot, M., Allen, J.: Health priorities and the social determinants of health. EMHJ-Eastern Mediterr. Heal. J. **21**, 671–672 (2015)
17. Allen, J., Balfour, R., Bell, R., Marmot, M.: Social determinants of mental health. Int. Rev. psychiatry. **26**, 392–407 (2014)
18. Shilton, T., Sparks, M., McQueen, D., Lamarre, M.-C., Jackson, S.: Proposal for new definition of health. BMJ **343**, d5359 (2011)
19. Huber, M., et al.: How should we define health? BMJ **343**, d4163 (2011)
20. Kharabsheh, M., Meqdadi, O., Alabed, M., Veeranki, S., Abbadi, A., Alzyoud, S.: A machine learning approach for predicting nicotine dependence. Int. J. Adv. Comput. Sci. Appl. **10**, 179–184 (2019)
21. Santani, D., Do, T.-M.-T., Labhart, F., Landolt, S., Kuntsche, E., Gatica-Perez, D.: DrinkSense: characterizing youth drinking behavior using smartphones. IEEE Trans. Mob. Comput. **17**, 2279–2292 (2018)
22. Wang, L., Christensen, J.L., Jeong, D.C., Miller, L.C.: Virtual prognostication: When virtual alcohol choices predict change in alcohol consumption over 6-months. Comput. Human Behav. **90**, 388–396 (2019)
23. Mazzoleni, M., Previdi, F., Bonfiglio, N.S.: Classification algorithms analysis for brain–computer interface in drug craving therapy. Biomed. Signal Process. Control **52**, 463–472 (2019)
24. Jiajie, L., Narasimhan, K., Elamaran, V., Arunkumar, N., Solarte, M., Ramirez-Gonzalez, G.: Clinical decision support system for alcoholism detection using the analysis of EEG signals. IEEE Access. **6**, 61457–61461 (2018)
25. Matsushita, G.H.G., Sugi, A.H., Costa, Y.M.G., Gomez-A, A., Da Cunha, C., Oliveira, L.S.: Phasic dopamine release identification using convolutional neural network. Comput. Biol. Med. **114**, 103466 (2019)
26. Chen, M., Jing, Y., Wang, L., Feng, Z., Xie, X.-Q.: DAKB-GPCRs: an integrated computational platform for drug abuse related GPCRs. J. Chem. Inf. Model. **59**, 1283–1289 (2019)
27. Huang, T., Elghafari, A., Relia, K., Chunara, R.: High-resolution temporal representations of alcohol and tobacco behaviors from social media data. Proc. ACM Human-Comput. Interact. **1**, 1–26 (2017)

28. Ellis, R.J., Wang, Z., Genes, N., Ma'Ayan, A.: Predicting opioid dependence from electronic health records with machine learning. BioData Min. **12**, 3 (2019)
29. Ruisoto, P., Vaca, S.L., López-Goñi, J.J., Cacho, R., Fernández-Suárez, I.: Gender differences in problematic alcohol consumption in university professors. Int. J. Environ. Res. Public Health. **14**, 1069 (2017)
30. Ruisoto, P., Cacho, R., López-Goñi, J.J., Vaca, S., Jiménez, M.: Prevalence and profile of alcohol consumption among university students in Ecuador. Gac. Sanit. **30**, 370–374 (2016)
31. Iavindrasana, J., Cohen, G., Depeursinge, A., Müller, H., Meyer, R., Geissbuhler, A.: Clinical data mining: a review. Yearb. Med. Inform. **18**, 121–133 (2009)
32. Szolovits, P.: Artificial Intelligence in Medicine. Routledge (2019)
33. Torres-Carrión, P.V., Reátigui, R., Valdiviezo, P., Bustamante-Granda, B., Vaca-Gallegos, S.: Application of techniques based on Artificial Intelligence for predicting the consumption of drugs and substances. A systematic mapping review. In: International Conference on Applied Technologies ICAT 2019, Quito, Ecuador (2019)
34. World Health Organization WHO: The alcohol, smoking and substance involvement screening test (ASSIST): development, reliability and feasibility. Addiction **97**, 1183–1194 (2002)
35. Hughes, M.E., Waite, L.J., Hawkley, L.C., Cacioppo, J.T.: A short scale for measuring loneliness in large surveys: Results from two population-based studies. Res. Aging. **26**, 655–672 (2004)
36. Moyano, N.C., Tais, M.M., Muñoz, M.P.: Propiedades psicométricas de la Escala de Satisfacción con la Vida de Diener. Rev. Argentina Clínica Psicológica. **22**, 161–168 (2013)
37. Steinberg, L., Sharp, C., Stanford, M.S., Tharp, A.T.: New tricks for an old measure: the development of the Barratt Impulsiveness Scale-Brief (BIS-Brief). Psychol. Assess. **25**, 216 (2013)
38. Cohen, S., Kamarck, T., Mermelstein, R.: A global measure of perceived stress. J. Health Soc. Behav. **24**, 385–396 (1983)
39. Reinert, D.F., Allen, J.P.: The alcohol use disorders identification test (AUDIT): a review of recent research. Alcohol. Clin. Exp. Res. **26**, 272–279 (2002)
40. Heatherton, T.F., Kozlowski, L.T., Frecker, R.C., Fagerstrom, K.: The Fagerström test for nicotine dependence: a revision of the Fagerstrom Tolerance Questionnaire. Br. J. Addict. **86**, 1119–1127 (1991)
41. Kroenke, K., Spitzer, R.L., Williams, J.B.W.: The PHQ-9: validity of a brief depression severity measure. J. Gen. Intern. Med. **16**, 606–613 (2001)
42. Ruiz, F.J., Luciano, C., Cangas, A.J., Beltrán, I.: Measuring experiential avoidance and psychological inflexibility: the Spanish version of the acceptance and action questionnaire-II. Psicothema **25**, 123–129 (2013)
43. Renau, V., Oberst, U., Gosling, S., Rusiñol, J., Chamarro, A.: Translation and validation of the ten-item-personality inventory into Spanish and Catalan. Aloma Rev. Psicol. Ciències l'Educació i l'Esport. **31** (2013)
44. Tibshirani, R., Walther, G., Hastie, T.: Estimating the number of clusters in a data set via the gap statistic. J. R. Stat. Soc. Ser. B (Stat. Methodol.) **63**, 411–423 (2001)
45. Witten, D.M., Tibshirani, R.: A framework for feature selection in clustering. J. Am. Stat. Assoc. **105**, 713–726 (2010)
46. Roldos, M.I., Corso, P.: The economic burden of intimate partner violence in Ecuador: setting the agenda for future research and violence prevention policies. West. J. Emerg. Med. **14**, 347 (2013)
47. Jothi, N., Husain, W.: Data mining in healthcare–a review. Procedia Comput. Sci. **72**, 306–313 (2015)

Copper Price Variation Forecasts Using Genetic Algorithms

Raúl Carrasco[1,2(✉)] (iD), Christian Fernández-Campusano[3] (iD), Ismael Soto[2] (iD),
Carolina Lagos[4,8] (iD), Nicolas Krommenacker[5] (iD), Leonardo Banguera[6] (iD),
and Claudia Durán[7] (iD)

[1] Facultad de Ingeniería, Ciencia y Tecnología, Universidad Bernardo O'Higgins,
Santiago, Chile
raul.carrasco.a@usach.cl
[2] Departamento de Ingeniería Eléctrica, Universidad de Santiago de Chile,
Santiago, Chile
[3] Department of Architecture and Computer Technology,
University of the Basque Country UPV/EHU, Donostia-San Sebastián, Spain
[4] Escuela de Ingeniería Eléctrica, Pontificia Universidad Católica de Valparaíso,
Valparaíso, Chile
[5] Université de Lorraine, CRAN CNRS UMR 7039,
54506 Vandoeuvre Les Nancy, France
[6] Facultad de Ingeniería Industrial, Universidad de Guayaquil, Guayaquil, Ecuador
[7] Faculty of Engineering, Department of Industry,
Universidad Tecnológica Metropolitana, 7800002 Santiago, Chile
[8] Facultad de Administración y Economía, Universidad de Santiago de Chile,
Santiago, Chile

Abstract. The use of genetic algorithms and techniques of big data support the decision-making of manner effective in problems, such as the variation in copper prices. Today, the price of copper and its variations represent a significant financial issue for mining companies and the Chilean government because of its high impact on the national economy. This paper reviews the forecast of volatility for the copper market over a period, which is of interest to different participants such as producers, consumers, governments and investors. To do this, we propose to apply genetic algorithms to predict the variation in copper prices, in order to improve the degree of certainty by incorporating of the inverse of the percentage of sign prediction *PSP*.

Keywords: Genetic algorithms · Forecasting · Directional accuracy · Copper

1 Introduction

The price of copper and its variations is a crucial financial problem for mining companies. In the case of Chile, this problem affects the Chilean government, due to the high impact on the results in the country's economy. The series of

© Springer Nature Switzerland AG 2020
M. Botto-Tobar et al. (Eds.): ICAT 2019, CCIS 1194, pp. 284–296, 2020.
https://doi.org/10.1007/978-3-030-42520-3_23

prices in the markets in general and of the commodities, as is the case of copper, have high volatility, dynamics and turbulence, due to this it is imperative to estimate their price.

Mathematical modelling and prediction of commodity values is the subject of constant research by private agents, insurers and government institutions to ensure free competition in the securities market [1–3]. In this paper, to apply genetic algorithms to predict the variation in copper prices.

Today, capital markets require real-time information (data) to support and enable short-term and long-term decision-making, allowing them to manoeuvre effectively in the face of the turbulent global economy (it is denoted as *latency reduction*).

Regarding the latency and the value of the data. The value of the data decreases rapidly, that is, the low latency data has more value than the high latency data. Reducing data and analysis latency depends essentially on solutions in Big Data. However, the decision to reduce latency requires changes in business processes. Thus, providing fresher data does not create business value unless it is used in a timely manner.

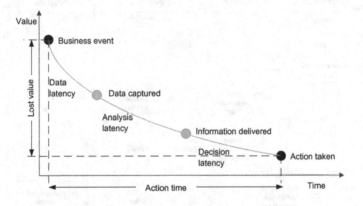

Fig. 1. Latency in decision-making.

In Fig. 1 proposed in [4], we can see the latency reduction curve. The longer the delay or latency of the response, the lower the value. Figure 1 shows the time of action (or distance of action), that is, the duration between the event and the action, and the net benefit is the value of the decisions (lost or gained) over some time.

2   Genetic Algorithm

The Genetic Algorithms (GA) refers to the evolutionary algorithms class, and its development is based on the natural genetic evolution of individuals in a defined environment. The initial work on GA was conducted by Holland [5] in 1975, which explores their use in the study of a wide range of complex, naturally occurring processes, concentrating on systems having multiple factors that interact in non-linear ways. Work that allowed investigations such as the related to modelling dynamic multivariate forecasting systems in [6].

For the model, a partial solution is represented by binary chains of constant length. This local solution is improved by using methods based search multi-point evolutionary theories, achieving solutions of better quality and speed compared to search algorithms previously investigated.

The algorithm was implemented in an object-oriented platform to obtain forecasts on copper prices. On the other hand, the data analysis is carried out with the statistical program R [7].

2.1   Phases of the Genetic Algorithm

Figure 2 shows four phases of the genetic algorithm, which are generation, population, actions and results. These are described below:

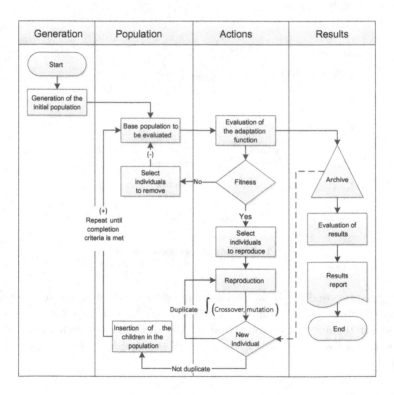

Fig. 2. Four phases of the genetic algorithm.

P1. Generation: It generates the initial population in a stochastic process of predefined size.

P2. Population: Determine a population of thirty individuals, which are renewed through the departure of unfit individuals and the incorporation of the new evolved genetic material. The steps in this phase are:

– To prepare the base population for evaluation.
– To eliminate individuals who do not qualify according to the aptitude function defined in the problem.
– To incorporate into the base population, the offspring spawned by the best individuals in the population, adding, in this way, new genetic material.

P3. Actions: Evaluate and select individuals according to aptitude function (fitness function), then reproduce and mutate. The steps of this phase are:

– To evaluate individuals according to the function of aptitude (fitness function).
– To select the individuals:
  • Capable (fit), those will remain in the base population and transfer their genetic material to the next generation;
  • Incapable (unfit), those will be eliminated from the base population.
– To reproduce capable individuals according to crossbreeding operators.
– To mutate the reproduced individuals according to the mutation operators and the restrictions of not duplicating the individuals, which must pass to the base population.

P4. Results: To archive the results of each generation and, once the algorithm is finished, evaluation and analysis of the results are made to produce a report of these.

3 Application: A Copper Price Forecast

The applicability of the genetic algorithm is to obtain a dynamic multivariate forecast model, which maximizes the predictive percentage of sign related to the daily variations of copper prices (Cu) presented by the London Metal Exchange—discovering a mathematical formula that generates approximately the historical patterns of copper time series.

Our time series corresponds to a sequence of values that measure the price of markets at equal intervals of time, maintaining the consistency in the activity and the method of measurement. In this case, the extra sample data correspond to daily closing values for the period from February 24 to August 29, 2016. For the intro-sample variability period, the days of lags corresponding to *Rolling* of 60 days was considered.

The dynamic multivariate forecast models used help to project the future prices of the time series [8,9], understanding the behaviour and what is happening with the data [8,9] of the Cu and Dow variables as a stock market, according to their values *Stragglers*.

3.1   Metrics and Data

Determination of Variations. To determine the price variations of the *Dow* and *Cu*, in general terms it is expressed with the difference operator ∇. This operator is used to express relations of type $\nabla Y_t = X_t - X_{t-1}$, where X_t is a balance variable and ∇Y_t will be the corresponding flow variable [10].

∇ is defined as:

$$\nabla Y_t = X_t - X_{t-1} \ \forall t, \ t \in Z \tag{1}$$

Where: ∇Y_t is the price variation, X_t is the price of the period, X_{t-1} is the price of the previous period.

We can represent the variations of Dow *and* Cu, *according to Eq. (1).*

$$\nabla Dow_t = Dow_t - Dow_{t-1} \ \forall t, \ t \in Z \tag{2}$$

Where: ∇Dow_t is the price variation of the *Dow*, Dow_t is the price of the *Dow* period, Dow_{t-1} is the price of the previous *Dow* period.

$$\nabla Cu_t = Cu_t - Cu_{t-1} \ \forall t, \ t \in Z \tag{3}$$

Where: ∇Cu_t is the price variation of Cu, Cu_t is the price of the Cu period, Cu_{t-1} is the price of the previous Cu period.

Determination of Projections. The projection of the variation of Cu will be denoted as a function of the lags variations of the *Dow* and *Cu* of Eqs. 2 and 3 and of forecast errors.

It is defined in this study as:

$$\nabla \overline{Cu}_t = \sum_{i=1}^{4} \theta(\nabla Dow_{t-i} \bullet \beta_{di} + \nabla Cu_{t-i} \bullet \beta_{ci} + \epsilon_{t-i} \bullet \beta_{ei}) \tag{4}$$

Where:

$\nabla \overline{Cu}_t$, is the projected variation,
$\theta()$ is the Heaviside function, which multiplies the betas calculated with the input of the variable,
∇Dow_{t-i} are the lags of the *Dow* variations,
∇Cu_{t-i} are the lags of the variations of Cu,
ϵ_{t-i} is the prediction error.

The projection is done by minimizing the squared error of the 60-day *Rolling* estimate with Newton Raphson, defined by:

$$\min \left(e^2(n) \right) = \min \left(\sum_{i=1}^{n} \left(\nabla Cu_t - \nabla \overline{Cu}_t \right)^2 \right) \tag{5}$$

Where:

$n = 60$,
$e^2(n)$ is the squared error of the estimate in the n periods,
∇Cu_t is the value of the variation of the period,
$\nabla \overline{Cu}_t$ is the value of the projected variation.

Note 1: The objective function is not subject to a series of constraints.

Note 2: The vector of decision variables corresponds to the calculated betas, which minimize the sum of the errors to the prediction table for *Rolling* of 60 days.

Determination of the Percentage of Sign Prediction (PSP). Moreover, we must determine the PSP [11]. To calculate the PSP, the sign of the projected variation is compared with the sign of the observed variation, in each period of $t+n$, where $t = 1, 2 \ldots, n$ starting from $t+1$. If the signs of the projected variation and observed variation coincide, the value "1", is obtained, which represents a success. In the opposite case, "0" indicates a model prediction error [12]. If the signs coincide, the effectiveness of the prediction increases, and in case of no coincidence, the prediction error of the model increases. The PSP of the model is defined by:

$$PSP = \frac{\sum_{j=1}^{n} \theta \left(\nabla Cu_{j,t+1} \bullet \nabla \overline{Cu}_{j,t+1} \right)}{n} \qquad \forall, 1 \le j \le n \qquad (6)$$

Where:

PSP is the percentage of sign prediction (PSP) presented in the equation,
∇Cu_t is the value of the period variation,
$\nabla \overline{Cu}_t$ is the value of the projected variation,
$\theta()$ is the dichotomous function of Heaviside; $\theta() = 1$ iff $\nabla Cu_{j,t+1} \bullet \nabla \overline{Cu}_{j,t+1} > 0$, or $\theta() = 0$ iff $\nabla Cu_{j,t+1} \bullet \nabla \overline{Cu}_{j,t+1} \le 0$,
n is the total number of predictions performed.

The $PSP_{\max}$ variable used will be the maximum between PSP and $(1 - PSP)$, as shown in Eq. 7.

$$PSP_{\max} = \max(PSP, (1 - PSP)) \qquad (7)$$

Test of Pesaran and Timmermann. The directional accuracy Test of Pesaran and Timmermann [13] was applied in order to measure the statistical significance of the predictive capacity of the dynamic multivariate prognostic model with genetic algorithms [14,15].

The directional correctness test is used to measure the statistical significance of the predictive ability of the models analyzed [11,16]. The directional correctness test tests the null hypothesis that the observed variations are distributed independently of the projected variations. Therefore, if the null hypothesis is rejected, it is said that there is statistical evidence that the model has the ability to predict the future evolution of the observed variable.

This test compares the sign of the projection with that of the observed value for each $j\text{-}esime$ observation of the sample set $(j = 1, 2, \ldots, n)$. Where the sign indicates the direction in which the stock market will move: up if it is positive, or down if it is negative. If the signs coincide, the prediction effectiveness increases, and in case of no coincidence, the prediction error of the model increases (same as the methodology used to calculate PSP).

In order to obtain the percentage of real positive changes observed, the following equation is represented by:

$$P = \frac{\sum_{j=1}^{n} \theta \left(\nabla P_{j,t+1} \right)}{n} \tag{8}$$

Where:

P is the percentage the real positive changes observed, and
$\theta()$ is the dichotomous function Heaviside; $\theta() = 1$ iff $\nabla P_{j,t+1} > 0$ or $\theta() = 0$ iff $\nabla P_{j,t+1} \leq 0$.

The percentage of positive projection variations is represented in the following equation:

$$\overline{P} = \frac{\sum_{j=1}^{n} \theta \left(\nabla \overline{P}_{j,t+1} \right)}{n} \tag{9}$$

Where:

$\overline{P}$ is the percentage of projected real positive changes, and
$\theta()$ is the dichotomous function Heaviside; $\theta() = 1$ iff $\nabla \overline{P}_{j,t+1} > 0$, or $\theta() = 0$ iff $\nabla \overline{P}_{j,t+1} \leq 0$.

In addition, the success ratio when the actual variations and projected variations are independently distributed for $\nabla P_{j,t+1}$ and $\nabla \overline{P}_{j,t+1}$, SRI, is given by:

$$SRI = P \,\overline{P} + (1 - P)(1 - \overline{P}) \tag{10}$$

To determine the variance of the SRI ratio, it is defined as:

$$\text{var}(SRI) = \frac{\left(\begin{array}{c} n \left(2\overline{P} - 1 \right)^2 P \left(1 - P \right) + \\ n \left(2P - 1 \right)^2 \overline{P} \left(1 - \overline{P} \right) + \\ 4P\overline{P} \left(1 - P \right) \left(1 - \overline{P} \right) \end{array} \right)}{n^2} \tag{11}$$

On the other hand, the variance of the SR ratio, it is defined as:

$$\text{var}(SR) = \frac{SRI \left(1 - SRI \right)}{n^2} \tag{12}$$

Finally, the Directional Accuracy (DA) by [15] is given by:

$$DA = \frac{(SR - SRI)}{\sqrt{\text{var}\left(SR \right) - \text{var}\left(SRI \right)}} \xrightarrow{d} N\left(0, 1 \right) \tag{13}$$

This test follows a standard normal distribution. The result of this equation is compared to a critical one, which will depend on the level of trust required to be tested. That is to say; If the DA value is between the rejection values, we do not reject the null hypothesis that the observed variations are distributed independently of the projected variations [17, 18]. From the latter, it is understood that we try to reject the null hypothesis. That is to say, that the value DA is not between the critical values mentioned and that, therefore, there is a predictive capacity.

3.2   Codification of Variables

Each chromosome has several genes, which correspond to the parameters of the problem. To work computationally with the genes is necessary to encode them in a string (i.e., in a sequence of symbols composed, in this case, of zeros and ones).

For correct codification and good resolution of the problem. It was constructed by matrix blocks of the Dow_t, Cu_t and ε_t lags. We used the heuristic rule called the building blocks rule, that is, related parameters that must be close to each other on the chromosome.

Chromosome [1010][1110][1101], would be represented by a three-block string, where the first four-gene block is represented by the Dow_t at $t-1, t-2, t-3$ and $t-4$ lags. The second block of four genes ranging from gene five to gene eight is represented by Cu_t lags in its lags from $t-1$ to $t-4$. For the third block of four genes ranging from gene 9 to gene 12 is represented by ε_t in the prediction for $t-1$ to $t-4$.

3.3   Generation of Initial Population and First Generation

A population of fixed-sized chromosomes of only thirty individuals has been chosen arbitrarily, considering the calculation times of our computational resources and privileging the search and optimization based on the Darwinian theory.

Our initial population, or first generation of chromosomes, is generated from a random generation. The genetic algorithms being tools to obtain approximate solutions to problems in which to evaluate the exact resolution would be very costly in time. So, It is necessary to program in a way that does not allow the twins (siblings alike) not to evaluate more than once the same model. Also, it is advisable not to generate or reproduce a null chromosome, composed of zeros for the three blocks, represented as [0000][0000][0000].

From an initial population of Cu chromosomes generated randomly from thirty models, as shown in Fig. 3. The ten best chromosomes are chosen by the aptitude assessment function (in an elitist manner), which are assigned to the second generation to be reproduced until it completes again thirty, as shown in the Fig. 2. In this way, elitism can rapidly increase the performance of a genetic algorithm since this avoids losing the best solution found [19]. Then, it is possible that this method quickly leads to a local optimum.

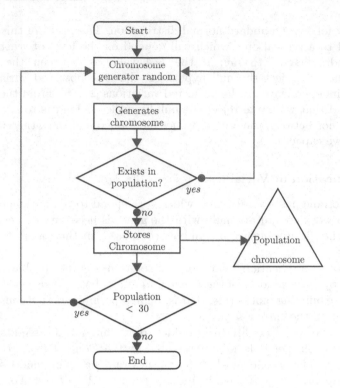

Fig. 3. Initial population generation.

3.4   Evaluation and Selection

The evaluation function fitness, according to Eq. 14, plays an essential role in the potential classification of solutions in terms of their characteristics. It is the criterion evaluation of the quality of the individuals.

$$Fitness = \frac{PSP_{\max} + Profitability}{2} \tag{14}$$

Where:

Fitness is the multi-objective evaluation function,
$PSP_{\max}$ is the maximum PSP between PSP and $(1 - PSP)$, as shown in Eq. 7,
Profitability is the individual's or chromosome's profitability.

It provides higher power and robustness to the search technique. And this operator fulfils the function of making a selection of the best individuals so that they are considered in the process of generation of the new population [19]. The genetic algorithm uses the elitist selection technique to select the individuals to be copied to the next generation. The technique ensures the selection of the most suitable members of each generation and preserves them to deliver their attributes to their descendants. That is, those selected as reproducers of the next generation. During the evaluation, the gene is decoded, turning it into a series of parameters (presented at the coding point of the variable). Finally, a solution is obtained with the best score based on the best performance.

3.5 Mutation, Reproduction and Stop

5% was assigned for the mutation function, with the restriction of not allowing duplicate individuals, which increases the effective mutation rate after each generation. In gene transfer, the father provides the first two genes of each block, and the remaining two are the mother, which are selected for random reproduction among the best individuals. The stopping criterion was applied upon completion of evaluating the seventeenth generation.

4 Results

4.1 The Best Models

According to Table 1, the best model [0100][1100][0111] are born in the twenty-first generation both with a predictive capacity (PSP) of 67.12% and a profitability of the period of 9.02%.

In the results of positions 2 to 5 can be observed in Table 1.

Table 1. Top five models.

Chromosome	PSP_{max}	Profitability	Generation
[0100][1100][0111]	67.12%	9.02%	21°
[0100][1000][0011]	65.75%	8.03%	17°
[0011][1000][0000]	61.64%	10.23%	5°
[0100][1100][0011]	64.38%	7.32%	19°
[0111][0001][0000]	61.64%	9.89%	5°

Figure 4 shows the evolution of the relative price of Cu and the best individual [0100][1100][0111], showing the best result of 9.02% of the active management of chromosome [0100][1100][0111] versus the Buy and Hold strategy of 0.45% represented by Cu. Active management with the support of genetic algorithms improves the profitability result. We can see that with active portfolio management, buying and selling according to the signal variation of the price of the positive or negative Cu, a yield of 9.02% is obtained at the end of the evaluation period. Represented by the best individual [0100][1100][0111].

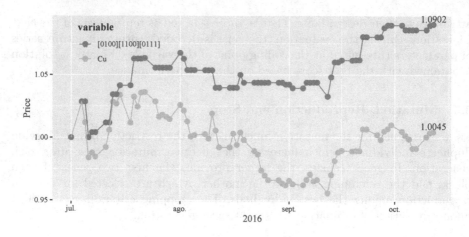

Fig. 4. Results comparison.

4.2   Findings

The $PSP_{\max}$ of the best results of the individuals selected for their best fitness in each generation, turned out to be maximum $(1 - PSP)$ on PSP as shown in Eq. 7 which is part of Eq. 14, and can be observed in Figs. 5a and b [7].

The importance of finding and using an individual with a low level of success (<50%), it allows being classified as a liar or inverse, which is shown in Fig. 5a. Identifying and measuring the liar as $(1 - PSP)$, has allowed us to obtain better results with some individuals. It can be seen in Fig. 5b, for the distribution of fitness for the inverse and normal groups.

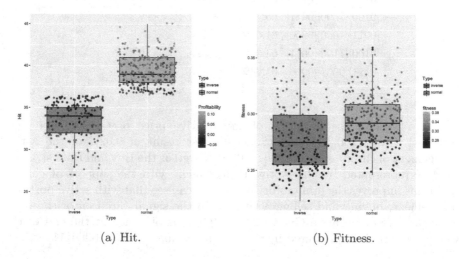

(a) Hit.   (b) Fitness.

Fig. 5. Normal and inverse.

5   Conclusions

The application of genetic algorithms to the Cu price prediction models effectively allowed an improved model. Our results showed that the chromosome [0100][1100][0111] obtained a maximum PSP of 67.12% and a yield of 9.02%, versus a "Buy and Hold" return of 0.45%, in 73 days for investment.

It is worth mentioning that the first generations do not show good results, however, reproduction and mutation achieve an evolution of generations with desired results. The above described makes it possible to obtain forecast models of the copper price variation with greater precision.

In addition, the models showed the highest accumulated profitability during the evaluation period. Thus, the models constructed from genetic algorithms presented a statistically significant predictive capacity, as demonstrated by the results of the directional accuracy test of Pesaran & Timmermann.

Comparison with other methods such as SVM (Support Vector Machines) and Neural Networks is proposed as future work.

Acknowledgement. The authors are grateful for the financial support of the projects Fondef/Conicyt IT17M10012 and STIC-AmSud 19-STIC-08.

References

1. Carrasco, R., Vargas, M., Soto, I., Fuentealba, D., Banguera, L., Fuertes, G.: Chaotic time series for copper's price forecast. In: Liu, K., Nakata, K., Li, W., Baranauskas, C. (eds.) ICISO 2018. IAICT, vol. 527, pp. 278–288. Springer, Cham (2018). https://doi.org/10.1007/978-3-319-94541-5_28
2. Carrasco, R., Astudillo, G., Soto, I., Chacon, M., Fuentealba, D.: Forecast of copper price series using vector support machines. In: 2018 7th International Conference on Industrial Technology and Management (ICITM), pp. 380–384. IEEE, Oxford (2018). https://doi.org/10.1109/ICITM.2018.8333979
3. Seguel, F., Carrasco, R., Adasme, P., Alfaro, M., Soto, I.: A meta-heuristic approach for copper price forecasting. In: Liu, K., Nakata, K., Li, W., Galarreta, D. (eds.) ICISO 2015. IAICT, vol. 449, pp. 156–165. Springer, Cham (2015). https://doi.org/10.1007/978-3-319-16274-4_16
4. Hackathorn, R.: The BI Watch Real-Time to Real-Value. DM REVIEW (2004)
5. Holland, J.H.: Adaptation in natural and artificial systems: an introductory analysis with applications to biology, control, and artificial intelligence. University of Michigan Press (1975)
6. Hsu, C.M.: A hybrid procedure for stock price prediction by integrating self-organizing map and genetic programming. Expert Syst. Appl. **38**(11), 14026–14036 (2011). https://doi.org/10.1016/j.eswa.2011.04.210
7. R Core Team: R: A Language and Environment for Statistical Computing (2016)
8. Carrasco, R., Vargas, M., Alfaro, M., Soto, I., Fuertes, G.: Copper metal price using chaotic time series forecating. IEEE Lat. Am. Trans. **13**(6), 1961–1965 (2015). https://doi.org/10.1109/TLA.2015.7164223
9. Guerino, M.: Técnicas avanzadas para la predicción de la variación de IFN. Tesis para optar al grado de magíster en finanzas, Universidad de Chile (2006)

10. Guerrero, V.: Análisis estadístico de series de tiempo económicas, 2a edn. International Thomson, México, D.F. (2003)
11. Parisi, A., Parisi, F., Díaz, D.: Forecasting gold price changes: rolling and recursive neural network models. J. Multinational Finan. Manag. 18(5), 477–487 (2008). https://doi.org/10.1016/j.mulfin.2007.12.002
12. Parisi, A., Parisi, F., Cornejo, E.: Algoritmos genéticos y modelos multivariados recursivos en la predicción de índices bursátiles de américa del norte: IPC, TSE, NASDAQ y DJI. Fondo de Cultura Económica 71(284(4)), 789–809 (2004)
13. Ahn, Y.B., Tsuchiya, Y.: Directional analysis of consumers' forecasts of inflation in a small open economy: evidence from South Korea. Appl. Econ. 48(10), 854–864 (2016). https://doi.org/10.1080/00036846.2015.1088144
14. Anatolyen, S., Gerko, A.: A trading approach to testing for predictability. J. Bus. Econ. Stat. 23(4), 455–461 (2005)
15. Pesaran, M.H., Timmermann, A.: A simple nonparametric test of predictive performance. J. Bus. Econ. Stat. 10(4), 461–465 (1992). https://doi.org/10.2307/1391822
16. Tsuchiya, Y.: Do production managers predict turning points? A directional analysis. Econ. Model. 58, 1–8 (2016). https://doi.org/10.1016/j.econmod.2016.05.019
17. García, I., Trigo, L., Costanzo, S., ter Horst, E.: Procesos gaussianos en la predicción de las fluctuaciones de la economía mexicana. El Trimestre Económico 77(3), 585–603 (2010)
18. Trigo, L., Costanzo, S.: Redes neuronales en la predicción de las fluctuaciones de la economía a partir del movimiento de los mercados de capitales. El Trimestre Económico 74(74), 415–440 (2007)
19. Marczyk, A.: Algoritmos genéticos y computación evolutiva. Departamento de Informática, Universidad de Colorado (2004)

A Survey on Hand Gesture Recognition Using Machine Learning and Infrared Information

Rubén Nogales[1,2]([✉]) [iD] and Marco E. Benalcázar[2] [iD]

[1] Facultad de Ingeniería en Sistemas Electrónica e Industrial,
Universidad Técnica de Ambato, Ambato, Ecuador
[2] Departamento de Informática y Ciencias de la Computación,
Escuela Politécnica Nacional, Quito, Ecuador
{ruben.nogales,marco.benalcazar}@epn.edu.ec

Abstract. The research consists in the hand gestures are movements that convey information and thus complement oral communication or by themselves constitute a form of communication between people. The function of a hand gesture recognition system is to identify the type of movement, from a given set of movements, and the instant when that movement is performed. Gesture recognition systems have multiple applications including sign language translation, bionics, human-machine interaction, gamming, and virtual reality. For this reason, hand gesture recognition is a problem where many researchers have focused their attention too. In this context, in this paper, we present a systematic literature review for hand gesture recognition using ma-chine learning and infrared information. This work has made because there is no work in the scientific literature that reviews gesture recognition systems based on machine learning and infrared information. In this work, we answer the research question: what is the architecture of the proposed models for hand gesture recognition based on machine learning and infrared information? For answering this research question, we used Kitchenham methodology. Finally, in this work, we also present trends and gaps with respect to the problem analyzed.

Keywords: Hand gesture recognition · Machine learning · Infrared information · Survey · Systematic literature review

1 Introduction

1.1 A Subsection Sample

The hand gesture recognition problem consists of quickly tracking the movements of the hands and retrieve information about its images, spatial positions such as finger position, velocity, and direction of movement, between other data. This makes the hand gesture recognition doesn't be a trivial problem. This problem has been tried to solve with computer-vision techniques based on RGB cameras. These techniques faced problems such as illumination, and impaired backgrounds, between others.

It has been also tried to solve using electromyography signals facing problems such as muscle developing, and injuries of the forearm. In some way, infrared sensors and

© Springer Nature Switzerland AG 2020
M. Botto-Tobar et al. (Eds.): ICAT 2019, CCIS 1194, pp. 297–311, 2020.
https://doi.org/10.1007/978-3-030-42520-3_24

depth cameras come to solve these problems. However, these sensors faced problems too such as the inaccuracy of spatial positions given and occlusion of the fingers.

In this sense, hand gesture recognition is a problem associated with pattern recognition and machine learning. The hand gestures recognition is necessary for interaction between people, especially when they present hearing or speech disabilities or between people that speaking different languages. In this field, hand gesture recognition has a great advanced, especially in sign language recognition.

In [1] mentioned that 93% of people use non-verbal communication in their daily lives. In human-machine interaction, non-verbal communication is an important challenge, especially in the development of the human-computer interface. In the medicine field, hand gesture recognition has taken very important advances, especially in the rehabilitation field, and in the operating room. In this context, the paper proposes a survey that permits identify which are the trends and the gaps with respect to the architecture of the models used in the hand gesture recognition using machine learning and infrared information.

2 Related Works

Considering the interest of the researchers for the hand gesture recognition problem, and the constant increasing of the technologies like machine learning, and retrieve data by infrared sensors, the present subsection proposed to make a review of surveys about hand gesture recognition using machine learning and infrared information.

To search for reviews about the problem, we used scientific databases such as Science Direct, Springer, ACM digital library, IEEE Xplorer and Willey. In the same sense, we used a search string that includes all keywords about the problem together with logic operators. The keywords are: hand gesture recognition, infrared, infrared information, machine learning, systematic literature review, state of the art, review and survey.

The search string is: *(((hand AND gesture AND recognition) AND "machine learning") AND (infrared OR ("infrared information"))) AND (("systematic literature review") OR ("state of the art") OR ("review") OR ("survey")).*

Using this search string over the databases above described, we don't obtain results, however, our survey will describe surveys that maintain relationships with the problem. From the articles found will retrieve information as such as: objectives, inclusion, and exclusion criteria and quality criteria assess used.

In [2] the authors presented a systematic literature review for hand gesture recognition on vision-based. For this systematic literature review the authors used Kitchenham methodology. The authors clearly show the objectives, databases used for retrieve information of the primary studies. In the study they clearly defined the inclusion and exclusion criteria. In the same way, they mentioned the quality of criteria assess used, how were data and data synthesis extracted. However, the authors have not been critical with the primary studies and the primary reviews.

In [3] the authors mention that the main objective is to understand what 3D mid-air hand gestures exist, and in which fields are applied. In the same way, the authors describe that the gestures sets have been develop according to faced phenomenon.

In this study they mentioned devices for data acquisition such as Microsoft Kinect and Leap Motion. For the primary studies they used electronic databases and specific journal well defined in the paper. Therefore, the authors defined the inclusion and exclusion criteria. The valid papers were discussed by each research and retrieved data was registered in a spreadsheet. Additionally, they detail why the other papers were discarded. However, their survey is based on types gestures and interfaces. They did not mention the methodology used.

In [4] the authors present a review of methods and databases for hand gesture recognition. They clearly identified the static and dynamic hand gestures. Some methods are mentioned and represented in Tables showing used features, classification methods and reported application. Did not mention the methodology, inclusion and exclusion criteria, quality of criteria assess used and how were data extracted and how were data synthesis. However, mentioned that the survey is very important due to unsolved challenges of reliable and speed of this phenomenon.

In [5] presented a survey about the use of depth images for hand tracking and gesture recognition. The main goal was obtaining papers that describe the phenomenon in terms of hand localization and gesture classification methods developed and used, where the applications has been tested and how Microsoft Kinect was used. In this sense, the survey describes depth sensors and methods for hand localization, gesture recognition and applications and environments. However, the survey does not mention what methodology is used, how primary studies where obtained, what were inclusion and exclusion criteria's, what criteria assess used, how were data extracted and how were data synthesis.

In [6] presents a study about seven most important patient monitoring application on vision based. This paper cites patient monitoring application, fall detection, action and activity monitoring, epilepsy monitoring, vital signs monitoring, facial expression monitoring. Each application cited present comprehensive studies about technologies used for solving, discussing and facing the problem. However, does not present the databases that retrieve information. In the same way does not present the inclusion and exclusion criteria for selecting the primary studies, how quality assess the papers, retrieve information and were data synthesis.

In [7] presented a study about automated hand gesture, and mentioned that is a recognition and segmentation problem. The main goals are based on analyze the application of computer methods for hand gesture analysis, and functional types and phases of gestures in theoretical terms. In this paper did not mention in which data bases the search was carried out, did not mention the inclusion and exclusion criteria. However, present how were data extracted. The data were retrieve in function the domains of application, means of data acquisition, data representation strategies, computational techniques or strategies used in data analysis and metric employed to evaluate the results. But did not mention how quality criteria assess where used for evaluating the papers.

In [8] present a comprehensive survey about soft computing approach for image segmentation. The objective of this survey is present applications for segmentation based on soft computing. This paper mentions that core of soft computing is Fuzzy logic, Artificial Neural Network and Genetic algorithms. The idea is present in the state of the art new techniques, new topics, new problems associated with segmentation of images. The paper presents a wide description, definitions of methods and sub methods used in soft computer valid for other researchers. Did not mention what databases were used for searching primary studies. Did not show how the primary studies are selected. However, each paper evaluated parameters such as: testing protocol, testing regime, performance indicators (accuracy, robustness, sensitivity, adaptability reliability, efficiency), performance metrics of the algorithms and image databases used for testing algorithms. The paper described the most widely used databases for image segmentation. The paper let opening future works, however did not mention a methodology used.

As we can see above reviewing the scientific literature shows that does not exist a systematic literature review with respect to hand gesture recognition based on infrared information using machine learning techniques. In this context, the present project proposes to identify trends, identify gaps for new studies and present analysis of previous studies of hand gesture recognition using machine learning and infrared information. In the scientific literature, there are architectures of models proposed for recognition and classification of static and dynamic hand gestures. These models are based on infrared information, depth and color images [9–15]. Real-time hand gesture recognition is a challenging problem [10, 12, 16, 17] due to the hardware and time constraints.

3 Methods

The present work is developing, because, we need to know what is the bench-marking about the architecture of models of hand gestures recognition using machine learning and infrared information. Additionally, neither of the works in the above section described what is the architecture, how to use the models of machine learning in the hand gesture recognition problem.

In this sense, we propose a generic model that permits the evaluation of the architecture of the articles reviewed in this work. The proposed architecture consists of the following modules: data acquisition, pre-processing, feature extraction, classification and post-processing.

In hand gesture recognition problems, data acquisition is composed of methods that allow us the extraction of data from the sensors use to implement the system for hand gesture recognition [18]. In the same problem, pre-processing consist of a set of operations that take as input a signal and return another signal with some of its properties enhanced [19].

For feature extraction we define as set of descriptors that represent the signal returned by the pre-processing module. Usually, these descriptors are arranged in a n-dimensional vector called feature vector [20].

The classification module labels feature vectors according to a classification rule.

The Post-processing module enhances the response obtained from the classifier.

3.1   Research Question

For this work, we propose a research question that guides our research. *What is the architecture of the proposed models for hand gesture recognition based on Machine Learning and Infrared information?*. This research question is proposed because is necessary to know how the modules together, and how the variables in each module are used in relation to our model propose. For answering this question, we propose searching all related works in the scientific literature. In the same sense, we proposed a search string for including all works that present the real-time hand gesture recognition using machine learning and infrared information. The search string was structured using key-words and logic connectors as AND, OR.

The key-words are: Hand gesture, Hand gesture recognition, Hand tracking, Hand poses recognition, Machine learning, Machine learning techniques, Machine learning algorithms, Infrared, Infrared information.

The search string is: *((((((hand gesture) OR (hand poses)) AND recognition) OR (hand tracking)) AND machine learning) AND ((infrared) OR (infrared information) OR (Leap Motion) OR (Kinect)))*.

3.2   Inclusion and Exclusion Criteria

The inclusion and exclusion criteria are defined for deciding what studies input our studied. The inclusion and exclusion criteria are defined in the Table 1.

Table 1. Inclusion and exclusion criteria.

Type	Description
Inclusion	Papers that using machine learning and infrared information for hand gesture recognition
	Only papers from databases mentioned in Sect. 2
	Only papers of congresses and journals
	Items that are peer-reviewed
	Works that present models or that compare models
	Publications between January 2015 and February 2019
Exclusion	Articles that are not related to hand gesture recognition
	Articles that not use infrared information and machine learning for hand gesture recognition
	Articles with years of publication earlier than 2015
	If the publications do not define population, intervention and outcomes (accuracy and time)
	All articles that do not be in English language
	Works that present only applications and that do not propose a model

Of applying the search string, we obtained 1156 related articles to the problem. To avoid biases, we decide to include database ArXiv in which finding 230 articles that referred to the problem. Next, we applied the inclusion and exclusion criteria in two phases. First, the search was constrained to the years 2015 and 2019, we select all articles that in its titles present the keywords above defined, also, the problem was searched in abstracts and conclusions. In this phase, we obtained 173 articles, 55 of IEEE, 30 of ACM, 42 of Science Direct, 40 of Springer, 6 of Willey, and 18 of ArXisv. Second, we excluded all articles that no be in English language and articles that present only applications, we included articles that present population, intervention, and outcomes, in this phase we obtained 62 articles, 27 of IEEE, 7 of ACM, 16 of Science Direct, 10 of Springer, and 2 of Willey. The process is shown in Fig. 1.

Fig. 1. Process for applying inclusion-exclusion criteria.

3.3   Quality Assessment of Articles

To the results obtained from the application of inclusion-exclusion criteria, we applied quality assessment criteria as: the experiments are well described, the research follows the goals of the work, the conclusions are supported by the results, and the results are well described, as shown in Table 2.

Table 2.   Quality assessment.

Items for quality assessment	Disagree	None of	Agree
The experiments are well described	1		
The research follows the goals of the work		1	
The conclusions are supported by the results			1
The results are well described			1

All possibilities are weighted in one. If the sum of columns 'agree and the nether agree nor disagree' is most than the sum of columns 'disagree and nether agree nor disagree', the paper pass. Finally, for our study, we selected 38 articles, 12 of IEEE, 5 of ACN, 13 of Science Direct, 7 of Springer and 1 of Willey. Of the works selected are 19 articles in Journals and 15 in Congress.

3.4 Data Extraction

The interest information will be retrieved from the primary studies for answering the research question, the information that extract is: data acquisition methods, preprocessing methods or algorithms, feature extraction algorithms, classifier algorithms and post-processing methods.

4 Analysis of the Studies

The hand gesture is a movement of the hand that carries any information. The hand gesture recognition is a process in which the machine intends to recognize a movement of the hand [2]. A lot of techniques are used for hand gesture recognition, one of these is machine learning, in the same sense, and the data origin for this processing is very varied, one of these is the infrared information with deep cameras. in this context, we presented the most influential primary studies in the Table 3.

Table 3. Most influential primary studies.

Article name	Ref.
Recognition of dynamic hand gestures from 3D motion data using LSTM and CNN architectures	[18]
Machine learning for hand gesture recognition using bag-of-words	[21]
Computer-vision-assisted palm rehabilitation with supervised learning	[22]
Recognizing arabic sign language gestures using depth sensors and a KSVM classifier	[23]
Exploiting recurrent neural networks and leap motion controller for the recognition of sign language and semaphoric hand gestures	[24]
Hand gesture recognition using kinect via deterministic learning	[25]
A comprehensive leap motion database for hand gesture recognition	[26]
A Multisensor technique for gesture recognition through intelligent skeletal pose analysis	[27]
Static and dynamic hand gesture recognition in depth data using dynamic time warping	[28]
Dynamic hand gesture recognition based on depth information	[29]
CNN+RNN depth and skeleton based dynamic hand gesture recognition	[30]
Hand gesture recognition based on wavelet invariant moments	[31]
A system for a hand gesture-manipulated virtual reality environment	[32]
Multi-layered gesture recognition with Kinect	[33]
Proactive sensing for improving hand pose estimation	[34]
Gestimator: shape and stroke similarity based gesture recognition	[35]
A real-time hand posture recognition system using deep neural networks	[36]
A hand gesture recognition system based on canonical superpixel-graph	[37]

(continued)

Table 3. (*continued*)

Article name	Ref.
Convolutional neural networks and long short-term memory for skeleton-based human activity and hand gesture recognition	[38]
Enhancement of gesture recognition for contactless interface using a personalized classifier in the operating room	[39]
A multimodal framework for sensor based sign language recognition	[40]
Domain adaptation with low-rank alignment for weakly supervised hand pose recovery	[41]
An image-to-class dynamic time warping approach for both 3D static and trajectory hand gesture recognition	[42]
Dynamic and combined gestures recognition based on multi-feature fusion in a complex environment	[43]
Depth sensor based automatic hand region extraction by using time-series curve and its application to Japanese finger-spelled sign language recognition	[44]
An Improved computer interface comprising a recurrent neural network and a natural user interface	[45]
American sign language alphabet recognition using convolutional neural networks with multiview augmentation and inference fusion	[46]
Hand gesture recognition from depth and infrared Kinect data for CAVE applications interaction	[47]
Hand gesture recognition with jointly calibrated Leap Motion and depth sensor	[48]
Hand gesture recognition using Leap Motion via deterministic learning	[49]
Virtual grasps recognition using fusion of Leap Motion and force myography	[50]
Automatic recognition of the American sign language fingerspelling alphabet to assist people living with speech or hearing impairments	[51]
Hand joints-based gesture recognition for noisy dataset using nested interval unscented Kalman filter with LSTM network	[52]
Kinect-based Taiwanese sign-language recognition system	[53]

In this work, matching the architecture of the models proposed with the architecture of found models in the scientific literature. Below presented Fig. 2 that represents the histogram of models.

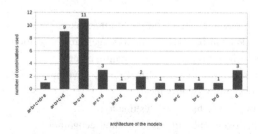

Fig. 2. The architecture of the model proposed vs. architecture of models found in the scientific literature; The parameters of the model structure proposed are: (a) data acquisition, (b) preprocessing, (c) feature extraction, (d) classifier and (e) post-processing.

According, the architecture proposed most papers used the pre-processing, feature extraction and classifier modules. However, to this architecture adds the data acquisition module. Both architectures are most used. In Fig. 2, we can see that don't all architectures mentioned the data acquisition module, however, all architectures used data. In Fig. 3, we can see what are the most type of data used for hand gesture recognition using machine learning and infrared information. The spatial positions of the hand are most used. These are relative position in the space with respect to a device that has the ability for recognizing these objects.

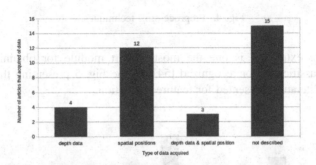

Fig. 3. Type of data used for intending resolve the problem of hand gesture recognition using machine learning and infrared information.

The second type of data most used is depth images. These types of images contain the distance between objects and devices that contain the depth cameras. These types of images typically used for forming 3D images. The scientific literature also presented articles that use both types of data spatial position and depth data. In the same sense, many articles don't mention what process used for data acquisition.

In this context, isn't possibly identify a trend. The devices most used are the Leap Motion Controller and Kinect. The Leap Motion has three led sensors and two depth cameras. This device is specialized for tracking hands and fingers. Frequently for the data acquisition module, its sample frequency is 200 frames by second. Themselves, the Kinect is a device that is imprecise with the hand and finger tracking, its sample frequency is 30 frames by second. These devices for interacting with a PC use their own SDK. With respect to the pre-processing module, the techniques used are dimensionality reduction, normalization, noise reducing filters, manual segmentation, images equalization, and articles that not reported the technique. How to shown in Fig. 4.

The papers that use dimensionality reduction, they do principal component analysis. themselves, the normalization carries the values between 0 and 1. For noise reduction the papers mentioned that used Kalman filter and the median filter. Also, for segmenting using canny filter and k-curvature.

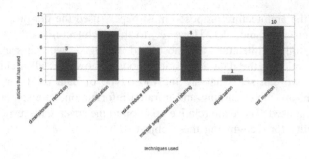

Fig. 4. Preprocessing techniques.

The feature extraction is one the most import module for obtaining a higher accuracy of classification or recognized [54]. In the Fig. 5 presenting the techniques that scientific literature presented for feature extraction.

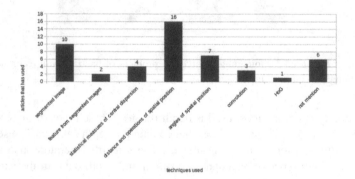

Fig. 5. Feature extraction techniques.

Between the feature extraction method reported in the scientific literature are angles of the spatial position, statistical measures of central dispersion, features from segmented images, distance and mathematical operations from spatial positions, and convolution. In Table 4 shown the techniques used when papers using angles of the spatial positions how feature vector, also shown the accuracy values reach. In this reported shown a value of 99.58, this accuracy was reached using a dataset of 5000 samples.

Table 4. Angles of spatial position.

Variables used	Accuracy reported
Accuracy reported Angles for each joint (3-angles) Angles between each two bones	93, 86, 96.41, 97.25, 99.58, 88.3

The techniques of statistical measures of central dispersion how feature vector presents values of accuracy less than the values of angles of spatial positions, how presented in Tables 4 and 5.

Table 5. Statistical measures of central dispersion.

Variables used	Accuracy reported
Arithmetic mean	
Standard deviation	
Root mean square	81, 90
Co variance	
Sign split window	

The techniques used with segmented images show in Table 6, also, show the accuracy reach in its studies. The reached values are high, an accuracy value reported is 98.2, obtained with a public dataset and this present 5040 samples.

Table 6. Feature from segmented images.

Variables used	Accuracy reported
SIFT	
SURF	
Isolate gestures	97, 98, 95, 98.44, 93.78, 91.56, 98.2 (public DS) 5040
Contour angles	
HoG	

In Table 7 shown the variables used with the distance and operations of spatial position, as well as, the accuracy obtained. With this technique, a value of 99.6 was reported, this value was reached with a dataset of 3000 samples.

Table 7. Distance and operations of spatial position.

Variables used	Accuracy reported
Distance center palm to finger tip	
Euclidean distance	
Distance between base of the fingers	
Distance between tips of fingers	90.8, 93.9, 82.5, 99.6, 97.85, 96.3
Direction of fingers	
Direction of the hand	
Relative position of the center palm (dynamic)	

The variables used with the convolutional techniques are presented in Table 8. This technique presents an accuracy value of 98.12 with a dataset of 507000 samples.

Table 8. Distance and operations of spatial position.

Variables used	Accuracy reported
6 layers 3 × 3 LSTM 256 units 16 layers 5 × 5 Image 32 × 32	85, 98.12, 88, 88.9

Finally, in this systematic literature review present how frequency the classifiers were used. In Fig. 6 shown what's classifiers were used and the most used. Clearly, you can see the SVM is used 14 times in papers analyzed, being the most used, however, the use of classifiers as neural networks and convolutional neural network have been growing.

In the same sense, papers that dynamic gesture recognition, reported that used deep neural networks and hidden Markov models.

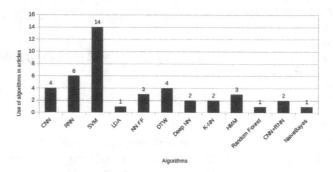

Fig. 6. Classifier methods.

5   Conclusions and Future Work

In this work we have analyzed the architecture of the proposed models for hand gesture recognition using the standard architecture composed of data acquisition, pre-processing, feature extraction, classification, and post-processing. Since the architectures of the proposed models differ a lot from one to the others, in this review we cannot define a trend because each study reported an architecture that fitted the most to its problem. However, the most popular architecture used is composed of algorithms of data acquisition, preprocessing, feature extraction and classification. The types of data most used for implementing the recognition systems include spatial positions and depth images. Additionally, many works combine both types of data. The algorithms that form the pre-processing and the feature extraction modules are usually designed heuristically. Finally, regarding classification, the most used algorithm is SVM; however, RNNs exhibit and increase in its use.

Acknowledgments. The authors gratefully acknowledge the financial support provided by Escuela Politécnica Nacional for the development of the research project PIJ-16-13, and the Universidad Técnica de Ambato by development research project PFISEI24 "Integración de Machine Learning y Visión por Computadora para la Manipulación de Objetos Aplicados al Youbot Kuka". Finally, to Doctoral program of the Departamento de Informática y Ciencias de la Computación of the Escuela Politécnica Nacional.

References

1. Rohith, H.R., Gowtham, S., Chandra, A.S.: Hand gesture recognition in real time using IR sensor. Int. J. Pure Appl. Math. **114**(7), 111–121 (2017)
2. Al-Shamayleh, A.S., Ahmad, R., Abushariah, M.A.: A systematic literature review on vision based gesture recognition techniques. Multimed. Tools Appl. **77**, 28121–28184 (2018)
3. Groenewald, C., Anslow, C., Islam, J., Rooney, C., Passmore, P.J., Wong, B.L.: Understanding 3D mid-air hand gestures with interactive surfaces and displays: a systematic literature review, pp. 1–13 (2016)
4. Pisharady, P.K., Saerbeck, M.: Recent methods and databases in vision-based hand gesture recognition: a review. Comput. Vis. Image Underst. **141**, 152–165 (2015)
5. Suarez, J., Murphy, R.R.: Hand gesture recognition with depth images: a review (2012)
6. Sathyanarayana, S.: Vision-based patient monitoring: a comprehensive review of algorithms and technologies. J. Ambient Intell. Humaniz. Comput. **9**, 225–251 (2015)
7. Madeo, R.C., Lima, C.A., Peres, S.M.: Studies in automated hand gesture analysis: an overview of functional types and gesture phases (2016)
8. Chouhan, S.S., Kaul, A.: Soft computing approaches for image segmentation a survey. Multimed. Tools Appl. **77**, 28483–28537 (2018)
9. Dominio, F., Donadeo, M., Marin, G., Zanuttigh, P., Cortelazzo, G.M.: Hand gesture recognition with depth data. In: Proceedings of the 4th ACM/IEEE, pp. 9–16 (2013)
10. Xu, Y., Wang, Q., Bai, X., Chen, Y.L., Wu, X.: A novel feature extracting method for dynamic gesture recognition based on support vector machine. In: Proceedings of the IEEE International Conference on Information, pp. 437–441 (2014)
11. Jais, H.M., Mahayuddin, Z.R., Arshad, H.: A review on gesture recognition using Kinect. In: The 5th International Conference on Electrical Engineering and Informatics 2015, pp. 594–599 (2015)
12. Plouffe, G., Cretu, A.-M.: Static and dynamic hand gesture recognition in depth data using dynamic time warping. IEEE Trans. Instrum. Meas. **65**(2), 305–316 (2016)
13. Czuszynski, K., Ruminski, J., Wtorek, J.: Pose classification in the gesture recognition using the linear optical sensor. In: Proceedings - 2017 10th International Conference on Human System Interactions, HSI 2017, pp. 18–24 (2017)
14. Park, S., Ryu, M., Chang, J.Y., Park, J.: A hand posture recognition system utilizing frequency difference of infrared light. In: Proceedings of the 20th ACM Symposium on Virtual Reality Software and Technology, pp. 65–68 (2014)

15. Jangyodsuk, P., Conly, C., Athitsos, V.: Sign language recognition using dynamic time warping and hand shape distance based on histogram of oriented gradient features. In: Proceedings of the 7th International Conference on PErvasive Technologies Related to Assistive Environments - PETRA 2014, pp. 1–6 (2014)

16. Doan, H. G., Vu, H., Tran, T.H.: Recognition of hand gestures from cyclic hand movements using spatial-temporal features. In: ACM International Conference Proceeding Series, pp. 260–267 (2015)

17. Wei, L., Tong, Z., Chu, J.: Dynamic hand gesture recognition with leap motion controller. IEEE Signal Process. Lett. **23**(9), 1188–1192 (2016)

18. Naguri, C.R., Bunescu, R.C.: Recognition of dynamic hand gestures from 3D motion data using LSTM and CNN architectures. In: Proceedings - 16th IEEE International Conference on Machine Learning and Applications, ICMLA 2017, pp. 1130 1133 (2018)

19. Sagayam, K.M., Hemanth, D.J.: A probabilistic model for state sequence analysis in hidden Markov model for hand gesture recognition. Comput. Intell. **35**, 59–81 (2019)

20. Du, Y., Liu, S., Feng, L., Chen, M., Wu, J.: Hand gesture recognition with leap motion and kinect devices, pp. 1–6 (2014)

21. Benmoussa, M., Mahmoudi, A.: Machine learning for hand gesture recognition using bag-of-words. In: 2018 International Conference on Intelligent Systems and Computer Vision, ISCV 2018, pp. 1–7 (2018)

22. Vamsikrishna, K.M., Dogra, D.P., Desarkar, M.S.: Computer vision assisted palm rehabilitation with supervised learning. IEEE Trans. Biomed. Eng. **9294**(i), 1–10 (2015)

23. Almasre, M.A., Al-Nuaim, H.: Recognizing arabic sign language gestures using depth sensors and a KSVM classifier, pp. 146–151 (2016)

24. Avola, D., Bernardi, M., Cinque, L., Foresti, G.L., Massaroni, C.: Exploiting recurrent neural networks and leap motion controller for the recognition of sign language and semaphoric hand gestures. IEEE Trans. Multimed. **PP**(8), 1 (2018)

25. Liu, F., Du, B., Wang, Q., Wang, Y., Zeng, W.: Hand gesture recognition using kinect via deterministic learning, pp. 2127–2132 (2017)

26. Khalifa, A.B.: A comprehensive leap motion database for hand gesture recognition, no. July 2013, pp. 514–519 (2016)

27. Rossol, N., Cheng, I., Basu, A.: A multisensor technique for gesture recognition through intelligent skeletal pose analysis. IEEE Trans. Hum.-Mach. Syst. **46**, 1–10 (2015)

28. Plouffe, G., Cretu, A.M.: Static and dynamic hand gesture recognition in depth data using dynamic time warping, pp. 1–12 (2015)

29. Bai, X., Li, C.: Dynamic hand gesture recognition based on depth information. In: 2018 International Conference on Control, Automation and Information Sciences (ICCAIS), (Iccais), pp. 216–221 (2018)

30. Lai, K., Yanushkevich, S.N.: CNN + RNN depth and skeleton based dynamic hand gesture recognition. In: 2018 24th International Conference on Pattern Recognition (ICPR), pp. 3451–3456 (2018)

31. Li, C.: Hand gesture recognition based on wavelet invariant moments, no. 61403302, pp. 459–464 (2017)

32. Clark, A., Moodley, D.: A system for a hand gesture-manipulated virtual reality environment. In: Proceedings of the Annual Conference of the South African Institute of Computer Scientists and Information Technologists on - SAICSIT 2016, pp. 1–10 (2016)

33. Jiang, F., Zhang, S., Wu, S., Gao, Y., Zhao, D.: Multi-layered gesture recognition with kinect. In: Escalera, S., Guyon, I., Athitsos, V. (eds.) Gesture Recognition. TSSCML, pp. 387–416. Springer, Cham (2017). https://doi.org/10.1007/978-3-319-57021-1_13

34. Hsiao, D.Y., Sun, M., Ballweber, C., Cooper, S.: Proactive sensing for improving hand pose estimation, pp. 2348–2352 (2016)

35. Ye, Y., Nurmi, P.: Gestimator - shape and stroke similarity based gesture recognition categories and subject descriptors, pp. 219–226 (2015)
36. Tang, A., Lu, K., Wang, Y., Huang, J., Li, H.: A real-time hand posture recognition system using. ACM Trans. Intell. Syst. Technol. (TIST) **6**(2), 1–23 (2015)
37. Wang, C., Liu, Z., Zhu, M., Zhao, J., Chan, S.: A hand gesture recognition system based on canonical superpixel-graph. Signal Process.: Image Commun. **58**, 87–98 (2017)
38. Núñez, J.C., Cabido, R., Pantrigo, J.J., Montemayor, A.S., Vélez, J.F.: Convolutional neural networks and long short-term memory for skeleton-based human activity and hand gesture recognition. Pattern Recognit. **76**, 80–94 (2018)
39. Cho, Y., Lee, A., Park, J., Ko, B., Kim, N.: Enhancement of gesture recognition for contactless interface using a personalized classifier in the operating room. Comput. Methods Programs Biomed. **161**, 39–44 (2018)
40. Kumar, P., Gauba, H., Roy, P.P., Dogra, D.P.: A multimodal framework for sensor based sign language recognition. Neurocomputing **259**, 21–38 (2016)
41. Hong, C., Zeng, Z., Xie, R., Zhuang, W., Wang, X.: Domain adaptation with low-rank alignment for weakly supervised hand pose recovery. Signal Process. **142**, 223–230 (2018)
42. Cheng, H., Dai, Z., Liu, Z., Zhao, Y.: An image-to-class dynamic time warping approach for both 3D static and trajectory hand gesture recognition. Pattern Recognit. **55**, 137–147 (2016)
43. Liang, W., Guixi, L.: Dynamic and combined gestures recognition based on multifeature fusion in a complex environment. J. China Univ. Posts Telecommun. **22**(2), 81–88 (2015)
44. Inoue, K., Shiraishi, T., Yoshioka, M., Yanagimoto, H.: Depth sensor based automatic hand region extraction by using time-series curve and its application to japanese finger-spelled sign language recognition. Procedia - Procedia Comput. Sci. **60**, 371–380 (2015)
45. Yang, J., Horie, R.: An improved computer interface comprising a recurrent neural network and a natural user interface. Procedia - Procedia Comput. Sci. **60**, 1386–1395 (2015)
46. Tao, W., Leu, M.C., Yin, Z.: American sign language alphabet recognition using convolutional neural networks with multiview augmentation and inference fusion. Eng. Appl. Artif. Intell. **76**(February), 202–213 (2018)
47. Leite, D.Q., et al.: Hand gesture recognition from depth and infrared Kinect data for CAVE applications interaction. Multimed. Tools Appl. **76**, 20423–20455 (2017)
48. Marin, G., Dominio, F., Zanuttigh, P.: Hand gesture recognition with jointly calibrated leap motion and depth sensor. Multimed. Tools Appl. **75**(22), 14991–15015 (2016)
49. Zeng, W., Wang, C., Wang, Q.: Hand gesture recognition using leap motion via deterministic learning. Multimed. Tools Appl. **77**, 28185–28206 (2018)
50. Jiang, X., Gang, Z., Carlo, X.: Virtual grasps recognition using fusion of leap motion and force myography. Virtual Reality **22**(4), 297–308 (2018)
51. Quesada, L., López, G., Guerrero, L.: Automatic recognition of the American sign language fingerspelling alphabet to assist people living with speech or hearing I pairments. J. Ambient Intell. Hum. Comput. **8**(4), 625–635 (2017)
52. Ma, C., Wang, A., Chen, G., Chi, X.: Hand joints-based gesture recognition for noisy dataset using nested interval unscented Kalman filter with LSTM network. Vis. Comput. **34**(6), 1053–1063 (2018)
53. Lee, G.C., Yeh, F.H., Hsiao, Y.H.: Kinect-based Taiwanese sign-language recognition system. Multimed. Tools Appl. **75**(1), 261–279 (2016)
54. Benalcázar, M.E.: Machine learning for computer vision: a review of theory and algorithms (2019)

Frontal Impact Analysis of an Interprovincial Bus Using the Finite Element Method: Case Study in Ecuador

Luis Santos-Correa[1,2(✉)], Diego Pineda-Maigua[1],
Fernando Ortega-Loza[1], Jhonatan Meza-Cartagena[1],
Ignacio Abril-Naranjo[1], Paola Cabrera-Zuleta[1], Jofre Díaz-Ayala[1],
and José Olmedo-Salazar[3]

[1] Instituto Superior Tecnológico Superior 17 de Julio, Urcuquí, Ecuador
lasc700@gmail.com
[2] Universidad Técnica del Norte, Ibarra, Ecuador
[3] Universidad de las Fuerzas Armadas ESPE, Sangolquí, Ecuador

Abstract. The present work shows the structural behavior of a bus body under frontal impact by explicit analysis of finite elements using LS Dyna software to propose a modification to the Ecuadorian Technical Regulations NTE INEN 1323. For this purpose, different tests were carried out to show the resistance to different types of Design gas proposed in national regulations. Therefore, they are reviewed national and international regulations that provide specific data to achieve this study. In addition, the results were validated by comparing a physical tensile with a computational one, where they were applied analysis criteria for the bus impact test. Convergence was obtained through results between the two tests with a calculated error of 1.89%. Finally, it has been observed that the national construction body of buses is not designed to withstand frontal impact loads.

Keywords: Computational mechanics · Finite elements · Explicit dynamics · Impact

1 Introduction

The Latin-American cities are in a stage of development. For instance, it is very important to have an efficient public transportation system that offers the necessary security guarantees for all occupants. However, this massive people mobility has been damaged by its poor development in infrastructure and body building regulations. The currently construction in Ecuador is largely handmade. Although, it has received a great knowledge injection and technical procedures from the field of engineering, It is said that in some cases the process continues to be empirical, because, there is not a clear criteria that allow builders to create safe structures with the necessary degree of deformability to absorb the greatest amount of energy released during an impact. Additionally, it isn't neither a certain degree of rigidity to protect the safety space where the occupants are located [2]. If the body is built with an excessive weight, it

© Springer Nature Switzerland AG 2020
M. Botto-Tobar et al. (Eds.): ICAT 2019, CCIS 1194, pp. 312–320, 2020.
https://doi.org/10.1007/978-3-030-42520-3_25

causes an overload to the frame and the engine of the vehicle, the structure becomes too rigid, and the bodywork is unable to dissipate kinetic energy of the bus, this produces a sharp deceleration that is transmitted directly to the occupants, causing serious injuries and increasing the death rate [3].

In Ecuador, at the end of 2018, there were 12460 sinister and 1058 deaths with a mortality rate in some provinces with the 11.1 people per 100,000 population, becoming the sixth cause of death in Ecuador. Where the excess of speed, disrespect for traffic signals, recklessness has been the main causes [4–6]. As can be seen then, the ability of a vehicle to resist an impact is measured from different technical criteria that must be placed on a balance to analyze the different commitments that result from stiffening the structure or make it more ductile, all this in order to prevent as much as possible cause injuries to passengers [7, 8]. For this purpose, it is necessary to have criteria of design and evaluation of manufacturing materials. Currently, this process can be assisted by simulation software that uses the finite element method as the basis Mathematics for calculation, of the effort and structural deformation in addition to many other physical possibilities [9, 10]. With this method, we avoid the cost of materials and manufacturing for the development of physical prototypes of tests, which if you think in a bus can become very expensive in time and money. In the same way you can improve the time of design and production [11].

Some works such as [2, 4, 7] have performed impact analysis on different vehicles, analyzing the structural consequences. However, it remains unsolved specific problems such as lack of security for passengers, the technical recommendations and improvement recommendations for the car construction technical process of body building. For this reason, this work has been carried out in Ambato city, in Tungurahua province, for being the city that accumulates most of the industry metalworking body of the country. With the support of different companies and organizations with those who allowed an approach to their industries, providing their knowledge, experience and all the work clothing required to perform the planned physical and computational tests. The national regulations were analyzed, dealing with what is stated in European and US regulations, where it stands out the Vehicle Regulations of the United Nations Economic Commission (UNECE) and the regulations of the National Traffic Safety Administration and Motorways (NHTSA), respectively. Consequently, the conditions were determined suitable for developing the impact test based on the 033 normative of the United Nations, was crash the bus frontally against to a rigid barrier height of 1500 mm, an impact speed of 15 m/s and a trajectory travel was assigned totally perpendicular to the impact barrier. As relevant results were obtained the magnitudes of dissipated energy, displacements of the bodywork, and slowdown in cabin.

The rest of the document is structured as follows: The Sect. 2, shows the Ecuadorian regulations and the proposed bodywork analysis scheme. Section 3 indicates the results obtained in the corresponding tests. Finally, Sect. 4 shows the conclusions and future work.

2 Regulations and Bodywork Analysis Scheme

This section shows the regulations in Ecuador and the necessary evidence for the validation of frontal impact tests. Otherwise, it presents the bodywork analysis of an interprovincial bus.

2.1 Ecuadorian Regulations

The Ecuadorian Technical Standard NTE INEN 1323: 2009 details the guidelines to follow to computationally validate a new bus body design through the application of three tests: (i) load test distributed on the roof that must be greater than or equal to half the weight the chassis can support, (ii) a quasi-static test that consisting of the analysis of the structure subjected to different static and dynamic charges [11] and (iii) dump the structure to a 800 mm pit and this is normalized according to United Nations regulation 66 [13].

This regulation also indicates how the security space is distributed within of the entire structure, even inside the cockpit, being this one ahead from the back of the driver's seat 600 mm away as indicated in Fig. 1. Thus, there are no guidelines that mark the behavior for the structure before frontal impacts, which allows to comply with this regulation, therefore, the purpose of this research is to simulate the dynamic event of frontal collision of an interprovincial bus using finite element software as a basis for the proposal for an amendment to the Ecuadorian Technical Standard NTE INEN 1323: 2009.

This study presents a great convenience and social relevance because the revision process of the bus construction regulations has not been modified since the year 2009, in order to improve national production and provide better security standards to the occupants of the cockpit, and in this way improve the overall quality of other national construction product [4, 14].

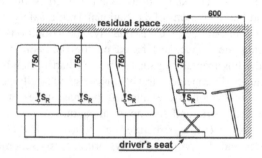

Fig. 1. Safety space for the driver's seat. NTE INEN 1323: 2009

2.2 Bodywork Analysis

To characterize the material, a tensile test was carried out in the laboratory of materials of the "Fomento Productivo Carrosero Metal Mecanico" center (Productive

Development Metal Mechanic Bodybuilder center). The test tubes of the material were obtained according to the specifications determined in the NTE standard INEN 0109: 2009. The material used was A grade steel according to NTE INEN 2415: 2011. This test also allowed to validate the calculated results. One of the test specimens submitted in Fig. 2.

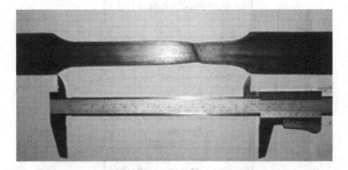

Fig. 2. Test tube of tensile material

The bus structure must be subjected to a frontal impact test to verify its resistance and degree of impact attenuation. Subsequently, it must be done including a model of the chassis and the engine; equal in mass and dimensions to the real components, in addition, the bus must be impacted frontally against a barrier of rigid construction that must have a minimum dimension of 1500 mm of height and 3000 mm of longitude. The bus speed and travel vector must be perpendicular to the plane formed by the impact barrier. The speed of the bus must be at least 14 m/s. No deformations should be obtained or displacements of any structural member that invade the security area in the cabin in order to safeguard the lives of the occupants.

Computational impact simulation is performed in LS Dyna software because it is Explicit dynamics specialist. For the execution of this test it has been mostly used two-dimensional elements with a mesh size of 30 mm, with who was obtained, optimum quality parameters. For the material the model is assigned 024 from the LS Dyna library, Piece wise linear plasticity, which allows you to enter a behavior curve based on the data generated in the laboratory [9]. A Belitshko Tsay type section and a Horu-energy control is assigned glass type Belytshko Flanagan, the established time step is 1 ms and 152 steps [5]. A total of 139826 elements were obtained, and a processing time close to 4 h on a Dell Precision ™ T5600 workstation with 128 GB of RAM and a 16-core Intel Xeon processor.

The mathematical equation that governs a presented system is as follows:

$$[M]\ddot{x} + [C]\dot{x} + [K]x = Fext \tag{1}$$

Where M corresponds to the mass matrix, C corresponds to the damping matrix and K corresponds to the stiffness matrix. In addition x, corresponds to displacement, $\dot{x}$ to velocity, $\ddot{x}$ to acceleration and F to external excitation force.

This mathematical problem is solved by the method of central differences, using an algorithm that runs iteratively, repeating in a finite loop, according to the number of calculation steps programmed in the study. It is presented below the calculation algorithm in Fig. 3.

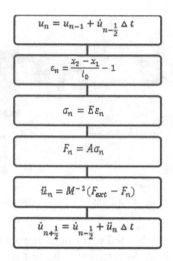

$$u_n = u_{n-1} + \dot{u}_{n-\frac{1}{2}} \Delta t$$

$$\varepsilon_n = \frac{x_2 - x_1}{l_D} - 1$$

$$\sigma_n = E \varepsilon_n$$

$$F_n = A \sigma_n$$

$$\ddot{u}_n = M^{-1}(F_{ext} - F_n)$$

$$\dot{u}_{n+\frac{1}{2}} = \dot{u}_{n-\frac{1}{2}} + \ddot{u}_n \Delta t$$

Fig. 3. Algorithm for calculating deformation tests

3   Evaluation of Results and Discussion

This section shows the general tests performed and their interpretation.

3.1   General Tests

It is observed that most of the elements in the bus cabin reach critical efforts, exceeding the maximum material breakage limit. These efforts are generated since the front part comes in contact with the rigid barrier. They become maximum spreading throughout the bodywork of the bus instantly t = 40 ms which is when It produces the largest exchange of kinetic force to internal energies.

Figure 4 shows where the maximum stresses of the model are located, focussing in addition to the front part exist on efforts in the structural supports roof. A maximum effort of 387.9 MPa is calculated in the red color zones.

Taking as a reference the back part of the bus, a total displacement of 720 mm is calculated, where 410 mm correspond to the deformation of the outside area of the bus cabin. As indicated in Fig. 4, the remaining 310 mm, correspond to an intrusion into the security zone reserved for occupants, which would probably be fatal or could cause serious damage to the physical integrity of the occupants of the cabin as it is shown in Fig. 5.

Currently the development and advancement of automotive technology, prepares its designs to dissipate the energy of an impact by programmed deformations of the

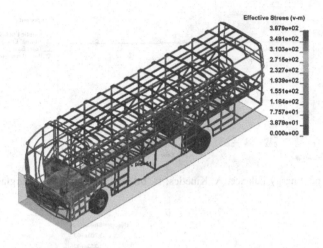

Fig. 4. Von Mises stress diagram on the front of the bus (Color figure online)

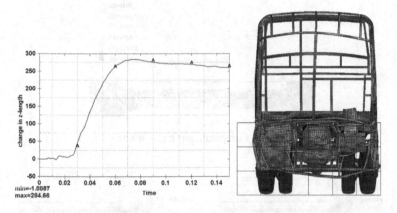

Fig. 5. Shift diagram for frontal nodes

structural members of the bodywork, or through structures that serve to dissipate the impact energy known as dimmers, however, the Ecuadorian designs do not have any of these mechanisms.

In Fig. 6, it can be seen that, while the total energy of the system remains constant, the kinetic energy of the movement decays completely at 40 ms. Conversely, the internal energy of the body suddenly increases, forming a crosslinking denoting, as structural members dissipate kinetic energy of impact.

With the objective of validate the study, the tensile test carried out in order to characterize the material, experience in the LS computing environment has been repeated Dyna, obtaining convergent data, determines the length of the calibrated area of the real test tube and the computational one after applying the assay, obtaining 168,234 mm and 171.48 mm respectively. Based on these results, it can be determined the range of study error, which is 1.89% (Fig. 7).

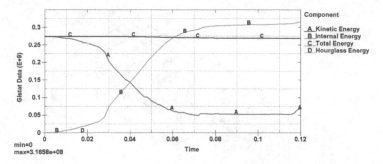

Fig. 6. Energy balance. A. Kinetics; B. Internal; C. Total; D. Hourglass.

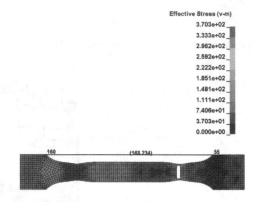

Fig. 7. Computational traction test.

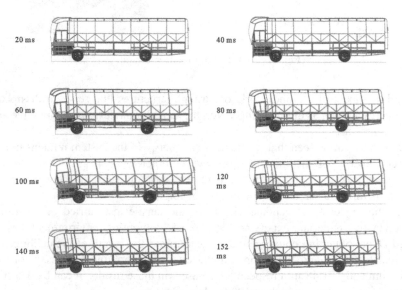

Fig. 8. Impact simulation

Taking as a reference the back part of the bus, a 720 mm displacement is calculated, of which 410 mm correspond to the deformation of the area that is outside the bus cabin. As indicated in Fig. 4, the remaining 310 mm, correspond to an intrusion into the security zone reserved for occupants, which would probably be fatal or could cause serious damage to the physical integrity of the occupants of the cabin. Figure 8 shows part of the results obtained in the analysis.

4   Conclusions and Future Works

The frontal impact of the body of a bus has been simulated using the package LS Dyna computational, since it is a software specially designed for analysis of events of explicit nonlinear dynamics, where there are large deformations of the structure and abrupt energy release in very short moments.

Optimal simulation parameters were established for this study through the documentary review, we worked with a mesh size of 30 mm for the whole part structural, with two-dimensional elements and a piece wise linear material model plasticity that allows entering data of a material characterized by Physical tests. Therefore, the results were validated by comparing a physical trial of traction with a computational one, where the same analysis criteria were applied for the bus impact test. Convergence of results was obtained between the two tests with a calculated error of 1.89%.

The kinetic energy of the moving vehicle is dissipated by the structure in 70 ms, that is we have an average deceleration of 22 G, however, there may be short deceleration peaks greater than 30 G that turns a potential dangerous bodywork, which can cause to the occupants bone lesions, loss of knowledge, rupture of blood vessels and even death.

The amendment to the Ecuadorian Technical Standard NTE INEN 1323: 2009 is proposed, to include an analysis of resistance to frontal impacts in the requirements that must approve bodywork to be homologated by the ANT.

Finally, it is recommended to design a suitable impact attenuator to be installed in the bus that lowers the level of deceleration and minimizes deformation and cabin intrusions.

Acknowledgment. Special thanks to Ing. Fernando Olmedo and Ing. Patricio Riofrio for their time and dedication in support of this project, in the same way to, the Laboratory of Materials Centro de Fomento Productivo Carrosero Metal Mecánico in Ambato city, and all other people who contributed positively to the development of this draft.

References

1. Rodríguez, P.L., Mántaras, D.Á.: Investigacin de Accidentes de Trfico. Manual de Reconstrucción. https://books.google.es/books?hl=es&lr=&id=1sv_AEi0qB8C&oi=fnd&pg=PA39&dq=Investigación+de+Accidentes+de+Tráfico.+Manual+de+Reconstrucción&ots=w5utleFjfx&sig=oJRrONDi978KCVcGZexTBIsnCuY#v=onepage&q=. Accessed 16 Sept 2019

2. Arias, M., Javier, F.: ESCUELA POLITÉCNICA NACIONAL FACULTAD DE INGENIERÍA MECÁNICA ANÁLISIS ESTRUCTURAL A CARGAS DE IMPACTO FRONTAL DE UN BUS TIPO INTERPROVINCIAL MEDIANTE EL MÉTODO DE ELEMENTOS FINITOS
3. Andrés, N., Bustos, A., Freddy, I., Palomeque, P.: UNIVERSIDAD DEL AZUAY FACULTAD DE CIENCIA Y TECNOLOGÍA ESCUELA DE INGENIERÍA EN MECÁNICA AUTOR: DIRECTOR
4. Cárdenas, D., Escudero, J., Quizhpi, S., Pinos, M.A.: Propuesta de diseño estructural para buses de carrocería interprovincial (2014)
5. Agencia Nacional de Tránsito: Descargables - Reporte Nacional de Siniestros de Tránsito Julio 2019 - Agencia Nacional de Tránsito del Ecuador - ANT. https://www.ant.gob.ec/index.php/descargable/file/6598-reporte-nacional-de-siniestros-de-transito-julio-2019. Accessed 16 Sept 2019
6. Agencia Nacional de Tránsito: Estadísticas sobre Siniestros de Tránsito - Agencia Nacional de Tránsito del Ecuador – ANT. Reporte Nacional de Siniestros (2018). https://www.ant.gob.ec/index.php/noticias/estadisticas. Accessed 13 Aug 2019
7. Li, Q., Yang, J.: Study of vehicle front structure crashworthiness based on pole impact with different position. In: Proceedings - 2013 5th Conference on Measuring Technology and Mechatronics Automation, ICMTMA 2013, pp. 1031–1035 (2013)
8. Wang, W., Hecht, M., Zhou, H.: Crashworthiness design of detachable cab. In: 2011 International Conference on Consumer Electronics, Communications and Networks, CECNet 2011 - Proceedings, pp. 170–173 (2011)
9. Dagdeviren, S., Yavuz, M., Kocabas, M.O., Unsal, E., Esat, V.: Structural crashworthiness analysis of a ladder frame chassis subjected to full frontal and pole side impacts. Int. J. Crashworthiness 21(5), 477–493 (2016)
10. Zhang, Y., Liu, C., Mu, X.: Robust finite-time H∞ control of singular stochastic systems via static output feedback. Appl. Math. Comput. 218(9), 5629–5640 (2012)
11. Altair University: Practical Finite Element Analysis (from Finite to Infinite). https://altairuniversity.com/learning-library/practical-finite-element-analysis-from-finite-to-infinite/. Accessed 16 Sept 2019
12. Wu, S.R., Gu, L.: Introduction to the Explicit Finite Element Method for Nonlinear Transient Dynamics. Wiley (2012)
13. Full text of 'NTE INEN 1323: Vehículos automotores. Carrocerías de buses. Requisitos.' https://archive.org/stream/ec.nte.1323.2009/ec.nte.1323.2009_djvu.txt. Accessed 16 Sept 2019
14. Full text of 'NTE INEN 0109: Ensayo de tracción para materiales metálicos a temperatura ambiente.' https://archive.org/stream/ec.nte.0109.2009/ec.nte.0109.2009_djvu.txt. Accessed 16 Sept 2019

Student Dropout Model Based on Logistic Regression

Blanca Rocio Cuji Chacha$^{(\boxtimes)}$, Wilma Lorena Gavilanes López,
Víctor Xavier Vicente Guerrero,
and Wilma Guadalupe Villacis Villacis

Faculty of Human and Education Sciences, Technical University of Ambato,
Ambato 180103, Ecuador
{blancarcujic, wilmalgavilanesl, vvicente7867,
wilmagvillacisv}@uta.edu.ec

Abstract. Student dropout is a phenomenon that affects the majority of higher education institutions in Ecuador. The objective of the research was to design a predictive model to detect possible dropouts before they decide to abandon their studies. This model is based on logistic regression, and the methodology used in this research is based on the Knowledge Discovery in Databases (KDD) Model; which has five stages: selection, processing, transformation, data mining and evaluation. The application of the Logit function of the R tool for the logistic regression helps the construction of the predictive model. This model evaluates possible dropout students and leads to the conclusion that grades have a greater influence on student dropout.

Keywords: Logistic regression · Predictive model · Student dropout

1 Introducción

Data mining tools have high prediction capabilities [1], the use of algorithms such as logistic regression and structural equation modeling allows predicting the dropout rate, through variables such as sex, race, academic performance, family social status, pre-university experiences, level of education of parents and social interaction. In addition, the educational environment is a relevant factor in student dropout.

The study is based on use of multivariate logistic regression models and optimization techniques. The Hausman test allows the author to select the most appropriate model for the mining application, determining with this mechanism that selected models are optimal; the test made to the models contrasts the coefficients of these and determines which will be the most efficient, in the search for the most statistically efficient. According to the research, the variables that influenced on student dropout were admission score, admission by second option, special quota programs and registration cost [2].

Moreover, research on academic dropout from the University of Atacama, Chile; proposes the creation of a predictive model developed in two stages, in the first is used

© Springer Nature Switzerland AG 2020
M. Botto-Tobar et al. (Eds.): ICAT 2019, CCIS 1194, pp. 321–333, 2020.
https://doi.org/10.1007/978-3-030-42520-3_26

one multiple linear regression based on a limited set of variable; In the second stage, logistic regression is used a dichotomous variable called dropout (=1) or not dropout (=0) is determined as a predictor. The prediction is checked taking into account the probability of occurrence of the event, while closer to (=1) it is a dropout but if it approaches (=0), it is a non-dropout. The author concludes that the rating directly influences on dropout, so that the population with lower averages is more likely to dropout, while those with higher grades do not [3].

In the study, related to the construction of an early warning model for the detection of students at risk of dropping out of the Metropolitan University of Education Sciences of Mexico, logistic regression is used to analyze to 457 male and 549 female students in the first year of study, which was determined as a year of high dropout. Regarding the most important variables, the study showed that those that refer to: grade point averages, language, mathematics, difference in years of admission and departure, age of admission and gender, were the most important at the time of predicting [4].

On the other hand [2], it is essential to find the gaps in the prediction of student dropout through the discovery of new attributes, using classification algorithms, regression, association rules, genetic algorithm, clustering, method of nearest neighbors, decision trees, with variables such as gender, level of study of the family, professional parents, parental occupation, basic services in the institution, methodology adopted by the teacher as well as marital status; all of these through the application of the process of data mining, which is beneficial to improve standards in education.

The creation of a predictive model of data mining, based on a decision tree and Bayesian networks (BN), increases the accuracy in predicting possible dropout students [3]. Studying how to provide knowledge and understanding of the different techniques of data mining, which have been used to predict student progress and academic performance, helps to find attributes for the prediction of possible dropout students. This can be done by using decision trees, neural networks with variables such as: gender, age, marital status, number of children, occupation, time spent on the computer and computer literacy [4].

The ideal strategies to prevent student dropout are the creation of flexible educational models that detect possible dropouts in time [5]; in this virtue, the application of mining techniques provides a deep insight into the student's data, and allows decision-making, through the analysis of variables such as: high school degree, admission test result, graduation grade, father's and mother's occupation as well as family financial condition [6].

Some of the most popular classification algorithms are: k-Nearest, nearby neighbors, random forest, AdaBoost, classification and regression trees (CART), support-vector machines (SVM) [7] and ordinary least square help to generate predictive models. Studying how to predict the number of times a student will repeat a course allows designing and implementing intelligent systems to predict whether a student will finish his studies [8]. This study is possible through the use of neural networks and variables such as attendance, course content, teaching method and evaluation system [9].

Variables such as the level in which students are found, academic performance, gender, marital status, ethnicity, place of birth and city of residence were used for the construction of a predictive model of student dropout. In the same sense, a three-level tree is created based on decision trees and the application of the KDD methodology.

This tree allows the detection of variables such the academic performance and level as those that have the greatest incidence on student dropout. Meanwhile, variables such as gender, marital status, and ethnicity, place of birth and city of residence were discarded. The prediction was made with a 94% probability of success [10].

On the other hand, there are studies in which psychological variables such as motivation, values and feelings are factors that intervene in student dropout. It is detected that these variables influence on grades [11]. This analysis is done through the application of k-means to detect possible future dropout students [12].

The creation of an algorithm for the prediction of student dropout from decision trees [13], allows to study the profiles of dropout students, taking into account the variables related to sex, age of admission, social status, marital status, health regime, place of birth and origin, parents' occupations, siblings, type of residence, family income, tuition cost, type of school, school day, area of program as well as grade point average [14]. As a result, it was determined that the socioeconomic factor has the greatest impact on student dropout.

The objective of the research was to design a predictive model of university student dropout based on logistic regression.

1.1 Regresión Logística

Logistic regression allows predicting the event through the analysis of a dichotomous variable; which can take a value (1.0), depending on whether the condition evaluated is met or not. First, the best model is evaluated, using the analysis of the changes that occur. If a variable is taken or discarded, a variable may or may not be related to the others through a causal, correlational or relational relationship, and if two models are equally efficient, the one that uses less number of variables is taken [15].

1.2 Logit

Logit, is a function used to estimate probabilities, the results of this determine the probability of the event. It also measures the influence of the selected attributes on the predicting variable [16], which is used in binary logistic regression, and is given by the function:

$$G(x = 1) = e^x/(1 + e^x) \tag{1}$$

Where x is equal to the values of each of the variables, R allows the implementation of this function using the logit (input data) command.

2 Metodología

The methodology used for the design of the predictive model is based on the KDD method, which allows finding patterns in a set of raw data. Afterwards, the patterns are to be analyzed using the data mining algorithm in order to generate a predictive model according to the predictive variables. It consists of five stages: data selection, processing, transformation, data mining and evaluation [10] (see Fig. 1).

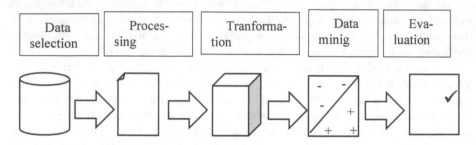

Fig. 1. KDD model.

2.1 Data Selection

In this phase, the data collection and integration must be carried out.

Data Collection. The set of selected data belongs to 712 students of the Basic Education Major of the Universidad Técnica de Ambato (UTA). The institution provided the information from the year 2010 in a matrix containing personal data such as, identification, which was replaced by a numerical identifier of 1... n, to maintain the confidentiality of the participants as set forth in the Law of the National System of Public Registries of Ecuador [17].

Moreover, more information was given such as: gender and academic data with attributes related to the academic period, course enrolled, subject, average of grades obtained in the first term (grade1p, where 1 = first term and p = first semester, identifier of the semester) and second term (grade1s, 1 = first term and s = second semester). This nomenclature is used to identify terms and all semesters included in the analysis. Finally, female = 1 and male = 2 were replaced (See Table 1).

Table 1. Matrix: general data - academic.

Identifier	Gender	grade1p	grade2p	grade1s	grade2s
1	1	8.5	8.5	8.5	8.3
2	2	7.0	6.8	7.2	7.0
3	1	8.5	8.5	8.5	10.0
4	1	6.7	7.0	6.5	8.0
5	1	8.5	6.5	8.4	7.0

Data Integration. At the beginning, there were 22304 records for the analysis, the attributes of the matrix were integrated into a spreadsheet to which the attributes, gender, student dropout, grade1p, grade2p were added. Prefixes representing the levels reached by the students were used (p = first, s = second, t = third, c = fourth, q = fifth, se = sixth, sp = seventh, or = eighth) (see Table 2).

Table 2. Integral matrix with attributes such as gender, grade1p, grade2p, dropout student.

Identifier	Gender	Initial level	Dropout	Dropout level
1	1	P	No	
2	1	P	Yes	S
3	1	P	No	
4	1	P	No	
5	1	P	Yes	S

The attributes, gender, dropout student, grade1p, grade2p... n, were determined as follows:

Gender: based on the participants' names.
Dropout student: based on the course and subject grades that reflect the student's historical data.
Grade1p, grade2p, grade1s, grade2s... n: average of grades per subject of all approved levels.

2.2 Data Processing

In this phase, data cleansing and data storage are taken into account.

After cleaning and integrating the data of 712 initial students, 643 final ones are selected for the study. Atypical data (extremely large or small data) [18] was detected, using SPSS with box diagrams (see Fig. 2), for the variables: grade1p, grade2p.

The variable grade1p shows atypical data, from students who have grades lower than 7.8. Thus, the records 197, 198, 200, 228, 380, 382, 384, 387, 395, 416, 417, 485,523, 531, 544 are in a range of grades between 7.8 and 0.

The same procedure was applied to all variables.

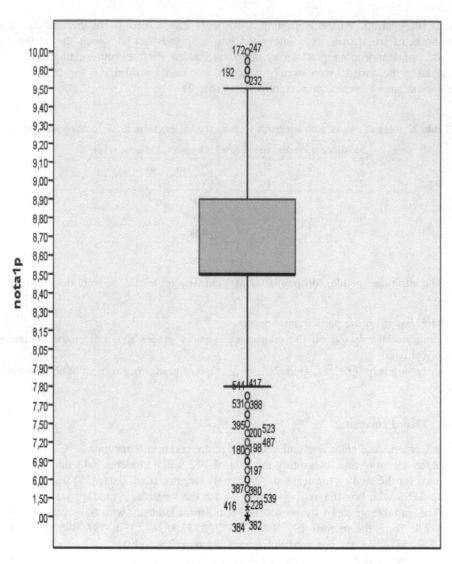

Fig. 2. A typical data variable grade1p.

2.3   Data Transformation

The original matrix was modified taking into account the following weights for qualitative variables such as, gender and dropout student (see Table 3).

Table 3. Weighing of qualitative variables.

Variable name	Weighing
Gender	1 = female
	2 = male
Dropout student	0 = no
	1 = yes

The data was transformed to an appropriate format and the yes and no answer was replaced by 1 and 0 respectively. In question yes, you lost a level and the records that did not contain one or more columns with data were deleted.

2.4   Data Mining

The Logistic Regression technique and the application of the logit algorithm were selected, depending on the literature reviewed, the great majority of studies show the creation of predictive models based on techniques such as decision trees, neural networks and classification rules.

Algorithm Selection. The Logistic Regression algorithm, Logit [19], was used for the data mining process.

Model Generation. The R software was used to generate the student dropout model, through logistic regression and the use of the Logit function, R; which is used to determine the influence of the variables on the prediction.

- The first generated model showed the variables grade1p, grade1t, grade2c, grade2se, grade1o as the least influential on the prediction for student dropout. Consequently, they were excluded (see Fig. 3).

```
Coefficients:
              Estimate  Std. Error  z value
(Intercept)  18.508153   5.444274    3.400
GENERO        0.077969   0.416609    0.187
nota1p       -0.003862   0.598001   -0.006
nota2p        0.446341   0.309675    1.441
nota1s       -1.795382   0.474391   -3.785
nota2s        0.855186   0.446310    1.916
nota1t        0.036066   0.242657    0.149
nota2t       -0.654547   0.232276   -2.818
nota1c       -0.388882   0.237430   -1.638
nota2c       -0.059700   0.173224   -0.345
nota1q        0.257728   0.229611    1.122
nota2q       -0.542207   0.169934   -3.191
nota1se      -0.350758   0.311726   -1.125
nota2se      -0.450074   0.295991   -1.521
nota1sp      -0.193046   0.454390   -0.425
nota2sp      -0.036522   0.392906   -0.093
nota1o        0.049071   0.188431    0.260
nota2o        0.262700   0.176255    1.490
             Pr(>|z|)
(Intercept)  0.000675  ***
GENERO       0.851541
nota1p       0.994847
nota2p       0.149494
nota1s       0.000154  ***
nota2s       0.055349  .
nota1t       0.881846
nota2t       0.004833  **
nota1c       0.101447
nota2c       0.730367
nota1q       0.261671
nota2q       0.001419  **
nota1se      0.260499
nota2se      0.128368
nota1sp      0.670949
nota2sp      0.925941
nota1o       0.794539
nota2o       0.136104
---
```

Fig. 3. Result of the first model.

The first model was generated based on 17 variables.

After the generation of the model, the variable of gender was excluded.

- Without taking into account the excluded variables, the model (model 2) is generated again, obtaining the following result (Fig. 4):

```
Deviance Residuals:
    Min       1Q     Median      3Q        Max
-2.20988  -0.44345  -0.28483  -0.09898   2.91182

Coefficients:
             Estimate Std. Error z value Pr(>|z|)
(Intercept)  18.7907     3.9694   4.734 2.20e-06 ***
nota2p        0.4502     0.2805   1.605 0.108460
nota1s       -1.7967     0.4502  -3.991 6.57e-05 ***
nota2s        0.8537     0.4394   1.943 0.051999 .
nota2t       -0.6220     0.1607  -3.870 0.000109 ***
nota1c       -0.4531     0.1288  -3.516 0.000438 ***
nota1q        0.2547     0.2261   1.127 0.259875
nota2q       -0.5433     0.1655  -3.282 0.001031 **
nota1se      -0.3774     0.3029  -1.246 0.212691
nota2se      -0.4392     0.2855  -1.538 0.123975
nota1sp      -0.2011     0.3324  -0.605 0.545302
nota2o        0.2839     0.1633   1.739 0.082122 .
---
Signif. codes:  0 '***' 0.001 '**' 0.01 '*' 0.05 '.' 0.1 ' ' 1
```

Fig. 4. Variables for the final model.

The new model is generated with 11 variables and the less relevant variables are eliminated such as grade1p, grade1t, grade2c, grade2se, grade1o and gender; for not significantly influencing the model.

- To estimate the probability of the variables prediction, the values obtained in the Logit function are replaced. The coefficients of the model are: (see Fig. 5).

```
        2           5            6           8          10
0.007741271 0.029723703 0.021443835 0.051245227 0.056496488
       15          16           17          18          20
0.036625014 0.027077413 0.075363736 0.052720433 0.048651009
       28          29           31          32          33
0.050287577 0.049850570 0.026263675 0.053106002 0.120857006
       38          41           42          44          46
0.144839302 0.074471138 0.125142604 0.130130842 0.018700825
       51          52           53          54          56
0.112847695 0.143190185 0.141034033 0.123642889 0.078134391
       64          65           67          68          69
0.109209987 0.093705684 0.101194638 0.101194638 0.112003134
```

Fig. 5. Model coefficients.

2.5   Evaluation

The model is evaluated based on the following procedure.

- It is determined that the probability of the predicting variable (dropout student) takes the values of 1 or 0 (1 = it is a dropout student; 0 = it is not a dropout student), the commands used in R are:
  confmatrix<-table (actual_value=test$dropout, predicted_value=res>0.5)
  confmatrix
- The model is right in predicting 167 non-dropout students and five are misclassified.
- According to the model, 32 students are detected as dropouts; while 11 are dropouts but the model has not detected them.
- The commands that show the effectiveness of the model are.
  (confmatrix[[1,1]] + confmatrix[[2,2]])/sum(confmatrix)
- A prediction effectiveness in the model of 92.56% is obtained (see Fig. 6).

```
                  predicted_value
actual_value FALSE TRUE
           0   167    5
           1    11   32
> (confmatrix[[1,1]]+confmatrix[[2,2]])/sum(confmatrix)
[1] 0.9255814
```

Fig. 6. Probability of model success.

3   Results

The results obtained during the investigation process are detailed below:

3.1   Variables Selected for Prediction

The variables that best predict the model are gender, grade1p, grade2p, grade1s, grade2s, grade1t, grade2t, grade1c, grade2c, grade1q, grade2q, grade1se, and grade2se. These variables were originated based on the application of the Logit function.

3.2   Predictive Model of Student Dropout

The generated model is as follows:

$$Glm(formula = dropout \sim grade2p + grade1s + grade2s + grade2t + grade1c$$
$$+ grade1q + grade2q + grade1se + grade1sp + grade2sp + grade2o, family = binomial,$$
$$data = students)$$

$$(2)$$

3.3   Experimental Results

Several models based on logistic regression with Logit algorithm were created, the models are obtained using R, open code predictive analysis platform, the data collection consists of 22304 student records, of which 80% of the records were used to train the model and the remaining 20% for the tests, the result indicates that the variable grade notes at all levels, is significant for the prediction of dropout. The predictions obtained through the model have a certainty probability of 92.56%.

Although logistic regression is a seldom used technique to predict student dropout, the basic idea is to create a robust model with a high level of prediction reliability.

4   Discussion

4.1   Summary of Findings

The objective of this study was to design a university student dropout model based on logistic regression. The results revealed that, even though the existence of different models built with techniques such as: Decision Tree, Naive Bayes, Neural Network and the use of algorithms such as ID3, Multilayer Perceptron among others, logistic regression is a technique that generates consistent models, on the basis of only on the data found stored in the institution's databases.

4.2   Theoretical Implications

A primary theoretical contribution of the present study in relation to the bibliographic review, is to determine the largest number of variables that have the highest incidence of dropout, determining sociodemographic variables (age, gender, ethnicity, education, disability, birth zone and origin), socioeconomic (social level, employment status), psychological factors (psychological well-being, integration, family support, behavior), and other variables such as average grades, lost semesters, study day, teacher methodology, father's and mother's educational level, note of the university admission test, among others.

4.3   Practical Applications

This study shows a significant strength in terms of the creation of a robust student dropout predictive model, with a low use technique, managing to demonstrate the high probability of success of the model reflected in the 92.56% reliability, in relation to others techniques such as: Decision Tree, Naive Bayes and Neural Network, which show a probability of success between 73.9% to 87%.

5   Conclusions

The significant reasons that lead university students to dropout the university are their academic performance reflected in the variables: note2p, note1s, note2s, note2t, note1c, note1q, note2q, note1se, note1sp, note2sp, note2o.

The probability of success of the predictive model is 92.56%, which implies a high degree of reliability in the prediction of the data. Predictive models generated with techniques such as Decision Tree, Naive Bayes and Neural Network, show a probability of success between 73.9% to 87%.

One of the biggest conflicts that came forward during the research was the poor quality of the data, which caused that during the preprocessing stage (cleaning and transformation), variables such as: place and date of birth, age, education level of parents are discarded and other variables due to the impossibility of obtaining their values and that somehow influence the results on student dropout.

Several models have been created to predict student dropout based on algorithms such as decision trees, neural networks, random forests, vector support machine, however, logistic regression is a technique that allows generating models consistent with the observed reality and support theoretical, based solely on the data found stored in the databases.

Future studies should examine psychological factors such as motivational orientations, personality, behavior, internalization disorders or problems, anxiety, shyness, emotional withdrawal and depression, as well as externalization disorders or problems, infra-control or dissocial.

References

1. Kerby, M.B.: Toward a new predictive model of student retention in higher education: an application of classical sociological theory. J. Coll. Stud. Retent. Res. Theory Pract. **17**, 138–161 (2015). https://doi.org/10.1177/1521025115578229
2. Kumar, M., Singh, A.J., Handa, D.: Literature survey on educational dropout prediction. Int. J. Educ. Manag. Eng. **7**, 8–19 (2017). https://doi.org/10.5815/ijeme.2017.02.02
3. Ercan, E.: At-risk students using machine learning techniques. Int. J. Mach. Learn. Comput. **2**(4), 476 (2012)
4. Kumar, M., Singh, A.J., Handa, D.: Literature survey on student's performance prediction in education using data mining techniques. Int. J. Educ. Manag. Eng. **7**, 40–49 (2017). https://doi.org/10.5815/ijeme.2017.06.05
5. Martelo, R.J., Herrera, K., Villabona, N.: Estrategias para disminuir la deserción universitaria mediante series de tiempo y multipol. Revista Espacios **38**, 4–6 (2017). ISSN 0798 1015
6. Kumar, M., Singh, A.J.: Evaluation of data mining techniques for predicting student's performance predicting students academic performance using data mining techniques: a case study view project evaluation of data mining techniques for predicting student's performance. Artic Int. J. Mod. Educ. Comput. Sci. **8**, 25–31 (2017). https://doi.org/10.5815/ijmecs.2017.08.04
7. Strecht, P., Cruz, L., Soares, C., et al.: A comparative study of classification and regression algorithms for modelling students' academic performance. In: Proceedings of 8th International Conference on Educational Data Mining, pp. 392–395 (2015)

8. Oyedotun, O.K., Tackie, S.N., Olaniyi, E.O., Khashman, A.: Data mining of students' performance: Turkish students as a case study. Int. J. Intell. Syst. Appl. **7**, 20–27 (2015). https://doi.org/10.5815/ijisa.2015.09.03

9. Ismail, S., Abdulla, S.: Design and implementation of an intelligent system to predict the student graduation AGPA. Aust. Educ. Comput. **30**, 7–9 (2015)

10. Cuji, B., Gavilanes, W., Sanchez, R.: Modelo predictivo de deserción estudiantil basado en arboles de decisión. Espacios **38**, 17 (2017)

11. Ramesh, V., Parkavi, P., Ramar, K.: Predicting student performance: a statistical and data mining approach. Int. J. Comput. Appl. **63**, 35–39 (2013). https://doi.org/10.5120/10489-5242

12. Iam-On, N., Boongoen, T.: Generating descriptive model for student dropout: a review of clustering approach. Hum.-Centric Comput. Inf. Sci. **7**, 1–24 (2017). https://doi.org/10.1186/s13673-016-0083-0

13. Altujjar, Y., Altamimi, W., Al-Turaiki, I., Al-Razgan, M.: Predicting critical courses affecting students performance: a case study. Procedia Comput. Sci. **82**, 65–71 (2016). https://doi.org/10.1016/j.procs.2016.04.010

14. Timaran, R., Jiménez, J.: Detection of student dropout patterns in undergraduate programs of higher education institutions with CRISP-DM. Form. University, pp. 1–19 (2014)

15. De la Fuente Fernández, S.: Logistic Regression. Santiago de la Fuente Fernández (2011)

16. Llano, R., Mosquera, V.: The logit model an alternative to measure probability of student permanence, Manizales (2006)

17. National Full Law of the National System of Public Records

18. Daniel, S., Cesar, P.: Data mining. Techniques and Tools, pp. 13–14 (2007)

19. Silva Ayçaguer, L.C.: Excursion to the logistic regression in health sciences. Díaz de Santos Editions (2000)

Prediction of the Incrustating Trend in Oil Extraction Pipelines: An Approach Based on Neural Decision Trees

B. Peralta[1(✉)], M. Salvador[1(✉)], O. Camacho[1(✉)], F. Escobar[1(✉)], and C. Goyes[2(✉)]

[1] Escuela Politécnica Nacional, Ladrón de Guevara E11-253, Quito, Ecuador
{bryan.peralta, marcelo.salvadorq, oscar.camacho, freddy.escobarf}@epn.edu.ec
[2] Baker Huges, Quito, Ecuador
cristian.goyes@epn.edu.ec

Abstract. The oil and gas industry assesses the tendency of mineral deposit formation based on the principle of chemical equilibrium of the fluid based on existing production data. Instead of using this approach, the present work has used artificial intelligence to develop predictions of the incrustating tendency within oil extraction pipes using physicochemical analyzes on the extracted oil, using the processing capacity of current computers and the use of artificial neural networks of deep learning with the objective of determining how reliable a prediction based on artificial intelligence can be. Simultaneously, contemporary evaluation methods require on-site inspections that mostly provide remediation measures involving the consumption of labor and financial resources. Consequently, a new method for predicting the embedded trend in pipes based on an artificial neural network using decision trees as classifiers is proposed. The neural network model is trained based on an extensive database of the characteristics of the oil and the incrustation generated in the pipeline to obtain a predictive model. Subsequently, the model generates a decision tree by selecting within the database that information relevant to the solution of the problem and excluding the rest. The results of the experimentation and simulation were satisfactorily compared, obtaining a success rate of 83,26% when evaluated with a dataset dedicated only to the validation phase. Finally, the incrustating trend detection model using decision trees proved to be an applicable technology in the field of engineering within the field of gas and oil belonging to the Ecuadorian industry.

Keywords: Artificial neural networks · Incrustating trend · Oil · Pipelines · Prediction

1 Introduction

Within the oil and gas industry, one of the problems that represents the greatest economic damage is deposition at scale. Oil and gas are generated from deposits of dolomite ($MgCO_3 \cdot CaCO_3$) and limestone ($CaCO_3$), from relatively pure states or in the form of siliceous sands (SiO_2) or cemented carbonates in the presence of limestone or dolomite material [1, 2]. The probability that deposits generated inside oil pipes and

© Springer Nature Switzerland AG 2020
M. Botto-Tobar et al. (Eds.): ICAT 2019, CCIS 1194, pp. 334–347, 2020.
https://doi.org/10.1007/978-3-030-42520-3_27

production equipment (valves, pumps, etc.) are linked to the properties of the oil, physicochemical characteristics of the formation water, the geology of the rocks, variations in pressure, temperature and hydrodynamic conditions of the well itself [3]. The incrustation present in the oil extraction pipes is a problem studied since 1930 that causes the reduction of the flow and influences the efficiency of the installation, which hinders oil exploration and production activities [4].

In mature wells, where the pressure is insufficient to raise the oil spontaneously, water must be injected to increase the pressure, since there are deposits inside the pipeline the pressure is even higher and therefore the oil production decreases, where a severe inlay is capable of strangling a production tunnel within 24 h [5]. Consequently, in order to prevent catastrophic damage and a total shutdown in production pipelines it is of utmost importance to apply the long-term detection of the tendency to the appearance of mineral deposits.

Traditional methods based on surveys in situ of the pipeline as destructive and non-destructive tests have been used to determine the incrustation trend. In 2013, da Silva, Cagnon and Saggioro used a focus on recurrent neural networks to solve problems of underground hydrology including the formation of mineral deposits [6]. A relevant advantage of information processing techniques is the ability to handle large amounts of data of different types (for example, quantitative values and qualitative information), although information processing operations are necessary.

In recent years in the field of gas and oil production, research on the analysis and prediction of the trend of pipe incrustation has grown gradually, focusing the research on the attempt to determine the trend through the relationship of physicochemical properties of the transported fluid and the operating conditions of the installation [7]. However, the existing prediction methods are based on the analysis of the formation of deposits under specific conditions of oil production wells, determined the degree of incrustation through equations and descriptive manuals. Due to the high demand for financial resources and labor and the consequent workload in conventional analysis, the method based on the inspection of characteristics to determine the incrustation trend often has little effect. Therefore, it is necessary to develop other alternatives for determining the incrustation trend.

With the development of computer resources and the increased processing capacity of large databases, researchers are paying more attention to artificial intelligence. Including machine learning and deep learning [8]. Machine learning induces the machine to learn each time its structure, program or data changes depending on the input information in response to external stimuli in order to improve its expected future performance [9]. Machine learning uses a wide range of algorithms that learn itera-tively from data to improve, describe data and predict results [10]. The greatest advantage of these learning structures is the generation of an algorithm that perceives, models its environment and calculates the appropriate actions trying to anticipate their effects [11]. Based on this, the application of machine learning in the field of prediction and the recognition of trends can greatly reduce the demand for resources, improving the efficiency and accuracy of the prediction.

This contribution proposes a new method for predicting the incrustating trend in oil extraction pipes using an artificial neural network based on decision trees. The paper is divided as follows: Sect. 2 shows some fundamentals. Section 3 presents the methodology. Section 4 presents the results; in Sect. 5 some discussions are offered and finally the conclusions.

2 Fundamentals

2.1 Basic Theory and Process

The machine learning technique is used to develop a model that provides a solution to a specific problem from a set of previously collected data. This learning scheme is also called supervised learning where an external agent, in this case the programmer, directs the artificial neural network in the generation of the classifier algorithm [12]. The model is suitable for problems that involve logical solutions, geometric projections or probabilistic analysis where mathematical equations and physical laws do not produce results that meet expectations.

When the machine learning technique develops the model from the training data, the actual observations data is applied to the model. The process is illustrated in Fig. 1 where the model, based on the raw data entered for training, calculates the appropriate actions to try anticipate the effects. Changes made at any stage of Fig. 1 can be defined as learning [13].

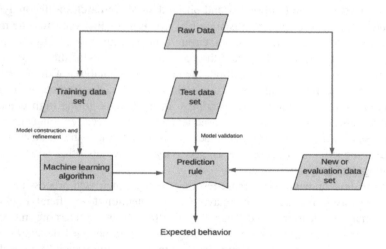

Fig. 1. Machine learning scheme for an artificial neural network

The neural decision tree (NDT) model developed by Lee and Yen combines an artificial neural network and a decision tree in a single entity [14], whose structure has been exported within Matlab software for use. It is possible to interpret the model as a set of learning criteria based on the theoretical foundations of the decision tree and the learning of the neural network. The main idea is to use an artificial neural network first to eliminate interdependencies between predictors, then the newly transformed input variables are entered into a learning process of the decision tree for classification [14].

The algorithm of the NDT model begins with the entry and acceptance of training data as input to the neural network model. Subsequently, the generated output is used for the construction of the decision tree, whose resulting rules will be the NDT model network [15].

The process of recognizing embedded trend in oil pipelines using a neural decision tree is shown in Fig. 2. First, the information collected is divided into two sets of information randomly. The first set is intended for training Neural network and constitutes 85% of the original database, the second set with the remaining database is intended for testing and validation of the neural network. The training data set is used to generate the classification algorithm that is adjusted by analyzing the importance of each of the predictors and their influence on the desired response. Finally, the validation of the model is carried out using the second set of test data that compare the response generated with the real trend registered, obtaining the loss or deviation when both are contrasted. A second external evaluation was carried out comparing the values generated by the model and the prediction made by the ScaleSoftPitzer software.

2.2   Neural Decision Tree Model

The training of the neural network using decision trees in this work uses the toolkit dedicated to the development of artificial intelligence focused on machine learning provided by Matlab [15]. The classification algorithm of the model is considered from the beginning as a decision tree, which consists of the typical architecture of the neural decision tree.

Architecture
The machine learning decision trees generated by Matlab are binary in nature. Each step in a prediction involves reviewing the value of a predictor as illustrated in Fig. 2, this being a simple decision tree [16].

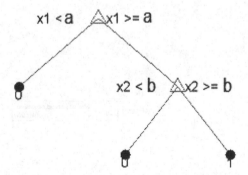

Fig. 2. Scheme of a simple decision tree

The classification tree in Fig. 2 predicts classifications based on two predictors, x1 and x2. The prediction begins with the upper node, represented by a triangle. The first decision is if x1 is less than a. If the condition is fulfilled, it is decided to follow the left branch and it is possible to visualize the classification of the types as type 0. However, if x1 exceeds the condition, then the right branch is followed. At this point the tree asks if x2 is less than b. If the answer is yes, the tree classifies the data as type 0, otherwise it classifies the data as type 1 [15].

3 Methodology

The characteristics of the neural decision tree developed in this work are presented in Table 1, which have been adjusted based on the particularity of the identification of the incrustation trend and the nature of the training process, using parameter optimization. Configuring the relationships between the predictor-response pair so that the deviation between the minimum objective observation generated and the estimated minimum objective observation is the smallest during the training process in each of the evaluations of the performance of the classification function as shown in the Fig. 3.

Table 1. Characteristics of the neural decision tree

Model parameters	TreeParams
Method	Tree
Type	Classification
Min leaf	6
Optimal leaf	15
Hyperparameters optimization results	Bayesian optimization
Min objective	0.2506
Min estimated objective	0.2568
Number objective evaluations	30

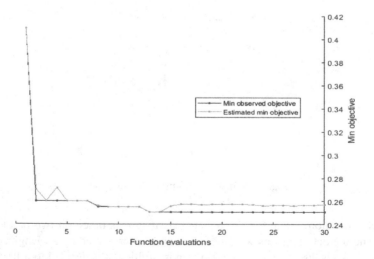

Fig. 3. Minimum objective vs. number of function evaluations

Figure 4 presents the behavior of the model function by tracking the estimated value of the objective function generated through the observable points or information collected, the average response of the model and the response generated for a next set of data as well as the deviations generated during the training process and a minimum feasible response value in each of the decision nodes within the neural network, until finding the minimum necessary number of nodes from which the response and observations the deviation of the answer begins to diminish.

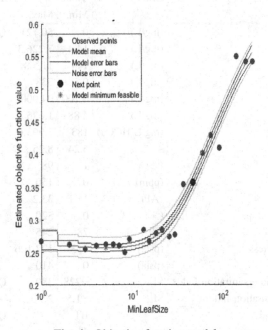

Fig. 4. Objective function model

Refinement of the Neural Decision Tree Model
Once the architecture of the neural network has been established and the classification algorithm is generated, it is subjected to a refinement process. The network improvement action is implemented using a function capable of removing zero-sized inputs (predictors), layers and outputs (responses) from it [17]. This generates a network with a smaller number of inputs and outputs, with the same competence to implement the same operations, since zero-sized inputs and outputs that do not transmit any information are dispensed [18].

3.1   Database and Resources

Table 2 presents the information collected from the physicochemical analyzes of the oil and the operating conditions, taking the denomination of predictors, together with the associated embedded trend for each set of predictors that is affected as the response.

The database includes 19 predictors and one response, out of a total of 488 tests performed. All the information was provided by many oil services companies that work in Ecuador.

Table 2. Database features

Predictors	Units	Values		Type
		Min.	Max.	
Na^+	(mg/L)	0	50,721	Quantitative
Mg^{2+}	(mg/L)	14.4	7,226.3	
Ca^{2+}	(mg/L)	144	11,528	
Fe^{2+}	(mg/L)	0.5	338	
Cl^-	(mg/L)	3,200	94,900	
SO_4^{2-}	(mg/L)	2.88	1,625	
Bicarbonates	(mg/L HCO_3)	183	2,903.6	
pH (st. c.)	–	5.24	8.25	
CO_2 gas	(%)	1	98	
H_2S gas	(ppm)	0	150	
API density	°API	15.1	33.2	
Bottomhole Temp.	(°F)	0	320	
Head Temp.	(°F)	78	245	
Bottomhole Press.	(psia)	313	127,106	
Head Press.	(psia)	0	410	
Sif – bottomhole saturation index	SI	−228	389	Quantitative
Sic - head saturation index	SI	−1.5	2.9	
Producing sand	–	BT		Qualitative
		TI		
		TS		
		UI		
		US		
		U		
Response				
Bottomhole incrustation trend	–	NSPD (No deposit occurs)		Qualitative
		LEVE (Mild)		
		MODERADA (Moderate)		
		SEVERA (Severe)		

The producing sand can be used alone or as a combination of the types shown above.

The software resources used for the development of the artificial neural network are presented in Table 3. Additionally, the hardware used and its characteristics are detailed in Table 4.

Table 3. Hardware configuration

Hardware	Brand and main parameters
Computer	Lenovo IdeaPad P500
CPU	Intel (R) Core (TM) i7-3520 M CPU 2.90 GHz
RAM	8 GB
Hard disk	ATA TOSHIBA DT01ACA2 SCSI Disk Device (1 TB)

Table 4. Software configuration

Proyect	Software name and version
Operating system	Microsoft Windows 8 Professional 64-bit
Matlab	Matlab R2019a
FrenchCreek	ScaleSoftPitzer V 13.0

4 Results

Figure 5 presents the analysis of the importance of the 17 quantitative predictors within the development of the classification algorithm represented in a bar chart with a unit value range, which is understood as the individual influence of the input information as a complement to the classification process to generate an answer. Of the 17 original predictors, 10 of them have been selected and grouped individually or in pairs based on their interaction with the qualifying algorithm. The value of the predictors acting together as sodium (Na) and magnesium (Mg) was 0.1248%, magnesium (Mg) and calcium (Ca) were 0.2097%, iron (Fe) and chlorine (Cl) were 0.9071% and the gaseous carbon dioxide ($CO_{2\ (g)}$) and gaseous hydrogen sulfide ($H_2S_{(g)}$) were 1.856%. Individually calcium (Ca) has an influence of 0.0046%. Finally, the rest of the predictors and their pairings have negligible influences.

The test data set previously separated from the original database, with a total of 73 observations, is used to test the quality of the prediction generated by the two-stage neural decision tree model, the first implemented on the algorithm classifier before refinement and the second after implementing refinement. The percentage of success obtained in both stages is presented in Table 5.

Once the refinement of the neural network is finished and with the best success rate of the predictions made, the decision tree shown in Fig. 6 is generated. The specifications and criteria values for the predictors are detailed in Table 6.

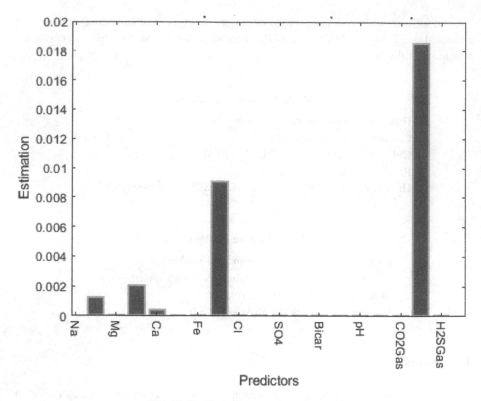

Fig. 5. Predictor importance estimates

Table 5. Performance of the neural decision tree model

Stage	Refinement	Success (%)
1	No	80.43
2	Yes	83.26

Finally, a second validation of the reliability of the predictions was performed by randomly separating 40 observations from the test data set, the information was entered into the ScaleSoftPitzer software to generate a prediction of the incrustation trend. The values of the real response of the observations were compared separately with the predictions made by both the neural decision tree and the ScaleSoftPitzer software. The success rate of both the software and the neural network are presented in Table 7.

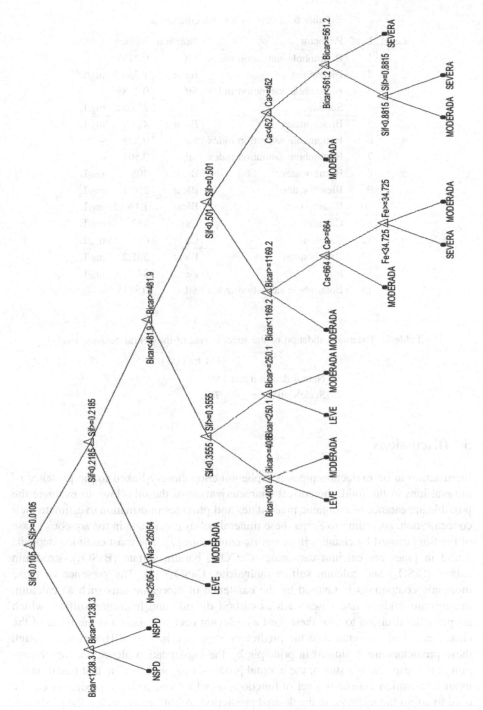

Fig. 6. Decision tree

Table 6. Decision tree specifications

Level	Leaf	Predictor	Notation	Value	Units
1	1	Bottomhole saturation index	Sif	0.2185	–
2	2	Bicarbonates	Bicar	1,238.3	mg/L
	3	Bottomhole saturation index	Sif	0.2185	–
3	4	Sodium	Na	25,054	mg/L
	5	Bicarbonates	Bicar	481.9	mg/L
4	6	Bottomhole saturation index	Sif	0.355	–
	7	Bottomhole saturation index	Sif	0.501	–
5	8	Bicarbonates	Bicar	408	mg/L
	9	Bicarbonates	Bicar	250.1	mg/L
	10	Bicarbonates	Bicar	1,169.2	mg/L
	11	Calcium	Ca	452	mg/L
6	12	Calcium	Ca	664	mg/L
	13	Bicarbonates	Bicar	561.2	mg/L
7	14	Iron	Fe	34.725	mg/L
	15	Bottomhole saturation index	Sif	0.8815	–

Table 7. External validation of the success rate of the neural decision tree

	Hit rate (%)
Neural decision tree	80
ScaleSoftPitzer	70

5 Discussions

Incrustation in oil extraction pipes is a phenomenon closely linked to the presence of mineral ions in the fluid, the correct characterization of the oil allows to evaluate the possible appearance of inorganic precipitates and provides information to estimate their concentration, according to Zerpa these mineral solids precipitate in the aqueous phase of the fluid caused by changes in operating conditions [7]. The most common deposits found in pipes are calcium carbonate ($CaCO_3$), barium sulfate ($BaSO_4$), strontium sulfate ($SrSO_4$) and calcium sulfate (anhydrite, $CaSO_4$) [2]. The presence of these inorganic compounds is favored by the existence of inorganic ions such as calcium, magnesium, sodium and gases such as carbon dioxide and hydrogen sulfide, which supports the decision to take these ions as relevant decision criteria at the time. of the classification of importance of the predictors as shown in Fig. 5. However, although these predictors are evaluated in principle by the classification algorithm, the generation of the response is a sum of the internal processes of the algorithm that based on the input information generates a set of functions in which one or more predictors can be used to adjust the response to the desired prediction. Additionally, within the predictors selected to be part of the classification nodes, the bottomhole saturation index representing the water and hydrocarbon content present in the porous space of the reservoir rock was added, because when this fluid is extracted the amount of mineral present in transported oil increases [3]. The neural decision tree was successfully developed following the architecture proposed by Lee and Yen where a decision node can lead to a response or connection with two other nodes at a lower level, but a response can never be based on two decision nodes [14]. The generated model had a success rate of 80.43%. Next, the algorithm was subjected to a refinement process in which decision nodes were removed that started from predictors with incomplete information or that generated an inconclusive response following the improvement process stipulated by Roe et al. [17], so that the success rate improved to 83.26%. The number of levels and decision nodes required by the final artificial neural network, as well as their characteristics are detailed in Table 6 and the generated decision tree is presented in Fig. 6. The results of the external validation using the ScaleSoftPitzer software and the actual responses of the observations are detailed in Table 7, in which it can be seen that the rate of correct predictions made by the neural decision tree remained above 80% and exceeded the prediction made by the external software, which implies that for the analysis of the incrustating trend of Ecuadorian crudes the classification algorithm generated by the artificial neural network has a greater precision than its counterpart, which presented a success rate of 70%. The lower success rate of the predictions made by the ScaleSoftPitzer software can be attributed to the fact that the classification criteria used are a generalization of the tendency of incrustation in oil extraction pipes located in the European continent, while the neural decision tree developed is closely linked to the characteristics of oil and operating conditions of the Ecuadorian industry.

6 Conclusions

This document proposed a method for predicting the incrustating trend in oil extraction pipes within the Ecuadorian industry based on the use of artificial intelligence, using a neural network model assisted by a decision tree, which seeks to implement the machine learning as a research tool in the prediction of mineral scale in the exercise of chemical engineering within the field of gas and oil.

The neural decision tree model was trained using petrochemical physicochemical analyzes and operating conditions for oil extraction. Training of the neural decision tree model requires a massive amount of data from different categories and conditions. The accuracy of the prediction generated is closely linked to the information fed from the tests, which implies that the absence of information from one or more predictors has a significant impact on the response generated by the classifying algorithm.

The decision tree generated by the artificial neural network provided a visual recognition tool for a given observation, facilitating visual inspection and obtaining prediction.

The success rate in predicting the incrustating trend of the neural decision tree model exceeded the performance of software dedicated to the same purpose within the scope of the Ecuadorian oil industry.

The reliability of the prediction by using the neural decision tree model may vary if the classification algorithm is used for the prediction of incrustation trends of an oil not extracted from Ecuadorian Amazon region.

Future research will propose a stricter data collection criterion in order to reduce the amount of missing information and prevent erroneous associations of the classification algorithm during the training process of the artificial neural network.

Further complementary research may use the scheme of the neural decision tree model developed to make predictions of the corrosive trend in oil extraction pipelines.

Weaknesses and strengths should be examined more rigorously of the different stages of the development of the artificial neural network, as well as the factors that influence the extraction of gas and oil, such as the volumetric flow of fluids that pass through the pipe and the measured and calculated amount of dissolved solids (TDS) present in the fluid.

References

1. Sandrine, P., Laurent, A., Vuataz, F.-D.: Review on chemical stimulation techniques in oil industry and applications to geothermal systems. Deep Heat Min. Assoc. **28**, 1–32 (2007)
2. Guo, B., Song, S., Ghalambor, A., Tian Ran, L.: Offshore Pipelines: Design, Installation, and Maintenance, no. 2 (2014)
3. Fink, J.: Petroleum Engineer's Guide to Oil Field Chemicals and Fluids, 1st edn. Elsevier, Waltham (2012)
4. Calixto, E.: Gas and Oil Reliability Engineering, 2nd edn. Elsevier, San Francisco (2016)
5. Vallejo, M.L.: Predicción y control de incrustaciones minerales en pozos petroleros. Instituto Politécnico Nacional de México (2011)
6. da Silva, I., Cagnon, J., Saggioro, N.: Recurrent neural network based approach for solving groundwater hydrology problems, vol. I, p. 13. Intech (2016). no. Deep learning
7. Zerpa, L.: A Practical Model to Predict Gas Hydrate Formation, Dissociation and Transportability in Oil and Gas Flowlines". Colorado School of Mines, Colorado (2013)
8. Gollapudi, S.: Practical Machine Learning. Tackle the Real-World Complexities of Modern Machine Learning with Innovative and Cutting-Edge Techniques, 1st edn. Packt Publishing, Birmingham (2016)
9. Smola, A., Vishwanathan, S.: Introduction to Machine Learning, vol. 1, 1st edn. Cambridge University Press, Cambridge (2009). no. July
10. Hurwitz, J., Kirsch, D., Wiley, J.: Machine Learning For Dummies, 1st edn. IBM, United States of America (2018). no. June
11. Nilsson, N.J.: Introduction to Machine Learning An Early Draft of A Proposed, 1st edn. Stanford University, Stanford (2005)
12. Kim, P.: MATLAB Deep Learning: With Machine Learning, Neural Networks and Artificial Intelligence, 1st edn. Apress, New York (2017)
13. Shalev-Shwartz, S., Ben-David, S.: Understanding Machine Learning: From Theory to Algorithms, vol. 9781107057, 1st edn. Cambridge University Press, Cambridge (2014)
14. Lee, Y.-S., Yen, S.-J.: Neural-based approaches for improving the accuracy of decision trees. In: Kambayashi, Y., Winiwarter, W., Arikawa, M. (eds.) DaWaK 2002. LNCS, vol. 2454, pp. 114–123. Springer, Heidelberg (2002). https://doi.org/10.1007/3-540-46145-0_12
15. Beale, M.H., Hagan, M.T., Demuth, H.B.: Neural Network Toolbox ^{TM} 7 User's Guide, 11th edn. MathWorks, Natick (2010)
16. Yao, Y., Yang, Y., Wang, Y., Zhao, X.: Artificial intelligence-based hull structural plate corrosion damage detection and recognition using convolutional neural network. Appl. Ocean Res. **90**, 101823 (2019)

17. Roe, B.P., Yang, H.J., Zhu, J., Liu, Y., Stancu, I., McGregor, G.: Boosted decision trees as an alternative to artificial neural networks for particle identification. Nucl. Instrum. Methods Phys. Res. Sect. A Accel. Spectrom. Detect. Assoc. Equip. **543**(2), 577–584 (2005)
18. Delete neural inputs, layers, and outputs with sizes of zero - MATLAB prune. https://www. mathworks.com/help/deeplearning/ref/prune.html?searchHighlight=prune&s_tid=doc_ srchtitle. Accessed 03 Oct 2019

Wind Energy Forecasting with Artificial Intelligence Techniques: A Review

Jorge Maldonado-Correa$^{(\boxtimes)}$ ⓘ, Marcelo Valdiviezo, Juan Solano ⓘ,
Marco Rojas, and Carlos Samaniego-Ojeda

Universidad Nacional de Loja, Facultad de Energía,
Ciudad Universitaria Guillermo Falconí, Loja, Ecuador
jorge.maldonado@unl.edu.ec

Abstract. The World Wind Energy Association (WWEA) forecasts that installed wind capacity worldwide will reach 800 GW by the end of 2021. Because wind is a random resource, both in speed and direction, the short-term forecasting of wind energy has become an important issue to be investigated. In this paper, a Systematic Literature Review (SLR) on non-parametric models and techniques for predicting short-term wind energy is presented based on four research questions related to both already applied methodologies and wind physical variables in order to determine the state of the art for the development of the research project "Artificial intelligence system for the short-term prediction of the energy production of the Villonaco wind farm". The results indicate that artificial neural networks (ANN) and support-vector machines (SVMs) were mainly used in related studies. In addition, ANNs are highlighted in comparison with other techniques of Wind Energy Forecasting.

Keywords: Forecasting wind energy · Artificial intelligence · Wind farm

1 Introduction

The latest annual report published by the Global Wind Energy Council (GWEC) indicates that installed wind capacity worldwide reached 599 GW by the end of 2018 [1], and capacity of 800 GW is projected for the year 2021 [2].

The wind is characterized as a random and intermittent resource, which can become highly variable even in the short term. Therefore, it is difficult to know in advance and accurately the amount of wind energy that we can count on at any given time. This variability makes the Wind Farms operation complex. Thus, its future energy production has to be estimated or predicted. Developing a wind energy prediction tool involves generating information on what will happen in the future from the data available on the present and past.

The most important models for wind speed forecasting are the physical methods, such as the Numerical Weather Prediction (NWP); the statistical methods, where the most popular is the ARIMA model; the intelligent models, where the most popular are based on ANNs; and the hybrid forecasting models that combine different types of algorithms. Physical methods are better for predicting wind speed in the long-term. Statistical methods and artificial intelligence models are efficient for short-term wind

© Springer Nature Switzerland AG 2020
M. Botto-Tobar et al. (Eds.): ICAT 2019, CCIS 1194, pp. 348–362, 2020.
https://doi.org/10.1007/978-3-030-42520-3_28

speed prediction. The forecasting of wind energy production is a complex task, but it is crucial to establish optimal planning by Wind Farms owners and operators [3].

In this work, we used the method involving a systematic review of Torres-Carrión's literature, which divides the relevant process into three main phases: planning, conducting, and reporting the review [4]. The authors reviewed 50 studies published in the last three years on wind energy prediction, including scientific articles and conference proceedings indexed in Scopus, which were classified and ordered with the help of the bibliographic manager Mendeley.

According to the literature reviewed, we have ascertained that the use of ANN for the prediction of wind speed in the short term (24 h) is more efficient and provides better results compared to the other methods, as demonstrated [3], a recent study with more than 180 references that have been published in the last five years. Another aspect to highlight is the fact that the majority of these studies focus on short-term prediction owing to the speed of response and the best percentages of accuracy, using different software for data analysis, among which MATLAB® is the most common one [5–7].

The objective of this paper is to determine the actual state-of-the-art on wind energy predictive models in the short-term, which will allow establishing the baseline of a project called: "Artificial intelligence system for the short-term prediction of the energy production of the Villonaco wind farm" [8]. In summary, the following contributions are made in this paper:

- An SLR using a methodology already developed, with the aim of knowing which are the methods of prediction of short-term wind energy using artificial intelligence (AI) tools.
- A methodology to make an SLR that can be applied to other fields of engineering and science.

This paper is organized as follows: In Sect. 2, a detailed description of the method used is provided, wherein we present the conceptual mindset, the research questions, and the semantic structure of the search. In Sect. 3, we describe the results obtained through these research questions, and finally, in Sect. 4, the relevant conclusions have been presented.

2 Method

This paper uses the methodology proposed by Torres-Carrión [4], in order to build an SLR, which is a method adapted from Kitchenham and Bacc [9, 10]. The steps of this methodology are shown in Fig. 1 and described below.

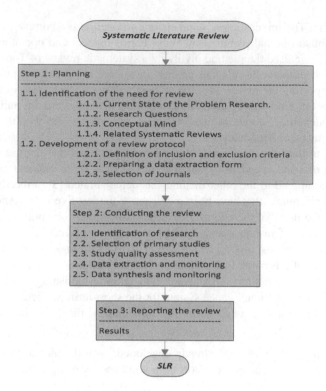

Fig. 1. Flow diagram of the methodology

2.1   Identification of the Need for Revision

According to [11] SCADA systems can measure from a few tens to thousands of endogenous and exogenous variables. The data provided by the SCADA system could be classified into the following categories:

- Weather parameters: The parameters related to the wind, e.g., speed, intensity, direction, etc., and other environmental parameters, e.g., pressure, temperature, etc.
- Energy conversion parameters: Parameters related to the energy conversion process, e.g., power output, voltage, pitch system, yaw system, etc.
- Temperature parameters: Temperatures of the turbine components and the air temperature close to the components.
- Vibration parameters: Vibrations from the drive train and tower.

The use of stored SCADA data from the wind turbine (WT) for the prediction of energy production and the prediction of possible failures in the different WT components is a current research topic. For example, in [12], a novel study on short-term wind energy prediction is presented, wherein wind speed and generated power data acquired by the SCADA system of wind turbines are used. In the aforementioned study, the authors propose a hybrid prediction model based on the combination of different tools and techniques such as the Hilbert-Huang Transform (HHT), genetic algorithms (GA), and ANN, thus achieving significant improvements in the accuracy of short-term wind energy predictions.

This literature review is based on the need to know the existing models and techniques for predicting the short-term energy production of Wind Farms, which will contribute to improving the operational efficiency of Wind Farms, reducing the uncertainty in energy production reports.

2.2   Research Questions

A research question is a question of interest to the scientific and/or professional community that allows focussing on a specific and concrete theme or phenomenon. According to [13] one of the main shortcomings of SLR is the lack of an explicit statement of the research questions to be investigated. Adequate formulation of research questions facilitates peer review and, therefore, contributes to the manuscript's acceptance.

In order to focus the literature search on our field of interest, and therefore be able to determine which are the wind energy forecasting models and the characteristics of the algorithms input data, we have asked ourselves the following research questions (RQ):

- RQ1: Which machine learning (ML) models are currently used for wind energy prediction?
- RQ2: Which methods are applied in wind energy for the evaluation of prediction models?
- RQ3: Which physical variables are applied for wind energy prediction models?
- RQ4: Which SCADA system variables are used for wind energy prediction models?

2.3   Conceptual Mind

The conceptual mind or the so-called "mentefacto" is regarded as one of the main instruments required to achieve good reading and learning [4]. The "mentefacto" tool has been created in pedagogy to represent concepts. According to its author, it is an ideogram or graphic sketch that represents something, assumes a complex idea, and conceptualizes it. Figure 2 depicts the conceptual mindset of this research. Construction of the conceptual mind requires one to respond to the following four intellectual operations:

- Supraordinate: The concept is included in a superior class that contains it.
- Isoordinate: It describes the qualities, properties, and characteristics of the given concept.
- Exclusion: Differences are established with the said concept.
- Infraordinate: The subclasses of the concept, its versions, and types are established.

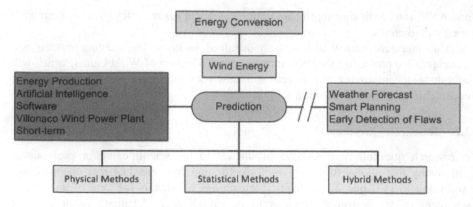

Fig. 2. The conceptual mind on wind energy prediction. Green: supraordinate; blue: isoordinate; yellow: exclusion; grey: infraordinate. (Color figure online)

For the development of the bibliographic review, a search script was developed, which can collect information from different databases. It has been structured in four levels, the first of which was taken from the conceptual mind, while the second level restricted the search to studies related to AI. Subsequently, the third level linked climate-related studies, whereas the fourth one directly linked the four research questions (Table 1). In addition, in each level, the terms provided by the thesaurus were taken into account, which made it possible to cover all those terms related by synonymy.

2.4 Related Systematic Literature Reviews

In order to identify previous studies conducted on SLRs in our field of interest, we conducted a search in Scopus employing the next search syntax:

((forecast* OR project* OR estimat* OR prevision OR prediction) AND ("AI" OR "machine learning" OR "artificial intelligence") AND (weather OR climate) AND (wind W/1 (farm OR turbine OR energy)) AND (SLR OR review OR survey OR "meta-analysis" OR mapping)) AND (LIMIT-TO (SUBAREA,"ENGI") OR LIMIT-TO (SUBJAREA,"ENER") OR LIMIT-TO (SUBAREA, "COMP") OR LIMIT-TO (SUBJAREA, "MATH") OR LIMIT-TO (SUBAREA, "ENVI")) AND (LIMIT-TO (PUBYEAR, 2019) OR LIMIT-TO (PUBYEAR, 2018) OR LIMIT-TO (PUBYEAR, 2017)) AND (LIMIT-TO (LANGUAGE, "English")))

The search script made it possible to find four articles catalogued as SLR. For instance, in [14] a detailed review of prediction wind energy with machine learning methods was presented. Subsequently, ten different models based on different influencing factors were compared. Then, the predicted results were combined with meteorological factors for the final prediction of wind energy. Lastly, the results of the prediction were compared showing that SVM's had the lowest error for the next 20 days forecasting. On the other hand, in [15] a new optimization algorithm called Fruit fly optimisation was proposed to predict short-term wind speed. The results of the simulation revealed that with the use of optimized extreme machine learning based on the new algorithm, the short-term wind speed forecasting was closer to the actual values, reflecting the effectiveness and practicality of the proposed algorithm. Similarly, a literature review on the use of spatial analysis in the implementation of renewable energy policies and a proposal for the use of a multi-criteria spatial decision support system based on geographic information systems was presented in [16]. Finally, [3] presented an exhaustive review of he NNAs used in wind energy systems, identifying the most commonly used methods for forecasting wind speed and demonstrating that NNAs may be an alternative to conventional methods.

However, the authors did not find literature reviews that allow us to answer the research questions posed, which forces us to undertake this study in order to achieve our goal. It is also necessary to point out that in the scientific literature consulted, all the case studies are related to wind farms located below 2700 m.a.s.l. Given that the Villonaco Wind Farm is located at an altitude higher than 2700 m.a.s.l, the authors have identified a knowledge gap that justifies the research and the present literature review.

Table 1. Semantic search structure

Level	Conceptual mind	Semantic structure
1st-level	Wind energy prediction	((Forecast* OR project* OR estimat* OR prevision OR prediction) AND (wind W/1 (farm OR turbine OR energy))
2nd-level	Artificial intelligence	AND ("AI" OR "machine learning" OR "artificial intelligence")
3rd-level	+Climate	AND (climate OR weather)
4th-level	Question	RQ1: (models) RQ2: (evaluation) RQ3: (physical methods) RQ4: (SCADA)

2.5 Selection of Journals

Table 2 presents the top ten the list of consulted journals and their primary metrics, such as the impact factor, location quartile according to the JCR (Journal Citation Report) and SJR (Scimago Journal Rank), and value according to the Google Scholar h5 indicator. We also detail the number of articles consulted in each journal and their order of importance calculated using the following expression:

$$Ord = (0.25\#papers)(IF - JCR)(IF - SJR)(h5) \tag{1}$$

Table 2. List of consulted journals. N: Number of papers; IF: Impact Factor; Qn: Quartile; h5: H Index

	N	JCR		SJR		h5	Value
		IF	Q_N	IF	Q_N		
Applied Energy	2	8.426	Q1	3.46	Q1	131	1909.58
Renewable and Sustainable Energy Reviews	1	10.55	Q1	3.29	Q1	174	1510.72
Renewable Energy	2	5.439	Q1	1.89	Q1	90	462.59
Energy Conversion and Management	1	7.181	Q1	2.73	Q1	88	431.29
Knowledge-Based Systems	1	5.101	Q1	1.46	Q1	74	137.78
Reliability Engineering and System Safety	1	4.039	Q1	1.94	Q1	59	115.58
Wind Energy	3	3.125	Q1	1.05	Q2	43	105.82
Energies	3	2.707	Q2	0.61	Q1	62	76.78
IEEE Access	1	4.098	Q1	0.61	Q1	89	55.62
IEEE Trasc. on Systems, Man, and Cyb. Syst.	1	1.34	Q1	2.15	Q1	71	51.33

2.6   Development of the Review Protocol

Definition of the Inclusion and Exclusion Criteria

For the development of the review protocol, certain criteria were considered, for the selection of articles related to the research questions, both collectively and individually. The general criteria for the review protocol are as follows:

- First, studies related to wind energy prediction that apply AI techniques were considered.
- Studies published in the last three years, between 2017 and 2019, were considered.

The specific criteria taken into account are as follows:

- Studies that evaluate ML models in wind energy prediction;
- Studies that describe the architecture of ML models as well as the AL techniques applied;
- Studies that present the variables used in the training of the models;
- Studies that compare different AL models applied for the prediction of wind energy;
- Studies that analyze the influence of climatic variables on wind energy prediction models;
- Studies that present techniques to evaluate wind energy prediction models uncertainty;
- Studies that use the SCADA system variables for wind energy prediction.

Certain exclusion criteria, such as the following, were also considered:

- Research work involving only climate prediction for wind power plants;
- Studies analyzing intelligent planning for wind power plants;
- Research focusing on the early detection of failures in wind power plants;
- Magazines that are not considered scientific articles.

Defining the Categories of Analysis

For RQ1, the following variables or categories of analysis were considered:

- ML models employed for wind energy prediction: Multilayer Perceptron (MLP), Nested Adaptive Neural Networks (NANN), Generalized Regression Neural Network (GRNN), Least Squares Support Vector Networks (LSSVN), Extreme Learning Machine (ELM), hybrid methods, among others.
- Datasets used in the training of the models created, which are related to Wind Farms around the world;
- The time variable that describes whether the studies have been performed for immediate, short-term, or long-term analysis;
- Frameworks and software used for the programming of wind energy prediction models.

For RQ2, we considered the evaluation metrics of prediction models, which are an important parameter to quantify uncertainty. The main ones include the following: Accuracy, Root-Mean-Square Deviation (RMSD), Mean Absolute Error (MAE), Mean Absolute Percentage Error (MAPE), and others such as Normalized Root Mean Square Error (nRMSE), Standard Deviation of Error (SDE).

For RQ3, the following elements of climate were considered: ambient temperature, atmospheric pressure, wind, and humidity. A few other variables considered include climate factors such as altitude, relief and solar radiation.

For RQ4, weather parameters, energy conversion parameters, temperature parameters, and vibration parameters collected by the SCADA system of the Wind Farms were taken into account.

3   Results

In this section, we present the results of the bibliographic search; we do so using tables comprising the research questions, the variables associated with each question, the bibliographic references, and the frequency of appearance f.

3.1   RQ1: Which ML Models Are Used for Wind Energy Prediction?

In relation to the first research question, we have been able to determine that hybrid ML models are frequently used, aiming to optimize the parameters for predicting wind energy. With regard to the first variable, the use of neural networks for the prediction of wind energy is frequent. Most research works use Wind Farm datasets corresponding with different parts of the planet, wherein those located in China are more frequent (Table 3).

Table 3. Articles on ML models for wind energy prediction

A. Models	Articles	f
A1- MLP	[12, 17]	2
A2- NANN	[18]	1
A3- GRNN	[19–21]	3
A4- LSSVN	[6, 9, 22, 23]	4
A5- ELM	[24–26]	3
A6- Hybrids (SVR, RBFNN, and RF, NN, EMD, KRR, RVFL)	[7, 17, 20, 23–25, 27–29]	9
A7- Others (NARX-NN, SVR, DNN)	[18, 22, 26, 30–38]	12
B. Datasets		
B1- Global energy forecasting competition 2012 - wind forecasting	[14]	1
B2- Others (ELIA wind power website)	[15, 19, 20, 22, 23, 27, 30, 31, 34, 35, 38–41]	14
C. Temporary		
C1-Short-term	[3, 7, 12, 14, 15, 17, 19, 21, 23–26, 28–30, 32, 34, 36, 37, 42, 43]	21
C2- Long-term	[21, 35, 41, 44, 45]	5
C3- Immediate	[6]	1
D. Framework		
D1- Keras	[34, 37]	2
D2- Tensorflow		0
E. Language		
E1- Python	[34]	1
E2- Php		0
F. Tools		
F1- Matlab	[5, 7, 12, 14, 29, 32, 40]	7
F2- R	[42, 44]	2

Each model employed uses a different dataset than that of the other studies, which correspond to different locations around the world, wherein the Wind Farms are located mainly in China and Europe, resulting in specialized models or those based on the climate characteristics specific to each region. MATLAB is the most commonly used software in this regard for its data analysis capacity. However, free software options including Python, R, and frameworks such as KERAS are also used for the development of neural networks.

3.2   RQ2: Which Methods Are Applied in Wind Energy for the Evaluation of Prediction Models?

Evaluation methods are statistical techniques that allow one to determine the degree of uncertainty that predictive models possess. In the bibliographic review, we can observe that the methods predominantly used for calculating errors in predictive models are the

RMSE and MAE. It is common to observe the simultaneous use of more than one of these methods, in order to validate the models more accurately (Table 4).

Table 4. Articles on the evaluation methods of wind energy prediction models

A. Evaluation	Articles	f
A1- Accuracy	[12, 15]	2
A2- RMSD	[6, 7, 12, 15, 17, 24, 27, 28, 30–33, 38–42, 44–46]	22
A3- MAE	[7, 12, 14, 15, 17, 21, 24, 25, 28–32, 36, 37, 39–42, 44, 45]	21
A4- MAPE	[15, 19, 25, 28, 30, 31, 46]	7
A5- Others: nRMSE, BDE	[17, 23, 26, 28, 32, 33, 46]	7

3.3 RQ3: Which Physical Variables Are Applied for Wind Energy Prediction Models?

As can be observed in Table 5, most of the scientific articles consulted show the use of prediction models that employ data gathered from physical variables associated with climate elements, specifically wind speed, ambient temperature, atmospheric pressure, and relative humidity, and in lower percentage, atmospheric pressure, humidity, altitude, and relief.

3.4 Which SCADA Variables Are Used for Wind Energy Prediction Models?

In the scientific literature consulted, few articles discuss the use of data collected from the SCADA system of wind turbines as input for the prediction models of wind energy. Moreover, in the articles consulted, it is described that the variables used in the SCADA to predict wind energy are electrical (active power), mechanical, and the nacelle's position, and/or the so-called yawing (Table 6).

Table 5. Articles on physical variables applied in wind energy prediction models

A. Components of the energy-climate	Articles	f
A1- Ambient temperature	[6, 12, 14, 19, 21, 22, 24, 27, 33, 37, 39, 40, 44, 45]	14
A2- Atmospheric pressure	[12, 14, 19, 22, 37, 39, 40, 45]	8
A3- Wind	[3, 5, 6, 12, 14, 17, 19, 21, 22, 24–29, 33, 35–40, 42, 44–48]	28
A4- Humidity	[6, 12, 22, 24, 39, 44]	7
B. Factors		
B1- Altitude	[19]	1
B2- Relief	[37, 45]	2
B3- Solar radiation	[6]	1

Table 6. Articles on SCADA variables applied in wind energy prediction models

A. SCADA variables	Articles	f
A1- Weather parameters	[12, 39, 42, 49–51]	6
A2- Energy conversion parameters	[12, 33, 34, 42]	3
A3- Temperature parameters	[39]	1
A4- Vibration parameters		0

4 Conclusions

This paper presents a review of wind energy forecasting with artificial intelligence techniques. In addition, the methodology can be used for other studies that need a systematic literature review. First, literature search needs are exposed, establishing research questions on the field of interest. Second, the "mentefacto" is presented as a hierarchical cognitive diagram for organizing knowledge. Here, fundamental ideas are used and secondary ideas are discarded. Then, both the relational search script and the semantic search structure are presented, where the inclusion and exclusion criteria detailed in the methodology are used. Finally, the responses of each research questions are presented.

At present, there is an abundance of scientific literature on wind energy prediction models, the most outstanding of which include physical models, advanced statistical models, CFD (Computational Fluid Dynamics) models. Moreover, there are models that use AI tools such as fuzzy logic, genetic algorithms, ANN, and so on. However, the use of deep neural networks (DNN) appears to be one of the most frequently used techniques for wind energy prediction, which when combined with optimization techniques and based on AI algorithms can adjust a large number of parameters of a network, thereby improving the systems' accuracy considerably.

The ANNs have a wide application in the field of wind energy. Specifically, numerous studies have shown that ANNs are efficient for predicting short-term wind speed. In addition, the hybrid method, based on ANN and SVM, provides better results for short-term predictions than other conventional techniques. On the other hand, an important aspect to consider regarding the different techniques used in this regard is the evaluation method, wherein the RMSE or MAE are the most common ways to determine the precision of established models.

According to the literature reviewed, wind speed is the variable with the greatest influence on predictive models of wind energy. Because all WTs have a built-in SCADA system, usually with a sampling time of 10 min, the data collected (such as wind speed, wind direction, air temperature, equipment temperature, vibrations, etc.) is used for training wind energy prediction algorithms.

Acknowledgement. The authors acknowledge the support of the 'Universidad Nacional de Loja' by means of the research project: Artificial intelligence system for the short-term prediction of the energy production of the Villonaco wind farm. 26-DI-FEIRNNR-2019.

References

1. Wang, H., Wang, H., Jiang, G., Li, J., Wang, Y.: Early fault detection of wind turbines based on operational condition clustering and optimized deep belief network modeling. Energies 12(6), 984 (2019)
2. Pandit, R., Infield, D.: Gaussian process operational curves for wind turbine condition monitoring. Energies 11(7), 1631 (2018)
3. Marugán, A.P., Márquez, F.P.G., Perez, J.M.P., Ruiz-Hernández, D.: A survey of artificial neural network in wind energy systems. Appl. Energy 228, 1822–1836 (2018)
4. Torres-Carrión, P.V., González-González, C.S., Aciar, S., Rodríguez-Morales, G.: Methodology for systematic literature review applied to engineering and education. In: 2018 IEEE Global Engineering Education Conference (EDUCON), pp. 1364–1373 (2018)
5. Tasnim, S., Rahman, A., Oo, A.M.T., Haque, M.E.: Wind power prediction in new stations based on knowledge of existing stations: a cluster based multi source domain adaptation approach. Knowl.-Based Syst. 145, 15–24 (2018)
6. Dong, W., Yang, Q.: Ultra-short term prediction model of wind power generation based on hybrid intelligent method. In: Chinese Control Conference, CCC, pp. 9148–9153 (July 2018)
7. Lu, H., Heng, J., Wang, C.: An AI-based hybrid forecasting model for wind speed forecasting. Lecture Notes in Computer Science (including Subseries Lecture Notes in Artificial Intelligence and Lecture Notes in Bioinformatics). LNCS, vol. 10637, pp. 221–230. University of Technology Sydney, Sydney (2017). https://doi.org/10.1007/978-3-319-70093-9_23
8. Hernandez, W., et al.: Modeling of a robust confidence band for the power curve of a wind turbine. Sensors 16(12), 2080 (2016)
9. Kitchenham, B.: Procedures for performing systematic reviews. Keele 33(2004), 1–26 (2004). Joint Technical Report
10. Bacca, J., Baldiris, S., Fabregat, R., Graf, S., Kinshuk: Augmented reality trends in education: a systematic review of research and applications. Educ. Technol. Soc. 17(4), 133–149 (2014)
11. Pliego Marugán, A., García Márquez, F.P.: Advanced analytics for detection and diagnosis of false alarms and faults: a real case study. Wind Energy 22(11), 1622–1635 (2019)
12. Zheng, D., Shi, M., Wang, Y., Eseye, A.T., Zhang, J.: Day-ahead wind power forecasting using a two-stage hybrid modeling approach based on scada and meteorological information, and evaluating the impact of input-data dependency on forecasting accuracy. Energies 10(12), 1988 (2017)
13. Velásquez, J.: Una Guía Corta para Escribir Revisiones Sistemáticas de Literatura. DYNA 82(189), 9–12 (2014)
14. Cao, Y., Hu, Q., Shi, H., Zhang, Y.: Prediction of wind power generation base on neural network in consideration of the fault time. IEEJ Trans. Electr. Electron. Eng. 14(5), 670–679 (2019). Shanghai Electric Power Design Institute Co., Ltd., No. 550, Xujiahui Road, 23rd Floor, Huangpu District, Shanghai, 200090, China
15. Chen, Y., Pi, D.: Novel fruit fly algorithm for global optimisation and its application to short-term wind forecasting. Connect. Sci. 31(3), 244–266 (2019)
16. Kazak, J., van Hoof, J., Szewranski, S.: Challenges in the wind turbines location process in Central Europe – the use of spatial decision support systems. Renew. Sustain. Energy Rev. 76, 425–433 (2017)

17. Dolara, A., Gandelli, A., Grimaccia, F., Leva, S., Mussetta, M.: Weather-based machine learning technique for day-ahead wind power forecasting. In: 2017 6th International Conference on Renewable Energy Research and Applications, ICRERA 2017, pp. 206–209 (January 2017)
18. Blanchard, T., Samanta, B.: Wind speed forecasting using neural networks. Wind Eng. **44**(1), 33–48 (2020). https://doi.org/10.1177/0309524X19849846
19. Li, N., He, F., Ma, W.: Wind power prediction based on extreme learning machine with kernel mean p-power error loss. Energies **12**(4), 673 (2019)
20. Cornejo-Bueno, L., Cuadra, L., Jiménez-Fernández, S., Acevedo-Rodríguez, J., Prieto, L., Salcedo-Sanz, S.: Wind power ramp events prediction with hybrid machine learning regression techniques and reanalysis data. Energies **10**(11), 1784 (2017)
21. Martín-Vázquez, R., Aler, R., Galván, I.M.: Wind energy forecasting at different time horizons with individual and global models. In: Iliadis, L., Maglogiannis, I., Plagianakos, V. (eds.) AIAI 2018. IAICT, vol. 519, pp. 240–248. Springer, Cham (2018). https://doi.org/10.1007/978-3-319-92007-8_21
22. Martín-Vázquez, R., Aler, R., Galván, I.M.: A study on feature selection methods for wind energy prediction. In: Rojas, I., Joya, G., Catala, A. (eds.) IWANN 2017. LNCS, vol. 10305, pp. 698–707. Springer, Cham (2017). https://doi.org/10.1007/978-3-319-59153-7_60
23. Qiu, X., Ren, Y., Suganthan, P.N., Amaratunga, G.A.: Short-term wind power ramp forecasting with empirical mode decomposition based ensemble learning techniques. In: 2017 IEEE Symposium Series on Computational Intelligence, SSCI 2017 - Proceedings, pp. 1–8 (January 2018)
24. Pan, C., Tan, Q., Qin, B.: A new method of wind speed prediction based on weighted optimal fuzzy c-means and modular extreme learning machine. Wind Eng. **42**(5), 447–457 (2018)
25. Tahir, M., El-Shatshat, R., Salama, M.M.A.: Improved stacked ensemble based model for very short-term wind power forecasting. In: Proceedings - 2018 53rd International Universities Power Engineering Conference, UPEC 2018 (2018)
26. Cocchi, G., Galli, L., Galvan, G., Sciandrone, M., Cantù, M., Tomaselli, G.: Machine learning methods for short-term bid forecasting in the renewable energy market: a case study in Italy. Wind Energy **21**(5), 357–371 (2018)
27. Zaunseder, E., Müller, L., Blankenburg, S.: High accuracy forecasting with limited input data: using FFNNs to predict offshore wind power generation. In: ACM International Conference Proceeding Series, pp. 61–68 (2018)
28. Lahouar, A., Slama, J.B.H.: Hour-ahead wind power forecast based on random forests. Renew. Energy **109**, 529–541 (2017)
29. Salfate, I., et al.: 24-hours wind speed forecasting and wind power generation in La Serena (Chile). Wind Eng. **42**(6), 607–623 (2018)
30. Jiao, R., Huang, X., Ma, X., Han, L., Tian, W.: A model combining stacked auto encoder and back propagation algorithm for short-term wind power forecasting. IEEE Access **6**, 17851–17858 (2018)
31. Wang, Y., Hu, Q., Meng, D., Zhu, P.: Deterministic and probabilistic wind power forecasting using a variational Bayesian-based adaptive robust multi-kernel regression model. Appl. Energy **208**, 1097–1112 (2017)
32. Zameer, A., Arshad, J., Khan, A., Raja, M.A.Z.: Intelligent and robust prediction of short term wind power using genetic programming based ensemble of neural networks. Energy Convers. Manag. **134**, 361–372 (2017)

33. Fischer, A., Montuelle, L., Mougeot, M., Picard, D.: Statistical learning for wind power: a modeling and stability study towards forecasting. Wind Energy **20**(12), 2037–2047 (2017)

34. Woon, W.L., Oehmcke, S., Kramer, O.: Spatio-temporal wind power prediction using recurrent neural networks. In: Liu, D., Xie, S., Li, Y., Zhao, D., El-Alfy, E.-S.M. (eds.) ICONIP 2017. LNCS, vol. 10638, pp. 556–563. Springer, Cham (2017). https://doi.org/10.1007/978-3-319-70139-4_56

35. Kumar, A., Ali, A.B.M.S.: Prospects of wind energy production in the western Fiji-an empirical study using machine learning forecasting algorithms. In: 2017 Australasian Universities Power Engineering Conference, AUPEC 2017, pp. 1–5 (November 2017)

36. Prasetyowati, A., Sudibyo, H., Sudiana, D.: Wind power prediction by using wavelet decomposition mode based NARX-neural network. In: ACM International Conference Proceeding Series, pp. 275–278 (2017)

37. Díaz-Vico, D., Torres-Barrán, A., Omari, A., Dorronsoro, J.R.: Deep neural networks for wind and solar energy prediction. Neural Process. Lett. **46**(3), 829–844 (2017)

38. Alonzo, B., Plougonven, R., Mougeot, M., Fischer, A., Dupré, A., Drobinski, P.: From numerical weather prediction outputs to accurate local surface wind speed: statistical modeling and forecasts. In: Drobinski, P., Mougeot, M., Picard, D., Plougonven, R., Tankov, P. (eds.) FRM 2017. SPMS, vol. 254, pp. 23–44. Springer, Cham (2018). https://doi.org/10.1007/978-3-319-99052-1_2

39. Yang, J.: A novel short-term multi-input–multi-output prediction model of wind speed and wind power with LSSVM based on improved ant colony algorithm optimization. Cluster Comput. **22**(2), 3293–3300 (2019). Electrical Engineering College, Northwest Minzu University, Lanzhou, 730100, China

40. Dong, W., Yang, Q., Fang, X.: Multi-step ahead wind power generation prediction based on hybrid machine learning techniques. Energies **11**(8), 1975 (2018)

41. Deo, R.C., Ghorbani, M.A., Samadianfard, S., Maraseni, T., Bilgili, M., Biazar, M.: Multi-layer perceptron hybrid model integrated with the firefly optimizer algorithm for windspeed prediction of target site using a limited set of neighboring reference station data. Renew. Energy **116**, 309–323 (2018)

42. Browell, J., Gilbert, C., McMillan, D.: Use of turbine-level data for improved wind power forecasting. In: 2017 IEEE Manchester PowerTech, Powertech 2017 (2017)

43. Gensler, A., Sick, B.: Probabilistic wind power forecasting: A multi-scheme ensemble technique with gradual cooperative soft gating. In: 2017 IEEE Symposium Series on Computational Intelligence, SSCI 2017 - Proceedings, pp. 1–10 (January 2018)

44. Díaz, S., Carta, J.A., Matías, J.M.: Performance assessment of five MCP models proposed for the estimation of long-term wind turbine power outputs at a target site using three machine learning techniques. Appl. Energy **209**, 455–477 (2018)

45. Labati, R.D., Genovese, A., Piuri, V., Scotti, F., Sforza, G.: A decision support system for wind power production. IEEE Trans. Syst. Man Cybern. Syst. **50**(1), 290–304 (2018). Department of Computer Science, Università degli Studi di Milano, 26013 Crema, Italy (e-mail: ruggero.donida@unimi.it)

46. Banerjee, A., Tian, J., Wang, S., Gao, W.: Weighted evaluation of wind power forecasting models using evolutionary optimization algorithms. Procedia Comput. Sci. **114**, 357–365 (2017)

47. Reyes, A., Ibargüengoytia, P.H., Jijón, J.D., Guerrero, T., García, U.A., Borunda, M.: Wind power forecasting for the Villonaco wind farm using AI techniques. In: Pichardo-Lagunas, O., Miranda-Jiménez, S. (eds.) MICAI 2016. LNCS (LNAI), vol. 10062, pp. 226–236. Springer, Cham (2017). https://doi.org/10.1007/978-3-319-62428-0_19

48. Zhu, Y., Chen, S., Luo, J., Wang, Y.: A novel wind power prediction technique based on radial basis function neural network. In: Qiao, F., Patnaik, S., Wang, J. (eds.) ICMIR 2017. AISC, vol. 690, pp. 180–184. Springer, Cham (2018). https://doi.org/10.1007/978-3-319-65978-7_27

49. Gonzalez, E., Stephen, B., Infield, D., Melero, J.J.: On the use of high-frequency SCADA data for improved wind turbine performance monitoring. J. Phys: Conf. Ser. **926**(1), 012009 (2017)

50. Bi, R., Zhou, C., Hepburn, D.M.: Applying instantaneous SCADA data to artificial intelligence based power curve monitoring and WTG fault forecasting. In: 2016 International Conference on Smart Grid and Clean Energy Technologies, ICSGCE 2016, pp. 176–181 (2017)

51. Vidal, Y., Pozo, F., Tutivén, C.: Wind turbine multi-fault detection and classification based on SCADA data. Energies **11**(11), 3018 (2018)

Regression Models Comparison for Efficiency in Electricity Consumption in Ecuadorian Schools: A Case of Study

Alejandro Toapanta-Lema[1,2]([⊠]), Walberto Gallegos[1],
Jefferson Rubio-Aguilar[1], Edilberto Llanes-Cedeño[1],
Jorge Carrascal-García[2], Letty García-López[2],
and Paul D. Rosero-Montalvo[3]

[1] Universidad Internacional SEK, Quito, Ecuador
alejandro_toapanta@hotmail.com
[2] Instituto Superior Tecnológico, 17 de Julio, Urcuquí, Ecuador
[3] Universidad Técnica del Norte, Ibarra, Ecuador

Abstract. Consumption forecast models with their proper billing allow establishing strategies to avoid overloads in systems and penalties for high consumption. This paper presents a comparison of multivariate data prediction models that allow detecting the final monthly cost of electricity consumption in relation to the different billing parameters. As relevant results, it was obtained that the models based on decision support machines have a better sensitivity when compared with different metrics that evaluate the prediction error with training set improved by backward elimination criteria.

Keywords: Regression models · Electric consume prediction

1 Introduction

The natural resources of the planet must be conserved to the maximum for their suitability in the coming years. For this reason, electricity generation is focused on renewable energies that meet the growing need of emerging cities [1]. Consequently, the consumption forecast is a planning task that allows all members of the energy market to cooperate effectively. It also allows the sectors with the highest consumption to avoid penalties [2]. In addition, electricity generating companies can supply the appropriate amount, avoiding overloads in the generation of the equipment. However, the use of electricity in homes and industries shows a high variation that depends on the lifestyle of the users and the type of good or service that is generated. To all this, we must consider climate change [3]. Therefore, there is a wide variety of studies of energy consumption. Among the most important is the implementation of sensors that always acquire data in real-time [4]. With this information, the implementation of machine learning algorithms that allow the recognition of patterns and the growth trend of the country's energy demand is sought. An example is the works [5–7] that propose the methodology for forecasting the electric charge curves for short-term planning technologies. For this, multivariate mathematical models and algorithms such as decision

© Springer Nature Switzerland AG 2020
M. Botto-Tobar et al. (Eds.): ICAT 2019, CCIS 1194, pp. 363–371, 2020.
https://doi.org/10.1007/978-3-030-42520-3_29

support machines (SVM), mathematical regressions, empirical decomposition (EMD), neural networks (ANN), among others, have demonstrated great accuracy in demonstrating the indicated phenomenon.

Ecuador has invested heavily in hydroelectric projects with the objective of changing its productive matrix and avoiding dependence on the sale of crude oil. In addition, it can reduce tons of CO_2 per year. Renewable energies under development are: (i) hydraulic energy, (ii) wind energy, (iii) bio-mass energy, (iv) biofuel energy, (v) Geothermal energy, (vi) Mareomotive energy, (vii) Photovoltaic energy and (viii) Concentrated solar energy. According to the effective power, the representative power plants in the generation of electricity are Hydraulic 58.53%, Thermal power stations 39.16%, Biomass 1.66%, Photovoltaic 0.32%, Wind 0.24%. However, there are no works that relate energy consumption to data models with the objective of counting forecasts of growing demand, especially in the public sector. Currently, works such as [8, 9] have presented proposals on consumption behavior and electricity demand in different cities. However, there is no robust analysis of information.

During this period of investment in hydroelectric plants in the country, the Ministry of Education subscribed to the Millennium Declaration, where the set of Millennium Development Goals (MDG) are established. Among the most important area in this field of education that is based on all children in the world complete primary education equally. For this reason. Ecuador established placement criteria in relation to the relegated sectors in conjunction with student demand and low academic results. All these infrastructures are called Schools of Education of Milenio (MES), which have a capacity of 1140 or 570 students per day (morning and evening) with blocks of classrooms, initial education, laboratories, multiple-use, bar, and administration, among others. Across the country, there are 97 active MONTHs [10, 11].

Each MONTH has laboratories, classrooms, sports fields, which have a significant installed load of 207.07 KW. However, there is currently no study of energy consumption in this education sector and its prognoses of future consumption for supply planning. For this reason, an analysis of the consumption forecast and its comparison with different mathematical models is presented in order to determine the appropriate one that allows regulating the electrical consumption and establishing consumption policies. As relevant results, it is determined that the multivariate regression model of decision support machines has a lower prediction error (around 3%) with a training matrix optimized under the criterion of backward elimination.

The rest of the document is structured as follows: Sect. 2 shows the methodology used and the mathematical models used. Section 3 presents the results of the models and their selection. Finally, Sect. 4 show the conclusions and future works.

2   Methodology and Mathematical Models

This section presents the description of the database obtained for the analysis and the subsequent use of the prediction algorithms.

2.1 Data Set Description

The data set is acquired from the collection of electricity consumption during the last 3 years of operation of the SUMAK YACHANA WASI MONTH, located in the Cotacachi-Imbabura canton. It has 997 students, 48 teachers in the morning and evening hours. It must be considered that the consumption is related to the active and reactive power in the hours of use. As a result, a matrix $Y \in R^{mxn}$ was obtained, where m is the number of samples collected and n are the attributes of each m. In this case $m = 36$ and $n = 8$. In addition, consideration should be given to taking data to validate the models. For this reason, 80% of the database for training and 20% for tests are planned. However, in order to find the attributes that are closely related to our objective variable, in this case the total payment item, the backward elimination criterion is used. The same that allows to eliminate variables that do not provide too much information to the mathematical model. In this sense, using *p-value* and *R-squared* provides information necessary for the selection of attributes [12]. The significance values for the model are shown in Fig. 1.

```
Coefficients:
                     Estimate Std. Error t value Pr(>|t|)
(Intercept)         34.938803  13.354928   2.616 0.017494 *
P..ACTIVE.8_18.Kwh   0.054831   0.007963   6.886 1.93e-06 ***
P.ACTIVE.18_22.Kwh   0.087445   0.023630   3.701 0.001637 **
P.ACTIVE.22_8.Kwh    0.053124   0.010780   4.928 0.000109 ***
P.REACTIVE.00_24.KVARh -0.031909 0.017829 -1.790 0.090324 .
DEMAND.8_18.KW       1.196235   0.906645   1.319 0.203570
DEMAND.18_22.KW      2.556066   0.856984   2.983 0.007981 **
---
Signif. codes:  0 '***' 0.001 '**' 0.01 '*' 0.05 '.' 0.1 ' ' 1

Residual standard error: 10.8 on 18 degrees of freedom
Multiple R-squared:  0.9581,    Adjusted R-squared:  0.9441
F-statistic: 68.55 on 6 and 18 DF,  p-value: 2.04e-11
```

Fig. 1. Significance importance for model prediction

As can be seen, the variables of reactive power and the demand in hours from 8 to 18 h do not provide valuable information to the model. This reassesses the significance value of each variable as shown in Fig. 2.

```
Coefficients:
                     Estimate Std. Error t value Pr(>|t|)
(Intercept)         38.098136  14.160240   2.691 0.014067 *
P..ACTIVE.8_18.Kwh   0.057864   0.005946   9.731 4.99e-09 ***
P.ACTIVE.18_22.Kwh   0.104660   0.023898   4.379 0.000290 ***
P.ACTIVE.22_8.Kwh    0.042576   0.010362   4.109 0.000545 ***
DEMAND.18_22.KW      3.198316   0.794444   4.026 0.000662 ***
---
Signif. codes:  0 '***' 0.001 '**' 0.01 '*' 0.05 '.' 0.1 ' ' 1

Residual standard error: 11.57 on 20 degrees of freedom
Multiple R-squared:  0.9465,    Adjusted R-squared:  0.9358
F-statistic: 88.52 on 4 and 20 DF,  p-value: 1.997e-12
```

Fig. 2. Significance importance for model prediction

With this information, two training databases of the models are proposed. On the one hand, the matrix $Y \in R^{mxn}$ discussed above with all the variables of electricity consumption. On the other hand, the matrix $X \in R^{mxp}$, where p only has 6 values.

2.2   Regressions Models

In order to generate a consumption forecast, there are different models to perform a multivariate linear and no-linear regression analysis. In this sense, the most important ones are planted in relation to the literature found as: (i) covariance matrix, (ii) similarity Functions, (iii) Frequency table and others [12]. The first criterion (covariance matrix) is a simple linear model based on the equation:

$$y = b_0 + \sum_{i=0}^{m} x_i b_i \tag{1}$$

Where the values of x_i are the attributes of the Y and X matrices. The graphical result of this model is found in Fig. 3.

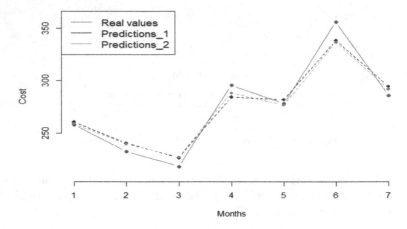

Fig. 3. Simple multivariant regression model. Predictions 1: Matrix Y, Predictions 2: Matrix X

To improve the fit of a linear model, you can square its values of the attributes of each matrix and find a multivariate regression of the equation:

$$y = b_0 + \sum_{j=0}^{r} \sum_{i=0}^{m} x_i^j b_i \tag{2}$$

In Fig. 4 the adjustment of the model with respect to the training set can be seen.

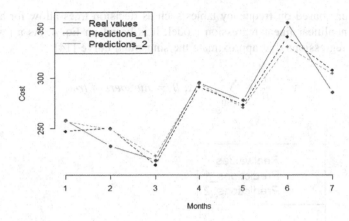

Fig. 4. Polynomial multivariant regression model. Predictions 1: Matrix Y, Predictions 2: Matrix X

Decision support machines (SVM) can be used as regression methods. SVM establishes tolerance margins to minimize error, individualizing the hyperplane that maximizes the prediction margin. This criterion is given by Eq. 3 when using a linear kernel method [13].

$$y = b_0 + \sum_{i=0}^{m} (\alpha_i - \alpha_i^*) . (x_i, x) + b \tag{3}$$

As a result of the implementation of this kernel method, its graphic representation is shown in Fig. 5.

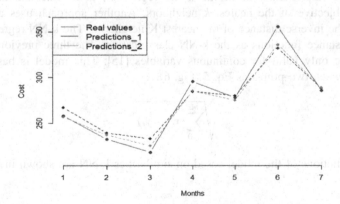

Fig. 5. SVM multivariant regression model. Predictions 1: Matrix Y, Predictions 2: Matrix X

Algorithms based on frequency tables such as decision trees allow for a complex and highly nonlinear linear regression model. It is based on Eq. 4. As a result, learn local linear regressions that approximate the sinusoidal curve [14].

$$y = \frac{1}{B} + \sum_{i=0}^{m} f_i * x, \, B = numbers \, of \, trees \tag{4}$$

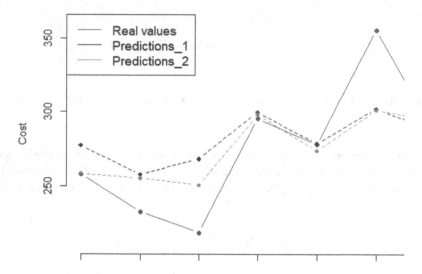

Fig. 6. Random Forest multivariant regression model. Predictions 1: Matrix Y, Predictions 2: Matrix X

A simple implementation of the k-NN regression is to calculate the average numerical objective of the nearest **k** neighbors. Another approach uses a weighted average of the inverse distance of the nearest K neighbors. The k-NN regression uses the same distance functions as the k-NN classification. The three previous distance measures are only valid for continuous variables [15]. This model is based on the distance between two points as Eq. 5 (Fig. 6).

$$\sqrt{\sum_{i=0}^{m} (x_i - y_i)^2} \tag{5}$$

The predictions of the model based on similarities k-NN are shown in Fig. 7.

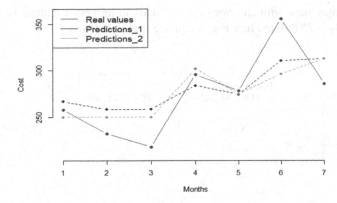

Fig. 7. k-NN multivariant regression model. Predictions 1: Matrix Y, Predictions 2: Matrix X

3 Results

With the models implemented, the different types of error allow to determine the appropriate one that represents the problem posed. For this reason, the cumulative sum of forecast errors (CFE) that provides additional information such as the deviation forecast was first analyzed. Second, the deviation from the mean error (MAD) that gives the absolute value of the difference between real and estimated value. Third, mean square error (RMSE) that gives the dispersion of the predicted error. Finally, average absolute percentage error (MAPE) that provides the error in percentage given by the model and the actual data (Table 1).

Table 1. Error analysis

Model	Linear		Polynomial		SVM		Random forest		k-NN	
Matrix Error	Y	X	Y	X	Y	X	Y	X	Y	X
CFE	0.36	−0.67	0.12	0.87	0.08	−2.71	6.84	1.13	6	1.75
MAD	1.23	1.05	1.59	1.49	1.43	0.99	3.16	2.55	2.55	3.32
RMSE	0.01	0.06	0.001	0.11	0.001	1.05	6.69	0.18	5.14	0.43
MAPE	4.3	3.7	4.8	5.2	5.1	3.4	11.9	9.2	12.3	11.1

According to the error analysis, the smallest forecast error is the linear model. However, the error dispersion (lower variability) presents the SVM model. In addition, the RMSE error allows the error to be squared and generates a greater analysis of the error. Finally, the MAPE error presents a lower prediction error as a percentage. As a result, SVM with the data matrix X has been shown to have better prediction performance. Consequently, the model is valid in the next 6 months, demonstrating a very significant adjustment with an error of 3.2%. This in monetary matters is an error of approximately 3.75 dollars. In a Fig. 8 can see this analysis.

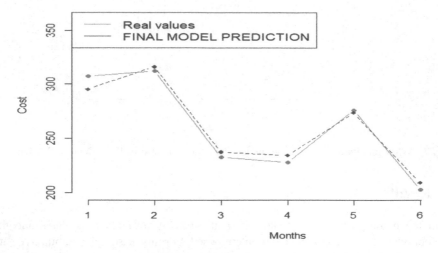

Fig. 8. Final regression model

As a discussion of this work, it can be deduced that these analyzes are very helpful for saving energy consumption in the education sector. Since these institutions are subsidized by the Ecuadorian state and generate a high cost nationwide. With this, energy-saving planning must be a priority for a country still dependent on oil. Consequently, the models proposed to depend on an adequate data reading process. In this sense, Ecuador has pending the automation of the consumption acquisition of electricity that causes errors in the payment of payroll. Finally, a local server is necessary to generate and process the data and models present in this investigation.

4   Conclusions and Future Works

This work establishes a solid basis for energy planning in education centers throughout the country. On the one hand, you can make consumption savings plans and validate it in real environments. On the other hand, in some sectors the reading of the meters is manual, where there are many errors in reading and improper charges to users. With the implementation of these prediction models, consumption monitoring can be carried out and problems or current leaks can be detected.

With respect to the implementation of prediction algorithms, it was possible to show that the linear models best fit other criteria established with their Random Forest or k-NN algorithms. In addition, the Backward Elimination criterion allowed reducing the training database and focusing on the attributes that provide more information to the model generation. With this, statistical criteria such as the p-value or the r-square help an adequate analysis of variables.

As future work, they propose installing current measurement sensors to generate their own consumption measurement and establish recommendations and power factor correction policies to have adequate items and subject to improvements.

References

1. Ishtiak, T., Orpon, R.M., Mashnoor, N., Ahmed, M., Nazim, M.A.: An advanced application to decrease household power consumption and save energy detecting the weather condition. In: 2017 8th IEEE Annual Information Technology, Electronics and Mobile Communication Conference, IEMCON 2017, pp. 622–627 (2017)
2. Dmitri, K., Maria, A., Anna, A.: Comparison of regression and neural network approaches to forecast daily power consumption. In: Proceedings - 2016 11th International Forum on Strategic Technology, IFOST 2016, pp. 247–250 (2017)
3. Zhang, X.M., Grolinger, K., Capretz, M.A.M., Seewald, L.: Forecasting residential energy consumption: single household perspective. In: Proceedings of 17th IEEE International Conference on Machine Learning and Applications, ICMLA 2018, pp. 110–117 (2019)
4. Sun, Y., Gu, W., Lu, J., Yang, Z.: Fuzzy clustering algorithm-based classification of daily electrical load patterns. In: 2015 12th International Conference on Fuzzy Systems and Knowledge Discovery, FSKD 2015, pp. 50–54 (2016)
5. Yildiz, B., Bilbao, J.I., Dore, J., Sproul, A.B.: Recent advances in the analysis of residential electricity consumption and applications of smart meter data. Appl. Energy 208, 402–427 (2017)
6. Liu, Y., Wang, W., Ghadimi, N.: Electricity load forecasting by an improved forecast engine for building level consumers. Energy 139, 18–30 (2017)
7. Zhao, H.X., Magoulès, F.: A review on the prediction of building energy consumption. Renew. Sustain. Energy Rev. 16(6), 3586–3592 (2012)
8. Baquero, M., Quesada, F.: Eficiencia energética en el sector residencial de la Ciudad de Cuenca, Ecuador. Maskana 7(2), 147–165 (2016)
9. Gabriel, D., Urgilés, T.: Universidad politecnica salesiana sede cuenca facultad de ingenierias carrera de ingeniería eléctrica
10. UEM en Funcionamiento – Ministerio de Educación. https://educacion.gob.ec/uem-en-funcionamiento/. Accessed 24 Sept 2019
11. Ministerio de Electricidad y Energía Renovable – Ente rector del Sector Eléctrico Ecuatoriano. http://historico.energia.gob.ec/. Accessed 24 Sept 2019
12. Rosero-Montalvo, P.D., et al.: Wireless sensor networks for irrigation in crops using multivariate regression models. In: 2018 IEEE Third Ecuador Technical Chapters Meeting (ETCM), pp. 1–6 (2018)
13. Huang, X., Wang, S.: Prediction of bottom-hole flow pressure in coalbed gas wells based on GA optimization SVM. In: Proceedings of 2018 IEEE 3rd Advanced Information Technology, Electronic and Automation Control Conference, IAEAC 2018, pp. 138–141 (2018)
14. Ye, X., Wu, X., Guo, Y.: Real-time quality prediction of casting billet based on random forest algorithm. In: Proceedings of the 2018 IEEE International Conference on Progress in Informatics and Computing, PIC 2018, pp. 140–143 (2019)
15. He, L., Song, Q., Shen, J.: K-NN numeric prediction using bagging and instance-relevant combination. In: Proceedings - 2nd International Symposium on Data, Privacy, and E-Commerce, ISDPE 2010, pp. 3–8 (2010)

Proposal for the Implementation of MLP Neural Networks on Arduino Platform

Kevin Andrés Suaza Cano⬤, Jhon Freddy Moofarry⬤, and Javier Ferney Castillo Garcia(✉)⬤

Universidad Santiago de Cali, Calle 5 No. 62-00 Barrio Pampalinda, Cali, Colombia
javier.castillo00@usc.edu.co

Abstract. This paper presents implementation MLP artificial neural networks on embedded low-cost microcontrollers that can be dynamically configured on the run. The methodology starts with the training process, goes through the codification of the neural network into the microcontroller format, and finishes with the execution process of the embedded NNs. It is presented how to compute deterministically the memory space require for a certain topology, as well as the required fields to execute the neural network. The training and verification was done with Matlab and programming with a IDE Arduino compiler. The results show statistical and graphical analysis for several topologies, average execution times for various transfer function, and accuracy.

Keywords: Artificial Neuronal Networks (ANN) · Multilayer Perceptron (MLP) · Matlab

1 Introduction

In the last years, the number and variety of applications using artificial neural networks have been increasing significantly thanks to their adaptability and robustness. Example of those applications is data fitting, pattern recognition, clustering, system identification, adaptive filters and inverse problems. In the field of control systems, the neural networks are suitable for complex process, multivariable systems, nonlinear systems, unknown models, hard-coupled systems, time-variant and dynamic control systems [1–3], here are even more software tools for off- and on-line training, simulation and validation making easier the work with NN. Most of the applications are implemented in commodity PCs. However, the implementation on embedded systems is still in discussion because of the reconfigurability and demanding computing resources, especially for low-cost applications.

This paper presents a methodological proposal for implementing MLP neural networks on low-cost embedded systems like microcontrollers, where the topology can be resized and implement different kind of activation functions.

The paper is organized as follows. First, Sect. 2 presents some previous works about the implementation of NN on embedded systems. Then, Sect. 3 explains basics of MLP networks and presents some software tools to work with neural networks. Next, Sect. 4 presents the methodology to implement MLP networks on embedded systems. Section 5

© Springer Nature Switzerland AG 2020
M. Botto-Tobar et al. (Eds.): ICAT 2019, CCIS 1194, pp. 372–385, 2020.
https://doi.org/10.1007/978-3-030-42520-3_30

presents the fitting of the peaks function as case of study. Section 6 presents results of some implementations on an Arduino Uno microcontroller running neural networks trained with Matlab. Finally, Sect. 7 draws some conclusions and purposes further work.

2 Previous Works

Despite the inheritable parallel nature of the neural networks, the most implementations use sequential single-core processors. It leads to establish a sampling time to let the processor carries out all the intra- and interlayer computations to get a valid output [4].

Computation can be sequentialized, the implementation of neural networks results relative easy on conventional processors (like those in microcontrollers) with respect to other traditional algorithms.

In 1997, a microcontroller system was implemented based on the architecture of an MLP to linearize a sensor. This research showed the technical feasibility for such implementations [5]. In 1999, a cooperative work presented the advantages and drawbacks when implementing neural network on microcontroller systems against implementing fuzzy systems 1997 [6]. In 2001, [7] presented new comparisons showing more advantages of neural networks over fuzzy systems. In [8], the author references different hardware implementations, specifically on microcontroller platforms using MLP neural networks.

3 Theoretical Framework

MLP Neural Networks. The artificial neural networks (ANN) are parallel systems for information processing, bio-inspired in the way as biological neural networks in the brain process.

The NNs have a wide number of different topologies. The most common is the multilayer perceptron (MLP) and the radial basis function (RBF) [4, 9]. This paper will focus in the MLP topology.

The MLP consists in an input layer, multiple hidden layers, and an output layer. Each node in the net is called a neuron (see Fig. 1), that includes a transfer function and an activation function.

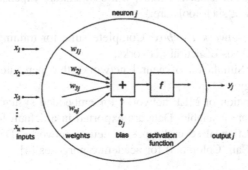

Fig. 1. Neural network node

The transfer function is the dot product between the inputs X and the weights W. Equation 1 describes the how the output is computed using an activation function f, whose input argument depends on the result of transfer function and the bias b.

$$y = f(x \cdot w + b) \tag{1}$$

If several nodes are connected in parallel, then it is denominated a layer. Figure 2 shows an example with one input layer, one hidden layer and one output layer, each with three neurons.

Some activation functions are for instance: tangential sigmoidal (*tansig*), the linear (*purelin*), and the hard-limits (*hardlim*).

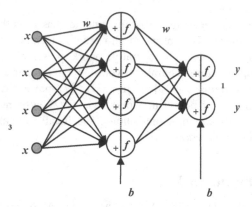

Fig. 2. Example of MLP.

The adaptation of a neural network is called the training. This is the process that changes the weights and biases according with some training parameters using a training function, in order to get the best fit with the training data. The Levenberg-Marquardt backpropagation training method is a common method to train MLPs.

Tools for Training and Generation of Neural Networks
Several tools are available in the market for the development with neural networks. Some of them are free or academic tools, while others are expensive but reliable and supported. Examples of such tools are:

- *Matlab's Neural Network Toolbox:* Complete suite for training, simulation, visualization, and analysis of neural networks.
- *NeuralGraphics*: Simulation of neural networks that generates an output of the simulated topology.
- *MLP Edit:* Generation of MLP networks for embedded systems. Training configuration and tests are available. Data are exported in a default format.
- *UV-SRNA*: Simulation of several types of neural networks. Developed at Universidad del Valle (Cali, Colombia) for academic purposes [4].

4   Methodological Proposal

For implementation of the *NN*, it has been divided in two parts. The first part consists in the steps to generate a neural network using a software tool. The second part describes the steps for the implementation and test on the embedded system.

The design flow for the generation of neural networks using a software tool is shown in Fig. 3.

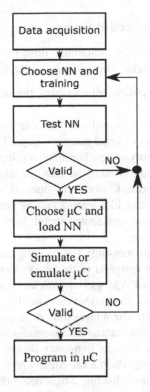

Fig. 3. Methodology's diagram for the NN implementation on a µC.

4.1   Methodology to Generate a Neural Network

The first step when modeling a system with neural networks is the data acquisition. It can be done evaluating the target function or sampling the system. Once a data collection is available, then it is used to train a neural network with the chosen topology (number of layers, delays, feedback).

It is very important to avoid preprocessing functions on the inputs and the outputs that cannot be executed in the embedded system. For instance, some tools scale inputs and outputs and then train the networks producing weight and bias matrices that need such scaled data when are evaluated. It implies that scaling parameters should be given to the embedded system as well as the scaling method must be hardcoded.

The activation functions are a very important issue for the hardware implementation, because some of them are very complex and time consuming, affecting the performance. Also, the accuracy can be affected when approximation methods are used to fit the nonlinearities, as well as loss of precision due to the sequence of rounding operations [10].

Once the network satisfies the requirements of accuracy, size; then the network parameters are encapsulated within a single array together with other parameters like the sample time. An example of such array will be presented in the next subsection.

4.2 Methodology for the Embedded System

The diagram in Fig. 4 presents the functional flow purposed for the execution of dynamic-resizable neural networks. These steps are explained as follows:

- **Initial configuration:** It initializes all variables, registers, modules and peripherals in the microcontroller.
- **Load default NN:** A default neural network can be loaded in RAM from the program memory or from an external memory, say a Flash memory or a SD-Card.
- **Update inputs and preprocessing:** The network inputs can adopt the value of external signals like those converted by an ADC or taken from the GPIO, or can be internal variables, e.g. feedback from the network outputs, delayed inputs in regressors. Some preprocessing like scaling can be performed in this step.
- **NN execution:** The dynamic-resizable neural network structure is evaluated according to the current inputs.
- **Update outputs and post-processing:** The recently computed network outputs are post-processed and then are assigned to the real outputs. For instance, the outputs can vary the duty cycle in a PWM module or can be out through the GPIO ports.
- **Idle state:** For real-time tasks, the system stays in idle state while the sample period *Ts* has still not been reached. The temporization *Ts* can be performed by the timer modules in the microcontroller, or can be externally done by real-time clocks and oscillators. The field *Ts* can define a frequency divider factor or a code to identify one predetermined sample periods in a set. Meanwhile, the system can be requested by the host in order to start downloading a new network configuration. If no request was received, the system goes back to step 3: *Update inputs and preprocessing.*
- **Set outputs to safe state:** If the system accepts the request for a new neural network, then it is strongly recommended to set all the outputs and other modules to a safe state, avoiding possible damages caused by real-time violation, by communication problems or by inconsistencies in the new network.
- **Load new NN:** The new neural network is load from the host into the data RAM. The communication can be done by RS-232 serial port, USB port, Ethernet, SPI, etc. The frame can contain information about the number of data to be send and other important information to keep the data integrity. Once the network has successfully loaded, the system restarts from step 3.

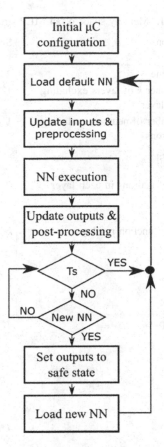

Fig. 4. Functional flow of the processor.

Most important thing when talking about a dynamic-resizable neural network lies in the memory space. While the input and output interface to the network is hardcoded, and therefore fixed, the core can be flexible adopting different hidden layers, number of neurons and transfer functions.

Table 1 presents a distribution of the linear memory space used to allocate an MLP structure that indicates the number of layers and neurons, their activation functions, and the values of all weights and biases. It also reserves space for the network inputs, as well as for the neurons outputs. A deterministic expression to compute the size of each item based on the data type is also given.

The variable that allocates the neural network should be declared in the source code as a byte vector so long as the memory lets it. Table 2 presents the size of some neural networks.

Table 1. Memory space o fan MLP structure.

Field	Description	Size (bytes)	Data type
Ts	Sample time	2	uint16
totalLayers = k	Total number k of layers excluding the input layer	1	uchar
totalNeurons = n	Total number of neurons excluding input neurons $$n = \sum_{i=1}^{k} m[i]$$	1	uchar
m[0] ... m[k]	Number of neurons in each layer	k + 1	uchar
f[1] ... f[k]	Activation function per layer	k	uchar
x[1] ... x[m0]	NN inputs	4*m[0]	float32
L[1].y[1] ... L[1].y[m1] ... L[k].y[1] ... L[k].y[m[k]]	Layer outputs	4*n	float32
L[1].b[1] ... L[1].b[m1] ... L[k].b[1] ... L[k].b[m[k]]	Biases	4*n	float32
L[1].w[1, 1] ... L[1].w[m[1], m[0]] ... L[k].w[1, 1] ... L[k].w[m[k], m[k − 1]]	Weights	$4 \sum_{i=1}^{k} (m[i] \cdot m[i-1])$	float32

Table 2. Size of different topologies.

Configuration	Size [bytes]	Example
[1, 2]	31	Basic perceptron
[100, 1]	815	Adaptive filter (tap 100)
[12, 10, 10, 1]	1147	12 inputs, 1 output and 2 hidden layers with 10 neurons each
[8, 8, 8, 8, 8]	1325	8 inputs, 8 outputs and 3 hidden layers with 8 neurons each
[16, 16]	1223	16 inputs and 16 outputs

At this point, it was designed a network structure in data RAM and now it has to be executed properly. Figure 5 show the C-code that carries out a sequential computation of each neuron in each layer according to its bias, weights and inputs.

The key is the use of pointers since the position of all data varies with the network configuration. Only two pointers remain always in the same position: the *totalLayers* and the *totalNeurons* pointers. Starting from their values, the function locates the beginning of the embedded vectors.

```
ALGORITHM 1. EXECUTION OF THE DYNAMIC NEURAL NETWORK.
float* executeNeuralNetwork(void){

    char layer = 0;
    char neuron = 0;
    char input = 0;

    // Initialize pointers
    unsigned char* n_inputs = (unsigned char*)
        (totalNeurons + 1);
    unsigned char* n_outputs = (unsigned char*)
        (totalNeurons + 2);
    unsigned char* f = n_outputs + *totalLayers;
    float* x = (float*) (f + *totalLayers);
    float* y = x + *n_inputs;
    float* b = y + *totalNeurons;
    float* w = b + *totalNeurons;

    // Pointer to the first neuron in the output layer
    float* last_y;

    // Main loop
    for(layer = 0; layer < *totalLayers; layer++){
        last_y = y;
        for(neuron = 0; neuron < *n_outputs; neuron++){
            y[neuron] = b[neuron];
            for(input = 0; input < *n_inputs; input++){
                y[neuron] = y[neuron] + (w[input])*(x[input]);
            }
            w = w + *n_inputs;
            y[neuron] = transferFunction(f[layer], y[neuron]);
        }
        x = x + *n_inputs;
        y = y + *n_outputs;
        b = b + *n_outputs;

        n_inputs++;
        n_outputs++;
    }
    return last_y;
}
```

Fig. 5. Algorithm for execution of the dynamic neural network.

After all pointers have been correctly initialized, then the main loop starts with the first layer. The inputs of this layer come from the X vector. It executes the dot product between the X vector and the corresponding weights of each neuron. Next the corresponding bias is added and then the activation function of the current layer is evaluated. All partial results in this process are stored in corresponding neuron output in the network structure.

Subsequently, when all neurons in the current layer have been updated, it proceeds with the next layer. In this case, the inputs will be the outputs of the predecessor layer.

At the end, the function returns a pointer to the first output in the output layer, so that the *updateOutputs()* function can locate the output values from the network.

In addition, when working with time-delayed inputs and regressors, it is recommended to implement circular buffers for each delayed input, which should be so long as the maximum delayed expected. The connection between a determined network input and a delayed input can be specified by an additional field in the network structure. Figure 6 shows how two circular buffers with maximum size 12 are used to feed the network inputs with a predefined format [12 0 6 2 1] indicating the delay of each input for the current network configuration.

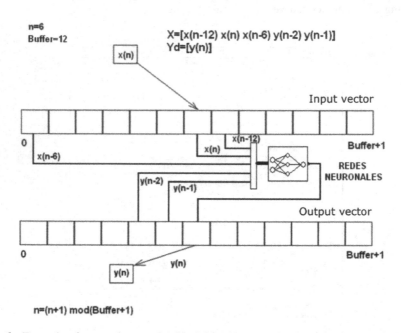

Fig. 6. Example of a neural network with delayed inputs. The network output is fedback.

4.3 Methodology for Loading a NN Structure in the Embedded System

Now that we know an appropriate way to build and run a neural network structure, then it is time to define how to export it to the embedded system (see Fig. 7).

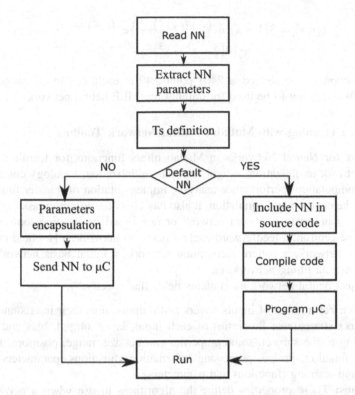

Fig. 7. Methodology for downloading the default and new neural networks into the embedded system.

5   Case of Study: Peaks Function

5.1   Target System: The Two-Variable Function Peaks

The function Peaks is a scalar, continuous, nonlinear function depending on two variables and is described by (2). Figure 8 shows a 3D plot of this function with a grid size of *1/8*.

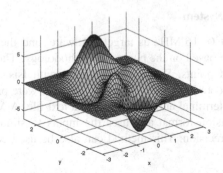

Fig. 8. Peaks function for 48 divisions in each (*x*, *y*) axis.

$$z(x, y) = 3(1 - x)^2 e^{-(x^2 + (y+1)^2)} - \frac{1}{3}e^{-((x+1)^2 + y^2)} \dots$$
$$-10(\frac{x}{5} - x^3 - y^5)e^{-(x^2 + y^2)} \dots \tag{2}$$

The function was evaluated at 2401 points, 49 in each axis in the range $[-3, 3]$. These points are going to be used for training the MLP neural network.

5.2 Offline Training with Matlab's Neural Network Toolbox

The toolbox for Neural Networks in Matlab offers functions for learning, training, plotting, network initialization, weight and bias initialization, topology configuration, net input computation, performance analysis, implementation of transfer functions, as well as for the creation and simulation. It also has GUIs for common tasks and demos.

The user can create a custom network or one based on existing topologies. For instance, a perceptron, a feedforward backpropagation network, a Hopfield network, a probabilistic network, a pattern recognition network, a radial basis network, a self-organizing map, a fitting network, etc.

The object neural network *nn* contains fields that specify:

- **Architecture:** number of inputs, layers and outputs; thus their interconnections.
- **Subobject structures:** Properties of each input, layer, output, bias and weights. According to the subject, some properties can be size, range, positions, topology, transfer function, pre/postprocessing information (functions, parameters and settings), and learning (functions and parameters).
- **Functions:** These properties define the algorithms to use when a network is to adapt, is to be initialized, is to have its performance measured, or is to be trained.
- **Weight and Bias values:** The *net.IW*, *net.LW* and *net.b* properties define the input weights, the interlayer weights and biases. This are cell matrices whose elements *(i, j)* define the weight or bias between the input/layer *i* with the layer *j*.
- **Creating network:** An MLP network is created by using the *newff()* function, passing the input matrix containing all the training points, and another matrix for the desired outputs. Other arguments are the vector that describes the number of neurons per hidden layer, and a cell containing strings indicating the transfer function for each hidden layer and for the output layer.

5.3 The Embedded System

Using an Arduino Uno @ 16 MHz as target hardware, and the IDE Arduino, the C source code was written based on the purposed methodology. The test code performs a swept over all the grid points. Simulations running with Isis Proteus were used to debug the system. The coordinates *(x, y)* and the output *z* were printed in hexadecimal format over the virtual terminal and then saved to a text file. A Matlab script casts the data to the single-precision format and uses the inputs to simulate the target neural network with double precision. This method can also be used combined with debugging on-system.

6   Performance and Error Analysis

Although the use of floating-point arithmetic is very accurate, its implementation is compiler dependant and is usually limited to single precision due to memory requirements and computational complexity that yields in longer execution times. It introduces additional errors that should be evaluated before the final implementation on the embedded system.

Execution Time and Memory Space

The execution time depends strongly of the operating frequency, but also of the cycles per instruction and the number of them. It becomes a little heuristic because of the architectural dependency; therefore, some tests should be run to estimate the maximum execution time of certain topologies in order to guarantee the real-time execution restricting the sample time Ts. Table 3 shows average execution time for the activation functions.

Table 3. Implementation of the activation functions.

Activation function	Code	Expression	Average execution time [μs]
Tansig(a)	0	tanh(a)	170.0
Purelin(a)	1	a	15
Logsig(a)	2	$\frac{1}{1+exp(-a)}$	100.0
Hardlim(a)	3	if(a $\geq$ 0.0) return 1.0; else return 0.0;	8.0
Hardlims(a)	4	if(a >= 0.0) return 1.0; else return -1.0;	8.0
Radbas(a)	5	exp($-$a*a)	172.0

6.1   Error Analysis

Table 4 shows the comparison between the hardware and software computations in a grid of 48 division for absolute error and relative error.

Table 4. Error analysis

Error	Mean	Variance	Standard deviation
Absolute	$-1.9144e-006$	$4.4471e-010$	$2.1088e-005$
Relative	$2.0076e-004$	$1.3519e-006$	$1.1627e-003$

7 Conclusion and Discussions

This paper has purposed a methodology for the implementation of dynamic-resizable MLP neural networks on low-cost microcontroller-based embedded systems. It has explained the whole procedure from the offline training starting with the data acquisition until building a running neural network structure in a microcontroller.

A study was presented fitting the function Peaks using the Matlab's Neural Network Toolbox. One MLP neural network structure was run and tested on an Arduino Uno, IDE Arduino was used to write and build the application C code.

Performance and error analysis were performed to validate the implementation, giving a relative error of order −4 when comparing with software results. The average execution time was computed for the activation functions and some network structures using Isis Proteus.

The computational cost of single-precision floating-point operations in commodity microcontrollers is affordable for systems that require low sample frequencies. For the example presented, a safe simple frequency is found below 100 Hz for 50 neurons with the tansig activation function. This is considered well enough for a wide range of industrial applications.

The tendency towards online neural networks training over low-cost embedded systems does still not offer enough robustness and reliability like does offline training. The main cause is that the user running offline training on PCs can validate much more data and perform more analysis than an embedded trainer. Also, the user enjoys other advantages like GUIs and huge data storage. Hence, this work offers engineers the method to download and run their neural networks designs on low-cost embedded systems in a fashion-like online soft programming.

The future work is oriented to explore other neural network topologies out of the MLP topology. Besides, not only neural networks can be implemented with this template, but also fuzzy systems. The team is currently working on the implementation of a neural-based inverse-control to control a nonlinear time-variant multivariable temperature plant with a low-cost microcontroller applying the presented methodology.

Acknowledgments. This work was supported by *Dirección general de Investigaciones* from Universidad Santiago de Cali through the project 829-621118-135 called *"Desarrollo de una plataforma robótica para estimulación cognitive y física en niños con discapacidad"*. We thank Aleksandra Obeso-Duque and Álvaro J. Caicedo-Beltrán for starting this research, the way was least difficult thanks for them.

References

1. Callinan, T.: Artificial Neural Network identification and control of the inverted pendulum. Dublin City University (2003)
2. Kim, I., Fok, S.: Neural network-based system identification and controller synthesis for an industrial sewing machine. Int. J. Control Autom. Syst. **2**(1), 83–91 (2004)
3. Medina, J., Parada, M.: A neural network-based closed loop identification of a magnetic bearings system. In: Proceedings of ASME Turbo Expo (2004)

4. Bravo Caicedo, E.F., López Sotelo, J.A.: Una aproximación práctica a las redes neuronales artificiales. Universidad del Valle (2009). ISBN 9789586707671
5. Dempsey, G., Alt, N.: Control sensor linearization using a microcontroller-based neural network. In: IEEE International Conference on Systems, Man, and Cybernetics 'Computational Cybernetics and Simulation', pp. 3078–3083 (1997)
6. Wilamowski, B.M., Binfet, J.: Do fuzzy controllers have advantages over neural controllers in microprocessor implementation. In: Proceedings of the Second International Conference on Recent Advances in Mechatronics, ICRAM 1999, pp. 342–347 (1999)
7. Binfet, J., Wilamowski, B.M.: Microprocessor implementation of fuzzy systems and neural networks. In: IEEE International Joint Conference on Neural Networks, IJCNN 2001, pp. 234–239 (2001)
8. Cotton, N.: A neural network implementation on embedded systems. Tesis Doctoral, Auburn University (2010)
9. Hu, Y., Hwang, J.N.: Handbook of Neural Network Signal Processing. CRC Press LLC, Boca Raton (2000). ISBN 0-8493-2359-2/01/$0.00 + $1.50
10. Cotton, N., Wilamowski, B.M.: Compensation of nonlinearities using neural networks implemented on inexpensive microcontrollers. IEEE Trans. Ind. Electron. 58(3), 733–740 (2011)

Design of an Intelligent Irrigation System Based on Fuzzy Logic

Fabián Cuzme-Rodríguez[✉], Edgar Maya-Olalla,
Leandro Salazar-Cárdenas[✉], Mauricio Domínguez-Limaico,
and Marcelo Zambrano Vizuete

Carrera de Ingeniería en Telecomunicaciones, Universidad Técnica del Norte,
Av. 17 de Julio 5-21 y Gral. José María Córdova, 100105 Ibarra, Ecuador
{fgcuzme,ljsalazar}@utn.edu.ec

Abstract. The present study contributes to the improvement of the processes of conventional agriculture that are still being carried out independently of the Information and Communication Technologies, which show shortcomings in the forms of irrigation carried out suffering an impact on the use of the water supply. From this point, precision agriculture becomes indispensable to improve the processes of agricultural production processes, allowing adequate management of agricultural plots supported by the use of technology to estimate, evaluate and understand the variations of the variables involved and offer quantities of water necessary for cultivation. This analysis covers the design and construction of an intelligent irrigation system based on fuzzy logic applied in vegetable crops. The fundamental mechanism of this system is to realize the control of irrigation through a scheme consisting of two modules, the data acquisition module, and the decision-making module. Considering that the section with the highest degree of responsibility is the integration of fuzzy logic as a control mechanism and that is part of the decision-making module. To achieve this, meteorological variables such as precipitation, temperature, the humidity of the environment and soil moisture are evaluated, which are considered as input variables for the diffuse system. The operation of the prototype is crystallized in a functional graphical interface and tested in two scenarios, where its efficiency in the proper use of the water supply is demonstrated.

Keywords: Fuzzy logic · Intelligent system · Irrigation system

1 Introduction

Agriculture is an activity that continues to be the main economic support in some countries, such as China that leads the ranking of countries with the highest production of vegetables worldwide with 39%, followed by India with 10% and the USA in third place with 4% [1]. In surveys carried out by the National Institute of Statistics and Censuses in relation to the surface and continuous agricultural production, they show that the 11.7 million hectares are being used for agricultural purposes, considering that 11.62% is formed by transitory crops, these crops are considered as the first source of economic income, whose contribution to GDP is 8% [2]. The transitory products are

© Springer Nature Switzerland AG 2020
M. Botto-Tobar et al. (Eds.): ICAT 2019, CCIS 1194, pp. 386–399, 2020.
https://doi.org/10.1007/978-3-030-42520-3_31

distinguished by vegetables, vegetables, and certain cereals, which are grown in the coast, mountains, and east of Ecuador; The main factor to enhance the production of these products is through efficient use of water.

In the production of crops (vegetables, vegetables, cereals) it is essential that an adequate irrigation technique be used, mainly characterized by distributing the amount of water and nutrients to the crop, these irrigation techniques may be accompanied by manual processes, automated and automated, depending on the use of Information and Communication Technologies (ICTs) in the application environment [3–5]. Some studies indicate that there is wear on water use due to the use of obsolete irrigation techniques, there being an uncontrolled waste of water [6], causing in some cases the generation of bacteria that directly affect an adequate growth of the plant. In this case, a drip irrigation study may cause waterlogging because the irrigation time is not adequately controlled [3, 6], preventing the increase in crop productivity and generating an obvious waste of water.

Several projects have been developed to improve the process of an irrigation system as in the case of [7], where an intelligent irrigation system is proposed that uses IoT and fuzzy Logic controller; another study is the one developed by [8], which proposes a low-cost intelligent irrigation system based on IoT; and finally the one developed by [5], which proposes an intelligent drip irrigation system and fertirrigation through wireless sensor networks (WSN). Although there are many other studies related to optimizing irrigation processes, manual techniques basically act similarly to automatic ones, however, manual or partially technified techniques do not integrate any level of automation or intelligence in decision making. Irrigation, as a livelihood element in production, has usually not been linked to the evolution of social, productive and environmental dynamics. For this reason, the irrigation techniques that were built for several years are being displaced in the current agricultural production systems [9].

Over the years, a fundamental aspect of conventional irrigation techniques that are applied to crops is the problems they present, including water management [5], which adds to other factors such as the adequate supply of fertilizers according to the soil conditions. Consequently, this can lead to poor production that affects low income for farmers [10]. In addition, there is concern about the dependence of water required on irrigation, a study reveals that 90% of the world's freshwater is retained in irrigation, and with the demographic increase each year, there is a need to expand the agricultural field and with it the use of water rises [7, 11]. Under this circumstance, the challenge is born to design and manage an adequate model of redistribution and integral management of irrigation systems, which favors small, medium and large farmers precisely, these aspects are achieved with the use of ICTs.

The evolution of ICTs has generated a successful boost in WSN wireless technologies, which allow the installation of an infrastructure that involves certain components such as sensors, actuators, computing and connectivity [12]. The purpose of these technologies in agricultural environments is to improve food production management by offering some advantages of constant monitoring of environmental parameters, data processing, uniform distribution of measurements, irrigation programming and information representation. This, in the end, makes it possible to sustainably maintain and increase agricultural production in a sustainable way, reducing the environmental impact established in intelligent irrigation systems [13].

An intelligent system is one that is designed with one or several artificial intelligence techniques [14], which are framed in: genetic algorithms, neural networks, robotics and fuzzy logic; whose peculiarities and functionalities resemble those of a human being, which allow systems to reason and act in situations as distinguished by a person. The system consists of a series of instructions, computational algorithms that allow you to generate a fast, automatic, immediate and more efficient response [15]. In this context, the present study is based on the use of fuzzy logic as a technique to be implemented in the irrigation system.

Diffuse logic is an efficient tool when implementing it in control systems, it is used in different areas of engineering such as chemical industry, robotics, biological processes, automation based systems; this with the purpose of replacing the repetitive activities that the man carries out, since sometimes these usually appear difficult and risky to control. It is considered as the way in which human beings perceive their surroundings in the real world, where situations or uncertain moments arise, for example, various criteria on the process of drying and wetting of clothing, speed of a vehicle on a road or the temperature prevailing in a room. Fuzzy logic is commonly expressed in a vague, distorted, inaccurate, fuzzy language and its conceptualization varies from person to person; In order to carry out all the timely treatment of blurred information, the elementary structure that constitutes the control design with fuzzy logic establishes certain stages such as fuzzy sets, membership functions, fuzzy operations, merger, fuzzy rules, inference, aggregate and defuzzification [16].

In the following sections, it is distributed as follows: Sect. 2 covers the materials and methods where the modules in which the system is divided are explained, as well as the definition of fuzzy logic for this study. In Sect. 3, it deals with the results obtained and discussion of the work and finally the conclusions of the study.

2   Materials and Methods

This section describes the architecture of the intelligent irrigation system, which is composed of two modules that try to suppress complexity when building the system. The first module used for data acquisition, and the second module of management and decision making the diffuse controller.

2.1   Data Acquisition Module

The data acquisition segment is represented by a WSN network with the purpose of obtaining and transmitting data on the temperature, humidity, and precipitation variables; In addition to integrating actuators that allows me to operate an electrovalve automatically that activates the water supply to the plot, and also controls the volume of the water with the help of a flowmeter. The hardware used in this module is an Arduino microcontroller that hosts sensors, actuators, and communication (tx/Rx) Fig. 1.

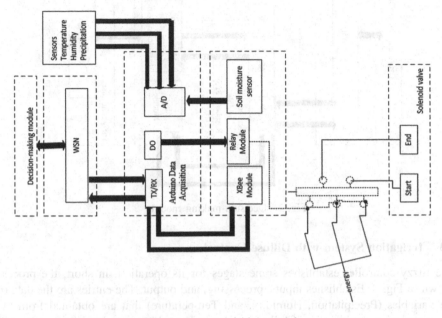

Fig. 1. Data acquisition module

2.2 Decision-Making Module

The elements that make up the hardware of the decision-making module are supported on a Raspberry Pi 3 board, where we find two communication systems (Tx/Rx), the first one through Xbee that links me to the module for acquiring data and the second one via WIFI that allows me to connect remotely from a terminal device to the Raspberry; the database where the information obtained from the data acquisition module is stored; and finally the programmable interface of API applications that is linked to the management and control system of the irrigation system and includes the fuzzy controller Fig. 2.

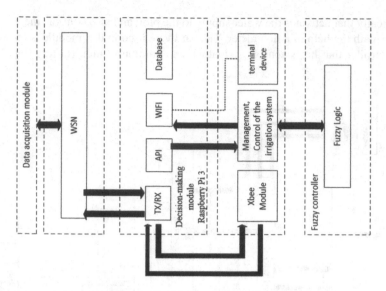

Fig. 2. Decision-making module

2.3 Irrigation System with Diffuse Control

The fuzzy controller establishes some stages for its operation, in short, the process shown in Fig. 3. Establishes inputs, processing, and output. The entries are the data of the variables (Precipitation, Humidity, and Temperature) that are obtained from the data acquisition module, specifically from the sensors; the processing interacts the API with the diffuse system that is composed of rules that will be verified to match any of them; the stable output will close the scope of the irrigation and the required time that can be nothing, very little, little, regular or enough. If the system detects that the plot needs water, the information will be transferred to the data acquisition module to activate the solenoid valve that will supply the water according to the data generated in the diffuse controller.

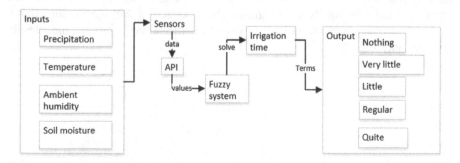

Fig. 3. Scheme of the fuzzy controller

The API is built in Python language exploiting some of its essential modules for the operation of the equipment, a collection of fuzzy logic algorithms called SCikit fuzzy is used [17], useful for solving environments where fuzzy logic is required.

In Fig. 4 we can see the flow chart of the operation of the system included in the diffuse controller.

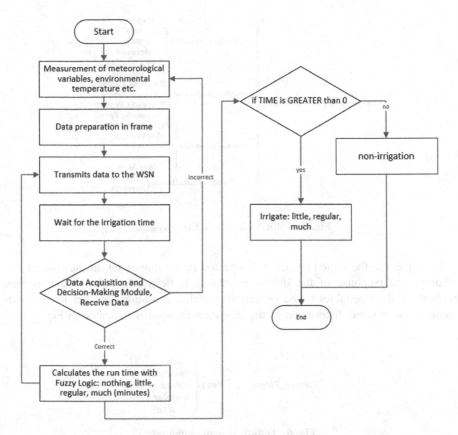

Fig. 4. Flowchart of the fuzzy controller

Within the fuzzy controller, some stages are set out below:

Fuzzification Phase. The inputs of the diffuse controller are given by the following meteorological factors: Temperature and Humidity of the Environment, Precipitation, and Humidity of the soil, which have been classified as diffuse input sets. The values of these factors are acquired by the data acquisition module and sent via the WSN network to the decision-making module, the time it takes to recover these values depends on the conditions present in the test environment, such as distance between modules, meteorological phenomena, obstacles and the physical properties of the hardware, this forces the API programming logic to work properly under these considerations.

The first diffuse control process is made up of the input sets, which have been segmented due to the operator's level of experience generating subsets; It is the stage known as blurred distribution and is the key that allows identifying the vague term conducive to each subset. Input subsets belonging to the fuzzy controller of this prototype are indicated in Fig. 5.

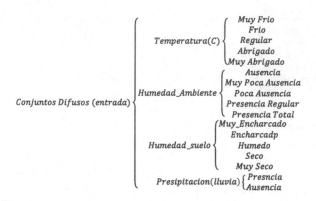

Fig. 5. Diffuse sets of the fuzzy system

Subsequently, the same procedure is repeated to the output set, in the case of this prototype the response of the diffuse controller is destined to the irrigation time required by the vegetables based on certain weather conditions present in the environment. The blurred distribution of the irrigation time set is indicated in Fig. 6.

$$Tiempo\_Riego(segundos) \begin{cases} Nada \\ Muy\ Poco \\ Poco \\ Regular \\ Mucho \end{cases}$$

Fig. 6. Diffuse system output set

Knowledgebase. It is based on the experience provided by a technical farmer with the particularity of expressing himself in an inaccurate language, which will later be transformed into fuzzy sets, this event is known as the basis of rules [16]. The number of rules can be very extensive and complicates the evaluation process, 30 rules have been considered for this case, and for each rule mathematically it is evaluated by the minimum method proposed by Mandami.

There are 30 rules proposed by this fuzzy controller and are extracted from a universe of rules, the format of the fuzzy rules takes on the aspect as specified in Table 1.

Table 1. Rule base for the intelligent irrigation system

N	Rules base
1	IF TA is MF AND HA is TP AND HS is E AND PR is AS then TR is N
2	IF TA is MF AND HA is TP AND HS is E AND PR is PS then TR is N
3	IF TA is MF AND HA is TP AND HS is S AND PR is PS then TR is N
4	IF TA is MF AND HA is TP AND HS is MS AND PR is PS then TR is N
5	IF TA is MF AND HA is TP AND HS is MS AND PR is AS then TR is R2
6	IF TA is F AND HA is PA AND HS is E AND PR is AS then TR is N
7	IF TA is F AND HA is PA AND HS is E AND PR is PS then TR is N
8	IF TA is F AND HA is PA AND HS is MH AND PR is AS then TR is N
9	IF TA is F AND HA is PA AND HS is MH AND PR is PS then TR is N
10	IF TA is F AND HA is PA AND HS is H AND PR is AS then TR is N
11	IF TA is F AND HA is PA AND HS is H AND PR is AS then TR is N
12	IF TA is F AND HA is PA AND HS is S AND PR is PS then TR is N
13	IF TA is R1 AND HA is PA AND HS is MH AND PR is AS then TR is N
14	IF TA is R1 AND HA is PA AND HS is MH AND PR is PS then TR is N
15	IF TA is R1 AND HA is PA AND HS is H AND PR is AS then TR is N
16	IF TA is R1 AND HA is PA AND HS is H AND PR is PS then TR is N
17	IF TA is R1 AND HA is PA AND HS is S AND PR is AS then TR is R
18	IF TA is R1 AND HA is PA AND HS is S AND PR is PS then TR is N
19	IF TA is R1 AND HA is PA AND HS is MS AND PR is AS then TR is M
20	IF TA is R1 AND HA is PA AND HS is MS AND PR is PS then TR is N
21	IF TA is A AND HA is AS AND HS is E AND PR is AS then TR is N
22	IF TA is A AND HA is AU AND HS is E AND PR is PS then TR is N
23	IF TA is A AND HA is AU AND HS is MH AND PR is AS then TR is N
24	IF TA is A AND HA is AU AND HS is H AND PR is AS then TR is P
25	IF TA is A AND HA is AU AND HS is H AND PR is PS then TR is N
26	IF TA is A AND HA is AU AND HS is S AND PR is AS then TR is R2
27	IF TA is MA AND HA is AU AND HS is MH AND PR is PS then TR is N
28	IF TA is MA AND HA is AU AND HS is H AN D PR is AS then TR is P
29	IF TA is MA AND HA is AU AND HS is S AND PR is AS then TR is M
30	IF TA is MA AND HA is AU AND HS is MS AND PR is AS then TR is M

These rules can be modified based on the experience and knowledge of the irrigation expert. The next step is to generate the rules with the "Scikit Fuzzy" module, where all the rules have the peculiarity of having a similar logical operation considered as AND (intersection); The solution to this type of operation is through the Min method and is defined as an intersection, however, the problem exists by having 4 subsets, therefore, this issue was solved by applying one of the fuzzy logic properties, which known as distributive and is represented by Eqs. 1 and 2.

$$\Delta i = A \cap B \cap C \cap D = (A \cap B) \cap (C \cap D) \tag{1}$$

$$\Delta = \min(\min(uA, uB), \min(uC, uD)) \tag{2}$$

Where A, B, C, D are the fuzzy input sets and Δi is the result of evaluating each rule.

Interference Motor. Based on the fuzzy logic math, the fuzzy inferences engine is added [16], where two types of elements are established: data (facts or evidence) and knowledge (the set of rules stored in the knowledge base), so the inference engine uses both to obtain new conclusions or facts. The mathematical equation used to evaluate each rule of the system is by the method of minimum (Min), as shown in Eq. 3.

$$\nabla j = min(\Delta i , Yi) \tag{3}$$

Where:

∇j it is the output of the fuzzy controller inference motor that throws the operation of each rule and the value is expressed in the fuzzy term.

Δi it is the result of evaluating each fuzzy rule.

Yi   is corresponding to each fuzzy rule

Defuzzification. As a final process, defuzzification is established, which consists of translating into a language in which the operator can interpret the final result since the response of the diffuse inference engine is manifested in diffuse language [16]. The method used by the prototype is by the centroid, as shown in Eq. 4.

$$time\_r = fuzz.defuzz(time, aggregated, 'centroide') \tag{4}$$

In the final stage, it is essential to get the output of the system, which contains the irrigation time in seconds; for that, the resulting value of defusing cation is transferred to its respective membership function.

3   Results and Discussion

The testing environment was developed with a focus on the proper use of water when considering a vegetable crop; the preparation of two plots of equal dimensions was established, the same ones that would serve to produce different types of vegetables, with the particularity of applying two irrigation methods, a drip irrigation without integrating any automation function that was suggested by a farmer technician, and another using the same drip irrigation technique but including the intelligent system. For the testing environment, time cycles were established during the vegetable production process in order to demonstrate the savings in water consumption required in a given period.

With the help of technical farmers of the farm where the tests were carried out, the following vegetables were included that included purple cabbage, lettuce, and chard on two similar plots, with dimensions of 4×1 m^2 in both cases, placing two rows of vegetables for each plot with a separation between vegetables of 30 cm. For this, the farmer technician recommended conditioning a suitable humidity on the land of the plots, the purpose is to ensure that the vegetables manage to adapt successfully to the land, which leads to starting with the installation of the irrigation channel for this scenario. In addition, a previous test was carried out supported by the agricultural technician in order to obtain the liter/hour water flow for each manual drip irrigation system and intelligent drip irrigation, establishing a 24-hour water flow rate for both systems.

In order to reflect a comfortable result in the consumption of water by the system, in the evaluation of water expenditure the smartest alternative was to establish comparisons in different periods of time. In the trial, two time periods were carried out, which are detailed in Table 2.

Table 2. Irrigation system test period

Period	Duration (days)
1	7
2	14

3.1 First Period

Figure 7 shows a real aspect of some of the characteristics of the plots described in previous sections prior to the operation of obtaining water consumption by each period.

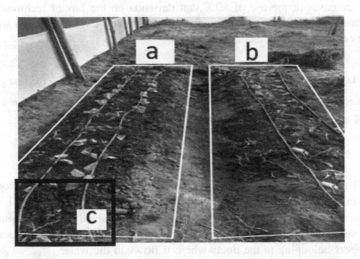

Fig. 7. First period; (a) irrigation plot using diffuse logic and drip; (b) plot using manual drip irrigation; (c) humidity sensor

Table 3 details the effects caused during the first test period.

Table 3. Effects caused during the first trial period

Type of irrigation	Time since planting (days)	Positive changes	Negative changes	Water consumption (liters)
Smart system	7	0% of vegetable mortality rate Leaves of the vegetable coloring, radiant. The soil in conditions suitable for development. Larger size	4 irrigations were made in 7 days	40
Manual system	7	Irrigation in less than 7 days	15.38% the mortality rate of the plot. Smaller size. Water stress Lack of soil moisture	48

The fundamental parameter for the calculation of water consumption in the case of the intelligent system comes from the system proposed in this study, whose value is recorded in the database managed by the system in the decision-making module.

When validating the amount of water with respect to the proposal of the farmer technician, assume that plot (b) of Fig. 7 where the manual drip irrigation method was established refers to the criteria of the farmer technician that 20 L/m^2 are sufficient for vegetables during the first period, which means that for every m^2 20 L of water is irrigated, since the surface of 4 m^2 is water consumption of the first period would be 80 L, however, this quantity can be modified according to the suggestion of the farmer technician, the suggested amount does not necessarily use 100%, his criteria express that there are cases of the use of 80% that depends on the farmer technical criteria. Table 4 shows the water consumption for the first seven days. compared to the recommendation of the farmer technician.

Table 4. Water consumption data of the types of irrigation applied in the first period

Irrigation techniques	Period (days)	Water consumption (liters)	Water required (%)
Manual drip system	7	48	60
Drip smart system	7	40	50

From Table 4, a saving of 10% is verified with an intelligent irrigation system technique in the 7-day cycle compared to the manual technique, the 48 L of water consumption referring to the manual technique were obtained from the measurement process recommended by the technical farmer calculated with a container placed in one of the drippers belonging to the ducts where it flows in the water.

3.2   Second Period

The second period was considered 14 days after the end of period 1 (7 days), with the priority of analyzing water consumption during this new cycle. Figure 8 shows the changes corresponding to this second period, where section A represents irrigation with diffuse control and section B with a manual drip technique.

Fig. 8. Second period; (A) irrigation plot using fuzzy logic; (B) plot with manual drip irrigation

In Fig. 8 it is evident to notice remarkable changes in the different plots, with better clarity in the technique with diffuse logic framed in the red box, obtaining better characteristics of the vegetables in relation to their growth, leaves, humidity ground. Water consumption for the second period is shown in Table 5.

Table 5. Water consumption data of the types of irrigation applied in the second period

Irrigation techniques	Period (days)	Water consumption (liters)	Water required (%)
Manual drip system	13	48	60
Drip smart system	13	20	41

Water consumption by the technique with diffuse logic in the second period achieved a saving of 20% compared to the manual drip technique. The behavior of the technique with fuzzy logic continues to demonstrate the powerful utility of benefiting in a certain way the consumption of water in the development of vegetables, as has been demonstrated in each of the periods in the testing environment, In addition, a better visual and colorful appearance is shown by the horizons in Fig. 9 which was not noticeable by the manual drip irrigation technique.

Fig. 9. Plot with irrigation applying fuzzy logic

The way that a saving in water supply can be guaranteed is by applying two irrigation techniques simultaneously, being able to demonstrate the savings in two different periods, this led to a continuous process of vegetable crops or any other type of cultivation can be better-obtained results in the proper use of water for crops, having a more efficient precision agriculture aligned to ICTs.

Most of the irrigation systems that have been developed in the study environment only established variable analysis parameters to know at what time irrigation was required or not, so the amount of water delivered to the Crops was in accordance with the criteria of the agricultural technician or empirical farmer. In the case of the proposed system that involves the decision-making process implemented, this logic process is done in an automated way considering the outputs that the diffuse controller gives me, where the intervention of the agricultural technician is not required only when necessary.

There are still challenges to complete, and one of them is to complement this system with other automated fertilization techniques to provide the necessary nutrients to the crops according to the needs they demand. So there would be significant savings for farmers who are engaged in this activity.

4   Conclusions

A manual drip irrigation system remains the most accessible alternative when implementing a vegetable irrigation system, however, over time we can see that these systems are being displaced by sophisticated techniques that involve ICTs, allowing Intensive management of irrigation that leads to economic savings and environmental conservation in a sustainable way. This study demonstrates that the integration of artificial intelligence such as fuzzy logic and WSN can solve problems in the environment easily and efficiently generating awareness in the use of natural resources.

References

1. Salcedo, S., Guzman, L.: Agricultura familiar en America Latina y el Caribe: Recomendaciones de Política (2014)
2. Instituto Nacional de Estadística y Censos: Encuesta de superficie y producción agropecuaria continua ESPAC 2017 (2017)
3. Organización de las Naciones Unidas para la Agricultura y la Alimentación FAO: Enfoques: Mejorar la tecnología de riego (2003). http://www.fao.org/ag/esp/revista/0303sp3.htm. Accessed 21 Nov 2019
4. Demin, P.: Aportes para el mejoramiento del manejo de los sistemas de riego. Inst Nac Tecnol Agropecu, vol. 1, pp. 1–24 (2014)
5. Mohanraj, I., Gokul, V., Ezhilarasie, R., Umamakeswari, A.: Intelligent drip irrigation and fertigation using wireless sensor networks. In: Proceedings - 2017 IEEE Technological Innovations in ICT for Agriculture and Rural Development, TIAR 2017, pp. 36–41, Institute of Electrical and Electronics Engineers Inc. (2018)
6. Ogasawara, J.: Estudio de los diferentes sistemas de riego agrícola utilizados en el Paraguay (2017)
7. Alomar, B., Alazzam, A.: A Smart irrigation system using IoT and fuzzy logic controller. In: ITT 2018 - Information Technology Trends: Emerging Technologies for Artificial Intelligence, pp. 175–179, Institute of Electrical and Electronics Engineers Inc. (2019)
8. Pernapati, K.: IoT based low cost smart irrigation system. In: Proceedings of the International Conference on Inventive Communication and Computational Technologies, ICICCT 2018, pp. 1312–1315, Institute of Electrical and Electronics Engineers Inc. (2018)
9. Consejo Nacional de Planificación: Plan Nacional de Desarrollo 2017–2021 - Toda una Vida (2017)
10. Ministerio de Agricultura Gandaría Acuacultura y Pesca (MAGAP): La Política Agropecuaria Ecuatoriana: Hacia el desarrollo territorial rural sostenible 2015–2025 (2016)
11. Organización de las Naciones Unidas para la Educación la Ciencia y la Cultura: Informe Mundial de Naciones Unidas sobre el Desarrollo de los Recursos Hídricos (2019)
12. Savić, T., Radonjic, M.: WSN architecture for smart irrigation system. In: 2018 23rd International Scientific-Professional Conference on Information Technology, IT 2018, pp. 1–4, Institute of Electrical and Electronics Engineers Inc. (2018)
13. Valero, J., Picornell, R.: El Riego y sus Tecnologías (2010)
14. Wangoo, D.P.: Artificial intelligence techniques in software engineering for automated software reuse and design. In: 2018 4th International Conference on Computing Communication and Automation, ICCCA 2018, Institute of Electrical and Electronics Engineers Inc. (2018)
15. Chen, G., Yue, L.: Research of irrigation control system based on fuzzy neural network. In: Proceedings 2011 International Conference on Mechatronic Science, Electric Engineering and Computer, MEC 2011, pp. 209–212 (2011)
16. Ross, T.J., Ross, T.J.: Fuzzy Logic With Engineering Applicationes, 3rd edn. Wiley, Hoboken (2010)
17. The scikit-image team: The scikit-fuzzy Documentation (2016)

Rehabilitation of Patients with Hemiplegia Using Deep Learning Techniques to Control a Video Game

Fabricio Tipantocta[1(✉)], Marcelo Zambrano Vizuete[2],
Ricardo Rosero[1], Wladimir Paredes[2], and Eduardo Velasco[1]

[1] Instituto Superior Tecnológico Sucre,
Av. 10 de Agosto N26-27 y Luis Mosquera Narváez, Quito, Ecuador
`ftipantocta@tecnologicosucre.edu.ec`
[2] Instituto Superior Tecnológico Rumiñahui,
Av. Atahualpa 1701 y 8 de FebreroSangolquí, Sangolquí, Ecuador

Abstract. This document presents the design and implementation of a video-game with the use of Deep Learning, which helps the rehabilitation of the upper limb in minor patients with hemiplegia. The video game was developed in the Unity silver-form and aims to incite the movement of the affected member, for this, the user's avatar performs different controlled actions through the gesticulation of the said limb. System tests were performed on a patient with left hemiparesis, which is a condition of the attenuated symptomatology of hemiplegia. The Neuronal Network was trained with 800 images of open hands and 800 images of closed hands of a child with left-hemiparesis. The results showed 99% reliability in the recognition of the hand, both open and closed. In this way, the patient with hemiparesis was motivated to open and close his hand continuously to carry out the control actions required by the videogame, promoting his rehabilitation and improvement of his fine motor skills.

Keywords: Deep learning · Hemiplegia · Neuronal networks · Videogame

1 Introduction

Daily technology advances and new rehabilitation methods for people with cerebral palsy are created, allowing them to improve their lifestyle. The generation of new technological resources that include robotic systems and video games for the rehabilitation of patients are the new fields of research [1].

Childhood cerebral palsy, globally, is between 2 to 3 per 1000 live births and continues to increase [2]. One way of presenting the PCI is hemiparesis, which affects a part of the body being congenital or acquired [3].

At present, there are many studies and technologies that allow the use of intelligent systems for the creation of assistance tools for people with disabilities [4]. The use of these types of solutions helps the rehabilitation of motor functions in patients in the recovery phase of a cerebrovascular accident, extending even in patients with cerebral palsy. There are several rehabilitation treatments, one of which is the "Mirror-Neuron" method [5]. This technique refers to the stimulation of the nervous system based on

© Springer Nature Switzerland AG 2020
M. Botto-Tobar et al. (Eds.): ICAT 2019, CCIS 1194, pp. 400–410, 2020.
https://doi.org/10.1007/978-3-030-42520-3_32

reflex learning work, which, through observation of motor activity, aims for the affected person to perform the same movements in arms and legs [6].

Based on the rehabilitation through the "Mirror-Neuron" method, a system was developed for children with hemiparesis to become interested and stimulate when performing the movements of the affected limb.

The objective of the project is to develop a videogame, implementing deep learning algorithms of a convolutional neural network, for the recognition of the movement of the hand in a patient with hemiparesis and thus perform control actions in order to rehabilitate the person.

2 Methodology

The methodology used for the development of the rehabilitation system is given by the following stages:

- Carrying out an analysis of patients with hemiparesis, for the determination of their causes and possible improvements.
- A study of the biomechanics of the hand, in people without the disease and with the disease.
- The development of a flexion-extension hand control system in a convolutional neural network algorithm.
- Design of an adventure video game with Unity software.
- Performance of the evaluation of functionality, with the patient and the video game and the control system of the convolutional network.

3 Proposed Software Architecture

For the application of the proposed convolutional neural network algorithm, the disorder (hemiparesis) to which it will be applied must be analyzed. Hand recognition is based on deep learning algorithms which are used in artificial intelligence applications. The video game reacts to the opening and closing of the hand, detected by the convolutional neural network.

3.1 Deep Learning

Deep learning is transforming and at the same time revolutionizing the field of artificial vision [7]. The use of images as input data is the opening to the theory of the convolutional neural network which handles a large amount of information for the input neurons. According to [8], a convolutional neural network (CNN) is a type of multi-layer network consisting of various convolutional and pooling layers (subsampling) alternated, which in the end has a series of full-connected layers such as a multilayer perceptron network. Figure 1 shows the model of the layers of the Deep Learning network.

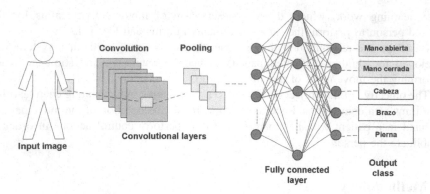

Fig. 1. Basic scheme of a Deep Learning network

For the project, it was planned to train the CNN with conventional methods such as the Google Tensorflow library and the anaconda 3.5 environment, on a Windows 10 computer and an Nvidia Gforce GPU to accelerate the network training process.

To train the network, it was estimated to place a set of 800 images of open hands and 800 images of closed hands of the patient with hemiparesis of 640 * 480 pixels with 3 RGB input channels to obtain the result of an open hand and a hand condition closed. To control the Tensorflow library, the code developed by Google in Python with the Anaconda 3.5 platform was used. Deep Learning is based on the theory of artificial neural network RAN, the following section justifies its model.

3.2   Artificial Neural Network (RAN) with Supervised Learning

RANs are parallel systems for information processing, inspired by the way in which the biological neuron networks of the brain process information, Fig. 2 shows the model of the artificial neuron consisting of the inputs (input), outputs (output), weights (w) and the functional scheme [7, 10].

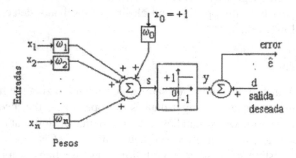

Fig. 2. Artificial neuron model.

Neural networks are organized in layers and each layer has a certain number of neurons that will connect with later ones. The neurons that contain the inputs become input layers and the neurons that contain the outputs become output layers and the intermediates become hidden layers as shown in Fig. 3 [7, 10].

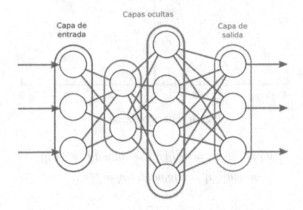

Fig. 3. Neural network with input, output and hidden layers.

The RAN used in the project is of the supervised type, since it has input and output neurons, training vectors will be compared to both input and output to calculate the error by iteration algorithms [7]. For the purpose of the project, three distance sensors were taken into account as inputs to the network and two tires in the configuration of a differential robot.

The supervised RAN is composed of two main algorithms that are feed-forward and the Backpropagation learning algorithm.

3.3 Feedforward Algorithm

For the first algorithm implemented, it is observed that each output neuron will have as input the sum of all the inputs, multiplied by the weights between the first layer and the subsequent layer [7].

$$input = \sum_i w_i * entrada_i \tag{1}$$

To calculate the output of the neurons in each layer, it is necessary to apply a mathematical activation function that will soften the output value between 0 and 1, therefore the function to be used in the application is known as sigmoid, Fig. 4 [10].

$$f(x) = \frac{1}{1 + e^{-x}} \tag{2}$$

The algorithm of the vector generated by each neuron will be multidimensional for which it operates with variables to interact between localities and the input applied to the sigmoid function [7].

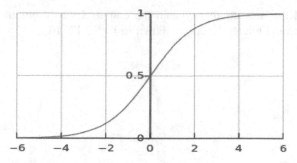

Fig. 4. Función sigmoide.

$$input[l][i] + = w[l][j][i] * output[l - 1][j];$$
$$output[l][i] = sigmoide(input[l][i]); \quad (3)$$

3.4 Learning Algorithm Back Propagation

When starting the neural network algorithm it is necessary to assign the weights of each of the neurons to be connected, so the easiest method is to assign them at random. This will make the desired output very different from what is needed.

The "backpropagation" algorithm takes the input vector and the desired output vector and this modifies the weights so that the error between the actual output of the network and the desired one is the minimum, it should be mentioned that the minimum does not mean 0 since the something -ritmo will search for the smallest of all generated values close to 0 [10].

If a training input with the assigned weights has an output of the same characteristic, this will be called a "t" vector and the desired output is called a "y" vector, then the following formula will represent the calculation of the error.

$$error_i = \frac{1}{2}(t_i - y_i)^2 \quad (4)$$

The technique used to find the corresponding weights and that the error of a function approaches 0 is known as the gradient descent method [7], which guarantees not the best 0 but the closest, so for each neuron i of the output layer its increase is calculated as:

$$\Delta i = f'(input_i) * (t_i - y_i) \quad (5)$$

Where f 'is the derivative of the activation function, and for each neuron j of the hidden layers it will be:

$$\Delta j = f'(input_j) \sum_i^{imax} W_{j,i} * \Delta i \tag{6}$$

Then each weight of the neural network starting from the output layer to the input layer "backpropagation" is updated according to the gradient descent method:

$$W_{k,j,i} + \alpha * a_j * \Delta i \tag{7}$$

Where α (alpha) will be the learning parameter between 0 and 1, which indicates how fast the neural network learns [7]. With these parameters the algorithm of the supervised neural network was designed.

3.5 Video Games with Unity

The video game industry has grown impressively in recent years and European countries like in Spain, enjoy this type of entertainment being this the sixth worldwide [11, 12].

With the new generation of digital natives, in which the protagonists are children and technology. The use and love of them to technological means, was used as motivation for the rehabilitation of children with hemiparesis, with the creation of a fantasy video game where avatar movements are carried out through the affected limb. For the creation of the video game, the Unity platform was selected for its usefulness when making 2D and 3D applications [13, 14].

For the design of the video game, the patient's age was taken into account, creating a 2D fantasy environment, consisting of an avatar, who has the ability to run and jump through different obstacles and enemies, to win, you have that pass the different scenarios increasing their points, as shown in Fig. 5.

Fig. 5. Game designed in unity.

4 Discussion and Results

4.1 Deep Learning Training

To train the convolutional neural network, 800 image samples were taken with the open hand (extension) and 800 images with the closed hand. These images were used as input values to execute the training, the tensor flow classifier executes the image learning algorithm step by step, until the error closest to 0 is found, in Fig. 6 shows the algorithm of learning with 400 steps and with 12000 steps; This is denoted when the algorithm tends to 0, and where the greatest number of learning steps are executed.

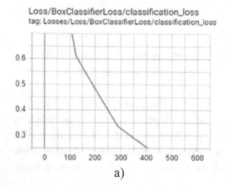
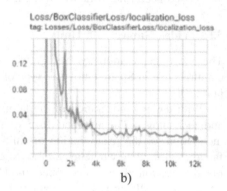

Fig. 6. Evolution of the learning algorithm: (a) 400 and (b) 12000 steps

With the error at 0.005%, after three hours and 12000 steps, the operation of the learning algorithm is cut and the network is trained; it should be specified that there is a maximum number of steps that is 200,000, but the error would be erasing the same values. As a result of the network training, it is presented in Fig. 7.

Fig. 7. Result of the convolutional neural network training.

As a result, there is a 99% probability percentage when the hand is open or closed. With these values you can perform control actions of the avatar in the video game.

4.2 CNN Application in the Video Game

In the execution of the project the softwares of: Python were used in the convolutional neural network and Unity for the creation of the videogame; to be able to connect the data between programs, a serial communicator was used through virtual COM ports. For this reason, the videogame must be run as the first instance and then the convolutional neural network is executed, with this the system would be ready for the patient to start playing, it can be seen in Fig. 8.

Fig. 8. Patient playing with the built-in system

Hand extension and closure are examined with the Tracker motion analysis software. The video game reacts to the opening of the hand, causing the player to jump into the created environment, taking into account that every 10 s an enemy will start to emerge that he must avoid with a jump (Fig. 9).

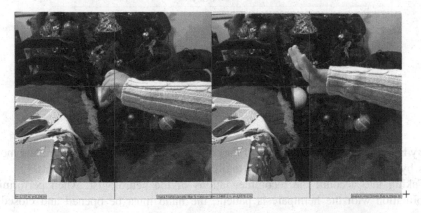

Fig. 9. Opening and closing of the hand so that the enemy jumps.

This first tracker analysis is taken into account observing the data collected in Table 1, which takes values of angular velocity and position.

Table 1. Angular velocity and hand position

t (seg)	ω (°/seg)	x(m)
1	−176,89	0
1,07	−151,43	−0,01
1,1	−149,36	−0,03
0,97	−135,51	0
1,03	−92,94	−0,01
0,43	−57,77	−0,1
0,93	−49,49	0,01
0,67	−27,73	−0,02
0,63	−8,34	−0,02
0,6	−1,89	−0,01
0,83	0	0,01

Observing how the patient, to the need for the player to jump, makes opening and closing movements and the maximum speed of 40 rev/sec can be appreciated with which he executes the action, it is observed in Fig. 10.

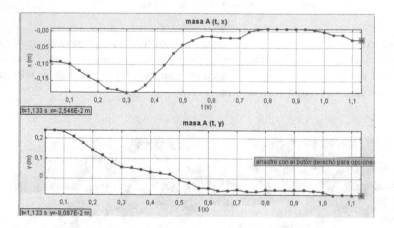

Fig. 10. Graph of angular velocity and left hand position.

By being in constant movement of the child's hand, he manages to stimulate motor skills by himself, since he must open and close his hand by his own means.

Taking into account the recommendations of the neurologist, when performing rehabilitation with the hemiparesis patient, the hand must be opened and closed at intervals of 10 s and in a time of 5 min, with a maximum of 5 tests per day, given this parameter an estimated 50 points are taken if the patient jumps all, Table 2 shows the tests performed on a random day.

Table 2. Funcionality test

Variable	Obs	ConteoPuntos
P1	50	25
P2	50	31
P3	50	46
P4	50	49
P5	50	47

As a result in this series of experiments, it is shown that between test 4 and test 5 it acquires more experience, obtaining a higher score. The work done by the hand in rehabilitation is observed, thus stimulating their motor skills.

5 Conclusions and Future Work

With the development of the project, it was observed that the patient with hemiparesis, was motivated to perform rehabilitations when playing with the videogame, making sudden movements to open the hand, which shows that the application worked correctly, but it should be noted the neurologist's recommendations as this could have an impact on patients who have a tendency to have epilepsy.

The system currently needs a large processing capacity, therefore, the computer to be used must have a video card so that the program can run normally, another option is to lower the quality of the images at the time of Train the network.

After performing the different tests of the system with the hemipareia patient, it is obtained that: the patient felt excited and tired, by the movements made with the hand and the videogame, managing to stimulate the limb and improving its fine motor skills.

For future works, other types of animations will be used, but above all virtual environments more attached to the taste of children's fantasy and compare the progress that has been obtained with respect to their rehabilitation.

References

1. Espin, S.: Hemiparesia izquierda a consecuencia de accidente de tránsito en adolescentes de 16 años, Repo.Uta.Edu.Ec, p. 115 (2014)
2. Echeverría, M.: Hemiparesia izquierda secuela de un traumatismo craneoencefálico infantil, Repo.Uta.Edu.Ec, no. 1, p. 130 (2011)
3. Castro-González, O.: Utilidad para terapeutas ocupacionales que trabajan en neurorehabilitación. Rev. Chil. Ter. Ocup. **6**, 1–10 (2006)

4. Giraldo, Á.N.G.M.I., Arango, M.I., Bermúdez, H.A.A.R.J.G.G.Z.G.: Geología de la Plancha 350 San Jose de Guaviare," vol. 19, no. 1, p. 3215 (2008)
5. Rizzolatti, G., Craighero, L.: The mirror-neuron system. Fam. Med. **48**(9), 731 (2016)
6. Monteagudo, P., Grado, J., Fisioterapia, E., Gimbernat Cantabria, E.U., Fernández Calle, C.: Eficacia De La Terapia Intensiva Bimanual En Niños Con Hemiparesia: Revisión Sistemática Effectiveness of Intensive Bimanual Therapy in Children With Hemiparesis: Systematic Review (2016)
7. Norris, D.J.: Beginning Artificial Intelligence with the Raspberry Pi (2017)
8. Durán, J.: Redes Neuronales Convolucionales en R: Reconocimiento de caracteres escritos a mano, Universidad de Sevilla, Sevilla (2017)
9. Quigley, M., Gerkey, B., Smart, W.D.: Programming robots with ROS a practical introduction to the robot operating system, vol. 53 (2015)
10. Maria Sopena, J.: Redes neuronales. Med. Cl. **122**, 336–338
11. Cuesta, D.B., Cuesta, B.G., Rodríguez-Osorio, J.J.: Videojuego 3D en Unity destinado al aprendizaje del Alfabeto dactilológico, pp. 1–162 (2014)
12. Ortuzar, M.: Desarrollo de un videojuego multijugador local con Unity3D : implementación de la Inteligencia Artificial (2017)
13. Martínez, J.S., Figueroa, P.: Guante háptico de realimentación variable para videojuegos y aplicaciones de realidad virtual. In: CEUR Workshop Proceedings, vol. 1957, pp. 176–188 (2017)
14. Camp, S.A., Final, P., Narrativas, G.M., Redondo, A.G., Mar, A.: Pirueto's Race : Videojuego destinado al aprendizaje dactilológico 3D en Unity del Alfabeto Créditos/Copyright (2016)

Application of a Heuristic-Diffuse Model to Support Decision-Making in Assessing the Post-seismic Structural Damage of a Building

Lorenzo Cevallos-Torres[1(✉)], Miguel Botto-Tobar[1,2],
Oscar León-Granizo[1], and Alejandro Cortez-Lara[3]

[1] Facultad de Ciencias Matemáticas y Físicas, Universidad de Guayaquil,
Víctor Manuel Rendón 429 e/Baquerizo Moreno y Córdova,
Guayaquil, Guayas, Ecuador
{lorenzo.cevallost,miguel.bottot,
oscar.leong}@ug.edu.ec
[2] Eindhoven University of Technology, Eindhoven, The Netherlands
m.a.botto.tobar@tue.nl
[3] Universidad Estatal de Milagro, Km. 2.5 vía a Virgen de Fátima,
Milagro, Ecuador
acortezl@unemi.edu.ec

Abstract. This research paper aims to evaluate structural damage of a building after a telluric movement occurred, by applying a heuristic-diffuse model to support the decision-making of experts in the seismic area. Ecuador is located in a highly seismic area, so it is necessary to study damage to the building structures after a telluric movement. Current damage assessment techniques are based on quantitative results, creating uncertainty. When an earthquake occurs, there is not enough certified staff, so professionals are attended without the experience to carry out the assessments and this involves the final decisions that are made not the most accurate. For the evaluation model proposed in this research, the AHP (Analytical Hierarchical Process) methodology is used in order to be able to determine the relative weights of the groups of variables that make up the building, this methodology consists of a heuristic method based on expert experience on a specific topic and in conjunction with fuzzy logic techniques will allow to work directly with the qualitative values of the data, thus providing subjective results easy to interpret that will serve as a decision-making aid. Once you have entered the data corresponding to the damage in the structural elements such as columns or beams, non-structural elements such as the facade, and data related to the type of soil, you get the levels or index of overall structural damage, non-structural and soil conditions, the level of habitability of the building is determined by applying fuzzy rules based on Mamdani's inference.

Keywords: Heuristics · Diffuse logic · Seismic damage · Building habitability · Post-seismic assessment

© Springer Nature Switzerland AG 2020
M. Botto-Tobar et al. (Eds.): ICAT 2019, CCIS 1194, pp. 411–423, 2020.
https://doi.org/10.1007/978-3-030-42520-3_33

1   Introduction

Ecuador is in a very seismic area due to geological failures that have caused disasters in recent years. In the face of the latest records of earthquakes, private or public institutions, including the University of Guayaquil, have been shown not adequately prepared for a rapid post-seismic recovery. This research refers to the problem that exists when conducting assessments of the structures of the building after a seismic movement occurred, the uncertainty that exists when the evaluation professional does not have sufficient experience in the area, generates as a consequence, that bad decisions are made, ranging from allowing the occupation of the building when it is unfit, to demolishing it.

The information handled in structural evaluations is subjective and depends on the judgement and perception of the evaluator, for this reason, an evaluation method is necessary to address the vagueness and uncertainty in this type of assessment, giving the mechanism support decision-making to determine the habitability of a building after a telluric movement has occurred. For this reason, a first work done by [1] analyzed a method for post-seismic evaluation using the MATLAB tool where the percentage of proximity to the visual method implemented by the expert when evaluating a building, the authors from our point of view, it also applies a hybrid model, in which it is very significant since they make use of another artificial intelligence model, which helps the expert in decision-making [2].

The second work corresponds to [3] where the authors propose a model that serves to support the damage assessment process after an earthquake. For the processing of this information a neural network of three layers of learning was designed, using Kohonen's algorithm where it is concluded that neural networks and fuzzy set theory are tools applicable to a damage assessment process seismic, due to the nature of the information used, subjective and incomplete. In contrast to the current case study, the AHP method could provide greater precision to the weights obtained, because the information was obtained directly from experts in the area who compared each criterion with each other and defined its importance [4].

2   Fuzzy Logic

Fuzzy logic is a multi-valued technique that allows mathematically to represent uncertainty and vagueness, providing formal tools for its treatment [5, 6]. A lot of problems can be solved given by a set of input values and producing an output value, atying to criteria of meaning and not precision. Among the most important terms regarding fuzzy logic are the theory of fuzzy sets, which is responsible for defining ambiguities and are an extension of classical set theory [7].

One of the main advantages of fuzzy logic [8], is the notion of partial pertinence that it has to deal with uncertainty, this allows to have a transition from total belonging to non-pertinence, allowing fuzzy or "confusing" entries to be correlated with actions based on linguistic rules.

2.1 Diffuse Sets

In general, you have that a classic set is a well-defined collection of elements, where it can be determined without an object belonging or not belonging to a particular set. The decision is defined as a "pertinence" or a total "non-pertinence", which the notion of set in this case is very narrow [2].

In a fuzzy set, however, each element of the universe is associated with a degree of pertinence, which is a number between 0 and 1, to that set. Under the above, a fuzzy set is defined as a correspondence (or function) that each element of the universe associates its degree of pertinence [9, 10]. If that pertinence level is close to 1, the closest the element will be to the corresponding set, otherwise, if the pertinence degree is close to 0, the element will be so far from the corresponding set [11].

Be X the universe of discourse, and its elements are denoted as x. In classical set theory, a C-on-X set is defined by the characteristic function of C as fc.

$$f_c(x) = \begin{cases} 1 \ when \ x \in C \\ 0 \ when \ x \notin C \end{cases} \tag{1}$$

If the function is generalized so that the values assigned to the elements in the set fall into a particular range to indicate the degree of belonging of the elements to that set, then a pertinence function of a particular fuzzy set will be maintained [12]. The pertinence function μ_A with which a fuzzy set is defined is given by:

$$\mu_A = X \to [0, 1] \tag{2}$$

Where $\mu_A(x) = 1$ yes x is totally in A, $\mu_A(x) = 0$ yes x is not in A y $0 < \mu_A(x) < 1$ yes x is partially in A. This value between 0 and 1 represents the degree of pertinence (pertinence value) of an element x to set A.

2.2 Integration of Heuristics with Fuzzy Logic

Heuristics is the use of empirical knowledge, practical rules and other ploys, to reduce or limit the problem of finding a solution [13]. For this reason a heuristic method was used to establish the weights of the variables involved as inputs to the heuristic-diffuse model, the heuristic method used was the hierarchical analysis process (HpA), a technique of support for decision-making multi-criteria which is based on the hierarchy, paired comparison, and weights of importance, of the criteria considered [14, 15]. Proposed by Thomas Saaty, it seeks to convert subjective assessments that are of relative importance into a set of global weights that will serve to select the best alternative.

The weights are obtained is based on an evaluation scale detailed in Table 1, and it is these weights that are entered into the model to determine structural, non-structural damage rates and soil conditions [16].

3 Post-seismic Evaluation Using Heuristics and Fuzzy Logic

For this case study, a building was chosen located within the campus of the University of Guayaquil, so it was taken as a study reference, the building of the Faculty of Philosophy and Letters Sciences, where there is currently uncertainty about its habitat of the building after the April 2016 earthquake in Ecuador. The habitability of the building will be evaluated from a set of selected variables with the help of an expert in the structural area (Table 1). Each variable will have an associated evaluation scale based on the damage index proposed by Benedetti (Table 2) because it conforms to setting the pertinence functions for the fuzzy logic model.

Scale applies for both structural element evaluation and non-structural evaluation.

Table 1. Hierarchy of evaluation criteria after consultation with the expert.

Level 1 criteria	Level 2 criteria
Structural elements	Columns
	Beams
	Beam-column connections
	Stairs
Non-structural elements	Facade
	Internal walls
Soil conditions	Settlement/liquefaction
	Soil type

Table 2. Scale for the assessment of the degree of damage proposed by Benedetti (1988).

Damage level	Degree of damage	Damage expected (%)	Average damage (%)
1	None	0 a 20	10
2	Light	20 a 40	30
3	Moderate	40 a 60	50
4	Severe	60 a 80	70
5	Total	80 a 100	90

For the scale of soil conditions, the following was determined (Table 3):

Table 3. Scale for the assessment of soil conditions.

Conditions soil	Damage expected (%)	Average damage (%)
Very good	0 a 20	10
Good	20 a 40	30
Regular	40 a 60	50
Bad	60 a 80	70
Very bad	80 a 100	90

The AHP method was used to establish the weights of relative importance of the variables, it was not deepened in the selection of decision alternatives.

4 Case Study Through the Analytical Hierarchical Process (AHP)

The relative importance weights of each level 2 criterion are inputs that are needed for the diffuse controller and in this study are obtained from the AHP methodology that uses comparisons between pairs of elements, building array from these Comparisons [17, 18]. These comparisons were made through a survey of 9 experts, using the Saaty evaluation scale, which assigns a numerical value between 1 and 9 depending on the relative importance between each of the Level 2 criteria [19].

4.1 Construction of the Comparison Array

The comparison array is a square array where the level 2 criteria are placed in their columns and rows to then make the comparison between each pair of criteria.

$$A = \begin{bmatrix} a_{1,1} & \cdots & a_{1,n} \\ \vdots & \ddots & \vdots \\ a_{n,1} & \cdots & a_{n,n} \end{bmatrix} \tag{3}$$

To calculate the priority level between peer comparisons, the scale of proportions or intensities named by Saaty is used. The main diagonal of the array will always be filled by 1 because the comparison of criteria to themselves in the evaluation will be "equally important" corresponding to numerical evaluation 1 and for cases where the criterion is less predominant a reverse rating is assigned, for example, 1/9.

4.2 Results Analysis

In order to continue the process, it is necessary to know if the comparison array is consistent. Consistency indicates that the decision shows consistent judgment in peer-to-peer comparison. A consistent array is considered when the consistency ratio is <0.1, if this condition is not met the entire process must be re-executed until it can have an acceptable consistency.

This step is performed only when the number of criteria is >2, for this reason it was only performed for the array of the structural elements that has 4 evaluation criteria.

Calculating the Consistency Index
The consistency of the array is calculated using the following formulas:

$$CI = \frac{\lambda_{max} - n}{n - 1} \tag{4}$$

Where n is the size of the array and λ_{max} it is obtained by adding the result of the multiplication between the comparison array and the resulting array of the average of each of the rows of this normalized.

Calculating the Consistency Ratio

The consistency ratio is calculated by dividing the consistency index obtained in the previous step by the random consistency index:

$$CR = \frac{IC}{ICA} \tag{5}$$

The random consistency index is obtained with the following formula:

$$ICA = \frac{1,98(n-2)}{n} \tag{6}$$

From the survey of 9 experts, 9 arrays were obtained for each group of variables, which were synthesized with the formula of the geometric average giving rise to the following comparison array (Table 4):

Table 4. Comparison array

Structural elements		C1	C2	C3	C4
C1	Columns	1	3,56	5,49	7,40
C2	Beams	0,28	1	1,16	3,52
C3	Beam-column connections	0,18	0,86	1	3,49
C4	Stairs	0,13	0,28	0,29	1

Once the arrays are synthesized, the normalized array is obtained to one with the formula (7). The following way is obtained at 0.1752 value located in row 2 column 1 of the data array:

$$\frac{0,28}{(1+0,28+0,18+0,13)} = 0,1752$$

In the same way it is done with all the elements of the array.

5   Post-seismic Evaluation-Fuzzy Logic

For the case study, fuzzy logic and fuzzy set theory were applied to establish the level of structural, non-structural damage, and ground condition level.

Once the variables and their weights (WDEi, WDNEi, WCSi) corresponding to the columns, beams, knots, stairs for the group of structural elements are obtained; facade and internal walls for non-structural elements; and settlement and soil type for soil conditions, it is important to define the linguistic values for each variable. Five linguistic values (none, light, moderate, severe, total) will be used for this study for structural and non-structural damage level; and (very good, good, medium, bad, very bad) for soil conditions.

Using fuzzy sets, one gets the pertinence functions of each variable; in this process, it is necessary to define the range of abscissae along with their degrees corresponding to the degree of pertinence. The following nomenclatures are presented below:

I_{DE}=*Structural Damage Index*

I_{DNE}=*Non-Structural Damage Index*

I_{CS}=*Soil Conditions Index*

$\mu(x) = pertinence\, function$

μ_{DE}=*membership function for Structural Damage*

μ_{DNE}=*membership function for non-Structural Damage*

μ_{CS}=*Membership function for Soil Conditions*

Once the input data is determined, it is compared to the values of the abscissae that have been defined in the pertinence functions to obtain a level of damage for each variable. This process is known as fuzification.

A new fuzzy set of output must then be generated using the implication, choosing a minimum degree of belonging between the retrieved diffuse value and the fuzzy set of the output variable.

Once the results of the fuzzy sets have been obtained, the outputs are added through the join operation and then the defuzification is done with the centroid method, with the aim of generating an index that will be the level of structural damage, non-structural and ground condition level. That's what the following formulas are defined for:

Damage to Structural Elements *i*

Union

$$DE_i = (DE_N \cup DE_L \cup DE_M \cup DE_S \cup DE_T) \tag{7}$$

$$\mu_{DE_i}(DE) = \max\left(w_{DEi} * \mu_{DE_N}(DE_{N,i}), \ldots, w_{DEi} * \mu_{DE_T}(DE_{T,i})\right) \tag{8}$$

This is the (max) join of fuzzy sets with respect to damage entries in structural elements.

Structural Damage Index

Union

$$DE = (DE_1 \cup DE_2 \cup DE_3 \cup DE_4) \tag{9}$$

$$\mu_{DE}(DE) = \max\left(\mu_{DE_1}(DE), \ldots, \mu_{DE_4}(DE)\right) \tag{10}$$

Defuzification

Through Defuzification, using the centroid method, you get a structural damage index.

$$I_{DE} = \left[\max\left(\mu_{DE_1}(DE), \ldots, \mu_{DE_4}(DE)\right)\right]_{centroid} \tag{11}$$

Table 5. Code developed in Matlab for the design of variable pertinence functions for structural elements (columns, beams. Knots, stairs).

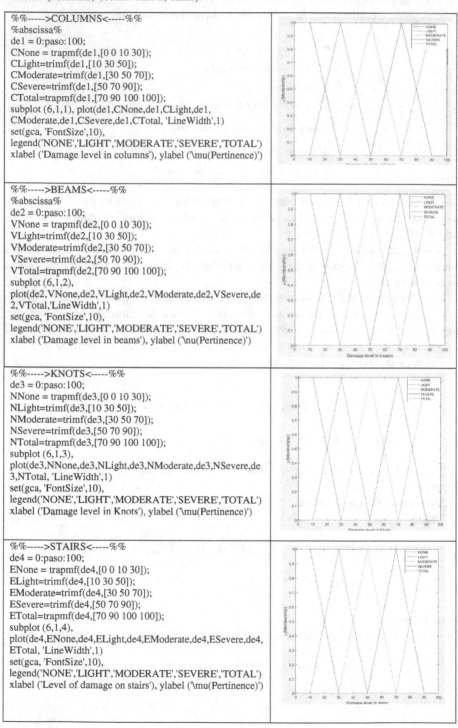

```
%%----->COLUMNS<-----%%
%abscissa%
de1 = 0:paso:100;
CNone = trapmf(de1,[0 0 10 30]);
CLight=trimf(de1,[10 30 50]);
CModerate=trimf(de1,[30 50 70]);
CSevere=trimf(de1,[50 70 90]);
CTotal=trapmf(de1,[70 90 100 100]);
subplot (6,1,1), plot(de1,CNone,de1,CLight,de1,
CModerate,de1,CSevere,de1,CTotal, 'LineWidth',1)
set(gca, 'FontSize',10),
legend('NONE','LIGHT','MODERATE','SEVERE','TOTAL')
xlabel ('Damage level in columns'), ylabel ('\mu(Pertinence)')
```

```
%%----->BEAMS<-----%%
%abscissa%
de2 = 0:paso:100;
VNone = trapmf(de2,[0 0 10 30]);
VLight=trimf(de2,[10 30 50]);
VModerate=trimf(de2,[30 50 70]);
VSevere=trimf(de2,[50 70 90]);
VTotal=trapmf(de2,[70 90 100 100]);
subplot (6,1,2),
plot(de2,VNone,de2,VLight,de2,VModerate,de2,VSevere,de
2,VTotal,'LineWidth',1)
set(gca, 'FontSize',10),
legend('NONE','LIGHT','MODERATE','SEVERE','TOTAL')
xlabel ('Damage level in beams'), ylabel ('\nu(Pertinence)')
```

```
%%----->KNOTS<-----%%
de3 = 0:paso:100;
NNone = trapmf(de3,[0 0 10 30]);
NLight=trimf(de3,[10 30 50]);
NModerate=trimf(de3,[30 50 70]);
NSevere=trimf(de3,[50 70 90]);
NTotal=trapmf(de3,[70 90 100 100]);
subplot (6,1,3),
plot(de3,NNone,de3,NLight,de3,NModerate,de3,NSevere,de
3,NTotal, 'LineWidth',1)
set(gca, 'FontSize',10),
legend('NONE','LIGHT','MODERATE','SEVERE','TOTAL')
xlabel ('Damage level in Knots'), ylabel ('\mu(Pertinence)')
```

```
%%----->STAIRS<-----%%
de4 = 0:paso:100;
ENone = trapmf(de4,[0 0 10 30]);
ELight=trimf(de4,[10 30 50]);
EModerate=trimf(de4,[30 50 70]);
ESevere=trimf(de4,[50 70 90]);
ETotal=trapmf(de4,[70 90 100 100]);
subplot (6,1,4),
plot(de4,ENone,de4,ELight,de4,EModerate,de4,ESevere,de4,
ETotal, 'LineWidth',1)
set(gca, 'FontSize',10),
legend('NONE','LIGHT','MODERATE','SEVERE','TOTAL')
xlabel ('Level of damage on stairs'), ylabel ('\mu(Pertinence)')
```

The columns have 0% damage according to the Expert Advisor assessment and processing this value with MATLAB determined that it has full pertinence in the fuzzy set of "none" (NONE) (Table 5).

For beams, knots and ladder, the expert also gave a value of 0% damage, which also has a total pertinence of the "NONE" set (Table 6).

Table 6. Code developed in MATLAB for the design of output pertinence functions.

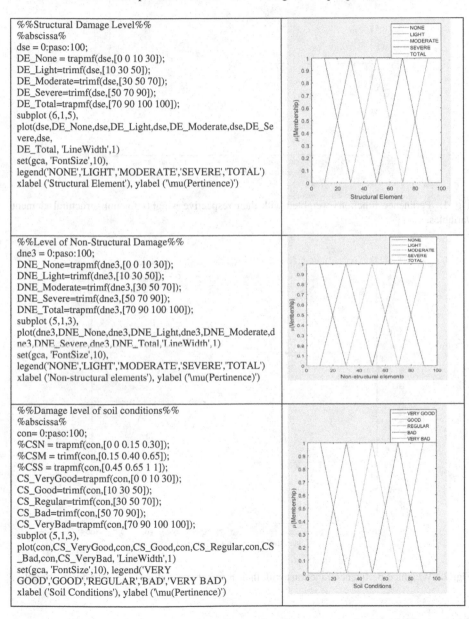

Code	Plot

`%%Structural Damage Level%%`
`%abscissa%`
`dse = 0:paso:100;`
`DE_None = trapmf(dse,[0 0 10 30]);`
`DE_Light=trimf(dse,[10 30 50]);`
`DE_Moderate=trimf(dse,[30 50 70]);`
`DE_Severe=trimf(dse,[50 70 90]);`
`DE_Total=trapmf(dse,[70 90 100 100]);`
`subplot (6,1,5),`
`plot(dse,DE_None,dse,DE_Light,dse,DE_Moderate,dse,DE_Se`
`vere,dse,`
`DE_Total, 'LineWidth',1)`
`set(gca, 'FontSize',10),`
`legend('NONE','LIGHT','MODERATE','SEVERE','TOTAL')`
`xlabel ('Structural Element'), ylabel ('\mu(Pertinence)')`	

`%%Level of Non-Structural Damage%%`
`dne3 = 0:paso:100;`
`DNE_None=trapmf(dne3,[0 0 10 30]);`
`DNE_Light=trimf(dne3,[10 30 50]);`
`DNE_Moderate=trimf(dne3,[30 50 70]);`
`DNE_Severe=trimf(dne3,[50 70 90]);`
`DNE_Total=trapmf(dne3,[70 90 100 100]);`
`subplot (5,1,3),`
`plot(dne3,DNE_None,dne3,DNE_Light,dne3,DNE_Moderate,d`
`ne3,DNE_Severe,dne3,DNE_Total,'LineWidth',1)`
`set(gca, 'FontSize',10),`
`legend('NONE','LIGHT','MODERATE','SEVERE','TOTAL')`
`xlabel ('Non-structural elements'), ylabel ('\mu(Pertinence)')`	

`%%Damage level of soil conditions%%`
`%abscissa%`
`con= 0:paso:100;`
`%CSN = trapmf(con,[0 0 0.15 0.30]);`
`%CSM = trimf(con,[0.15 0.40 0.65]);`
`%CSS = trapmf(con,[0.45 0.65 1 1]);`
`CS_VeryGood=trapmf(con,[0 0 10 30]);`
`CS_Good=trimf(con,[10 30 50]);`
`CS_Regular=trimf(con,[30 50 70]);`
`CS_Bad=trimf(con,[50 70 90]);`
`CS_VeryBad=trapmf(con,[70 90 100 100]);`
`subplot (5,1,3),`
`plot(con,CS_VeryGood,con,CS_Good,con,CS_Regular,con,CS`
`_Bad,con,CS_VeryBad, 'LineWidth',1)`
`set(gca, 'FontSize',10), legend('VERY`
`GOOD','GOOD','REGULAR','BAD','VERY BAD')`
`xlabel ('Soil Conditions'), ylabel ('\mu(Pertinence)')`	

Weighted pertinence functions are triangular and trapezoidal functions (Figs. 1 and 2).

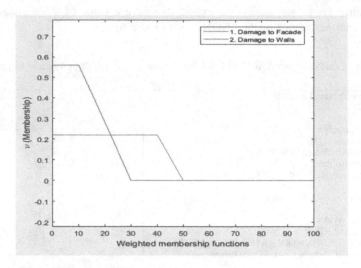

Fig. 1. Pertinence functions weighted with their respective weights for non-structural element variables.

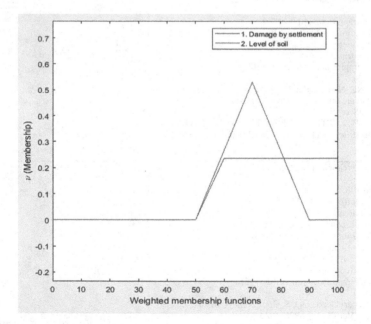

Fig. 2. Weighted pertinence functions with their respective weights for soil condition variables.

Once 10.83 structural damage, 17.55 non-structural damage and 75.47 soil conditions are obtained, 2 fuzzy controls are developed using MATLAB's "Fuzzy Logic Toolbox" tool.

The first fuzzy control receives 2 inputs (structural damage index and non-structural damage index), to which the Fuzification process, rule evaluation, output aggregation and defuzification will be applied. For Fuzification, as for damage to individual variables, it is also necessary to establish pertinence functions; in this study, the pertinence functions for the global damage level will be of the trapezoidal and triangular type, resulting in the overall damage level of the building, the calculation of which will be performed by evaluating the levels of each entry with fuzzy rules Table 7.

Table 7. Diffuse rules for overall building damage.

Structural	Non-structural				
	None	Light	Moderate	Severe	Total
None	None	Light	Light	Moderate	Moderate
Light	Light	Light	Moderate	Moderate	Moderate
Moderate	Moderate	Moderate	Moderate	Severe	Severe
Severe	Severe	Severe	Severe	Total	Total
Total	Total	Total	Total	Total	Total

Entering data for structural damage (10.83) and non-structural damage (17.55) into the fuzzy controller results in a result of 20. This overall damage data (20) is then entered along with the ground conditions (75.47) to the next controller that will ultimately determine the habitability of the building based on a set of rules. Which gave a value of 50, indicating that the building is Habitable with repairses, the result can be seen in Fig. 3.

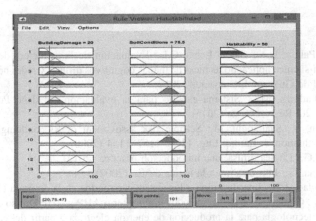

Fig. 3. Assessing rules and obtaining habitability.

To obtain the level of habitability, this diffuse controller receives as inputs the overall damage of the building and the level of ground conditions, the numbers in the figure indicate each rule, in total there are 13 fuzzy rules. For each rule, fuzification must be applied to obtain a degree of pertinence for each set of the rule antecedent, in addition to applying the minimum or maximum method to the antecedent depending on the connector of each rule, as most rules have AND connector, the minimum method is performed to get the true value of the antecedent which is needed to cut the diffuse set of output of the consequential. Once this process is performed for each rule, the aggregation of outputs is performed by maximum methods to obtain a single fuzzy set for all rules, a set that will subsequently be defuzified by centroid method to obtain a living value.

Conclusions and Future Works
For the study of the habitability of a building of the University of Guayaquil, first, by using fuzzy logic, fuzzy set theory and an AHP heuristic method to obtain relative weights, the structural damage index was determined, index no and finally the index of soil conditions for in a second instance to obtain the overall damage of the building and habitability.

As a final result a value of 50 was determined, indicating that the building is Habitable with repairses. The factor that influenced habitability was that the building has a settlement of approximately 80 cm, being in an area close to the ester and with the heavy rains causes the study center to flood at certain times.

It was further determined that the relative weights obtained through the hierarchical analysis process (HPA) had a significant impact on the outcome of the non-structural damage index and soil conditions, while, for the structural damage index, the weights relatives did not change the outcome of that index.

For future work, based on experience in this study, test data from previous evaluations is collected to then apply a heuristic method to optimize the selection or generation of fuzzy rules, so that the rules are less and allow to get the same result.

References

1. Cajamarca Palma, L.A., García Lema, C.E.: Evaluación de la estructura de daños de un edificio post-sísmico aplicando técnicas de mapas cognitivos difusos y redes neuronales de la Universidad de Guayaquil, Guayaquil (2017)
2. Carreño Tibaduiza, M.L.: Sistema experto para la evaluación del daño Post-sísmico en edificios, p. 20. ResearchGate (2015)
3. Tesfamaraim, S., Saatcioglu, M.: Seismic risk assessment of RC buildings using fuzzy synthetic evaluation. J. Earthq. Eng. **12**(7), 1157–1184 (2008)
4. Morcillo, C.G.: Lógica difusa, una introducción práctica (2012)
5. Morales Luna, G.: Introducción a la lógica difusa (2002)
6. Fortuna Lindao, J.M.: Introduccióon a los sistemas expertos en la empresa (1990)
7. Yajure, C.A.: Comparación de los métodos multicriterio AHP y AHP Difuso en la selección de la mejor tecnología para la producción de energía eléctrica a partir del carbón mineral. Sistema de Información Científica Redalyc **20**(3), 6 (2015)

8. Ministerio de Desarrollo Urbano y Vivienda: Vivienda, Guía práctica para evaluación sísmica y rehabilitación de estructuras, Quito (2016)
9. Carreño, M.L., Cardona, O.D., Barbat, A.H.: Evaluación de la habitabilidad de edificios afectados por sismo utilizando la teoría de conjuntos difusos y las redes neuronales artificiales. Rev. internacional de métodos numéricos para cálculo y diseño en ingeniería **27**(4), 278–293 (2011)
10. Duarte, O.G.: Sistemas de lógica difusa: fundamentos. Ingeniería e Investigación **42**, 22–30 (1999)
11. Eduardo, C., De Vito, E.L.: Introducción al razonamiento aproximado: lógica difusa. Rev. Americana de Medicina Respiratoria **6**(3), 126–136 (2006)
12. Cevallos-Torres, L.J., Guijarro-Rodriguez, A., Valencia-Martínez, N., Tapia-Celi, J., Naranjo-Rosales, W.: Evaluation of vulnerability and seismic risk parameters through a fuzzy logic approach. In: Valencia-García, R., Lagos-Ortiz, K., Alcaraz-Mármol, G., Del Cioppo, J., Vera-Lucio, N., Bucaram-Leverone, M. (eds.) Technologies and Innovation. CCIS, vol. 749, pp. 113–130. Springer, Cham (2017). https://doi.org/10.1007/978-3-319-67283-0_9
13. González, M.P., Pascual, A., Lorés, J.: Evaluación heurística. Introducción a la Interacción Persona-Ordenador. AIPO: Asociación Interacción Persona-Ordenador (2001)
14. Gómez, J.C.O., Cabrera, J.P.O.: El proceso de análisis jerárquico (AHP) y la toma de decisiones multicriterio. Ejemplo de aplicación. Scientia et technica **14**(39), 247–252 (2008)
15. Graupera, E.F.: Gestión de la información en la utilización del proceso analítico jerárquico para la toma de decisiones de nuevos productos. In: Anales de documentación, vol. 3 (2000)
16. Cevallos-Torres, L., Botto-Tobar, M.: The system simulation and their learning processes. Problem-Based Learning: A Didactic Strategy in the Teaching of System Simulation. SCI, vol. 824, pp. 1–11. Springer, Cham (2019). https://doi.org/10.1007/978-3-030-13393-1_1
17. Moreno-Jiménez, J.M., et al.: Validez, robustez y estabilidad en decisión multicriterio. Análisis de sensibilidad en el proceso analítico jerárquico. Rev. de la Real Academia de Ciencias Exactas, Físicas y Naturales **92**(4), 387–397 (1998)
18. Martínez, E., et al.: Apoyo a la selección de emplazamientos óptimos de edificios. Localización de un edificio universitario mediante el Proceso Analítico Jerárquico (AHP). Informes de la Construcción **62**(519), 35–45 (2010)
19. García, L., Fernández, S.J.: Procedimiento de aplicación del trabajo creativo en grupo de expertos. Ingeniería Energética **29**(2), 46–50 (2008)

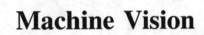

Machine Vision

Osteosynthesis Device Evaluation Using the Boundary Elements Method

Brizeida Gámez[1]([⊠]) [iD], David Ojeda[1] [iD], Marco Ciaccia[1] [iD],
and Iván Iglesias[2] [iD]

[1] Design, Simulation and Manufacturing Research Group (GIDSIM),
Técnica del Norte University, Ibarra, Ecuador
{bngamez,daojeda,mciaccia}@utn.edu.ec
[2] Intelligent Systems Research Group (GISI), Técnica del Norte University,
Ibarra, Ecuador
iiglesias@utn.edu.ec

Abstract. The elastic analysis of a dynamic compression plate (DCP) used for
the forearm bone fracture reduction is presented. For this propose is employed a
tool based on the Boundary Element Method (BEM) and an iterative domain
decomposition technique with which is possible to develop 3D models and non-
homogeneous materials. The numerical results obtained for the analysis has
been validated establishing a comparison with an experimental test employing a
DCP made of steel 316L with real dimensions. The results demonstrate that it is
possible to use the BEM for the osteosynthesis devices design.

Keywords: Dynamic compression plate · Elastostatic analysis · Boundary
element method · Iterative domain decomposition

1 Introduction

Plates are implants placed on the bone surface, fixed through screws, for the internal
reduction of fractures. They have several holes and are made of stainless steel or
titanium. There are different models, classified according to the shape, design of the
holes, site chosen for the fixation, and application mode. Generally, the plates operation
is based on three biomechanical principles: dynamic compression, neutralization, and
containment [1, 2]. The DCP is a plate designed to exert a variable compression force
on the bone segments; they can be identified by the oval holes for the eccentric
insertion of the screws [1–3].

Computer tools based on numerical techniques have been used for the design of
these implants to simulate the operation conditions of the plate on the bone surface,
evaluate the stresses and displacements to which they are subjected, and establish
device optimization criteria. The finite element method has provided a response to the
multi-region analysis between the plate and the bone, determining the displacements
and stresses generated throughout the device [4, 5]. In many of these papers it is
observed that plate displacement analyses are performed with an excessive separation
between the bones [6, 7]. This separation must be avoided because it prevents the bone
callus formation; therefore, in this work a 1 mm separation will be set for the

© Springer Nature Switzerland AG 2020
M. Botto-Tobar et al. (Eds.): ICAT 2019, CCIS 1194, pp. 427–438, 2020.
https://doi.org/10.1007/978-3-030-42520-3_34

simulation. On the other hand, due to the finite element method, often the analyses take excessive time computational because discretization must be performed throughout the domain.

The boundary element method is considered appropriate for performing complex simulations since computational time is lowered because a domain dimension is reduced [8]. We could not find any literature reference of an analysis of the mechanical behavior of an osteosynthesis plate using the BEM; this circumstance has become one of the motivation of this investigation.

The BEM has been extensively employed to solve problems in the elasticity field [9–14], since it allows to obtain the stresses and displacements of an elastic solid. In the biomechanics field, the BEM has been widely used producing excellent results [15–19]. In the case of the DCP, it is required to represent the problem as a plate-bone-screw set, i.e., it is considered as a multi-region problem. Therefore, it is proposed to apply the Iterative Domain Decomposition (IDD) technique, based in the BEM, since it is very effective for the solution of elasticity problems in 2D and 3D domains. Moreover, the parallel computation variant of the IDD method is ideal to solve large-scale problems [20–25], therefore it can be employed. In this work, the simulation of a DCP for forearm bone fractures using a robust and time-saving domain decomposition procedure, in addition to an iteration algorithm designed for parallel computation, is presented.

2   Materials and Methods

The analyzed model corresponds to a DCP with six holes for fixation screws. The geometry of the plate is 3 mm thick, 10 mm wide and 73 mm long (see Fig. 1). The fixation screws are AO/ASIF 3.5 mm diameter [24].

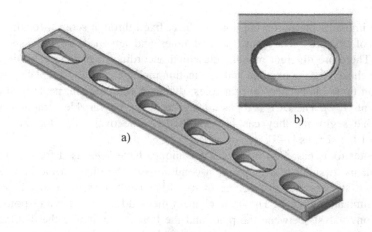

a)   b)

Fig. 1. A forearm DCP. (a) Plate isometric view, (b) Plate hole detail.

In order to analyze the performance of the plate, the critical condition is considered. This is the case where the two bone portions are not fully brought into contact during surgery, causing the plate to completely absorb the compression exerted on the bone [26]. This condition is considered by leaving a 1 mm separation between the bone models, see Fig. 2. The compression effect represents the process that happens after bone remodeling, where the bone begins to absorb loads by itself and the plate only resists axial load, this appears during the process of fracture consolidation.

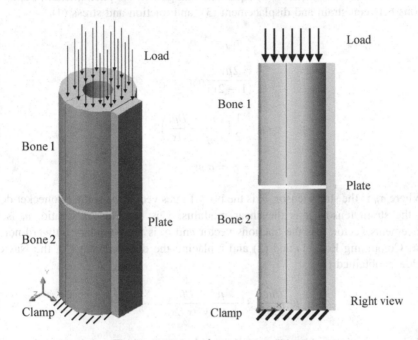

Fig. 2. DCP-bone boundary conditions

In the developed model, the fractured bone is represented by two separated cylinders. The cylinder's dimensions are external diameter, 23 mm; internal diameter, 9 mm; length, 73 mm. All degrees of freedom at the proximal end of Bone 2 were restricted to simulate the connection with the elbow. The load was applied at the opposite side (over Bone 1) to cause the transference of stresses from the bone to the plate.

The material used for the plate is stainless steel 316L, which has and elastic modulus of $E = 1.93 \cdot 10^{11}$ Pa and a Poisson ratio of $v = 0.3$. The bone is compressed with a pressure of 0.14 MPa. The plate is discretized with 56 biquadratic elements, whereas each bone is discretized with 80 biquadratic elements. The model converges with an iterative norm of $1.08 \cdot 10^{-03}$.

3 Results

3.1 Boundary Element Method in Elasticity

The boundary element method allows the determination of tractions, displacements and stresses on the boundary and in the internal points of a domain [4]. To establish the boundary integral equation in elasticity, it is important to indicate previously the equilibrium (1) and constitutive (2) equations that govern the phenomenon, beside the relations between strain and displacement (3), and traction and stress (4).

$$\sigma_{ij,j} = -b_i \tag{1}$$

$$\sigma_{ij} = \frac{2\mu v}{(1 - 2v)} \delta_{ij} e_{ii} + \mu e_{ij} \tag{2}$$

$$e_{ij} = \frac{1}{2} \left(\frac{\partial u_i}{\partial x_j} + \frac{\partial u_j}{\partial x_i} \right) \tag{3}$$

$$t_i = \sigma_{ij} n_j \tag{4}$$

Where σ_{ij} is the stress tensor, b_i is the body forces vector, δ_{ij} is the Kronecker delta, e_{ij} is the strain tensor, μ is the shear modulus, v is the Poisson's ratio, u_i is the displacements vector, t_i is the tractions vector and n_j is the boundary outward normal vector. Combining Eqs. (1) and (2) and replacing the displacement (3), the Navier's equation is obtained (5).

$$\mu \frac{\partial u_i}{\partial x_j \partial x_j} + \frac{\mu}{1 - 2v} \frac{\partial u_j}{\partial x_i \partial x_j} = -b_i \tag{5}$$

To have a well-posed boundary value problem, on each part of the boundary either the displacement $u_i = \overline{u}_i$ on boundary Γ_u or the traction $t_i = \overline{t}_i$ on boundary Γ_t is prescribed, where $\Gamma_t \cup \Gamma_u = \Gamma$ is the solution boundary of the domain Ω [12]. Based in the Somigilana equation, a formulation where the displacements in a collocation point p and the displacements u_i are related through an integral relation, and the tractions t_i on all boundaries Γ are prescribed, is set up. Furthermore, the body forces b_i remain related to the formulation through a domain integral Ω, as shown in (6).

$$c_{ij}^p u_i^p + \int_\Gamma H_{ij} u_i d\Gamma = \int_\Gamma G_{ij} t_i d\Gamma + \int_\Omega G_{ij} b_i d\Omega \tag{6}$$

Where H_{ij} and G_{ij} are fundamental solutions in terms of displacement and traction, respectively, and c_{ij}^p is a geometric constant with $c_{ij}^p = 1$ if $p \in \Omega$, $c_{ij}^p = 0$ if $p \notin \Omega$, and $c_{ij}^p = \frac{1}{2}$ on a smooth boundary [8]. Equation (7) shows the fundamental solutions for the two-dimensional case.

$$G_{ij} = \frac{1}{8\pi\mu(1-v)} \left[(4v-3)\delta_{ij} \ln r + \frac{\partial r}{\partial x_i} \frac{\partial r}{\partial x_j} \right] \tag{7}$$

$$H_{ij} = \frac{-1}{2\pi(1-v)r} \left(\frac{\partial r}{\partial n} \frac{\partial r}{\partial x_i} \frac{\partial r}{\partial x_j} \right) + \left[\frac{-(1-2v)}{4\pi(1-v)r} \left(\delta_{ij} \frac{\partial r}{\partial n} + \frac{\partial r}{\partial x_i} n_j - \frac{\partial r}{\partial x_j} n_i \right) \right]$$

Where r (radial dimension) is defined as $|x_i - x_i^p|$.

3.2   Iterative Domain Decomposition

The parallel IDD technique based in the BEM has been effectively applied in models of linear and non-linear heat conduction problems [20–23] with applications to turbo-machinery in homogeneous and piecewise non-homogeneous media [24, 25], as well as in meshless modeling of conjugate heat transfer [26, 27] and piecewise non-homogeneous material properties [28].

In this work, the application of the parallel IDD technique is shown for the design of a DCP. In order to apply the IDD technique, it is necessary to know in advance the requirements of memory for solving the problem in a single region. Figure 3 shows a simple domain Ω, its boundary conditions and the distinctive discretization of the BEM.

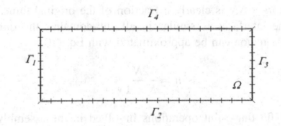

Fig. 3. BEM single region discretization.

The discretization of a single region of spatial dimension d (2 or 3), with NE boundary elements, having NN independent nodes each element, results in a $N \times N$ system of linear equations, where $N = NE \times NN \times d$, as shown in (8).

$$\Omega \Rightarrow [A]_{NxN} \{x\}_{Nx1} = \{b\}_{Nx1} \tag{8}$$

Where $\{x\}$ represents the unknown values of the tractions and displacements at the boundary. The number of floating-point operations involved in the assembly of the linear system (8) and the required memory are proportional to N^2. Direct solutions methods, such as the LU decomposition, require $\mathcal{O}(N^3)$ floating-point operations, while indirect techniques, such as the Bi-conjugate Gradient or Generalized Minimum Residual methods require, in general, $\mathcal{O}(N^2)$ floating point operations, to achieve convergence.

Instead of solving the problem on the whole domain in a single process, a multi-region solution procedure can be implemented. In this approach, artificial boundaries are created to divide the original domain Ω into K sub-domains, Ω_l ∴ $l = 1 \cdots K$, and each one is independently discretized, as shown in Fig. 4 for $K = 4$. If the conditions in the artificial boundaries are known, each sub-region can be independently solved in a standard process.

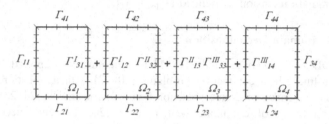

Fig. 4. BEM domain decomposition and discretization.

The application of the BEM in each of the sub-domains shown in Fig. 4, with ne boundary elements, generates a linear equations system as shown in (9).

$$[A]_{nxn}\{x\}_{nx1} = \{b\}_{nx1} \tag{9}$$

Where $n = d \times ne \times NN$ is clearly a fraction of the original dimension N. If the discretization of the interfaces is similar in all sub-domains, the dimension of the boundary of a single region can be approximated with Eq. (10).

$$n = \frac{2N}{K+1} \tag{10}$$

The number of floating-point operations involved in the assembly of the linear system (9) and the required memory are proportional to n^2. The memory savings proportion can be computed using Eq. (11).

$$\frac{n^2}{N^2} = \left(\frac{2}{K+1}\right)^2 \tag{11}$$

For $K = 4$, the required memory for each sub-domain corresponds to 16.6% of the memory needed for the whole region; however, only one region need to be in the computer memory at a time, since the rest of the sub-regions can be easily stored in disc for later use. If the linear system is solved by a direct method, $\mathcal{O}(n^3)$ floating-point operations are needed, and the savings in computer operations can be calculated with Eq. (12).

$$\frac{n^3}{N^3} = \left(\frac{2}{K+1}\right)^3 \tag{12}$$

For $K = 4$, the direct solution of the linear system for each sub-domain would consume only 6.4% of the floating-point operations count of the whole region problem; the complete number of operations can be estimated taking into account that a total of K solutions must take place. This technique is perfectly implemented in parallel computations and therefore theoretical savings can be closely achieved.

Since the interfaces between sub-domains are artificially prescribed, their boundary conditions are not known; therefore, the scheme shown in (13) is applied to guarantee the continuity of displacements u_i and the equilibrium of tractions t_i between the sub-domains.

$$u_i^a = u_i^b \tag{13}$$

$$t_i^a = -t_i^b$$

Where a and b denote the corresponding side of the studied interface. Adequate initial estimations of interfaces values are required to achieve conditions in (13) and fast convergence to the equilibrium solution. These initial values are obtained through a pre-process, using a preliminary discretization of low resolution in a simple region, consisting of an iterative procedure progressing from the initial estimation to a final solution that satisfies a norm or convergence level, adapted to the equilibrium and continuity conditions. There are two stages involved in the iterative process: first, boundary conditions are imposed individually with prescribed displacements $\bar{u}_i$ on all interfaces; thus, tractions t_i are obtained through the application of the standard BEM in each sub-domain. Usually, the computed tractions do not agree in both sides of each interface since the original supposition commonly differ from the correct values. Therefore, it is necessary to force these tractions to satisfy the equilibrium conditions using Eq. (14).

$$\bar{t}_i^a = t_i^a - \frac{t_i^a + t_i^b}{2} \tag{14}$$

$$\bar{t}_i^b = t_i^b - \frac{t_i^a + t_i^b}{2}$$

In this way, the fundamental equilibrium condition are satisfied since tractions $\bar{t}_i^a$ and $\bar{t}_i^b$ on both sides of the interface have the same magnitude, but opposite signs. The second step consists of the imposition of boundary conditions with prescribed tractions t_i, computed in the previous step, obtaining a new displacement field at the interfaces. Like the tractions before, displacements are different on both sides of each interface, requiring an update to comply with the continuity conditions, as shown in (15).

$$\bar{u}_i^a = \frac{u_i^a + u_i^b}{2} \tag{15}$$

$$\bar{u}_i^b = \frac{u_i^a + u_i^b}{2}$$

The norm displayed in Eq. (16) measures the distance between the current displacements and the previous ones, and it can be used to stop the iterative process when the norm e is smaller than a prescribed deviation L.

$$e = \sqrt{\frac{1}{NNI} \sum_{i=1}^{NNI} (\bar{u}_i - u_i)^2} \tag{16}$$

Where NNI is the number of nodes in the interface.

Additional efficiency can be attained by using the LU factorization method to compute the solution. The LU factors of the coefficient matrices for all sub-domains need to be computed only at the first step, and can be stored on disk for subsequent use during the iteration process, for which only a forward and a backward substitution will be required to solve the corresponding linear systems. This feature allows a significant reduction in the operational count of floating-point and the computation time.

It should be noted that, for the development of the formulation established in this part (DDI) it was codified using MPICH. This was used by hardware and operating system established by Gámez et al. [28, 29]. Additionally, genetic algorithms are used to optimize the results, as set forth by Ojeda et al. [19] and Divo and Kassab [24]. A key step in the domain decomposition is to keep each sub-domain discretization to a number of elements that allows the problem to be stored in the available RAM to avoid disk paging [21].

3.3   Numerical Example

Contour plots of the displacements are shown in Fig. 5. Additionally, a displacement convergence analysis is shown in Fig. 6.

Figure 5(a) and (c), the greater displacement happens in the region Bone 1 and they are falling as they descend towards the end of this region, is observed. The Fig. 5b shows that the compression load effect generates an approach from Bone 1 to Bone 2, in the zone opposed to the DCP. It is possible to emphasize that for the critical conditions, for which the DCP has been simulated, the plate works like a bridge between Bone 1 and Bone 2.

On the other hand, in the Fig. 6 the mesh sensibility analysis shows as the numerical results converges efficiently obtaining a value to the displacement of $U_{Numerical} = 5.768 \cdot 10^{-5}$ m that correspond to 216 biquadratic boundary elements.

Finally, it is important to mention that in this case the regions that simulate the fractured bone portions have been considered of the same material of the DCP, because one will settle down a comparison with an experimental model, with the same typical, which will be presented in the following section.

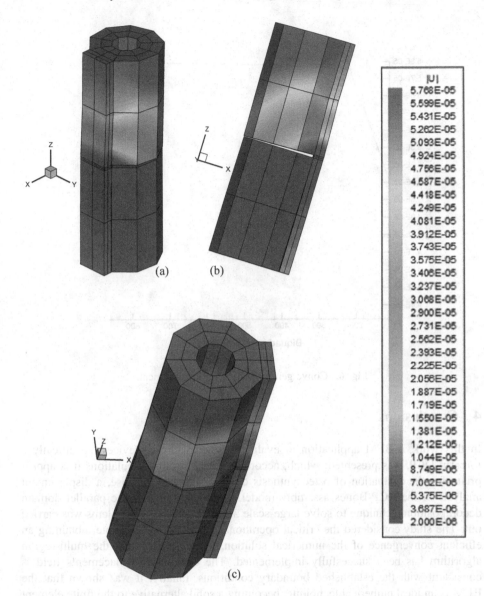

Fig. 5. Contour plot of displacements (m) for the elastic analysis, (a) View 1, (b) View 2, (c) View 3

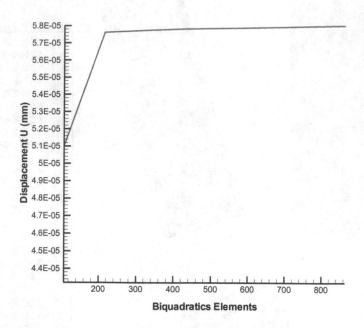

Fig. 6. Convergence analysis for displacement.

4 Conclusions

In this work, a BEM application to evaluate ostheosynthesis devices, specifically a forearm DCP, was presented; which, according to the results calculations it is appropriate for the evaluation of osteosynthesis devices. For this purpose, a displacement analysis of the DCP-Bones assembly model, employing an iterative parallel domain decomposition technique to solve large-scale 3D BEM elasticity problems was carried out. The study considered the critical operational condition for the plate, obtaining an efficient convergence of the numerical solution, demonstrating that the multi-region algorithm has been succesfully implemented. The computed displacements field is consistent with the established boundary conditions. Finally, it was shown that the BEM is an ideal numerical technique, becoming a solid alternative to the finite element method, for the evaluation of ostheosynthesis devices.

References

1. Taljanovic, M.S., Jones, M.D., Ruth, J.T., Benjamin, J.D., Sheppard, J.E., Hunter, T.B.: Fracture fixation. RadioGraphics **23**(6), 1569–1590 (2003)
2. Slone, R.M., Heare, M.M., Van der Griend, R.A., Montgomery, W.J.: Orthopedic fixation devices. RadioGraphics **11**(5), 823–847 (1991)

3. Chew, F.S., Pappas, C.N.: Radiology of the devices for fracture. Treatment in the extremities. Clin. Radiol. North America **33**(2), 375–389 (1995)
4. Tarnita, D., et al.: Numerical simulations of human tibia osteosynthesis using modular plates based on Nitinol staples. Romanian J. Morphol. Embryol. **51**(1), 145–150 (2010)
5. Baharnezhad, S., Farhangi, H., Allahyari, A.: Influence of geometry and design parameters on flexural behavior of dynamic compression plates (DCP): experiment and finite element analysis. J. Mech. Med. Biol. **13**(3), 1350032 (20 p.) (2013)
6. Snow, M., Thompson, G., Turner, P.: A mechanical comparison of the locking compression plate (LCP) and the low contact-dynamic compression plate (DCP) in an osteoporotic bone model. J. Orthop. Trauma **22**(2), 121–125 (2008)
7. Wieding, J., Souffrant, R., Fritsche, A., Mittelmeier, W., Bader, R.: Finite element analysis of osteosynthesis screw fixation in the bone stock: an appropriate method for automatic screw modelling. PLoS ONE **7**(3), e33776 (2012). https://doi.org/10.1371/journal.pone.0033776
8. Alarcon, E., Brebbia, C., Dominguez, J.: The boundary element method in elasticity. Int. J. Mech. Sci. **20**(9), 625–639 (1978)
9. Brebbia, C.A., Telles, J.C., Wrobel, L.C.: Boundary Element Techniques. Springer, Berlin (1984). https://doi.org/10.1007/978-3-642-48860-3
10. Kane, J.: Boundary Element Analysis in Engineering Continuum Mechanics. Prentice-Hall, New Jersey (1994)
11. Aliabadi, M.H.: The Boundary Element Method, Vol. 2: Applications in Solids and Structures. Wiley, Chichester (2000)
12. Cheng, A.H., Chen, C.S., Golberg, M.A., Rashed, Y.F.: BEM for thermoelasticity and elasticity with body force-a revisit. Eng. Anal. Bound. Elements **25**, 377–387 (2001)
13. Brebbia, C.A., Domínguez, J.: Boundary Element: an Introductory Course. Computational Mechanics, Boston (1989)
14. Becker, A.: Boundary Element Method in Engineering. McGraw-Hill Co., New York (1992)
15. Annicchiarico, W., Martínez, G., Cerrolaza, M.: Boundary elements and β-spline surface modeling for medical applications. J. Appl. Math. Model. **31**(2), 194–208 (2007)
16. Gámez, B., Ojeda, D., Divo, E., Kassab, A., Cerrolaza, M.: Crack analysis and cavity detection in cortical bone using the boundary element method. In: APCOM 2007 – EPMESCXI, Japan (2007)
17. Müller-Karger, C., González, C., Aliabadi, M.H., Cerrolaza, M.: Three dimensional BEM and FEM stress analysis of the human tibia under pathological conditions. J. Comput. Modeling Eng. Sci. **2**(1), 1–13 (2001)
18. Ojeda, D., Divo, E., Kassab, A., Cerrolaza, M.: Cavity detection in biomechanics by an inverse evolutionary point load BEM technique. Inverse Probl. Sci. Eng. **16**(8), 981–993 (2008)
19. Ojeda, D., Gámez, B., Divo, E., Kassab, A., Cerrolaza, M.: Singular superposition elastostatics BEM/GA algorithm for cavity detection. In: Proceeding of 29th International Conference on Boundary Elements and Other Mesh Reduction Methods, 04–06 June 2007. The New Forest, UK (2007)
20. Divo, E., Kassab, A.J.: A generalized BIE for transient heat conduction in heterogeneous media. AIAA J. Thermophys. Heat Transf. **12**(3), 364–373 (1998)
21. Divo, E., Kassab, A.J.: A boundary integral equation for steady heat conduction in anisotropic and heterogeneous media. Numer. Heat Transf. Part B: Fundam. **32**(1), 37–61 (1997)
22. Kassab, A.J., Divo, E.: A general boundary integral equation for isotropic heat conduction problems in bodies with space dependent properties. Eng. Anal. Bound. Elements **18**(4), 273–286 (1996)

23. Divo, E., Kassab, A.J.: Boundary Element Method for Heat Conduction with Applications in Nonhomogeneous Media. Wessex Institute of Technology (WIT) Press, Southampton (2003)
24. Erhart, K., Divo, E., Kassab, A.: A parallel domain decomposition boundary element method approach for the solution of large-scale transient heat conduction problems. Eng. Anal. Bound. Elements **30**, 553–563 (2006)
25. Divo, E., Kassab, A., Rodríguez, F.: Parallel domain decomposition approach for large-scale threedimensional boundary-element models in linear and nonlinear heat conduction. Numer. Heat Transf. Part B **44**, 417–437 (2003)
26. Divo, E., Kassab, A.J.: A meshless method for conjugate heat transfer problems. Eng. Anal. Bound. Elements **29**, 136–149 (2005)
27. Divo, E.A., Kassab, A.J.: An efficient localized RBF meshless method for fluid flow and conjugate heat transfer. ASME J. Heat Transf. **129**, 124–136 (2007)
28. Gámez, B., Divo, E.A., Kassab, A.J., Cerrolaza, M., Ojeda, D.: Parallelized iterative domain decomposition boundary element method for thermoelasticity in piecewise non-homogeneous media. Eng. Anal. Bound. Elements **32**(12), 1061–1073 (2008)
29. Müller, M.E., Allgöwer, M., Schneider, R., Willenegger, R.: AO Manual of Internal Fixation, 3rd edn. Springer, Berlin (1991)

From a Common Chair to a Device that Issues Reminders to Seniors

Orlando Erazo (ID), Gleiston Guerrero-Ulloa (✉) (ID), Dayana Guzmán,
and Carlos Cáceres

Facultad de Ciencias de la Ingeniería, Universidad Técnica Estatal de Quevedo,
Quevedo 120501, Ecuador
{oerazo, gguerrero, dayana.guzman2015,
carlos.caceres2015}@uteq.edu.ec

Abstract. Over the years, people tend to fail to recall different activities of everyday life, especially those that are performed less frequently. Forgetting a medical appointment, a family member's birthday, or to take a medicine, is a common problem for many older adults. Although Information and Communication Technologies provide several options to help older adults remember activities like these, many of them could feel discouraged by not being able to properly use these tools and take full advantage of them. In other cases, elderly people may feel intimidated and even refuse to interact with technological devices. For these reasons, this paper proposes the use of a conventional object, which can be found in any home, to help older adults remember certain activities. Specifically, an ordinary chair has been selected to be employed as a device to provide the necessary reminders. The reminders implemented in the designed prototype are provided by audio, lights and vibrations, things that users do not notice at first sight the presence of technology in the chair. This prototype was evaluated through a user study with the collaboration of older adults. The results of the evaluation were positive, which concludes that the proposal has a favorable reception. Thus, this proposal could provide an important contribution to the major goal of helping to improve the quality of life of the elderly population.

Keywords: Older adults · Reminders chair · Ubiquitous computing · Everyday objects

1 Introduction

Aging is a normal stage that all of us will reach. This stage is not a limitation to having a totally normal life and to be able to continue doing day to day activities However, there are age-related impairments that could interfere with the performance of these activities; one of the most frequent problems being memory loss.

Many experts have developed different technological devices and applications that provide help to older adults who have trouble remembering. Some examples of these solutions are the interactive dashboards [1], intelligent pill dispensers [2, 3], software that shows important episodes in the lives of older adults in order to keep their memory alive [4], and application of assistance for the elderly [5].

© Springer Nature Switzerland AG 2020
M. Botto-Tobar et al. (Eds.): ICAT 2019, CCIS 1194, pp. 439–448, 2020.
https://doi.org/10.1007/978-3-030-42520-3_35

All of these solutions have been evaluated with the participation of older adults, resulting in a great acceptance. Even solutions such as smart pill dispensers are already marketed. However, there are limitations: not all users have an economic solvency that can afford these devices; they have a limited amount of reminders that they can store and their functionality requires them to be transportable, but they do not have a convenient size for mobilization.

Based on the above, this paper analyzes the possibility of using everyday objects to provide reminders to older adults in a transparent way. The proposal aims to take advantage of the Internet of Things (IoT) [6] to enable the used objects to communicate, for example, with a mobile application. This application would make it possible, on the one hand, to set up reminders and, on the other hand, to deliver alerts to a caregiver or a relative of the older adult. However, one of the previous steps is the analysis of candidate objects to provide reminders. For this reason, this article describes the evaluation of a conventional chair "turned" into a "ubiquitous device" that can help older adults remember day-to-day activities. This chair provides reminders through auditory, vibratory and light alerts, without older adults feeling that they are interacting with a technological device; they will only see the chair in which they are sitting.

The developed prototype was evaluated through a user study. The study involved the type of people for whom the device is intended in order to assess its usefulness and functionality. The obtained results showed that participants liked the device. They expressed an understanding of the functionality of the chair and, most importantly, they did not notice that there was an extra device to send reminders. Participants suggested that this device could be used to provide everyday reminders, and in addition, it may provide some kind of companionship, for example, speaking to them.

The rest of the paper is organized as follows. In Sect. 2, some work related to the proposed prototype system has been written. Section 3 describes the materials and methods used in the development of the proposal: (1) prototype design and (2) its evaluation. The results and discussion are presented in Sect. 4. Finally, Sect. 5 presents the conclusions and future work.

2   Related Work

Today there is a remarkable growth of the elderly population. This growth causes an imbalance in the costs of public health services and the probable shortage of direct care workers such as nurses, paramedics and health aides. In addition, there are a significant number of older adults with progressive health problems who prefer to live in their homes where they have been established for many years [7].

According to Saracchini et al. [5], memory loss in older adults is one of the most common progressive health problems. They propose that the most viable solution to this problem is the use of technology through intuitive software assistance for the elderly. This proposal promotes the brain activity of the elderly; it helps to remember appointments and pending daily tasks. The software interface tries to simplify the preferences of the elderly. However, several of the subjects who participated in the evaluation said they had trouble managing the reminder's agenda because of its difficulty.

Situations of complexity such as the one mentioned above, plus others such as text size, size of buttons, lack of adaptability, among others, led a group of researchers to evaluate various mobile applications to determine which are suitable for use by older adults [8]. The applications Med Helper Pill [9], MediSafe [10], Pill Reminder [11] were recommended for their friendly interfaces, efficiency in health control at individual and population levels. These applications can encourage healthy behaviors to prevent or reduce health problems, offer reminders to take medications, scheduled appointments, and warning when medications are running out.

Other related work [12], which evaluates the acceptance of assistive technology in older adults, indicates that it is not only necessary to create the technological devices; designers of devices should also ensure that those devices are pleasing and easy for seniors to understand them. This study [12] found that users prefer an intelligent device that performs its functions autonomously. In addition, participants suggested that devices should offer company while supervising them periodically, "be an advisor" to maintain a healthy physical condition and have alarm or reminder functions to take the right medication at the right time.

Remembering to take medication is one of the most important daily activities that older adults forget to do. These omissions can accelerate the progression of a disease and subsequently increase medical expenses [2]. The solution to this problem can be found in a pillbox [2]. It requires to be fed by the medicine with its respective schedule, administered by an assistant that must have knowledge of the use of the device. This pillbox fulfills its function, but it necessitates the use of batteries and has a limitation on the maximum number of alarms that can be stored.

A similar but more updated design is proposed by Muñoz González [3]. It consists of a dispenser of pills for older adults with reduced autonomy. This dispenser has a touch screen on which the user can enter the medication schedule. In addition, it has a built-in NFC bracelet which is used to remind the elderly whenever it is time to take his/her medication. Despite its optimal functionality, the dispenser has a significant size that makes it difficult to move from one place to another.

In order to maintain stable health, it is not only necessary to take care of oneself physically or to take one's medication on the established schedules; but emotional states of older adults also have an influence on their health. For this reason, Etchemendy et al. [4] developed an application called Mayordomo that collects photos of happy moments from childhood to the present with a short story that narrates the events, photos of loved ones, relevant information of their lives, and so on. However, the interface of this software needs improvements to attract users, since many people do not want to involve technology in their lives.

Considering that it is difficult to involve technology in the lives of older people, Veldhoven et al. designed an interactive display of messages and reminders in the form of a board [1]. The designers wanted the screen to be nice so that seniors feel comfortable and do not think they are using technology, but they are using a simple dashboard to show the ads that people must remember daily. In addition, it has a website where family members can place the reminders that have to appear on the board. Nevertheless, there are other ideas that could be explored, particularly taking into consideration household objects that often go unnoticed.

3   Materials and Methods

3.1   Prototype Design

Faced with the problem of the decreasing memory in older adults, which in many cases may lead to dependence on another person to cope with their lives normally, this paper proposes to use an "intelligent" chair that can issue reminders semi-transparently. The idea of using a chair is based on the fact that it is a very popular object to rest and, in certain cases, older people tend to have a preference for a chair in their home. The chair, according to the proposal, will provide the necessary reminders combining voice, lights and vibrations. This prototype will work for people who still react to stimuli and understand verbal communication. In other cases it could serve as an alert or reminder device to the people who are responsible for their care. Figure 1 shows the front and back views of the proposed chair.

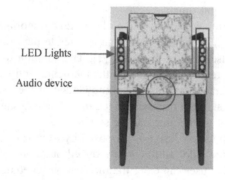

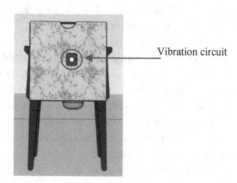

Fig. 1. Proposed prototype

For the development of the prototype to evaluate our proposal, a chair with armrests similar to the one in Fig. 1, was selected. Preliminary tests were also carried out planning scenarios of how an older adult would act if he/she were using the chair. These tests helped to define the location of the different components/circuits.

The lights circuit (see Figs. 1 and 2) was implemented in the front of the chair to allow the user of this device to have visibility of the color reflections emitted by the employed LED strips [13]. This circuit was previously simulated using Proteus [14]. It is composed of three LED strips (red, blue and green), buttons and a battery connected as shown in Fig. 2 (above).

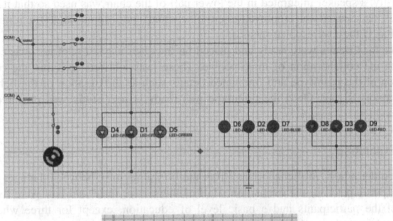

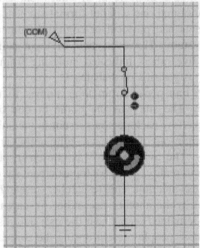

Fig. 2. Diagrams of the simulated circuits in Proteus: vibrating motor (bottom) and LED lights (top). (Color figure online)

The use of colors is very common to alert people about events that take place. Some of these events are scheduled (e.g., traffic lights), and others are unexpected (e.g., traffic accidents). In [15], the author's two present studies, (1) about the appropriate values of intensity, color, frequency of changes and brightness of lights to transmit information, and (2) how lights affect context consciousness in people. The favorable results obtained in these studies increased the motivation to use colored lights in our prototype.

The vibration alert circuit was also simulated by using Proteus software as shown in Fig. 2 (above). A power supply, a switch, a pushbutton and a motor were employed for its implementation. The switch was utilized to control the general circuit and provide the necessary voltage. Finally, the location of the circuit was determined. The back of the chair was chosen, because people usually lay back to support their backs. This decision allows older adults to better receive the vibratory alerts.

Finally, a speaker integrated in the lower part of the chair was used so that it is not visible. This speaker was employed to emit the sounds corresponding to the reminders emitted as if they were voice memos.

Once the circuits were ready and placed on the chair used, a pilot test was carried out with the collaboration of three people. These tests allowed verifying the correct operation of the prototype and fine-tuned details such as the duration of the lights on and vibratory signal. Once the prototype was ready, a user study was prepared and performed.

3.2 Evaluation

Twelve older adults, all Ecuadorians (from the littoral region), between the ages of 65 and 87, participated in the evaluation. Half of them were male and half were female. Most of the participants had a basic level of education, except for three who had completed secondary school and one without formal studies.

None of the participants had physical or cognitive ailments that prevented them from taking part in the study. However, two of them self-reported pain in their back/spinal column and five others indicated suffering from a disease that did not influence the study, such as gastritis or diabetes.

All participants were transported from their homes to the study site with the researchers' collaboration. The site was the home of one of the authors. Figure 3 illustrates the spots occupied by each person present in the study. The chair prototype was positioned in the living room of that house, eliminating any type of distractor that could influence the study. The living room was chosen in order to use a room of a house environment closer to reality. Next, each participant sat down on the chair while a researcher conducted the evaluation. This researcher (researcher 1 in Fig. 3) sat down in front of the participant, while another researcher (researcher 2 in Fig. 3) acted as observer. This researcher remained seated near the user and researcher 1 took notes. In addition, there were no more people present during the evaluation. This decision also avoided distractions during the study.

The evaluation began with a brief introduction of the functionality of the prototype reminders chair to each participant. Then, researcher 1 provided some instructions to the participant to proceed to sit down on the chair and to begin a dialogue about different topics in his/her life. As the conversation unfolded, reminders were issued in random time lapses. These reminders were as follows:

- You have to take your medicine (health reminder).

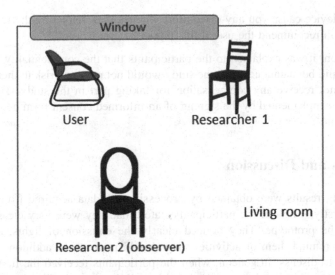

Fig. 3. Distribution of spots for people present in the study.

- Today is your son's birthday (birthday reminder).
- You have a medical appointment today at 16:00 (appointment reminder).
- It's time for your favorite TV show (leisure reminder).

Since some older adults tend to separate activities according to gender, a final reminder of domestic chores was made for women and another for men. These reminders were as follows:

- You have to turn off the stove burners (woman).
- You have to turn off the bedroom light (man).

The reminders were emitted through a combination of lights, vibration and audio. The visual alerts were sent by means of LED lights of different colors. The color red meant a family alert, the color blue meant a health alert, and the color yellow meant an alert for a leisure activity. Both the lights and the vibrations were emitted twice by reminder. Each alert lasted two seconds, with an interval of one second, based on the pilot test mentioned above. The used audio consisted of a female voice indicating the above reminders.

After the evaluation with the prototype, a questionnaire was applied to collect additional data. The participants first answered questions of a demographic nature and then about the evaluation of our proposal. The first part consisted of questions about age, sex, educational level, health condition, among others. In the second part, the following questions were raised:

1. Do you have a clear idea of the purpose of the device?
2. Do you think the service offered by the device is useful for your personal case?
3. Which was the aspect of the device that most caught your attention, positively or negatively?

4. Did the device cause you any discomfort when it was delivering the reminder?
5. Would you recommend the use of this device?

In addition, it was explained to the participants that the confidentiality of the data collected would be maintained, that the study would not pose any risk to them, and that they would not receive any compensation for taking part in the study. These explanations were supplemented by the signing of an informed consent form before starting data collection.

4 Results and Discussion

The following results were obtained by processing the data acquired from the questionnaire. Firstly, 100% of the participants stated that they were very clear about the purpose of the prototype. They noticed clearly the emission of lights, audios and vibrations to remind them of activities they should perform. In addition, there were other details to discuss. In general, when the participants received the first reminder, they were amazed because it was something they did not expect. The second reminder caused less amazement than the first one. By the time the third reminder appeared, the participants had already realized the objective of the prototype. There was no novelty in the reaction of the users in the reminders provided later. In the end, all participants made reference to the fact that the device had mechanisms for remembering activities or tasks to be performed while they are resting quietly.

Similarly, referring to the second question, all participants responded that in their personal case this prototype would be useful.

Thirdly, results about what caught the attention to participants positively or negatively from the device are shown below in Table 1. According to this table, the use of vibrations was the preferred kind of alert to catch their attention, followed by sound. On the contrary, particularly in comparison to [15], the lights went practically unnoticed. One possible motive could be the clarity of the evaluation site because the study was conducted during the day and near a window (bearing in mind participants' comfort). Also, there was a person who was not struck by any of the ways of issuing the prototype reminders. However, this answer could be due to the fact that the participant showed little interest during the evaluation, which could have derived in not being adequately aware of the different elements.

Table 1. Kind of alerts that most caught participants' attention.

Alert	Number of participants	Percentage
Vibration	7	58.4
Sound	4	33.3
Lights	0	0.0
None	1	8.3
Total	12	100

Considering that the participants did not know how the reminders were going to be sent, when we asked them if the alerts caused some discomfort, all of them replied that it did not. Several of them even went further, saying that it would be a very useful chair because, at their age, it is very common to forget many things. In addition, some participants pointed out that the chair could be "a good companion", as it could speak to them when they were alone.

Finally, the participants were asked whether they would recommend the use of the device or not. All the older adults who participated responded that they would recommend it. Two of them provided additional comments. One said that while he agreed with the proposal, the reminders "interrupted him during the conversation". Another participant stated that the reminders "awakened his mind" and that they would help him greatly to improve his memory.

5 Conclusions and Future Work

Considering that older people are becoming "technologically excluded", a significant amount of effort has been made to improve this situation. Even though some of them may be against the use of technological tools, they can be provided with ubiquitous solutions for the home that help seniors in their daily activities. For this reason, in this work we have analyzed the possibility of using a conventional chair to provide reminders to older adults about day-to-day activities that they may have a tendency to forget.

For this purpose, a prototype was built and an evaluation was carried out with senior citizens. In general, the prototype was well accepted by the older adults. Most of them said they understood the functionality of the prototype, and only one was indifferent.

Referring to the type of alerts used, it was evident that the LED lights were unnoticed. On the other hand, the sound alerts and vibrations that were emitted were clearly noticeable to the participants. However, these facts did not influence the acceptance of the prototype.

While the evaluated reminder device successfully fulfills its purpose, it may evolve to offer greater functionality. The chair may be suitable to act as a companion device for older adults. This aspect could eventually be addressed in future work.

In the future, the reminder chair will be supplemented with an application (mobile and/or web) that allows family members, or staff caring for an elderly person, to manage the necessary reminders. This application should allow configuring the needed reminders, with their schedule and even select the type or types of alerts with which an older adult would prefer receiving the reminders. Furthermore, in order to reduce the chances of seniors ignoring reminders, a wireless sensors network [16] (e.g., using motion sensors [17]) could be implemented so that alerts cease when the elderly joins in to comply with what the reminder suggests (similar to [18]). This could be notified to the relative or caregiver of the older adult. In addition, taking advantage of this sensors network and IoT, reminders related to events that could cause damage to the elderly or even to trigger a catastrophe could be incorporated (e.g., forget a plugged up iron or a stove left on). Finally, we plan to use other common things available at homes to work together with the proposed chair, taking advantage of IoT. All in all, we hope to contribute in this manner to help to improve the quality of life of the elderly population.

Acknowledgments. This work was partially supported by FOCICYT 2018-2019, Sexta Convocatoria (Universidad Técnica Estatal de Quevedo, Quevedo, Ecuador).

References

1. van Veldhoven, E.R., Vastenburg, M.H., Keyson, D.V.: Designing an interactive messaging and reminder display for elderly. In: Aarts, E., et al. (eds.) AmI 2008. LNCS, vol. 5355, pp. 126–140. Springer, Heidelberg (2008). https://doi.org/10.1007/978-3-540-89617-3_9
2. Gupta, K., Jain, A., Vardhan, P.H., Singh, S., Amber, A., Sethi, A.: MedAssist: automated medication kit. In: Proceedings - 2014 Texas Instruments India Educators Conference, TIIEC 2014, pp. 93–99 (2017)
3. Muñoz González, M.T.: Proyecto de Dosificador de Pastillas para Personas con Autonomía Reducida. Universitat Politècnica de Catalunya (2016)
4. Etchemendy, E., Baños, R.M., Botella, C., Castilla, D.: Program of life review support on the new technologies for elderly people: an positive psychology application. Escritos de Psicología **2**, 1–7 (2010)
5. Saracchini, R., Catalina, C., Bordoni, L.: Tecnología Asistencial Móvil, con Realidad Aumentada, para las Personas Mayores. Comunicar **23**(45), 65–74 (2015)
6. Atzori, L., Iera, A., Morabito, G.: The internet of things: a survey. Comput. Netw. **54**(15), 2787–2805 (2010)
7. Weitz, D., María, D., Lianza, F., Schmidt, N., Nant, J.P.: Smart home simulation model for synthetic sensor datasets generation. Sistemas y Telemática **14**(39), 71–84 (2016)
8. Helbostad, J., et al.: Mobile health applications to promote active and healthy ageing. Sensors **17**(3), 622 (2017)
9. Med Helper Pro Pill Reminder. https://med-helper-pro.mx.aptoide.com. Accessed 25 Sept 2019
10. Medisafe - Patient engagement platform solving medication management. https://www.medisafe.com. Accessed 25 Sept 2019
11. Rxremind. https://rxremindme.com. Accessed 25 Sept 2019
12. Heerink, M., Kröse, B., Evers, V., Wielinga, B.: Assessing acceptance of assistive social agent technology by older adults: the almere model. Int. J. Soc. Robot. **2**(4), 361–375 (2010)
13. Cunha, M., Fuks, H.: AmbLEDs collaborative healthcare for AAL systems. In: Proceedings of the 2015 IEEE 19th International Conference on Computer Supported Cooperative Work in Design, CSCWD 2015, pp. 626–631 (2015)
14. Proteus. https://www.labcenter.com. Accessed 25 Sept 2019
15. Davis, K., Owusu, E.B., Marcenaro, L., Feijs, L., Regazzoni, C., Hu, J.: Effects of ambient lighting displays on peripheral activity awareness. IEEE Access **5**, 9318–9335 (2017)
16. Akyildiz, I.F., Su, W., Sankarasubramaniam, Y., Cayirci, E.: Wireless sensor networks: a survey. Comput. Netw. **38**(4), 393–422 (2002)
17. Basu, D., Moretti, G., Sen Gupta, G., Marsland, S.: Wireless sensor network based smart home: sensor selection, deployment and monitoring. In: 2013 IEEE Sensors Applications Symposium, SAS 2013 - Proceedings, pp. 49–54. IEEE (2013)
18. Guerrero-Ulloa, G., Rodríguez-Domínguez, C., Hornos, M.J.: IoT-based system to help care for dependent elderly. In: Botto-Tobar, M., Pizarro, G., Zúñiga-Prieto, M., D'Armas, M., Zúñiga Sánchez, M. (eds.) CITT 2018. CCIS, vol. 895, pp. 41–55. Springer, Cham (2019). https://doi.org/10.1007/978-3-030-05532-5_4

Hand Exercise Using a Haptic Device

Paulo A. S. Mendes[1]([✉]), João P. Ferreira[1,2], A. Paulo Coimbra[1,3],
Manuel M. Crisóstomo[1,3], and César Bouças[1]

[1] ISR - Institute of Systems and Robotics, University of Coimbra, Coimbra, Portugal
`33paulomendes@gmail.com`, `{mcris,cesar.boucas}@isr.uc.pt`
[2] Department of Electrical Engineering,
Superior Institute of Engineering of Coimbra, Coimbra, Portugal
`ferreira@isec.pt`
[3] Department of Electrical and Computer Engineering, University of Coimbra,
Coimbra, Portugal
`acoimbra@deec.uc.pt`

Abstract. It is known that the brain uses the sense of touch, in different parts of the body, to acquire information to react to the environment. With nowadays technology, it is possible to create distinct virtual environments and to feel them with haptic devices. Using haptic devices, it is possible to train and develop different parts of the human body, including the brain. These devices allow users to feel and touch virtual objects with a high realism. The present paper proposes different controller methods to use a haptic device to help the user to exercise their hands. The hand exercises proposed are the straight-line, square, circle and ellipse follow-up. In this work four different types of controllers are compared: proportional, proportional-derivative and logarithmic and sigmoid function based controllers. Each one of the used controllers were tested with the hand exercises mentioned. The sigmoid and logarithmic function based controllers achieves more suitable results for the user haptic perception and trajectory follow-up.

Keywords: Haptic device · Digital control · Hand rehabilitation

1 Introduction

Touch is a fundamental aspect of interpersonal communication. It is a very powerful sense, and presently, new technologies allow computer systems' users to feel virtual objects.

In 2005, Brewster mentioned that State-of-the-art haptic (or force-feedback) devices allow users to feel and touch virtual objects with a high degree of realism [3]. Brewster defines haptics as a general term relating to the sense of touch.

The Fundação para a Ciência e Tecnologia (FCT) and COMPETE 2020 program are gratefully acknowledged for funding this work with PTDC/EEI-AUT/5141/2014 (Automatic Adaptation of a Humanoid Robot Gait to Different Floor-Robot Friction Coefficients).

In 2019 Seim *et al.* advise that haptic technology can be used as a tool for learning new skills [17]. Aijaz *et al.* mention that the haptic sense (sense of touch) establishes a link between humans and unknown environments in a similar way as the auditory and visual senses [1]. Martinez *et al.* instructs that haptic technology has the potential to expand and transform the ways that students can experience a variety of science, technology, engineering, and math topics [13].

"Haptic technology, also known as kinesthetic communication or $3D$ touch, refers to any technology which can create an experience of touch by applying forces, vibrations, or motions to the user"[1]. "In brief, a haptic device provides position input like a mouse but also stimulates the sense of touch by applying output to the user in the form of forces" (Brewster [3], p. 2). Vardar explored the mechanisms underlying haptic perception of electrovibration [20]. Electrovibration is one approach to generate realistic haptic feedback on touch screens. Dhiab *et al.* with piezoelectric actuators of various shapes and a vibration motor were capable to provide a local vibrotactile feedback that enables multi-users and multi-touch interactions [5]. The created vibration is felt by the user fingers. Haptic devices are progressively being more used and maturated. Xu *et al.* concluded work to reduce difficulties associated to communication delays. Xu *et al.* progress result in a most stable teleoperation [22].

Presently, there are many devices developed for Haptic applications. The Haptic Device (HD) mechanical stimulation can be applied in many areas, such as surgery, education, games, virtual reality and human-computer interaction. With evolution of new technologies, haptic devices bring the possibility to accelerate the research in those areas. A HD transmits visual and physical cues to the user to realize an intended purpose such as handwriting. With some software control, using that hardware, it is possible to help medical staff in the patient rehabilitation of a given health complication, to reach the highest possible level of independence. "Rehabilitation is greatly needed by those who have sustained severe injury, often due to trauma, stroke, infection, tumor, surgery, or progressive disease. Especially for the stroke patients, rehabilitation helps improve the chances of successful recovery" (Kim [9], p. 1).

Sebastian *et al.* developed a soft robotic haptic interface with variable stiffness for rehabilitation of neurologically impaired hand function, this device can help, in a near future, patients to regain functional hand control [16]. Aijaz *et al.* developed work refers the merge of haptics with the next generation 5G cellular network. Noticing, that haptic communication provides an additional dimension over traditional audiovisual communication for truly immersive steering and control in remote environments [1]. Gao *et al.* developed work to enable a better tactile understanding for robots, resulting in a purely visual haptic prediction model [6]. All this, by classifying surfaces with haptic adjectives from visual and physical interaction data, inspired by the human visual prediction and feedback cognitive pattern. Quaid *et al.*, in March 2019, had made a patent of a new

---

[1] Wikipedia haptic technology website (2019, November 19): https://en.wikipedia.org/wiki/Haptic_technology.

haptic guidance system and method for joint replacement in the human body [15]. Vetter applies haptic feedback in physical representation and modification of digital sound data [21]. "... development of the Haptic Wave has been initiated for visually impaired sound engineers and was appreciated as a valuable tool in sound production" (Vetter [21], p. 2). Dandu *et al.* used a single actuator to generate tactile stimuli [4]. Dandu's *et al.* method is based on the frequency-dependent damping of propagation waves in the skin. Grigorii *et al.* developed a method for real-time closed loop-rendering of surface stiction on an electroadhesive surface haptic display [7]. The promising results revealed a robust function to a variety of contact conditions such as a range of normal forces, velocities and swipe directions.

Hannaford *et al.* developed work to control haptic interfaces, with the defined passivity controller (PC) and observer (PO), reached good results [8]. "The PC has several desirable properties for applications including haptic interface control. The PO and PC can both be implemented with simple software in existing haptic interface systems" (Hannaford *et al.* [8], p. 9). Srimathveeravalli *et al.* designed a shared control (SC) assistance algorithm, derived from the haptic attributes concept [19]. The SC assistance algorithm was used to train a candidate's non dominant hand with trajectory follow-up [18]. Srimathveeravalli *et al.* concluded that for a person using their hand, to follow a given trajectory, the SC technique results in the best performance. Kong *et al.* shown results, with the developed position and force control, achieves good position and force tracking performance [10]. Kong *et al.* work is based in the Proportional (P) Integral (I) Derivative (D) controller.

In this paper it is presented the work developed with a haptic device, the $3D$ Systems' "touch". It is studied which one of the controllers have the best position and force control performance and which one results in the best usability for the HD user. The main purpose of this work is to select a position controller and compare, usual controllers (P, PD controller) with distinct one's, achieved results with the haptic device for trajectory follow-up. The conclusions achieved will be applied in future work, such as handwriting and Parkinson's diseased people rehabilitation.

This paper is organized as follows. First section is the Introduction, after is the Material and Method in Sect. 2, containing the hardware and software set-up, the device forward kinematics, control mechanism and controllers presentation. In Sect. 3 are presented the results obtained with the haptic device controllers for the developed hand exercises. In Sect. 4 is the discussion about the developed work. The last section is the conclusion about the developed work.

2   Material and Method

Hardware Set-Up. The hardware used is a Haptic Device (HD) with a developed workstation to facilitate the data/exercises acquirement/follow-up. That workstation increases height to the HD Y axis in 52 mm. To control the device is used a common computer with a Universal Serial Bus (USB) port to connect the device. The used HD is the $3D$ Systems' "touch". This HD provides force feedback and allows the user, for example, to freely sculpt $3D$ clay, improve scientific or medical simulations and increase productivity with interactive training[2]. Figure 1 shows the device.

Fig. 1. 3D Systems "touch" haptic device.

Software Set-Up. The software was developed using C/C++, Matlab and Simulink. The C/C++ code is based on OpenHaptics developer software and it communicates directly with the device. OpenHaptics enables software developers to add haptics and true $3D$ navigation to a wide range of applications including $3D$ design and modeling, medical applications, games, entertainment, visualization, and simulation[3]. This software allows the developer to know the joint and gimbal angles and the position of the last HD link.

The control applied in the device was developed using MATLAB and Simulink. To make the communication between Simulink and the HD, the C/C++ software were developed as a MATLAB MEX file function[4] (based on the solution developed by Mohammadi *et al.* [14]), providing a powerful mechanism for extending the capabilities of the Simulink environment. The software flowchart can be seen in Fig. 2.

2.1 $3D$ Systems' "Touch" Kinematics

Based on the Denavit Hartenberg [2] representation (following Koul *et al.* [11] and Martin *et al.* [12]), the homogeneous transformation matrix, T_{link2}^{base}, which yields the position and orientation of the second link end (where the gimbal is

---

[2] $3D$ Systems webpage (2019, November 19): https://www.3dsystems.com/haptics-devices/touch.

[3] $3D$ Systems OpenHaptics website (2019, November 19): https://www.3dsystems.com/haptics-devices/openhaptics.

[4] MATLAB website introducing MEX Files (2019, November 19): https://www.mathworks.com/help/matlab/matlab_external/introducing-mex-files.html.

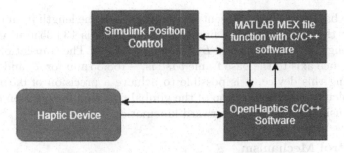

Fig. 2. Hardware and software flowchart.

connected) with respect to the base frame *base*, is presented as follows.

$$T_{link2}^{base} = \begin{bmatrix} c_1 & s_1s_3 & -c_3s_1 & -s_1(L_1c_2 + L_2s_3) \\ 0 & c_3 & s_3 & L_1s_2 - L_2c_3 \\ s_1 & -c_1s_3 & c_1c_3 & c_1(L_1c_2 + L_2s_3) \\ 0 & 0 & 0 & 1 \end{bmatrix}$$

Mentioning, θ_1, θ_2 and θ_3 identify the rotation joint angles of the haptic device, L_1 and L_2 identify the robotic link lengths shown in Fig. 10. In the previous matrix c_1 shall be understood as $cos(\theta_1)$, s_2 as $sin(\theta_2)$.

The homogeneous transformation from the second link end to the gimbal tip is given by the matrix T_{Gtip}^{link2}.

$$T_{Gtip}^{link2} = \begin{bmatrix} c_4 & -s_4s_5 & c_5s_4 & -L_3c_5s_4 \\ 0 & c_5 & s_5 & -L_3s_5 \\ -s_4 & -c_4s_5 & c_4c_5 & -L_3c_4c_5 \\ 0 & 0 & 0 & 1 \end{bmatrix}$$

Using the T_{link2}^{base} matrix, the position of the second link end in relation to the world reference axis, is given by Eqs. 1, 2 and 3.

$$p_x = -s_1(L_1c_2 + L_2s_3); \tag{1}$$

$$p_y = L_1s_2 - L_2c_3 + d_y; \tag{2}$$

$$p_z = c_1(L_1c_2 + L_2s_3) + d_z; \tag{3}$$

The position of the gimbal tip in relation to the world reference axis, is given by Eqs. 4, 5 and 6, resulting from the multiplication of T_{link2}^{base} with T_{Gtip}^{link2}.

$$P_x = -s_1(L_1c_2 + L_2s_3) + L_3c_3c_4c_5s_1 - L_3c_1c_5s_4 - L_3s_1s_3s_5; \tag{4}$$

$$P_y = L_1s_2 - L_2c_3 - L_3c_3s_5 - L_3c_4c_5s_3 + d_y; \tag{5}$$

$$P_z = c_1(L_1c_2 + L_2s_3) + L_3c_1s_3s_5 - L_3c_5s_1s_4 - L_3c_1c_3c_4c_5 + d_z; \tag{6}$$

It is necessary to add d_y and d_z to the position equations due to the world reference axis being translational, in d_y under the Y axis and d_z under the Z axis, to the base axis. In the presented equation, L_3 is the length from the second link end to the gimbal tip. The first link (L_1) length is 133.35 mm, the second link (L_2) length is 133.35 mm and L_3 is equal to 30 mm. The translation from the world reference axis to the base frame *base* is, -168.35 mm for d_z and 23.35 mm for d_y. Using this device, it is possible to achieve a precision of 0.5 mm in the position calculation. The position of the gimbal tip, used for position control, is calculated using the presented forward kinematics.

2.2   Control Mechanism

The data from the HD is obtained in the Simulink as an output of the HD Simulink block. This data are the joint and gimbal angles and the second link end position. The input of that block, and the HD basis of control, are the forces intended to apply in the gimbal tip to follow the intended trajectory.

The HD is controlled with forces as an input of the HD Simulink block. The forces to be applied should always be defined for the the X, Y and Z directions of the world reference axis. The applied force is calculated using a controller with negative feedback of the position ($P_{hd}(x_{hd}, y_{hd}, z_{hd})$). The intended position ($P_{intended}(x, y, z)$), is the input in the control mesh. The difference between P_{hd} and $P_{intended}$ multiplied by the controller gain function results in the vector force $[F_x \ F_y \ F_z]^T$. The input in the HD Simulink block is defined by Eq. 7.

$$F = E \times C_{function} \ \forall \ E = P_{intended} - P_{hd} \tag{7}$$

In this equation, the $C_{function}$ is one of the controller functions presented following.

2.3   Controller Functions

Proportional Gain Controller. The proportional gain controller follows Eq. 8 (described as a continuous time signal). The controller diagram is shown in Fig. 3.

$$f(t) = e(t) \times K_P \ \forall \ e(t) = P_{intended}(t) - P_{hd}(t) \tag{8}$$

The simple proportional gain position controller result in a good position control in the real world. The gimbal tip of the HD has an error associated to this control technique, however, that error is minimal for the intended purpose. The proportional gain is $K_P = 0.1755$.

Proportional Derivative Gain Controller. The proportional derivative gain controller follows Eq. 9 (described as a continuous time signal). The controller diagram is shown in Fig. 4.

$$f(t) = e(t) \times K_P + K_D \frac{d\, e(t)}{dt} \ \forall \ e(t) = P_{intended}(t) - P_{hd}(t) \tag{9}$$

The proportional derivative gain position controller result in a good position control in the real world. The gimbal tip of the HD shows excellent results using this controller. The position controller diagram is shown in Fig. 4. The proportional gain is $K_P = 0.05$ and the derivative gain is $K_D = 0.7$.

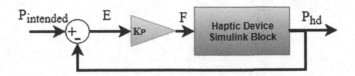

Fig. 3. Proportional (P) position controller diagram, used in Simulink. The error (E) is in millimeters and the Forces (F) in Newtons.

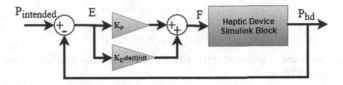

Fig. 4. Proportional derivative (PD) position controller diagram, used in Simulink.

Logarithmic Function Based Controller. The logarithmic function based controller follows Eq. 10. The controller diagram is shown in Fig. 5.

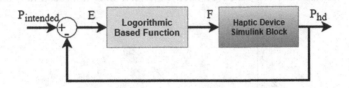

Fig. 5. Logarithmic function based position controller diagram, used in Simulink.

$$f(t) = e(t) \times Logarithmic_{function} \ \forall \ e(t) = P_{intended}(t) - P_{hd}(t) \qquad (10)$$

The force to be applied in each direction, the X, Y and Z of the reference world axis, have to be positive or negative. It was developed a logarithmic based function to obtain, following Eq. 10, positive and negative forces. The resulting force calculated using the developed function can be seen in Fig. 6, for an error smaller than 10 mm.

Sigmoid Based Function Controller. The sigmoid function based controller follows Eq. 11. The controller diagram is shown in Fig. 7.

$$f(t) = e(t) \times Sigmoid_{function} \ \forall \ e(t) = P_{intended}(t) - P_{hd}(t) \qquad (11)$$

The resulting actuation force calculated from the logarithmic function is high for lower error values, the sigmoid function was developed to achieve lower forces for

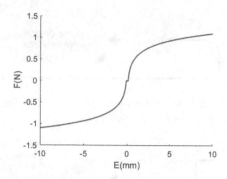

Fig. 6. Resulting force calculated with the logarithmic based function for a position error smaller than 10 mm.

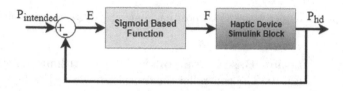

Fig. 7. Sigmoid function based position controller diagram, used in Simulink.

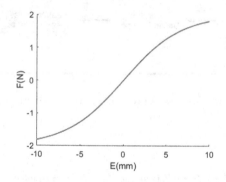

Fig. 8. Resulting force calculated with the sigmoid based function for a position error smaller than 10 mm.

the lower error values. The resulting force, calculated from the applied sigmoid based function, could be seen in the Fig. 8. Comparing Fig. 6 with Fig. 8, it could be concluded that the force curve slope is much lower for the force calculated with the sigmoid based function.

2.4 Developed Hand Exercises

The hand exercises were developed in Simulink. It is used, as mentioned in Sect. 2.2, the HD simulink block diagram to obtain the intended variables as an output (in this case the joint angles to calculate the gimbal tip position using the

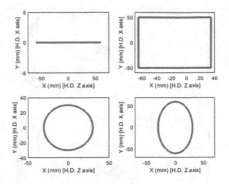

Fig. 9. Developed trajectories for the user to follow-up with the haptic device.

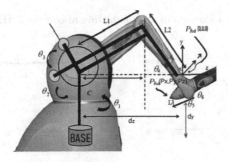

Fig. 10. HD initial conditions. The HD positioning axis (world reference axis) is denoted as the X, Y and Z near the gimbal.

presented forward kinematics). The input of the control mesh is an array of data (MATLAB Data Structure) or mathematical functions to reach the intended purposes. The hand exercises, defined as trajectories, are composed of (x, y, z) coordinates. The developed exercises (trajectories) can be seen in Fig. 9.

It is necessary to understand the world reference axis that is shown in Fig. 10. The calculated gimbal tip position, using the presented forward kinematics, result in coordinates in the world reference axis.

3   Results

3.1   Straight Line Follow-Up

In this hand exercise it is applied forces in the HD with the intent that the gimbal tip follows a straight line trajectory under the Z HD axis. The result of the various controllers in this exercise could be seen in Fig. 11. The best result is achieved with the logarithmic function based controller.

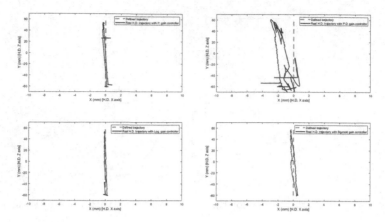

Fig. 11. HD trajectory follow-up of a straight line above the Z HD axis for the various controllers.

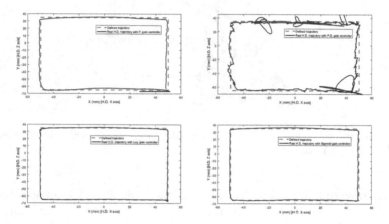

Fig. 12. HD trajectory follow-up of a square geometric figure for the various controllers.

3.2   Square Follow-Up

In this hand exercise it is applied forces in the HD with the intent that the gimbal tip follows a square trajectory in the XZ HD plane. The result of the various controllers in this exercise could be seen in Fig. 12. The best result is achieved with the logarithmic and sigmoid function based controller.

3.3   Circle Follow-Up

The purpose to this exercise is the user to follow a circular trajectory perpendicular to the Y HD axis plane. The input in the control mesh is a sinusoidal wave in the X and a cosine wave in the Z HD axis. Geometrically, the input in this exercise characterizes a circle with the center coordinates in $(0, 0, 0)$ with a

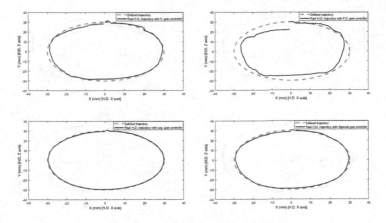

Fig. 13. HD trajectory follow-up of a circle geometric figure for the various controllers.

30 mm radius. The result of the various controllers in this exercise could be seen in Fig. 13. The best result is achieved with the logarithmic and sigmoid function based controller.

3.4   Ellipse Follow-Up

In this exercise the user has to follow an elliptical trajectory with the HD gimbal tip. Geometrically, the input in this exercise characterizes an ellipse with a center in the $(0, 0, 0)$ coordinate position, with a semi-major axis of 60 mm and a semi-minor axis of 30 mm. The result of the various controllers in this exercise could be seen in Fig. 14. The best result is achieved with the sigmoid function based controller.

4   Discussion

It is known that the proportional and proportional derivative controllers could achieve greater result, nonetheless, the presented results were obtained with the same conditions for all controllers (sample time and sample size).

In works developed with haptic devices, not only the position precision, but also the user perception of the forces applied by the device should be taken in consideration. Since the position precision obtained with the used controllers is high, it is useful to compare the results obtained with each controller in the trajectory follow-up and force feedback to the device user. Comparing Fig. 6 with Fig. 8, it could be concluded that the force curve slope is much lower for the force calculated with the sigmoid based function. With this characteristic it is achieved a best perception of the applied force for the haptic device user.

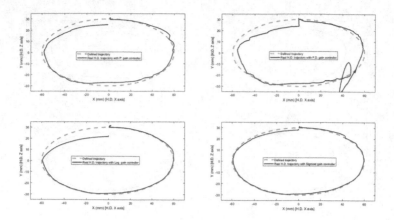

Fig. 14. HD trajectory follow-up of an ellipse geometric figure for the various controllers.

5 Conclusion

The logarithmic and sigmoid function based controllers result in a best force control applied by the device. Verifying the results, those two controllers reach the best position control under the trajectory follow-up. The resulting position is almost the same as the planned trajectory position (in millimeters).

Using a haptic device for force feedback, position control and trajectory follow-up it is concluded that the optimal results are achieved with a sigmoid or logarithmic function based controller. These two controllers, with a non-linear gain, result in a smoother force variation. This smoother force variation results in a best guidance and perception to the device user.

The characteristics of each controller should be tuned for the intended purpose. Moreover, it should be achieved the best ratio of position precision and device force feedback for the work in development.

The logarithmic and sigmoid function based controllers achieve the lowest position error and finest force feedback smoothness. It is concluded that those should be the position controllers used with haptic devices in future work.

Future work, based in the presented information, will be composed of Chinese alphabet learning, rehabilitation for post-stroke and Parkinson's diseased people and handwriting rehabilitation. To achieve a best force perception to the HD user, the Gudermannian, the Hyperbolic tangent, the Smoothstep and the generalized logistic functions shall be tested.

References

1. Aijaz, A., Dohler, M., Aghvami, A.H., Friderikos, V., Frodigh, M.: Realizing the tactile internet: haptic communications over next generation 5G cellular networks. IEEE Wirel. Commun. **24**(2), 82–89 (2016)

2. Beckman, J.A.: The PHANTOM Omni as an under-actuated robot (2007)
3. Brewster, S.: Impact of haptic 'touching' technology on cultural applications (2005)
4. Dandu, B., Shao, Y., Stanley, A., Visell, Y.: Spatiotemporal haptic effects from a single actuator via spectral control of cutaneous wave propagation, pp. 425–430 (2019)
5. Dhiab, A.B., Hudin, C.: Confinement of vibrotactile stimuli in narrow plates. In: 2019 IEEE World Haptics Conference (WHC), pp. 431–436. IEEE (2019)
6. Gao, Y., Hendricks, L.A., Kuchenbecker, K.J., Darrell, T.: Deep learning for tactile understanding from visual and haptic data. In: 2016 IEEE International Conference on Robotics and Automation (ICRA), pp. 536–543. IEEE (2016)
7. Grigorii, R., Peshkin, M., Colgate, J.: Stiction rendering in touch, pp. 13–18 (2019)
8. Hannaford, B., Ryu, J.-H.: Time-domain passivity control of haptic interfaces. IEEE Trans. Robot. Autom. **18**(1), 1–10 (2002)
9. Kim, Y.K., Yang, X.: Hand-writing rehabilitation in the haptic virtual environment. In: 2006 IEEE International Workshop on Haptic Audio Visual Environments and their Applications (HAVE 2006), pp. 161–164, November 2006
10. Kong, R., Dong, X., Liu, X.: Position and force control of teleoperation system based on PHANTOM Omni robots. Int. J. Mech. Eng. Robot. Res. **5**(1), 57 (2016)
11. Koul, M., Kumar, P., Singh, P., Muniyandi, M., Saha, S.: Gravity compensation for PHANTOM Omni haptic interface, May 2010
12. Martín, J.S., Triviño, G.: A study of the manipulability of the PHANToM OMNI haptic interface (2006)
13. Martinez, M.O., et al.: 3-D printed haptic devices for educational applications. In: 2016 IEEE Haptics Symposium (HAPTICS), pp. 126–133. IEEE (2016)
14. Mohammadi, A., Tavakoli, M., Jazayeri, A.: Phansim: a simulink toolkit for the sensable phantom haptic devices. In: Proceedings of the 23rd CANCAM, Canada, vol. 11, pp. 787–790 (2011)
15. Quaid, A., et al.: Haptic guidance system and method. US Patent App. 10/231,790, 19 March 2019
16. Sebastian, F., Fu, Q., Santello, M., Polygerinos, P.: Soft robotic haptic interface with variable stiffness for rehabilitation of neurologically impaired hand function. Front. Robot. AI **4**, 69 (2017). https://www.frontiersin.org/article/10.3389/frobt.2017.00069
17. Seim, C., et al.: Towards haptic learning on a smartwatch. In: Proceedings of the 2018 ACM International Symposium on Wearable Computers, pp. 228–229. ACM (2018)
18. Srimathveeravalli, G., Gourishankar, V., Kesavadas, T.: Comparative study: virtual fixtures and shared control for rehabilitation of fine motor skills. In: Second Joint EuroHaptics Conference and Symposium on Haptic Interfaces for Virtual Environment and Teleoperator Systems (WHC 2007), pp. 304–309, March 2007
19. Srimathveeravalli, G., Kesavadas, T.: Motor skill training assistance using haptic attributes, pp. 452–457, April 2005
20. Vardar, Y.: Tactile perception by electrovibration. Ph.D. thesis, Koç University (2018)
21. Vetter, J.: Tangible signals-physical representation of sound and haptic control feedback. In: Proceedings of the Thirteenth International Conference on Tangible, Embedded, and Embodied Interaction, pp. 741–744. ACM (2019)
22. Xu, X., Steinbach, E.: Elimination of cross-dimensional artifacts in the multi-dof time domain passivity approach for time-delayed teleoperation with haptic feedback, pp. 223–228 (2019)

Human Activity Recognition Using an Accelerometer Magnitude Value

Jhon Ivan Pilataxi Piltaxi$^{(\boxtimes)}$, María Fernanda Trujillo Guerrero,
Vanessa Carolina Benavides Laguapillo,
and Jorge Andrés Rosales Acosta

Escuela Politécnica Nacional, Ladrón de Guevara E11·253, Quito, Ecuador
jhon.pilataxi@epn.edu.ec

Abstract. Human activity recognition (HAR) is important for many applications to help healthcare and support systems due to fast increase of senior population worldwide. This paper describes a human activity recognition framework based on feature selection techniques from a waist single accelerometer. The objective is to identify the most important features to recognize static and dynamic human activities based on module acceleration, since a public database. A set of time and frequency features are getting from the module, so to analyze the impact of the features on the performance of the recognition system, a ReliefF algorithm is applied. Finally, a multiclass classification model is implemented thought Support Vector Machine (SVM). Experimental results indicate that the accuracy of the propose model is over of 85%, this percentage is like other works in which use each axes accelerometer. The advantage of this work is the use of the module value that allow identify the activity independently of the sensor position, also it reduces the computer resources.

Keywords: Waist single accelerometer · Accelerometer module value · Time and frequency features · Support Vector Machine (SVM) · Human Activity Recognition (HAR)

1 Introduction

Globally, the number of older persons is growing faster than the number of people in all younger age groups. The projections indicate that in 2050 there will be more older persons aged 60 or over than adolescents and youth at ages 10–24 years (2.1 billion versus 2.0 billion). The number of people at very advanced ages is increasing too: the global population aged 80 years or over is projected to triple between 2017 and 2050, increasing from 137 million to 425 million [1]. For these reasons is important improve the healthcare and support systems focused in senior population.

Nowadays, there are different researches based in Human Activity Recognition (HAR) due to its substantial contributions in the way to assistive elderly health care technologies [1–4], because that provides information about people's behavior and actions. One aim of the recognition is to provide valuable context aware information of users through the interpretation of sensory information collected from users and the surrounding environment. The activity information related to users are collected from

M. Botto-Tobar et al. (Eds.): ICAT 2019, CCIS 1194, pp. 462–472, 2020.
https://doi.org/10.1007/978-3-030-42520-3_37

users using body-worn sensors, and then retrieved information are exploited to carry out the desired recognition task [5, 6]. Depending the duration there are postural transitions and basic activities. The basic activities have a long duration, and they can be either dynamic or static [7].

The way to acquire data to this purpose has changed how is showed in works like [8–10], these developments use video cameras, microphones, GPSs, accelerometers and inertial measurement unit (IMU), acoustic sensors or force plate. The accelerometers and the IMUs have a lot of popularity for people's privacy is a downside of vision technologies that limits their use in every location. A small inertial measurement unit (IMU) has been development with the purpose to be comfortable, no invasive, reliable, and wearable like MetaWear or Arduino. In this way, a common smart phone with a three axial accelerometer is able to provide enough data to recognize activities like sitting, lying, standing, walking, and upstairs how the WISDM Activity Prediction [11], UCI Machine Learning Repository [12], Kaggle [13], or ExtraSensor [14] datasets show. The treatment of signal sensor depends of aspects such as kind the sensor, signal characteristics to establish, and the activities will recognize.

A static reference system is used with the sensor because it allows to obtain acceleration axes characteristics independently, for to reduce the error most studies the sensors are quiet in a specific place in the body like Fig. 1 shows. On the contrary we propose to use the acceleration module that is few used to HAR for it allows change the reference system to be an independent system, and it could reduce the number of features to research.

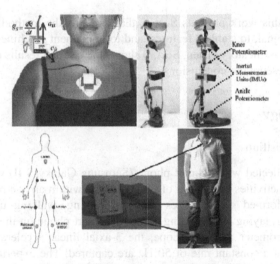

Fig. 1. Different sensor location [26–29]

The noise inside the data could interfere in the signal for this reason a low pass/high cut filter implementation is important. When the noise is reduced the sensor signal must be divided to extract the features, so sliding window approach to divide the sensor signal into smaller time windows is used works like [15, 16]. The most common features for activity recognition were calculated from time and frequency domain. In time domain, features include mean, standard deviation, median, maximum, minimum values of magnitude data, squared sum of magnitude data below 25 percentile, and squared sum of magnitude data below 75 percentile. Features in frequency domain include the peak frequency in the spectrum of magnitude data below 5 Hz frequency, number of peaks in the spectrum data, and integral of the spectrum of magnitude data.

For the recognition of human activities, machine learning is the most uses because it has a lot of classification techniques like SVM [17, 18]. The objective of the support vector machine algorithm is to find a hyperplane in an N-dimensional space (N—the number of features) that distinctly classifies the data points like Fig. 2 shows [19].

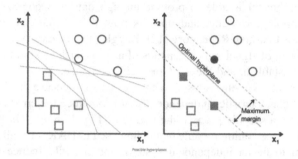

Fig. 2. Possible hyperplanes of SVN [19]

In this order this work presents Sect. 2 that explains the methodology applied to pre-process the signal, to stablish features, and to implement machine learning technic while the SVM classificatory. Then, Sect. 3 shows the test and results of this work, and finally Sect. 4 presents the conclusions.

2   Methodology

2.1   Data Acquisition

The data were collected with a smart phone (Samsung Galaxy S II) while performing certain everyday activities in a group of 30 volunteers within an age bracket of 19–48. Each person performed six activities such as walking, walking upstairs, walking downstairs, sitting, laying and standing wearing a smart phone on the waist. Using its embedded accelerometer and gyroscope, the 3-axial linear acceleration and 3-axial angular velocity at a constant rate of 50 Hz are captured. The experiments have been video-recorded to label the data manually. The obtained dataset has been randomly partitioned into two sets, where 70% of the volunteers was selected for generating the training data and 30% the test data. The sensor signals (accelerometer and gyroscope) were pre-processed [12].

2.2   Features Extraction

The features extracted from the sliding windows signals for activity recognition were calculated from time and frequency domain. For the time domain is used magnitude of acceleration defined by Eq. 1, while the frequency features employed relies on the Short Time Fourier Transform response defined by Eq. 2.

$$a_{xyz}(t) = \sqrt{a_x(t)^2 + a_y(t)^2 + a_z(t)^2} \tag{1}$$

Where $a_i(t)$ represent the acceleration in x, y and z-axis respectively.

$$A_h = \sum_{n=1}^{N} a_{xyz_n} e^{-i2\pi\frac{n}{hN}} \tag{2}$$

Where h = 1, ..., N, A_h^W is the output of Fourier Transform for each windows, composed for a set N complex number that represent the amplitude and phase of the harmonics, and N is the length of windows, in this case 128.

The Table 1 shows the features extracted.

Table 1. Features set

Features	Description
Mean	The average value of the acceleration module over the window
Maximum	The maximum value of acceleration module
Minimum	The minimum value of acceleration module
Median	The median signal value
Standard deviation	Measure of the spreadness of the signal
Variance	The square of standard deviation
Root mean square	The average of quadratic mean value
Range	The difference between maximum and minimum value of acceleration module
Skewness	The degree of asymmetry of acceleration module distribution
Kurtosis	The degree of peakedness of acceleration module distribution
Interquartile range	The difference between the 75th and the 25th percentiles of the acceleration module
Mean Jerk	The average value of Jerk (acceleration derivative)
Energy_0.68 Hz	Sum of the harmonics between 0.1 and 0.69 Hz
Energy_3 Hz	Sum of the harmonics between 0.7 and 3 Hz
Energy_5 Hz	Sum of the harmonics between 3.1 and 5 Hz
Mean harmonics	Average value of harmonics, without DC component
Pos. max. harmonic	Frequency of the maximum harmonic components
Harmonics kurtosis	The degree of peakedness of harmonic components
Harmonics skewness	The degree of peakedness of asymmetry components
Harmonics interquartile range	The difference between the 75th and the 25th percentiles of harmonics components

2.3   Features Normalization

Due to the units of the features described in Table 1, it is necessary normalize all the features. In this work, we use mean value $(\overline{X})$ and standard deviation (σ), according the Eq. 3, to normalize the features.

$$X_{normalized} = \frac{X - \overline{X}}{\sigma} \tag{3}$$

2.4   Features Selection

The 20 features extracted were ordered according their importance using Relieff algorithm. For each windows a set of 20 features are extracted and sorted as shown in the following vector.

Features = [Pos. Max. Harmonic, Range, Maximum, Energy_3 Hz, Standard Deviation, ...
Minimum, Median, Mean harmonics, Harmonics skewness, Harmonics kurtosis, ...
Interquartile Range, Harmonics interquartile range, Energy_0.68 Hz, Energy_5 Hz, ...
Root Mean Square, Mean, Variance, Skewness, Kurtosis, Mean Jerk]

The Fig. 3 shows the weight of each feature extracted, which represent their importance on the performance of classification. Since all features have a positive weight, all of them can be used in the training.

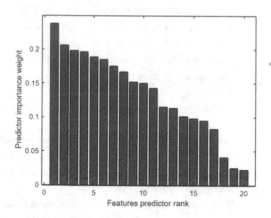

Fig. 3. Predictor importance weight of features

2.5   Training

The Fig. 4 shows the data used for the training (70% of database) in the plane formed by the features with greater weight: in Fig. 4a the x-axis represent the position of maximum harmonic, while in Fig. 4b represent maximum value of magnitude of acceleration, and y-axis in both correspond to the range.

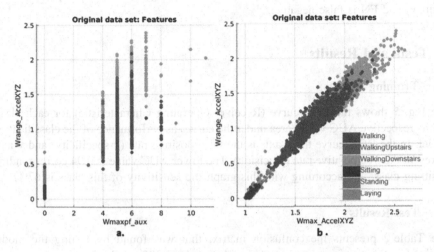

Fig. 4.   Training data

In this work, we used Support Vector Machine (SVM) binary classifier, which are generalized to the multiclass, case through a One-Vs-All (OVA) approach. The parameter for SVM are 10-fold cross validation, Gaussian kernel function. We have six class, which represent six principal human activities: walking, walking upstairs, walking downstairs, sitting, standing and lying.

2.6   Test

In order to quantify the performance of the model in the face of unknown data, statistical indicator such sensitivity, specificity, precision of each class, and the general accuracy of the model are used.

$$Sensitivity(\%) = \frac{TP}{TP + FN} * 100 \tag{4}$$

$$Specificity(\%) = \frac{TN}{TN + FP} * 100 \tag{5}$$

$$Precision(\%) = \frac{TP}{TP+FP} * 100 \tag{6}$$

$$Accuracy(\%) = \frac{TP+TN}{TP+TN+FP+FN} * 100 \tag{7}$$

Where TP represent true positive, TN mean true negative, FP correspond to false positive, and FN is false negative.

3 Tests and Results

3.1 Training Results

The Fig. 5 shows the ROC curve (Receiver Operating Characteristic) for each class, and its respective AUC value (area under the curve) as performance of the classification model. In the ROC curve the x-axis is the false positive rate (1-specificity) and y-axis represent the true positive rate (precision). The lower AUC value is 0.94 corresponding to sitting class, and according with this graph the sensitivity of this class is 67%.

3.2 Test Results

The Table 2 presents the confusion matrix, that was found by testing the model obtained in the training process tested with test data (30% of database).

Table 2. Confusion matrix

True class	Predicted class						Sensitivity (%)
Activity	Walking	Upstairs	Downstairs	Sitting	Standing	Laying	
Walking	**743**	20	3	0	0	0	95.36
Upstairs	38	**416**	17	0	0	0	88.32
Downstairs	1	34	**385**	0	0	0	91.67
Sitting	0	0	0	**328**	76	85	67.07
Standing	0	0	0	62	**469**	1	88.16
Laying	0	0	0	54	12	**471**	87.71
Precision (%)	92.38	88.51	95.06	73.87	84.2	84.56	
Specificity (%)	98.41	97.82	99.21	95.25	96.35	96.43	

The precision of the proposed model is calculated by the Eq. 7, and results 86.3%.

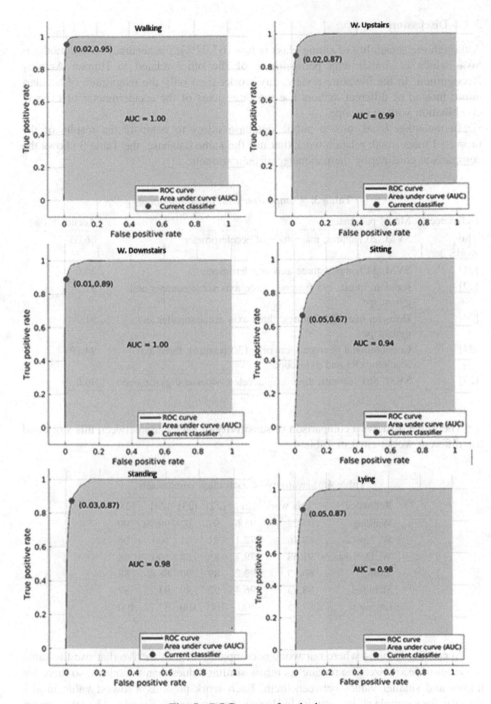

Fig. 5. ROC curve of each class

3.3   Discussion

Although the sensibility of sitting class is low (67.07%), the accuracy of the model is 86% which is similar with performance of the other related to Human Activity Recognition. In the literature review, any works used only the magnitude of acceleration; instead of different authors use the three axes of the accelerometer or even in combination with gyroscope.

In the other hand, uses a public database allow to compare the results of the proposed model with relation work that use the same database, the Table 3 shows the comparison considering the parameter of their models.

Table 3. Comparison with other authors

Reference	Model parameter	Accuracy (%)
This works	SVM, 20 features, magnitude of accelerometer	86.03
[21]	SVM, 34 features, three axis accelerometer	89.0
[22]	Random forest, 195 features, three axis accelerometer and gyroscope	91.3
[23]	Decision tree, 561 features, three axis accelerometer and gyroscope	91.7
[24]	Convolutional neuronal network, 130 features, three axis accelerometer and gyroscope	94.79
[25]	SVM, 561 features, three axis accelerometer and gyroscope	96.0

The Table 4 shows a comparison of sensitivity of each class between this work and the different works cited in Table 4.

Table 4. Sensitivity of each class comparison

Reference	This work	[21]	[22]	[23]	[24]	[25]
Walking	95.27	95.6	97	92	98.99	99
W. Upstairs	88.26	72.1	91	91	100	96
W. Downstairs	91.68	79.7	85	88	100	98
Sitting	66.95	92.2	89	90	88.8	88
Standing	88.13	96.4	92	89	93.23	97
Laying	87.75	100	100	100	87.71	100

According Table 4 where our work is compared with other five that use the same activities to recognize. The researches throw similar values than the others, so there are higher and smaller values between them. Each work presents a lowest value in any activity, for example sitting activity in Ronao, Anguita and this works. The table shows different works where applied machine learning technics and a variety of features.

4 Conclusion

In this paper, the problem of HAR focused to senior people is studied with three axial accelerometer sensor smart phone data located on the waist. We proposed a method that uses a filter to reduce the noise, after that a slide windows to extract the features in time and frequency domain. Then the SVM classificatory is applied to obtain the activity recognition. The advantage is the independence of the system because the module acceleration value is token like a reference.

The results show a precision over 74%, and a sensitivity over 66.95% between laying and sitting activities, and a precision over 83%, and a sensitivity over 87% between the others. Of those results we concluded than similar activities like sitting and laying need to be analyzed with more features or more sensors. On the other hand, our work achieves 86% of accuracy while other works have a higher percentage, but our work employs only acceleration module value beside it gives independency in the position use of the sensor.

In the future, the work will be improved if we collect more body activities data to analysis the transition postures of senior population. It will extend the ability of recognition more daily activities, and prediction complex daily activities.

References

1. Abhayasinghe, N., Murray, I.: Human activity recognition using thigh angle derived from single thigh mounted IMU data. In: 2014 International Conference on Indoor Positioning and Indoor Navigation (IPIN), pp. 111–115. IEEE (2014)
2. Chen, L., Hoey, J., Nugent, C., Cook, D., Yu, Z.: Sensor-based activity recognition. IEEE Trans. Syst. Man Cybern. Part C Appl. Rev. 42, 790–808 (2012)
3. Cook, D.J., Das, S.K.: Pervasive computing at scale: transforming the state of the art. Pervasive Mob. Comput. 8, 22–35 (2012)
4. Campbell, A., Choudhury, T.: From smart to cognitive phones. IEEE Pervasive Comput. 11 (3), 7–11 (2012)
5. Kallur, D.C.: Human localization and activity recognition using distributed motion sensors. Ph.D. thesis, Oklahoma State University (2014)
6. Motion capture systems - optitrack. https://www.naturalpoint.com/optitrack/
7. Bao, L., Intille, S.S.: Activity recognition from user-annotated acceleration data. In: Ferscha, A., Mattern, F. (eds.) Pervasive 2004. LNCS, vol. 3001, pp. 1–17. Springer, Heidelberg (2004). https://doi.org/10.1007/978-3-540-24646-6_1
8. Yatani, K., Truong, K.N.: Bodyscope: a wearable acoustic sensor for activity recognition. In: Proceedings of the 2012 ACM Conference on Ubiquitous Computing, pp. 341–350. ACM (2012)
9. Yu, C.-R., Wu, C.-L., Lu, C.-H., Fu, L.-C.: Human localization via multi-cameras and floor sensors in smart home. In: Proceedings of the IEEE International Conference on Systems, Man and Cybernetics (SMC), vol. 5, pp. 3822–3827, October 2006
10. Cheok, A.D., Li, Y.: Ubiquitous interaction with positioning and navigation using a novel light sensor-based information transmission system. Pers. Ubiquitous Comput. 12(6), 445–458 (2008)

11. Kwapisz, J.R., Weiss, G.M., Moore, S.A.: Activity recognition using cell phone accelerometers. In: Proceedings of the Fourth International Workshop on Knowledge Discovery from Sensor Data (at KDD 2010), Washington DC. WISDM Activity Prediction (2010). http://www.cis.fordham.edu/wisdm/dataset.php
12. Reyes-Ortiz, J.L., Anguita, D., Ghio, A., Oneto, L., Parra, X.: UCI Machine Learning Repository. https://archive.ics.uci.edu/ml/datasets/Human+Activity+Recognition+from+Co ntinuous+Ambient+Sensor+Data
13. Kaggle. https://www.kaggle.com/malekzadeh/motionsense-dataset
14. ExtraSensor. http://extrasensory.ucsd.edu/
15. Hong, Y.J., Kim, I.J., Ahn, S.C., Kim, H.G.: Mobile health monitoring system based on activity recognition using accelerometer. Simul. Model. Pract. Theory 18(4), 446–455 (2010)
16. Jimenez, A.R., Seco, F., Prieto, C., Guevara, J.: A comparison of pedestrian dead-reckoning algorithms using a low-cost MEMS IMU. In: Proceedings of the IEEE International Symposium on Intelligent Signal Processing (WISP), pp. 37–42, August 2009
17. Chernbumroong, S., Atkins, A.S., Yu, H.: Activity classification using a single wrist-worn accelerometer. In: 2011 5th International Conference on Software, Knowledge Information, Industrial Management and Applications (SKIMA), pp. 1–6. IEEE (2011)
18. Guo, G., Li, S.Z.: Content-based audio classification and retrieval by support vector machines. IEEE Trans. Neural Netw. 14(1), 209–215 (2003)
19. Gandhi, R.: Support Vector Machine — Introduction to Machine Learning Algorithms, Towards Data Science (2018). https://towardsdatascience.com/support-vector-machine-introduction-to-machine-learning-algorithms-934a444fca47. Acceded 12 Oct 2019
20. Anderson, I., et al.: Shakra: tracking and sharing daily activity levels with un-augmented mobile phones. Mob. Netw. Appl. 12(2–3), 185–199 (2007)
21. Anguita, D., Ghio, A., Oneto, L., Parra, X., Reyes-Ortiz, J.L.: Human activity recognition on smartphones using a multiclass hardware-friendly support vector machine. In: Bravo, J., Hervás, R., Rodríguez, M. (eds.) IWAAL 2012. LNCS, vol. 7657, pp. 216–223. Springer, Heidelberg (2012). https://doi.org/10.1007/978-3-642-35395-6_30
22. Bulbul, E., Cetin, A., Alper, I.: Human activity recognition using smartphones. In: 2nd International Symposium on Multidisciplinary Studies and Innovative Technologies (ISMSIT), Ankara, Turkey (2018)
23. Anguita, D., Ghio, A., Oneto, L., Parra, X., Reyes, J.: A public domain dataset for human activity recognition using smartphones. In: ESANN 2013 Proceedings on European Symposium on Artificial Neural Networks, Computational Intelligence and Machine Learning, Bruges (2013)
24. Uddin, M.T., Billah, M.M., Hossain, M.F.: Random forests based recognition of human activities on smartphones. In: 5th International Conference on Informatics, Electronics and Vision (ICIEV), Dhaka, Bangladesh. IEEE (2016)
25. Ronao, C., Cho, S.: Human activity recognition with smartphone sensors using deep learning neural networks. Expert Syst. Appl. 59, 235–244 (2016)
26. Najafi, B., Aminian, K., Paraschiv-Ionescu, A., Loew, F., Bula, C.J., Robert, P.: Ambulatory system for human motion analysis using a kinematic sensor: monitoring of daily physical activity in the elderly. IEEE Trans. Biomed. Eng. 50(6), 711–723 (2003)
27. Hemmati, S.: Detecting postural transitions: a robust wavelet-based approach. In: Annual International Conference of the IEEE Engineering in Medicine and Biology Society (2016)
28. Ganea, R., Paraschiv-Ionescu, A., Aminian, K.: Detection and classification of postural transitions in real-world conditions. IEEE Trans. Neural Syst. Rehabil. Eng. 20(5), 688–696 (2012)
29. Doulah, A., Shen, X., Sazonov, E.: Early detection of the initiation of sit-to-stand posture transitions using orthosis-mounted sensors. Sensors 17(12), 2712 (2017)

Semantic Segmentation of Weeds and Crops in Multispectral Images by Using a Convolutional Neural Networks Based on U-Net

Miguel Ángel Chicchón Apaza[1](✉) (iD),
Héctor Manuel Bedón Monzón[1](✉) (iD), and Ramon Alcarria[2] (iD)

[1] Exponential Technology Group (GITX-ULIMA),
Institute of Scientific Research (IDIC), University of Lima, Lima, Peru
chicchon.am@gmail.com, hbedon@ulima.edu.pe
[2] Telematics Engineering Department, Technical University of Madrid (UPM),
Madrid, Spain

Abstract. A first step in the process of automating weed removal in precision agriculture is the semantic segmentation of crops, weeds and soil. Deep learning techniques based on convolutional neural networks are successfully applied today and one of the most popular network architectures in semantic segmentation problems is U-Net. In this article, the variants in the U-Net architecture were evaluated based on the aggregation of residual and recurring blocks to improve their performance. For training and testing, a set of data available on the Internet was used, consisting of 60 multispectral images with unbalanced pixels, so techniques were applied to increase and balance the data. Experimental results show a slight increase in quality metrics compared to the classic U-Net architecture.

Keywords: Deep learning · Convolutional neural network · U-Net · Semantic segmentation · Precision agriculture

1 Introduction

The sustainability of agriculture is one of Sustainable Development Objectives (SDO) of the United Nations. To achieve this objective, new smart farming methods are required to increase or maintain crop yields minimizing environmental impact. Precision agriculture techniques achieve this goal through the spatial study of key indicators of crop health and the application of treatments such as herbicides, pesticides and fertilizers, only in relevant areas [1].

Conventional weed control systems apply uniformly with the same dose of herbicide in the entire field. In contrast, the new perception-controlled elimination systems offer the potential to perform a treatment for each plant, for example, by selective spraying or mechanical weed control. However, this process requires a plant classification system that can analyze the image data recorded in the field in real time and label

The original version of this chapter was revised: the author's name has been changed as "Ramon Alcarria". The correction to this chapter is available at
https://doi.org/10.1007/978-3-030-42520-3_52

individual plants as crops or weeds [2]. Field images acquired with these new systems can provide abundant information, however, their natural environment with different plants that grow together in a messy scene present many challenges [3]. Among the challenges are vegetation segmentation (vegetation in the first layer and ground in the background), segmentation of individual plants, segmentation of crops and weeds, and phenotyping of individual plants. The first three challenges are addressed directly by machine learning. The fourth challenge includes the growth stage, the position of the plant stem, the amount of biomass, the leaf count, the leaf area, among others. In addition, the crop/weed coverage index, crop spacing, crop plant counts and other derived measurements are of special interest to farmers.

This article focuses on the design, implementation and evaluation of deep learning algorithms based on the U-Net convolutional network architecture for crop and weed segmentation in multispectral images used in precision agriculture. The main contribution is the evaluation of modifications on the U-Net network in order to make it more optimal for the recognition of weeds and crops. For this reason, three variants of the U-Net convolutional network architecture are presented and its performance is evaluated using metrics such as the Jaccard index or Intersection over Union (IoU) and recall. The rest of this article is structured as follows: Section 1 presents a synthesis of the contribution of the main articles focused on this problem. Section 2 describes the methodology to follow. Experimentation and results are presented in Sect. 3 and finally in Sect. 4 discussions and future work.

1.1 Related Work

Image filters are used using the computational vision approach [4]. Søgaard [5] uses active shape models to classify weed types.

For weed discrimination, models were used for real-time detection [6] using the Haar wavelet transform (HWT) for image decomposition and the k-nearest neighbors (KNN) method obtaining 94% of precision improving the used baselines. Random Forest and support vector machines (SVM) are used for detection [7]. Also, semi-supervised approaches were used [8].

In recent years several studies have been carried out for the application of deep learning in agriculture, among them we have the works [9, 11] of the techniques used for deep learning. Convolutional neural networks (CNN) are studied in [2, 11]. In other investigations, unsupervised models of labeling is used first, then apply CNN based on ResNet18 [12]. Another approach uses a CNN with sliding windows [13], where from the calculation of a relationship between weed detection rate (WD) and crop waste (CW), it was discovered that the size of the sliding window of [80 80] results in an effective detection of weeds with 63.28% and a minor cause of crop damage with 13.33%. Lottes [14] uses fully convolutional networks (FCN) with an encoder-decoder structure achieving a level of completeness of 92.4% for weeds and 96.1% for cultivation. In other investigations, 86.2% accuracy is achieved for 22 types of weeds with crops [15] and 94% accuracy at pixel level [16].

1.2   U-Net

In biomedical image segmentation context, it is assumed that thousands of training data are required for successful training of a deep learning network. Ronneberger [17] presented the U-Net model based in CNN with a training strategy that focus primarily on data augmentation and contraction-expansion to use the available data more efficiently (see Fig. 1). The network can be trained from few images and its performance is remarkable. U-Net was also used in other applications such as radiofrequency [18]. The use of U-Net for this problem is explain in [10], where it is compared with other neural networks. There are other alternatives such as SegNet [19] applied to weed detection [1] or WeedMap that has been used in precision agriculture [20].

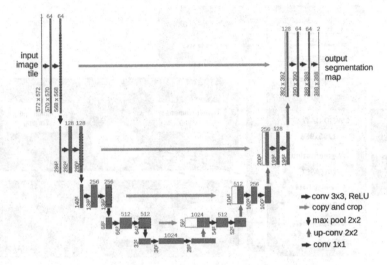

Fig. 1. Architecture U-Net [17].

1.3   Dataset

For the labeling data process, it is necessary Human intervention, which can be a very tedious task, initiatives [21] are proposed for the automatic generation of data sets based on a series of key features. Other several investigations use their own set of data taken on drones or cameras [12, 15]. Huag [3] proposes a data set of 60 images called CWFID (Crop Weed Field Image Dataset) which is complemented in [1]. This data set is used in investigations [7, 10] and in this article.

2   Methods

The objective of this article is to answer the questions: Is the U-Net convolutional network architecture effective for the segmentation of weeds and crops? Is it possible to improve the effectiveness of the U-Net convolutional network architecture by adding residual and recurrent layers for weed and crop segmentation? To answer these questions, the methodology described below is followed (see Fig. 2):

- Acquisition of the data set containing masks of weeds, soils and crops.
- Pre-processing through data augmentation explained above.
- Separation of test sets, validation and tests.
- Reduced tests (less steps) of the model using the hyper-parameters chosen in order to choose the best values.
- Training with the chosen hyper-parameters, using the set of tests and validation.
- Obtaining the metrics defined by validating the model with the set of tests.

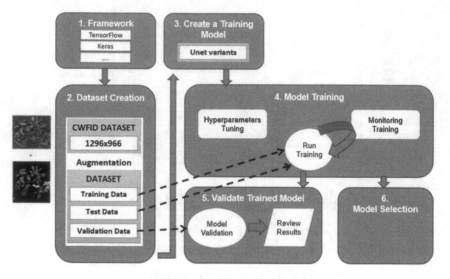

Fig. 2. Proposed process.

2.1   Pre-processing

Generally, these data sets contain very few images, so augmentation was performed with the following strategies:

- Reflection of images horizontally and vertically.
- Sliding images.
- Noise by altering the RGB channels.
- Elastic deformation.
- Gaussian noise.
- Cropping.

Additionally, the size of the images was reduced in order to have sufficient computational capacity to perform the tests.

2.2 Quality Metrics

The Jaccard index or Intersection Over Union (IOU) was used, since it is a metric widely used in object detection and allows measuring the degree of similarity between the predicted image and the mask image.

Another metric used is recall due to the interest in controlling the proportion of real positives correctly identified. In the case of the problem, it is of interest to keep the number of crops identified as weeds (negative faults) as low as possible [7].

Additionally, precision and F1 score were used as complementary metrics in order to make comparisons with the baseline.

2.3 Proposed Model

The models evaluated are variants of the U-Net convolutional network architecture, which is one of the most popular architectures in segmentation applications.

First, a recurrent convolutional neural network based on U-Net was evaluated, since the accumulation of characteristics with recurrent residual convolutional layers guarantees a better representation of the characteristics for segmentation tasks. Secondly, a residual convolutional neural network based on U-Net models was evaluated, because a residual unit helps the training of a deep architecture. Thirdly, a recurrent residual convolutional neuronal network was evaluated in order to use the advantages already mentioned. In Fig. 3 the U-Net base architecture is observed, where the blocks in red are convolutional units modified according to the variants shown in Fig. 4.

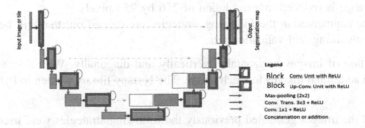

Fig. 3. Architecture variants U-Net [17]. (Color figure online)

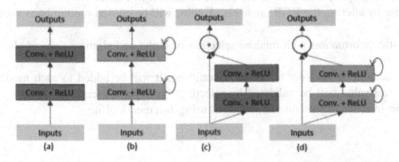

Fig. 4. Variants of convolutional units: (a) front convolutional units, (b) recurrent convolutional block (c) residual convolutional unit, and (d) recurrent residual convolutional units [22].

3 Results

3.1 Dataset

The data set used in this investigation is the Crop Weed Field Image Dataset (CWFID) [3], which consists of a set of 60 images of 1296 × 966 pixels, labeled with 3 classes (soil, weed, crop) that are shown in Fig. 5. Scaling was performed to reduce images to 246 × 256 pixels in order to improve computational capacity.

Fig. 5. Right: multispectral image. Left: labeled Image [3].

A data set with the following characteristics has been prepared: 40 images randomly chosen as training set. From the training set we will take the images number 11, 20, 41 and 52 to be aligned with the baseline. The set of tests will be the remaining 20 images.

The image is reduced to a resolution of 256 by 256 pixels.

For the augmentation the following strategies are carried out that will be applied only to the training and validation sets:

- Reflection of images horizontally, vertically and diagonally. With this we would have 3 additional images for each image. The Numpy library written in Python was used.

On all the images generated previously the following strategies were used:

- Sliding of the images: The sliding was done by random values and filling the remaining space with part of the image as shown in Fig. 6.
- Noise by altering the RGB channels. A color will be chosen randomly as shown in Fig. 6.
- Elastic deformation with random selection of alpha and sigma values as shown in Fig. 6.
- Gaussian noise in order to prevent overfitting. It will be added to each model and the best value will be validated by selecting hyper parameters.
- Crop in order to generate new images using fragments of it.

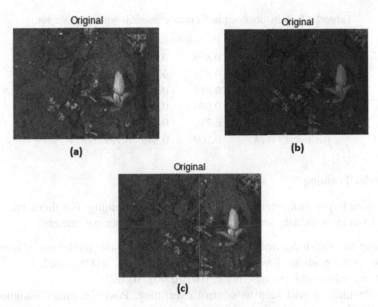

Fig. 6. (a) Sliding images. (b) Noise with channel alteration. (c) Elastic deformation. (Color figure online)

After making these modifications we have the following sets of data: 1560 images as a training set, 520 as a validation set and 20 as a test set.

3.2 Experimentation Environment and Baseline

Google Collaboratory has been used as a cloud platform. It allows us to carry out a collaborative and distributed work. It uses an Intel (R) Xeon (R) CPU @ 2.30 GHz processor with 12 GB of RAM. A Tesla P100 GPU with 16 GB of memory. Experiments were performed using Anaconda as a development environment and Python 3.6 as a programming language. The neural network models were developed using the Keras library on Tensorflow 2.0.

The strategies used were aligned to the research carried out by Cereda [10] which contains experimentation with U-Net and he uses the chosen metrics. The proposed models were developed from [22]. Cereda [13] conducted the experiment with 10 classifiers, which were evaluated with the indicators: Accuracy, Precision, Recall, F1 and Jaccard. This evaluation was performed at the pixel level in full size of the images extracted for the test set. The results of the neural network models are shown in Table 1. It can be seen in the results of Table 1 that the U-Net classifier has better performance levels in the majority of quality indicators used in the evaluation of the investigation.

Table 1. Results obtained in Cereda's research with the data set.

Model	Accuracy	Precision	Recall	F1	Jaccard
U-Net	**0.897**	0.886	**0.972**	**0.922**	0.879
U-ReNet	0.863	0.873	0.942	0.898	0.879
U-ReNet2	0.873	**0.895**	0.915	0.897	0.879
ReSeg	0.854	0.860	0.946	0.894	0.879
ReConv	0.737	0.770	0.883	0.809	0.879
ReConcat	0.743	0.769	0.846	0.799	0.879

3.3 Model Training

The following hyper parameters have been used during training. For them, each model was run 10 times with data set 1 using the following hyper-parameters.

- Learning rate (lr): It controls how much the weights of our model are adjusted with respect to the gradient. Possible values assigned: 0.01, 0.005, 0.001.
- L2: Assigned possible values: none, 0.01, 0.001, 0.0001.
- Gaussian filter: it will help us control overfitting. Possible values assigned: 0.5, 0.05, 0.005.
- Dropout that will be added to each of the convolutional layers. Possible values assigned: none, 0.1, 0.2.
- Batch normalization that will be added to each of the convolutional layers.

After performing the tests, the following hyper parameters have been chosen for each model in Table 2.

Table 2. Better hyper parameters.

Model	Jaccard
U-Net	lr = 0.001, Gaussian filter = 0.005, batch normalization
ResU-Net	lr = 0.001, Gaussian filter = 0.005, batch normalization
RU-Net	lr = 0.001, Gaussian filter = 0.005, batch normalization
R2U-Net	lr = 0.001, Gaussian filter = 0.005, batch normalization

Next, the training was carried out using the selected hyper parameters. Each model was executed in 200 periods using as a loss function: categorical crossentropy, Adam as an optimizer and a batch size of 30 for U-Net and ResU-Net, and 10 for RU-Net and R2U-Net. Table 3 shows the most relevant configurations.

Table 3. Description and relevant settings.

Model	Total parameters	Trainable parameters	Non-trainable parameters
U-Net	7768099	7764131	3968
ResU-Net	8117827	8113859	3968
RU-Net	24281347	24269443	11904
R2U-Net	24631075	24619171	11904

The metrics obtained are shown in Table 4.

Table 4. Results obtained with the data set.

Model	Accuracy	Precision	Recall	F1	Jaccard
U-Net	0.8825	0.8939	0.9307	0.8351	0.9760
ResU-Net	0.8724	0.9397	0.9669	0.8505	0.9749
RU-Net	0.9027	0.9212	0.9675	0.8917	0.9783
R2U-Net	0.8660	0.9311	0.9649	0.8376	0.9740

The learning curves for the training and validation sets are shown in Fig. 7. The execution times are shown in Table 5.

Table 5. Runtime in milliseconds per image.

Model	Time
U-Net	12
ResU-Net	14
RU-Net	37
R2U-Net	40

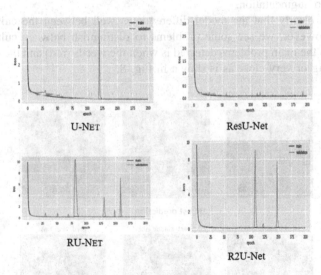

U-NET

ResU-Net

RU-NET

R2U-Net

Fig. 7. Learning curve for data set.

4 Discussion

As part of this chapter, based on the results, some topics will be discussed to interpret the experiments performed and find opportunities for improvements.

In Table 6 shows the best values of each data set. The results obtained with respect to the baseline are better except in the recall metric and F-1. The RU-Net model obtained the best results in all metrics except Precision.

The research proposes three additional models that don't use the baseline, and U-Net is the only one presented in both. When comparing the results, the baseline has been exceeded in precision and Jaccard.

Table 6. Better values in the data set.

Model	Accuracy	Precision	Recall	F1	Jaccard
Base line	U-Net	U-ReNet2	U-Net	U-Net	All
Base line	0.897	0.925	0.972	0.922	0.879
Best model	RU-Net	ResU-Net	RU-Net	RU-Net	RU-Net
Best model	0.9027	0.9397	0.9675	0.8917	0.9783

The learning curve of the models used is shown in Fig. 7. It is important to note that all graphs have a similar shape. Note that some models show temporary fluctuations in the loss function in the validation set. This could be due to a possible noise in the data due to augmentation.

It is show clearly that the model differentiates well between the cultivation and vegetation. However, we see some problems to distinguish between cultivation and weeds. One of the main problems detected is when the weeds (red) and the crop (green) are overlapping or very close as is shown in Fig. 8.

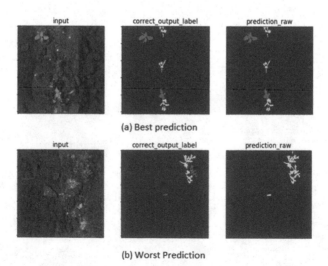

Fig. 8. Predictions with RU-Net. (Color figure online)

During the experiment, data augmentation has been used, as part of the improvement opportunity it is proposed to increase the amount of data augmentation. For example, in the baseline It is used up to 25,000 images with the almost 1000 used. Also, perform more tests by making changes in hyperparameters. Perform tests with a larger batch size, although this requires greater computational capacity. Additionally, Try other optimizers like RMSProp. Tests were performed using dropout layers and batch normalization layers where better results could be observed. It is necessary to improve the architecture using and/or proposing improvements to the layers such as attention mechanisms that allow efficient location of objects and an increase in performance in general.

4.1   Conclusions

From a practical point of view, this work should be expanded to be able to distinguish different types of weeds and to estimate the growth status of the crops. This implies extending the manual annotation to include this new data. For the weed detection problem, it is necessary to obtain a larger data set than the used for the present investigation.

The main objective of the present work was to carry out the experimentation of architectures of neural networks based on U-Net applied to the segmentation of crops and weeds having as base line an experimentation already carried out [10]. It is concluded from the results that using recurrent layers within the U-Net architecture allows to improve the effectiveness in the problem of crop and weed segmentation with multispectral images of the data set used. In contrast, the residual layers did not add any improvement.

From the evaluation analysis of the segmentation, it was observed that the same metric result can be obtained in different ways, therefore, it should be interesting to investigate which of the metrics is most suitable for resolving this type of problem. Finally, it is proposed to deepen research on topics such as data augmentation, the choice of hyper-parameters and assembly models in order to achieve better results. Additionally, Perform the experiment with other architectures and different data sets.

Acknowledgments. This research was supported by National Agriculture Innovation Program (PNIA) of Peru and the Institute of Scientific Research (IDIC) of the University of Lima.

References

1. Sa, I., et al.: WeedNet: dense semantic weed classification using multispectral images and MAV for smart farming. IEEE Robot. Autom. Lett. 3(1), 588–595 (2017). https://doi.org/10.1109/LRA.2017.2774979
2. Milioto, A., Lottes, P., Stachniss, C.: Real-time semantic segmentation of crop and weed for precision agriculture robots leveraging background knowledge in CNNs. In: 2018 IEEE International Conference on Robotics and Automation (ICRA), pp. 2229–2235. IEEE (2018). https://doi.org/10.1109/icra.2018.8460962

3. Haug, S., Ostermann, J.: A crop/weed field image dataset for the evaluation of computer vision based precision agriculture tasks. In: Agapito, L., Bronstein, M.M., Rother, C. (eds.) ECCV 2014. LNCS, vol. 8928, pp. 105–116. Springer, Cham (2015). https://doi.org/10. 1007/978-3-319-16220-1_8

4. Molina-Villa, M.A., Solaque-Guzmán, L.E.: Machine vision system for weed detection using image filtering in vegetables crops. Revista Facultad de Ingeniería Universidad de Antioquia **80**, 124–130 (2016). https://doi.org/10.17533/udea.redin.n80a13

5. Søgaard, H.T.: Weed classification by active shape models. Biosyst. Eng. **91**(3), 271–281 (2005). https://doi.org/10.1016/j.biosystemseng.2005.04.011

6. Ahmad, I., Siddiqi, M.H., Fatima, I., Lee, S., Lee, Y.K.: Weed classification based on Haar wavelet transform via k-nearest neighbor (k-NN) for real-time automatic sprayer control system. In: Proceedings of the 5th International Conference on Ubiquitous Information Management and Communication, p. 17. ACM (2011). https://doi.org/10.1145/1968613. 1968634

7. Voorhoeve, L.: Machine Learning for Crop and Weed Classification (2018)

8. Lottes, P., Stachniss, C.: Semi-supervised online visual crop and weed classification in precision farming exploiting plant arrangement. In: 2017 IEEE/RSJ International Conference on Intelligent Robots and Systems (IROS), pp. 5155–5161. IEEE (2017). https://doi.org/10. 1109/iros.2017.8206403

9. Kamilaris, A., Prenafeta-Boldú, F.X.: Deep learning in agriculture: a survey. Comput. Electron. Agric. **147**, 70–90 (2018). https://doi.org/10.1016/j.compag.2018.02.016

10. Cereda, S.: A comparison of different neural networks for agricultural image segmentation (2017)

11. Potena, C., Nardi, D., Pretto, A.: Fast and accurate crop and weed identification with summarized train sets for precision agriculture. In: Chen, W., Hosoda, K., Menegatti, E., Shimizu, M., Wang, H. (eds.) IAS 2016. AISC, vol. 531, pp. 105–121. Springer, Cham (2017). https://doi.org/10.1007/978-3-319-48036-7_9

12. Bah, M.D., Hafiane, A., Canals, R.: Deep learning with unsupervised data labeling for weeds detection on UAV images (2018). arXiv preprint arXiv:1805.12395

13. Kantipudi, K., Lai, C., Min, C.-H., Chiang, R.C.: Weed detection among crops by convolutional neural networks with sliding windows. In: 14th International Conference on Precision Agriculture, Quebec (2018)

14. Lottes, P., Behley, J., Milioto, A., Stachniss, C.: Fully convolutional networks with sequential information for robust crop and weed detection in precision farming. IEEE Robot. Autom. Lett. **3**(4), 2870–2877 (2018). https://doi.org/10.1109/LRA.2018.2846289

15. Dyrmann, M., Karstoft, H., Midtiby, H.S.: Plant species classification using deep convolutional neural network. Biosyst. Eng. **151**, 72–80 (2016). https://doi.org/10.1016/j. biosystemseng.2016.08.024

16. Dyrmann, M., Mortensen, A.K., Midtiby, H.S., Jørgensen, R.N.: Pixel-wise classification of weeds and crops in images by using a fully convolutional neural network. In: Proceedings of the International Conference on Agricultural Engineering, Aarhus, Denmark, pp. 26–29 (2016)

17. Ronneberger, O., Fischer, P., Brox, T.: U-Net: convolutional networks for biomedical image segmentation. In: Navab, N., Hornegger, J., Wells, W.M., Frangi, A.F. (eds.) MICCAI 2015. LNCS, vol. 9351, pp. 234–241. Springer, Cham (2015). https://doi.org/10.1007/978-3-319-24574-4_28

18. Akeret, J., Chang, C., Lucchi, A., Refregier, A.: Radio frequency interference mitigation using deep convolutional neural networks. Astron. Comput. **18**, 35–39 (2017)

19. Badrinarayanan, V., Kendall, A., Cipolla, R.: SegNet: a deep convolutional encoder-decoder architecture for image segmentation. IEEE Trans. Pattern Anal. Mach. Intell. **39**(12), 2481–2495 (2017)
20. Sa, I., et al.: Weedmap: a large-scale semantic weed mapping framework using aerial multispectral imaging and deep neural network for precision farming. Remote Sens. **10**(9), 1423 (2018)
21. Di Cicco, M., Potena, C., Grisetti, G., Pretto, A.: Automatic model based dataset generation for fast and accurate crop and weeds detection. In: 2017 IEEE/RSJ International Conference on Intelligent Robots and Systems (IROS), pp. 5188–5195. IEEE (2017)
22. Alom, M.Z., Hasan, M., Yakopcic, C., Taha, T.M., Asari, V.K.: Recurrent residual convolutional neural network based on U-Net (R2U-Net) for medical image segmentation (2018). arXiv preprint arXiv:1802.06955

Validation of an Algorithm for the Detection of the Image of a Person Using Multiple Cameras

Washington Garcia-Quilachamin[1,2](✉), Luzmila Pro Concepción[1](✉),
Jorge Herrera-Tapia[2](✉), Richard José Salazar[2](✉),
and Wellington Toala-Mero[2](✉)

[1] Universidad Nacional Mayor San Marcos,
Av. German Amezaga No. 375, Lima, Peru
profegarcia50l@gmail.com, lproc2003@hotmail.com
[2] Universidad Laica Eloy Alfaro de Manabi, Via San Mateo, Manta, Ecuador
jorge.herrera.tapia@gmail.com, salazarrjt@yahoo.com,
saulkslayer@gmail.com

Abstract. Among the various contexts of technological evolution, there is the application of algorithms in cameras for the detection and identification of people and their utilization in various areas, for example surveillance in smart home environments and companies. This is considered as a complex recognition task. Therein, this research presents an algorithm that was designed and implemented in Matlab, the same that through a camera allows to detect the image of a person. The objective in this investigation is to corroborate the process in the detection of the image of a person in multiple cameras through the application of an algorithm. The validation of the data obtained was carried out through a mathematical model, which allowed us to substantiate the detection of the image of a person through five cameras, considering as parameters time and distance. As a result of the study and the application of the algorithm, its functionality was verified in the positive detection of the image of a person, a mathematical model was also obtained to confirm its effectiveness and validity through different tests.

Keywords: Algorithm detection · Image acquisition · Mathematical model · Definite integral · Cameras

1 Introduction

Among the contexts of technological evolution, there is the application of algorithms in cameras for object detection, which is of great importance in development and innovation. Thus, the recognition tasks are considered complex for a computer according to [1], because they have problems in the processes of object classification, edge detection, movement tracking, etc.

In [2], object recognition is defined as the process of identifying a specific object or the class of objects in an image or video. It is considered that from the study of the detection of objects the innovation in the detection of people can be started. As the

© Springer Nature Switzerland AG 2020
M. Botto-Tobar et al. (Eds.): ICAT 2019, CCIS 1194, pp. 486–501, 2020.
https://doi.org/10.1007/978-3-030-42520-3_39

author [3] states, the type of detection not only focuses on finding the kind of object to be detected, but also locating the extent of the object in the image.

According to [4], several researchers have developed methods and techniques to detect people in images, considering that the study of an effective method for recognition continues. There are several factors that motivate this research, one of them being the application for the safety of people, which are considered as the most important and according to [5], mention that facial recognition systems play a vital role in many applications, including surveillance, biometrics and safety, so the authors in [6] establish that the Viola-Jones algorithm is currently one of the most used to solve problems of searching a person's face.

Considering the security field is a topic widely taken into account in smart cities, offices and homes. According to the authors [7], they state that the safety of a home and family is very important for everyone, so that the systems of recognition of human silhouettes can be connected to the Internet of Things (IoT).

The objective of this investigation is to validate the process and information obtained by applying the algorithm in the detection of the image of a person in multiple cameras. The validation was carried out through a mathematical model, which allowed us to base the detection of the image of a person through these cameras, considering as parameters time and distance. The area as a function of time, considering a definite integral starting from a point zero.

As a result of the study in the application of the algorithm in the computer, for the detection of the image of a person using five cameras, the data were obtained in relation to the time and distance in which a person is detected with each camera. These data were used in the analysis through the application of a definite integral, considering the author in [8]. It concludes with the effective validity of the algorithm applied in the detection of the image of a person in a distance range of 3 m to 9 m, also the speed of image capture and its feasibility determine that the greater distance from the person in relation to the camera, the smaller the angle for detection.

This document is organized as follows, in Sect. 2 a review of the related works on the subject of our research, developed framework, applied algorithm, tools used and selection criteria is described. Section 3 describes the results obtained and the analysis of data in the application of mathematical models for the validation of the process in the detection of the image of a person with the five cameras. Finally, there are the conclusions of this investigation.

2   Related Works

We consider the authors [9], who state that there are several ways to detect and recognize a person, also [10] argue that there are studies with algorithm-based models to identify faces, people, traffic signs, tumors and aspects of visual data. Convolutional Neural Network (dCNN), is a model used to classify images, group them by similarity and recognize objects within scenes, as the authors [11] argue.

The authors [12–14] and [15] discuss Hidden Markov (HMM) as a model to describe the characteristics of a stochastic process based on computational methods for classifying the Human physical activity and for the recognition of different activities that the human performs, being appropriate for the application in pattern recognition.

According to [16] and [17], they state that Support Vector Machines (SVM) is an algorithm model associated with learning that analyzes the data used for classification and recoil analysis in relation to artificial intelligence. The models described are based on algorithms that can identify faces, people, traffic signals, tumors and many other aspects of visual data, as they considered by [11] and [18].

Based on the research work developed, Table 1 describes the algorithms found that allow the recognition of an image with their advantages and disadvantages. This information is considered of great importance since there are no similar research papers.

Table 1. Recognition algorithms

Algorithm	Description	Advantage	Disadvantages	Ref.
Markov chain Monte Carlo	Applied to the HMM model, it is used for the recognition of human activities	Latent movements that are selectively shared between multiple paths and activities efficiently	It has several limitations, for example the assumption of independence	[21]
Weakly Supervised Multi-Type Attribute Learning	It considers contextual signals and progressively increases accuracy using a limited number of data labeled by people	It divides human attributes into multiple types, contains incompatible attributes and only one of them can be positive	It does not need more training on target data sets. It is not so accurate	[11]
Viola-Jones	It is based on a series of weak classifiers called Haar-like-feactures	It stands out for its low computational cost and it allows its use in real time	Occlusion is a problem for this algorithm	[6, 22, 23]
EigenFaces	It is a method of facial recognition based on the component analysis approaches	It evaluates images of size 23×28 pixels in an order to compare interpolarization	It is not coupled with Support vector machine and Artificial Neural Network	[24, 25]
Adaboost	Adaboost is a boosting algorithm presented by Freund and Sachapire in the generation of online learning	Used to select features and train classifiers. Dramatically increase detector speed	It takes weak classifiers and multiple tries each time	[22, 26]

According to the authors [19], they state that the Viola-Jones algorithm is based on a series of classifiers called Haar-like-features, which were used in this investigation and that allowed the efficient application starting from an integral real time image considering that it presents a problem in occlusion when detecting an image.

2.1 Workframe

The purpose is to validate the algorithm that allows the detection of the image of a person, for which it is necessary to understand the process, the same that was based on the flowchart shown in Fig. 1, considering that the algorithm developed in Matlab will be employed and using the image acquisition tool applications, computer visio system toolbox to capture a pattern.

This work is largely motivated by efficient detection, considering that the objective is to enable the detection of a person's image in real time through a computer camera, on which the calculation of the integral image is based, as [6] manifest it.

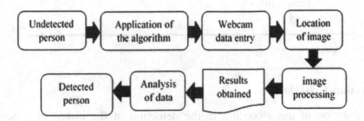

Fig. 1. Flow chart for image detection

2.2 Tools Used

The hardware used for the purpose of this research was a Toshiba Satellite C55-C5240 Laptop, Intel® Core ™ Intel Core i5-5200U 2.2 GHz processor, 8 GB RAM, 1 GB HDD, Toshiba HD Camera, 64-bit Windows 10, Intel HD Graphics 5500.

EasyCap device, used in the video signal. Toshiba Webcam HD camera, IR-Max 250 camera, IR-Hilook THC-T110-P camera, DS-2CE16D0T-IR Turbo HD camera, Genius Facecam 320X camera and Matlab 2017b was used as software.

2.3 Applied Algorithm

The algorithm used to detect a person's image in real time is based on Viola-Jones, which was improved and described below.

The data obtained in relation to the detection of the image of a person are imported to a local Excel file, in the directing route of data saving.

Algorithm: Image Detection

```
Data: video, peopleDetector, obj, frame, bboxes
peopleDetector = vision.PeopleDetector( )
obj=imaq.VideoDevice ( )
    set (obj);
    figure ( );
    while (true)
            frame=step(obj);
            bboxes=step (peopleDetector,frame);
            obj_p=insertObjectAnnotation (frame,bboxes);
            imshow(obj_p,)
            f=findobj( );
            pause (0.05)
    end
    release (obj)
    end
```

2.4   Selection Criteria

For the validation of the algorithm in the detection of the image of a person, five different types of cameras were selected in their size and class, for which their brand or specifications are not considered to limit this investigation.

Among the cameras selected and that were used for the development of the proposed objective, there are the computer webcam, Genius camera and cameras for infrared security type Domo and Bullet.

Table 2 describes the characteristics of the cameras considered in this study. The image resolution of the Toshiba and Genius camera was considered as 640×480 pixels considering that their configuration is not supported, while the resolution of the image on the Hilook, Max and Turbo cameras is assigned by default when used with the EasyCap device, which allows conversion in the resolution of the image.

Table 2. Camera features

Camera	Night vision	Image format	Image resolution
Toshiba	No	YUY2	640×480
Genius	No	YUY2	640×480
Hilook	Si	YUY2	720×576
Max	Si	YUY2	720×576
Turbo	Si	YUY2	720×576

With each of the five cameras described in Table 2, 30 images related to a person were detected in a distance range of 3 m to 9 m, with an interval of 1 m. Additionally, 210 images were obtained per camera and for this validation and 1050 data were considered as the total result of the five cameras used in this investigation.

2.5   Mathematical Model Implemented

With the information obtained, the respective mathematical calculations were developed, considering the authors [8] and [20].

The procedures described are based on the equation to obtain the area of a triangle (1), the velocity (2) and the definite integral (3). With the application of the equation, the respective calculations of the triangular waveform were made, which were feasible to validate the process in this investigation.

For this process, the graphs of Fig. 2 are considered, which have the shape of a right triangle between the projection axes and the stroke of the function.

$$A = \frac{1}{2}b * a \tag{1}$$

Where b is the base as a function of distance and a, the height of our data as a function of time being values (x, y) respectively. The speed (2) of the cameras used in this investigation was determined based on data related to time and distance.

$$v = \frac{d}{t} \tag{2}$$

Where "d" represents distance and "t" time, which are considered as (x, y) respectively. The definite integral (3) was considered, which was applied to evaluate the values obtained in relation to the time and distance parameters of each camera used for the detection of the image of a person. Where "d" will be the distance value and t is the time value obtained from each camera respectively

$$\int_0^d \frac{1}{2}(x/t).dx \tag{3}$$

3   Results and Analysis

3.1   Results of the Detection in Several Cameras

In Table 3, the person's height, detection angle, camera height and distance are described, which are the parameters used and the data obtained in relation to the detection of a person's image.

Table 3. Parameters and measurement data

Person stature (m)	Detection angle	Camera height (m)	Distance (m)
1.67	28°	0.88	3.00
1.67	24°	0.88	4.00
1.67	20°	0.88	5.00
1.67	16°	0.88	6.00
1.67	12°	0.88	7.00
1.67	8°	0.88	8.00
1.67	4°	0.88	9.00

Table 4 shows the average of the 1050 data obtained by the five cameras, considering the time in seconds detected by the image of a person in relation to the distance from 3 m to 9 m, with an interval of 1 m.

Table 4. Detection time per camera in seconds

Distance (m)	Types of cameras				
	Hilook (s)	Max (s)	Turbo (s)	Genius (s)	Toshiba (s)
3.00	0.4035898	0.3966187	0.3969076	0.3740454	0.3778240
4.00	0.3996954	0.4138642	0.4262454	0.3771838	0.3748886
5.00	0.4319552	0.4117347	0.4056770	0.3741177	0.3747536
6.00	0.4126833	0.4128826	0.4129655	0.3722851	0.3741777
7.00	0.4101977	0.4164330	0.4111486	0.3792567	0.3813519
8.00	0.4089741	0.4166274	0.4148233	0.3783192	0.3798207
9.00	0.4103726	0.4115626	0.4095482	0.3796371	0.3774584

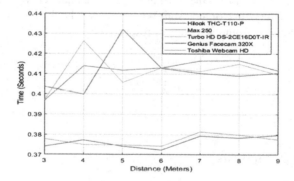

Fig. 2. Time-distance strokes

The values in Table 4 are shown graphically in Fig. 2, which are the result of the application of the algorithm in the detection using the five different cameras in a distance range from 3 m to 9 m. In addition, the strokes related to the average time and distance at which a person's image was detected are displayed.

The stroke values of the Hilook THC-T110-P camera are shown, which shows variations and this demonstrates that this camera is less efficient as it is displayed in the range of 3 m to 9 m, showing the largest peak of all the strokes in 5 m as well as a low speed at 4 m and low continuity from 6 m to 9 m progressively climbing presenting a high peak at 5 m. The Max 250 camera stroke shows a low starting peak at 3 m and displays more continuous speeds of all data in a distance range from 4 m to 9 m. The stroke of the Turbo HD DS-2CE16D0T-IR camera shows a high peak at 4 m with a drop peak at 5 m and a continuity in its speed in the range of 6 m to 9 m.

The strokes of the Genius FaceCam 320x camera represent a slight peak at 4 m with a slight scalloped fall from 4 m to 6 m and a moderate climb to 7 m maintaining continuity in the range of 7 m to 9 m. and finally, the strokes of the Toshiba Webcam HD camera range from a drop peak to 4 m, maintaining its continuity in the range of 4 m to 6 m, to climb a peak at 7 m and a slight stepped drop in the range of 7 m to 9 m.

The results obtained and shown in Fig. 2 display how strokes of the five cameras allowed to demonstrate the effectiveness of the algorithm in the detection of the image, considering the speed performance of the different cameras used in this investigation.

It must be considered that the proposal is not to evaluate the cameras, but the efficiency of the algorithm and that these results allow to demonstrate that the algorithm works, independent of the type and brand of the cameras, of course the values vary according to their specifications.

The application of the algorithm in the detection of the image of a person was feasible to validate its effectiveness in the five cameras. Table 5 shows the results of the detection at different focal angles and their respective accuracy classified as positive, false positive and false negative.

Table 5. Focal angle and image detection

Camera	Focal angle		Detection
	Inclination	Rotation	
Hilook	0° a 180°	0° a 360°	Positive/False-Positive/False-Negative
Max	0° a 180°	0° a 360°	False-Positive/Positive
Turbo	0° a 90°	0° a 360°	Positive/False-Positive/False-Negative
Genius	0° a 180°	0° a 360°	Positive/False-Positive
Toshiba	0° a 180°	0°	Positive

To determine the effectiveness of the algorithm in the detection of the image of a person, the shorter detection time of each camera was considered as an indicator, which was represented with a 100% effectiveness, taking into account that at a lower time the effectiveness in detection will be greater.

Tables 6 and 7, describes the effectiveness rate in percentages based on the time and distance range of 3 m to 9 m, in relation to the detection of an image through the five cameras used in this investigation for the purpose of validating the algorithm.

Table 6. Detection time camera effectiveness

Distance (m)	Types of cameras					
	Hilook (s)	Effective	Max (s)	Effective	Turbo (s)	Effective
3	0.4035898	99.0%	0.3966187	100.0%	0.3969076	100.0%
4	0.3996954	100.0%	0.4138642	95.8%	0.4262454	93.1%
5	0.4319552	92.5%	0.4117347	96.3%	0.4056770	97.8%
6	0.4126833	96.9%	0.4128826	96.1%	0.4129655	96.1%
7	0.4101977	97.4%	0.4164330	95.2%	0.4111486	96.5%
8	0.4089741	97.7%	0.4166274	95.2%	0.4148233	95.7%
9	0.4103726	97.4%	0.4115626	96.4%	0.4095482	96.9%

Table 7. Detection time camera effectiveness

Distance (m)	Types of cameras			
	Genius (s)	Effective	Toshiba (s)	Effective
3	0.3740454	99.5%	0.3778240	99.0%
4	0.3771838	98.7%	0.3748886	99.8%
5	0.3741177	99.5%	0.3747536	99.8%
6	0.3722851	100.0%	0.3741777	100.0%
7	0.3792567	98.2%	0.3813519	98.1%
8	0.3783192	98.4%	0.3798207	98.5%
9	0.3796371	98.1%	0.3774584	99.1%

Table 8. Image detection of a person

Type of cameras	Distance range							
	3m		4m		5m		6m	
	Detection	Effective	Detection	Effective	Detection	Effective	Detection	Effective
Hilook	Positive	99.0%	False Negative	100.0%	False Positive	92.5%	False Positive	96.9%
Max	Positive	100.0%	Positive	95.8%	Positive	96.3%	Positive	96.1%
Turbo	False Negative	100.0%	Positive	93.1%	False Negative	97.8%	False Positive	96.1%
Genius	Positive	99.5%	Positive	98.7%	False Positive	99.5%	Positive	100.0%
Toshiba	Positive	99.0%	Positive	99.8%	Positive	99.8%	Positive	100%

Similarly, in Tables 8 and 9 images of the detection of a person in real time enclosed in a yellow frame are shown, which were captured using five different cameras through the application of the algorithm to be validated. These visualized images are a sample of the total of 1050 detections made to the same person in a range of 3 m to 9 m with an interval of 1 m, considering that the image of the person in each interval was captured 30 times.

Table 9. Image detection of a person

Type of cameras	Distance range					
	7m		8m		9m	
	Detection	Effective	Detection	Effective	Detection	Effective
Hilook						
	Positive	97.4%	False Positive	97.7%	False Positive	97.4%
Max						
	Positive	95.2%	False Positive	95.2%	Positive	96.4%
Turbo						
	False Negative	96.5%	Positive	95.7%	Positive	96.9%
Genius						
	Positive	98.2%	Positive	98.4%	False Positive	98.1%
Toshiba						
	Positive	98.1%	Positive	98.5%	Positive	99.1%

3.2 Analysis in the Detection of the Image of a Person

Based on the results obtained in the detection of the image of a person considering a distance range of 3 m to 9 m, with the application of the algorithm configured in Matlab and the analysis of the five cameras, it was determined that the Hilook THC-T110-P camera has an effectiveness of 99.0% in the detection of an image in a time of 0.4035898 s, considering that 100% represents a time of 0.3996954 s for its detection analysis within the range of 3 m to 9 m, which is positive at 3 m and 7 m, false negative at 4 m, false positive from 5 m to 6 m and 8 m to 9 m.

With the Max 250 camera, image detection is 96.4% effective with a time of 0.4115626 s, considering that 100% equals 0.3966187 s. The detection is positive in the range of 3 m to 7 m and 9 m, at 8 m the detection is false positive.

The Turbo HD DS-2CE16D0T camera, has a 97.8% detection effectiveness with a time of 0.4056770 s, 100% being in relation to 0.36969076 s and the image detection is false negative at 3 m, 5 m and 7 m, positive at 4 m and from 8 m to 9 m, false positive at 6 m.

The 99.5% effectiveness was considered with a time of 0.3740454 s with respect to the Genius camera, 100% being the equivalent of a time of 0.3722851 s and the positive detection of 3 m to 4 m and 6 m to 8 m, false positive at 5 m and 9 m.

Finally, the Toshiba Webcam HD camera detects the image, with an effectiveness of 99.8% in relation to a time of 0.3747536 s, considering that 100% represents a time of 0.3741777 s and the detection is positive in the range of 3 m to 9 m.

Considering the five cameras and establishing the shortest detection time of each camera, it was determined the Genius Camera was the best with a time of 0.3722851 s and 100% in the detection of the image according to data in Table 6.

It is established that the most Effective Camera in the detection of the image is the Toshiba with a time of 0.3741777 s and an effectiveness of 99.5%. In relation to video surveillance cameras, it is established that the Max camera has an effectiveness of 93.9% with a time of 0.4117347 s and a positive detection in the range of 3 m to 7 m and 9 m, false positive at 8 m.

3.3   Results with the Mathematical Model

The result is the speed in the detection of the image of a person, by substituting a value assigned to d in the integral (3), starting from 3 m to 9 m. For which the distance is considered to calculate the speed of each camera, substituting the values in t according to the distance and in which the calculation is made with the described values.

The speeds were obtained by clearing the values as variables in the integral and applying the property of the exponent in the resulting integral. Subsequently, the integral is evaluated considering the values from 0 as a constant in the lower limit and from 3 m to 9 m in the upper limit. Applying the definite integral (4), the speed was obtained in relation to distance and time.

$$\vec{v} = d/t \Rightarrow \int\limits_{0}^{d} (dx)/dt \qquad (4)$$

As a result of the calculations developed below, the obtaining of the speed in relation to the different cameras used is described considering the range of distances from 3 m to 9 m. The calculation of the speed in the detection of the image of a person in the Hilook Chamber considering a limit of 0 to 3, allowed to obtain as a result the speed of 11.149935 in m/s, in the distance range of 0 m at 3 m. In relation to the Max Camera, the limit was considered as 0 to 3 meters, for which the speed of 11.345910 in m/s is obtained as a result, in the distance range from 0 m to 3 m.

Considering a limit of 0 to 3, a speed of 11.337651 in m/s was determined, in a distance range of 0 m to 3 m with respect to the Turbo Camera. The result is the speed of 12.030625 in m/s, in relation to the Genius camera in the distance range of 0 m to 3 m. and finally, the speed of 11.910307 in m/s is the result with respect to the Toshiba Camera considering a limit of 0 to 3, in a distance range of 0 m to 3 m.

Table 10. Image detection speed in m/s

Distance (m)	Types of cameras				
	Hilook (s)	Max (s)	Turbo (s)	Genius (s)	Toshiba (s)
3.00	11.149935	11.345910	11.337651	12.030625	11.910307
4.00	20.015242	19.330012	18.768531	21.209819	21.339673
5.00	28.938186	30.359355	30.812691	33.411945	33.355250
6.00	43.616982	43.595928	43.587176	48.350041	48.105486
7.00	59.727297	58.832994	59.589161	64.600045	64.245124
8.00	78.244564	76.807238	77.141279	84.584658	84.250279
9.00	98.690800	98.405443	98.889459	106.680828	107.296592

Table 10 shows the results obtained, in relation to the calculations developed considering a lower limit of 0 m, up to an upper limit of 4 m to 9 m.

Figure 3 displays the relationship between the speed in "m/s" and the x-axis and the distance in "meters", of the five different cameras used for the detection of a person's image on the y-axis.

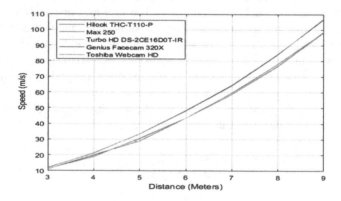

Fig. 3. Speed in relation to distance

3.4   Analysis of the Application of the Mathematical Model

Considering the results obtained in Table 10, applying as a mathematical model a definite integral equation in relation to the speed of the detection of the image of a person, it was determined that, for the Hilook camera in a distance range of 3 m to 9 m, the Average detection speed is 48.6261 m/s.

The Max 250 camera detects the image of the person at an average speed of 48.3824 m/s in a distance range of 3 m to 9 m.

With the Turbo camera, the detection of the image in a range of 3 m to 9 m, the average speed is 48.5894 m/s.

The Genius Camera detects the image in the range of 3 m to 9 m, with an average speed of 52.9811 m/s.

Finally, with the Toshiba camera, the detection of the image of the person was performed in the range of 3 m to 9 m and its average speed of 52.9289 m/s.

The speed and distance parameters of the frame in Fig. 3 are considered in relation to the detection of the image in a range of 3 m to 9 m. The strokes shown in the figure allow the Toshiba camera to be the most optimal because of its average speed of 52.9289 m/s, but it must be considered that this camera works with the computer processor and the algorithm runs without any interface.

This allows to establish the Hilook video surveillance camera with an average speed of 48.6261 m/s. It is considered optimal with the application of the algorithm, considering that in this investigation a device was used as an interface for the detection of the image of a person.

4 Conclusions

The contribution of this research work is related to the detection of the silhouette of the image of a person in contribution to security through video surveillance, considering that the studies that exist are focused on the face of a person. So, the implementation of the algorithm for the detection of the image of a person using multiple cameras, allowed to validate its effectiveness by obtaining as a result a positive detection in the functional field at the time of being used. The validation of the algorithm was carried out with five different types of cameras in their size and class, for which the brand or specifications that limit this research were not considered.

The use of the algorithm in the cameras determined that the video surveillance camera Max detects the image of a person with a 93.9% effectiveness in a time of 0.4117347 s and considers the Toshiba camera as the most feasible with a time of 0.3741777 s and an effectiveness of 99.5%. Therefore, the Max camera is considered the best in detecting the image of a person when using the algorithm and a device in the detection interface in relation to the Toshiba camera, which works with a direct interface with the computer.

The values obtained were validated through the mathematical model and allowed to determine that two of the five cameras used in this investigation show a higher speed than the others, demonstrating that they are the fastest at the time of capturing and detecting the image of a person.

In this research the application of the algorithm in the computer through a device, allowed the interface with the Hilook video surveillance camera, which is considered as an optimal displacement based on an average speed of 48.6261 m/s, faster in relation to the Max and Turbo cameras, in a distance range of 3 m to 9 m. The Toshiba camera with an average speed of 52.9289 m/s, is faster than the Genius, considering that the

Toshiba camera is integrated in the computer and allows the application of the algorithm to be direct and more optimal.

Applying the algorithm in the detection of the image of a person through the cameras allowed to determine that the fastest camera in the detection reduces the detection time and optimizes the software as well as the hardware used, being a contribution to video surveillance and a feasible energy efficiency in the use of materials.

These results obtained and analyzed strengthen the research developed for the validation of the algorithm and its subsequent implementation in contribution to the safety of people through video security, considering to link its application with the Internet of Things in a Smart Grid Home in the future.

References

1. Lin, X.J., Wu, Q.X., Wang, X., Zhuo, Z.Q., Zhang, G.R.: People recognition in multi-cameras using the visual color processing mechanism. Neurocomputing **188**, 71–81 (2016)
2. Mathworks: Image category classification using bag of features (2016)
3. Sinhal, K.: Object detection using deep learning for advanced users (2017)
4. Santos, M.Y., et al.: A big data analytics architecture for Industry 4.0. In: Rocha, Á., Correia, A.M., Adeli, H., Reis, L.P., Costanzo, S. (eds.) WorldCIST 2017. AISC, vol. 570, pp. 175–184. Springer, Cham (2017). https://doi.org/10.1007/978-3-319-56538-5_19
5. Matai, J., Irturk, A., Kastner, R.: Design and implementation of an FPGA-based real-time face recognition system. In: Proceedings of IEEE International Symposium on Field-Programmable Custom Computing Machines, FCCM 2011, pp. 97–100 (2011)
6. Alyushin, M.V., Lyubshov, A.A.: The Viola-Jones algorithm performance enhancement for a person's face recognition task in the long-wave infrared radiation range. In: Proceedings of the 2018 IEEE Conference of Russian Young Researchers in Electrical and Electronic Engineering, ElConRus 2018, Janua, vol. 2018, pp. 1813–1816 (2018)
7. Aydin, I., Othman, N.A.: A new IoT combined face detection of people by using computer vision for security application, p. 5 (2017)
8. Duffy, D.G.: Advanced Engineering Mathematics with MATLAB, 4th edn. Chapman and Hall/CRC, New York (2016)
9. Kamaruzaman, F., Shafie, A.A.: Recognizing faces with normalized local Gabor features and Spiking Neuron Patterns. Pattern Recogn. **53**, 102–115 (2016)
10. Wang, L., Tan, T., Ning, H., Weiming, H.: Silhouette analysis-based gait recognition for human identification. IEEE Trans. Pattern Anal. Mach. Intell. **25**(12), 1505–1518 (2003)
11. Su, C., Zhang, S., Xing, J., Gao, W., Tian, Q.: Multi-type attributes driven multi-camera person re-identification. Pattern Recogn. **75**, 77–89 (2018)
12. Nefian, A.V., Hayes, M.H.: Face detection and recognition using hidden Markov models, pp. 141–145 (2002)
13. Baca, A.: Methods for recognition and classification of human motion patterns – a prerequisite for intelligent devices assisting in sports activities. In: IFAC, vol. 45, no. 2 (2012)
14. Mahapatra, A., Mishra, T.K., Sa, P.K., Majhi, B.: Human recognition system for outdoor videos using hidden Markov model. AEU - Int. J. Electron. Commun. **68**(3), 227–236 (2014)
15. Jurafsky, D., Martin, J.H.: Hidden Markov models. In: Speech and Language Processing, chap. 9 (2017)

16. Lu, Y., Boukharouba, K., Boonært, J., Fleury, A., Lecœuche, S.: Application of an incremental SVM algorithm for on-line human recognition from video surveillance using texture and color features. Neurocomputing **126**, 132–140 (2014)
17. Kleinsmith, M.: Zero to hero: guide to object detection using deep learning: faster R-CNN, YOLO, SSD (2016). http://cv-tricks.com/object-detection/faster-r-cnn-yolo-ssd/
18. Kasim, S., Hassan, R., Zaini, N.H., Ahmad, A.S., Ramli, A.A., Saedudin, R.R.: A study on facial expression recognition using Local Binary Pattern. Int. J. Adv. Sci. Eng. Inf. Technol. **7**(5), 1621–1626 (2017)
19. Viola, P., Jones, M.: Robust real-time face detection. In: Proceedings Eighth IEEE International Conference on Computer Vision, ICCV 2001, vol. 2, p. 747 (2001)
20. Yolanda, L., Martín, M., Gutiérrez Mendoza, L., Mary, L., Nieves, A.: Guidelines to design of virtual learning objects (VLO). Application to the Teaching-Learning Process of Area under the Integral Calculus Curve Lignes directrices pour la conception d' objets d'apprentissage virtuels (OAV). Application au processus d' ense. Rev. Científica Gen. José María Córdova, vol. 14, pp. 127–147 (2016)
21. Sun, S., Zhao, J., Gao, Q.: Modeling and recognizing human trajectories with beta process hidden Markov models. Pattern Recogn. **48**(8), 2407–2417 (2015)
22. Klette, R.: Concise Computer Vision. Springer, London (2014). https://doi.org/10.1007/978-1-4471-6320-6
23. Tavallali, P., Yazdi, M., Khosravi, M.R.: An efficient training procedure for Viola-Jones face detector. In: Proceedings of the 2017 International Conference on Computer Science and Computational Intelligence, CSCI 2017, pp. 828–831 (2018)
24. Rodavia, M.R.D., Bernaldez, O., Ballita, M.: Web and mobile based facial recognition security system using Eigenfaces algorithm. In: Proceedings of 2016 IEEE International Conference on Teaching, Assessment and Learning for Engineering, TALE 2016, December 2017, pp. 86–92 (2017)
25. Gabriel, F., Julián, C., Reyes, M.V., López Sánchez, A., Alberto, C., Ríos, J.: Reconocimiento Facial Por El Método De Eigenfaces. Pist. Educ. **127**(04), 66–81 (2017)
26. Febrero, P.: Adaboost con aplicación a detección de caras mediante el algoritmo de Viola-Jones Néstor Paz, Febrero de 2009 (2009)

Unmanned Aerial Vehicle for Rescue and Triage

Darwin Armando Mora Arias[1] (ID), Juan Carlos Ortega Castro[2] (ID),
Carlos Flores-Vázquez[2] (ID), Daniel Icaza[2] (ID),
and Juan-Carlos Cobos-Torres[2(✉)] (ID)

[1] TelComSistema Cia. Ltda., J. Jaramillo 1-16, Cuenca 010110, Ecuador
[2] Catholic University of Cuenca, Av. Américas y G. Torres,
Cuenca 010101, Ecuador
juan.cobos@ucacue.edu.ec

Abstract. In recent years, the rescue of natural disaster victims has included the support of robotic systems to search for trapped people. However, the victims that are found by robots do not have their vital signs evaluated until the rescue team reaches their location. This can complicate matters in difficult-to-access locations and places affected by toxic waste or radiation, where the physical integrity of rescue teams is at risk. This research proposes the use of an unmanned aerial vehicle in the search for victims and performing basic triage (heart and respiratory rate measurement) through a contactless method to support rescue efforts. The main contribution is a decrease in response time in case of a search-and-rescue emergency. The system consists of navigating over a certain area designated as the disaster zone for the search of possible disaster victims that are lying on the ground. Once the victim is located, the navigation is reprogrammed to carry out the search and face recognition. Finally, by automatically selecting a skin area, the heart and respiratory rates are measured. The measurement is carried out through the photoplethysmography imaging technique, without any contact sensor. The comparison of the basic triage results with and without contact confirms to us the efficacy of the proposed method. The Bland-Altman data analysis shows a close correlation of heart and respiratory rates measured with both approaches (correlation coefficient of 0.90 for heart rate and 0.84 for respiratory rate).

Keywords: Recue UAV · Triage UAV · Search and rescue · Vital signs · Photoplethysmography imaging

1 Introduction

Natural and manmade disasters can pose great risks to health as well as to life. Ecuador, for example, is situated in what is referred to as the "Ring of Fire." It is located next to a large tectonic plate, and volcanoes are found throughout its mountain ranges. Therefore, this zone has a high level of seismic danger [1]. Because of its geographical and geologic features, this region is exposed to adverse circumstances of natural and anthropogenic causes, such as active volcanoes, geologic faults, tsunamis, as well as a

© Springer Nature Switzerland AG 2020
M. Botto-Tobar et al. (Eds.): ICAT 2019, CCIS 1194, pp. 502–516, 2020.
https://doi.org/10.1007/978-3-030-42520-3_40

possible terrorist risk due to its proximity to Colombia. The latter could trigger chemical and/or nuclear attacks, putting the population's health at risk.

On April 16, 2017, a series of earthquakes hit the coastal region of Ecuador. This was one of the greatest tragedies in decades, with 671 fatalities [2], many of whom could have survived, but due to the loss of road access, many victims died without even receiving medical attention despite the fact that many of them had managed to escape from the rubble of the collapsed buildings. Areas affected by natural disasters often collapse completely. Tackling floods, earthquakes, landslides, and forest fires becomes increasingly complicated given the inaccessibility of the places where there are still people alive. In the case of terrorist attacks of a chemical or nuclear nature, locating survivors becomes difficult or even unfeasible for health workers.

The seconds and minutes following a catastrophe are critical in minimizing the damage caused. Normally, there is a clear protocol of action for the coordination of specialized personnel and the management of resources, and this facilitates an efficient response to a catastrophe.

However, human intervention can prove very difficult in certain circumstances due to the very consequences of the catastrophe and the real risk to the people in charge of managing the crisis. In this type of setting, technology can contribute to the improvement of said management using robot support. Robots can serve as the first line of action to gather information in difficult-to-reach places in a quick and precise way. If this intervention is performed in a coordinated manner, the chances of minimizing the damage suffered from the catastrophe increase drastically.

Currently, there are various alternatives, such as teleoperated robots, [3] that through various detection systems can perform a search for human bodies, which is a great benefit over the majority of ground-based robots. However, teleoperated robots have limitations to achieving their goal that are inherent to their form of locomotion [4–6]. Wheels and caterpillar tracks impede avoiding and overcoming obstacles in the trajectory of the robot. Most existing robots can perform a search for people, but they do not provide information about their vital signs in a non-contact manner. Consequently, the task of the robots becomes even more complicated, for finding the right position to place a sensor on the victim is often impossible.

It is undeniable that a correctly performed triage will help health workers to take good decisions about the priority of attention and transport of the patients [7]. In this way, the rescue efforts can be more efficient and effective. Currently, various proposals for rescue robots exist. The Souryu IV robot [8] is designed to move through debris to carry out a search. It is equipped with RGB and IR cameras as well as a lighting system of high-power light-emitting diodes (LED). This robot is still in an experimental phase, and as of yet there is no information as to whether or not it will have the ability to measure vital signs.

Park et al. [9] propose a victim-recognition system using infrared images for a search-and-rescue robot in dark environments. The robot is equipped with a RGB-D, Microsoft Kinect, and IR camera. The Kinect utilizes a structured light source. In any case, the authors present a system for the recognition of a human victim in dark environments, but not vital sign detection. Another interesting robot is Vital-Bot [10], a small mobile robot that has a 5.8 GHz radar receiver for non-contact vital sign detection. The robot is capable of detecting human heartbeats and is designed for search-and-rescue missions. It is a project on which its authors are still working. Recently, another research group has proposed the miniaturization of the radar [11].

Other techniques to detect victims and measure their vital signs from a distance exist [12–14]. Using the Ultra-Wideband radar, the heart and respiratory rates of victims under debris can be detected. There are proposals to process thermal images of the victims and in this manner detect heart and respiratory rates. Thermal cameras can help to determine whether a victim has vital signs, but the cost still remains high. Additionally, they often suffer interference from the environment.

One mobile robot [15] for critical care in the medical attention area is designed with various sensors (GPS, ultrasound, and a heart rate sensor). The problem with this robot is that the heart rate sensor needs to be placed on the victim to carry out an evaluation; therefore, if the victim is unconscious or immobilized, the evaluation is rendered impossible.

Unmanned aerial vehicles (UAV) or drones are small robots that are extremely useful for a variety of applications. Generally, they boast low response times, provide new opportunities for rescue teams, and work despite adverse conditions such as high temperatures, smoke, pollutants, and toxic materials.

This investigation aims to develop a system for the search, detection, and basic triage of victims employing a UAV, the characteristics of their mounted cameras, and their capacity to reach places that are difficult to access for other robots and even men. Once the victim is found, the UAV will proceed to the remote physiological measurement of heart and respiratory rates utilizing the photoplethysmography imaging technique.

2   Materials and Methods

2.1   Hardware

To achieve the aims of the investigation, a drone (UAV) will be used. The brand is Parrot, and the model is Anafi 4k (Fig. 1). Anafi is a drone equipped with a 180-degree tilting gimbal that can capture images from difficult angles, a zoom without loss of resolution up to 2.8X, and a three-axis image stabilization to take stable photos and videos. Furthermore, it has a 1/2.4" 21MP CMOS Sony Sensor, GPS and GLONASS satellite navigation systems, and a flight time of 25 min.

Parrot provides the library Olympe as the programming interface to connect and control the drone from a remote Python script executed on a computer. Olympe sends command messages to the drone (take off, land, camera orientation, flight plans, etc.), checks the current status of the drone, waits for event messages, starts and stops the video streaming, and records the video stream and the associated metadata.

Fig. 1. Parrot Anafi 4k unmanned aerial vehicle

2.2 System Architecture

The system is designed under the Robot Operating System (ROS). For its development and testing, a UAV with an onboard camera was used. The objective of the system is to perform the navigation with a UAV over a disaster zone to perform basic triage on the detected victims. Once the navigation has commenced, by means of the onboard camera, the UAV carries out the online video stream for the search and detection of people lying on the ground. If a person is detected during its general flight plan, the location of the person is calculated and the person detected is positioned in the center of the frame of the video. The flight navigation then changes to victim navigation. It approaches and descends until it is positioned one meter away from the person lying on the ground and at a 45-degree angle from the face of the victim. Once the drone is in position, the detection of the face of the victim is performed. With this information, the location of the face of the victim is calculated, and the flight navigation changes to face navigation to position the camera over the region of interest (ROI) on the face of the victim.

Once the UAV is in position, the measurement of the heart and respiratory rates of the victim is carried out utilizing the photoplethysmography imaging technique as part of the basic triage. Once the geo-positioning coordinates, heart rate, and respiratory rate information is sent, the UAV continues its general navigation until the entire area of its initial flight plan is covered. Upon completion of the general flight plan, the UAV returns to its takeoff position. The general system architecture (Fig. 2) consists of four phases: video stream, navigation, positioning, and detection. Each one of these phases is made up of subsystems that interact with each other.

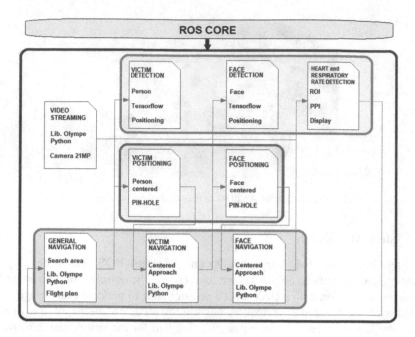

Fig. 2. System architecture.

2.3   Video Stream

The video stream phase provides us the required information for the analysis of the environment, person detection and positioning, and navigation. From the initiation of the system, the video stream starts and throughout the whole process keeps working in parallel with the rest of the phases.

The obtaining and streaming of the video of the UAV is carried out by means of the Olympe library by Parrot through a Python script for its integration with the ROS. The UAV connects to the computer through a wireless WiFi interface. It constantly sends information as to its status and battery charge level. If its battery charge level reaches 20%, the UAV returns to the initial takeoff location. The algorithms in the detection, positioning, and navigation stages will work upon the images obtained from the video.

2.4   Navigation

This phase of the system is focused on the navigation of the UAV (Anafi drone) with a flight plan. Upon takeoff, the "home" position is determined; that is, the point to which the drone will return at the end of the planned flight.

To carry out the UAV flight plan, we used the Olympe library. With its integrated GPS satellite navigation system, we determined its P0 position for the start and end of the route. The area of the planned flight for the general navigation is 15 by 30 m at a height of 10 m (Fig. 3).

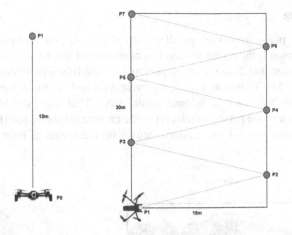

Fig. 3. General route flight navigation.

When the UAV is at position P1 at an altitude of 10 m, it starts on its route to cover the area, passing through points P2, P3, P4, P5, P6, and P7 (Fig. 3). When the UAV reaches position P7, it returns automatically to position P0 and lands.

During the general UAV navigation, the streamed video is processed to carry out the recognition of people lying on the ground. In this manner, if the UAV identifies a person during its route, its flight plan is modified. The UAV will center the image on the person, carrying out the victim navigation (Fig. 4a) to approach the victim to up to 2 m of distance at a 45-degree angle. Once the UAV is in this position, it carries out face identification and positioning. With the help of face navigation (Fig. 4b), it positions itself above a region of interest on the face to carry out the triage. Upon completion, it continues its flight plan with general navigation.

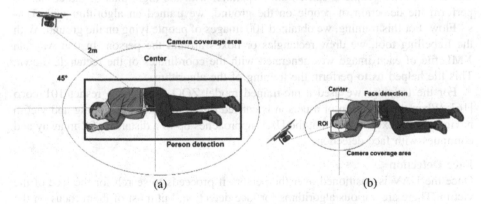

Fig. 4. (a) Victim navigation, (b) face navigation

The navigation phase consists of three subsystems: general navigation, victim navigation, and face navigation.

2.5 Positioning

The positioning phase works in parallel with detection and navigation. When the detection of a person lying on the ground has been carried out, by means of the Pin-hole method, the location and distance of the person are established in relation to the camera of the UAV (Fig. 5a). To initiate the victim navigation and approach the person for face detection, the Pin-hole method is once again used. That way, the location and the distance of the face of the person in relation to the camera is established (Fig. 5b), and by means of the face navigation, the camera focuses on the region of interest for analysis.

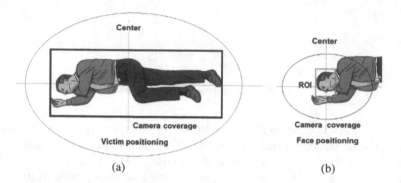

(a)   (b)

Fig. 5. (a) Victim positioning, (b) face positioning

2.6 Detection

This phase consists of three important processes: detection of people lying on the ground, face detection, and detection of heart and respiratory rates.

Detection of People
The detection of people is carried out in parallel with the flight plan of the drone. To perform the detection of people on the ground, we trained an algorithm with TensorFlow. For this training, we obtained 100 images of people lying on the ground. With the Labelling tool, we drew rectangles or boxes around the person. In that way, an XML file of each image was generated with the coordinates of the rectangle drawn. This file helped us to perform the training of the algorithm.

For the training, we used a pre-trained model ZOO faster_rcnn_resnet 101_coco [16]. When the detection of a person is carried out, the victim positioning and victim navigation are executed so that the UAV approaches up to a distance of 2 m away and continues with face detection.

Face Detection
Once the UAV is positioned over the person, it proceeds to search for the face of the victim. There are various algorithms for face detection, but most of them focus on the detection of front-facing human faces. The majority of facial features have a certain similarity, regardless of sex. This allows for the development of techniques for face detection and recognition. The image of the body of the victim captured by the UAV is converted to grayscale and is resized. This makes for a reduction in the computation

needed for face detection. The face detection is performed using the Dlib library [17] because of its features and precision. Subsequently, using the coordinates of the face, the program looks for facial marks (Fig. 6). The facial marks in question are the beginning and end of the eyebrows. With this information, the position of the head can be estimated (roll, yaw, and pitch).

Fig. 6. Detection of face and facial marks using Dlib

The beginning and end of the eyebrows define the area of analysis on the forehead of the victim. The imaginary line that connects the eyebrows help define the pitch. The size of the imaginary line, which is the distance between the eyebrows, defines the size of the area of analysis. This distance varies depending on the distance between the face and the camera. Finally, the ROI (Fig. 7) increases horizontally in the opposite direction to the movement of the face (yaw) and in both directions (roll). If the facial marks corresponding to the eyebrows cannot be found, the marks used will be the forehead and the chin. In this case, the cheek of the victim will be the ROI for analysis. Finally, once the region of interest has been estimated, it is analyzed for vital signs detection.

Fig. 7. ROI identification

Vital Signs Detection

The photoplethysmography imaging technique is used for the estimation of the heart and respiratory rates. It is a technique that allows for remote, non-contact measurements of vital signs. It is an alternative to conventional photoplethysmography on humans, as the photoplethysmography imaging technique is performed by means of a video camera instead of a photodetector. The camera can be RGB or multi-spectral, but it is important that the spectral sensitivity of the cameras and the absorption spectrum of the oxygenated and deoxygenated hemoglobin coincide. The technique has a basic layout (Fig. 8). The images of the identified region of interest (forehead or cheek) are analyzed by channel spacing.

All the pixels of each frame of each one of the channels are averaged. With each new average of each region of interest, a FIFO memory structure is generated (The memory structure has a dimension of 6 s). Inside the memory structure, a photoplethysmographic signal is stored that contains at least six heartbeats. Once the memory structures are full, the method of least squares described [18] is used to filter the trend signal of the averaged signal. Upon subtracting the average signal and the tracking signal, a low-frequency filter is generated. A filter pass whose band type is IIR Butterworth filter pass is then used with the signal without trend, with the heart and respiratory limits of a normal person. Subsequently, the Fourier transform is applied to find the signal spectrum. The matching peak with the heart and respiratory rates is searched for. The respiratory rate is detected through the sinus arrhythmia in the heart rate.

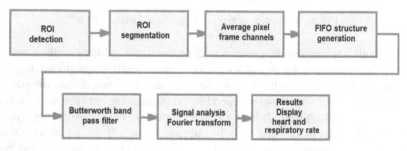

Fig. 8. Basic layout for heart and respiratory rate detection

3   Results

A three-stage plan was carried out. The three stages were the detection stage, the navigation and positioning stage, and the complete system stage. Each stage was analyzed separately to make the needed adjustments and to ensure the proper functioning of the entire system.

The trials were conducted in two different places, the parking lot of the Catholic University of Cuenca and a concrete court. Trials were performed with three people, who lay on the floor simulating the victims, to measure the response times and how long each process took. The detection stage worked as intended, correctly detecting the people, the faces, and the heart and respiratory rates, with a small variation in the trial times due to the impact of the wind on the stability of the UAV. The navigation and positioning trial helped us to determine the correct distance and angle for the location of the UAV in the face detection and the detection of heart and respiratory rates.

The complete system trial was carried out successfully, with all the stages of the architecture working properly and as a whole. The UAV set its position with the GPS geo-positioning coordinates upon takeoff, started the video stream, climbed to a height of 10 m, and commenced the general navigation. The navigation continued until the first person was detected, whereupon the positioning was carried out to center the person in the frame of the video. Subsequently, using the victim navigation, it approached up to a distance of 2 m with the camera at a 45-degree angle (Fig. 9a). When it reached that position, it carried out the face detection and positioned the face of the victim in the center of the frame of the video. The face navigation was performed to be in the best position and to detect the region of interest. Finally, it carried out the heart and respiratory rate detection (Fig. 9b). The vital signs measurements were correctly validated with a high degree of correlation. Upon completion of this process, the UAV returned to its general navigation route, completed its trajectory, returned to its takeoff position, ended the video stream, and landed seamlessly.

To confirm that the basic triage had worked properly, the data obtained by the photoplethysmography imaging technique were compared with data gathered using a heart rate monitor with a finger sensor and the help of a mobile application that helps one to control one's respiratory rate.

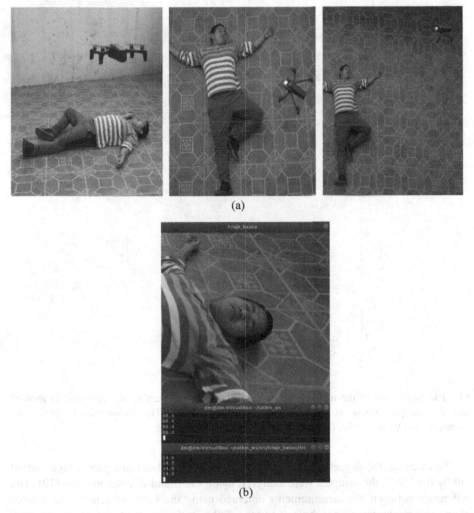

Fig. 9. (a) Photographic report of system operation, (b) image of the system in operation performing heart and respiratory rate detection

To calculate the degree of agreement, the measurements taken during 60 s on the three people acting as the victims were used as a sample. The results that were obtained by means of the camera of the UAV were compared with the results obtained by means of the heart rate monitor with a finger sensor (Fig. 10) that was used to obtain the heart rate of the person acting as a victim. The victims controlled their respiratory rates with the help of the previously mentioned application.

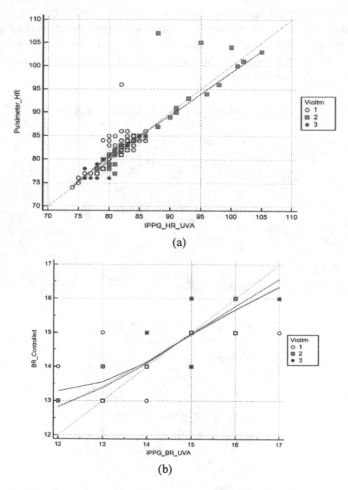

Fig. 10. Scatterplot of the relationship between the samples taken by the non-contact method and the samples taken by the method involving contact. The measurements performed correspond to three people. (a) Heart rate and (b) respiratory rate.

To calculate the degree of agreement of the samples taken during the triage carried out by the UAV, the samples were analyzed using the Bland-Altman method [19]. The difference between the measurements obtained using the UAV system by the photo-plethysmography imaging technique and those obtained using the heart rate monitor are represented as a function of the average of the two systems for heart rate and respiratory rate (Fig. 11). The averages are represented by dotted lines. The limits of agreement of 95% (±1.96 SD) are represented by continuous lines. The average bias was 0.4 for heart rate with limits of agreement of 95%, −4.3 to 5.2 bpm, and for the respiratory rate, the average bias was 0.11 with limits of agreement of 95%, −1.14 to 1.36 bpm.

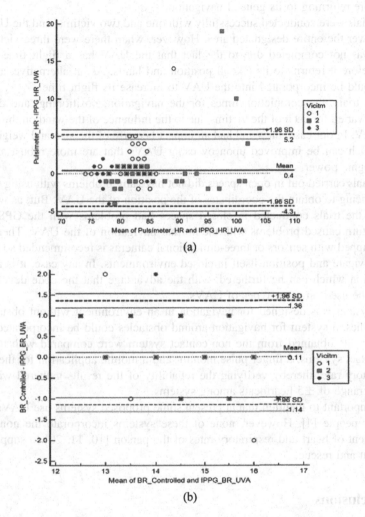

Fig. 11. Bland-Altman plots that show the degree of agreement between the measurements taken using the proposed non-contact method and the method involving contact. The performed measurements correspond to three people. (a) Heart rate, (b) respiratory rate.

4 Discussion

During the system testing, some points for future improvement were found. For example, when the face detection was carried out on a person lying face down, the system did not recognize any face. This means that the basic triage would not be performed and that the UAV would return to its general navigation route. In this case, the system should report the presence of a victim and send the coordinates to the rescue team before returning to its general navigation.

The trials were conducted successfully with one and two victims, and the UAV was able to cover the entire designated area. However, when there were three victims, the process was not completed due to the fact that the UAV has a flight time of only 25 min before it returns to its takeoff position and lands. As an alternative, a backup battery could be incorporated into the UAV to increase its flight time.

In the trials, the completion times for the navigation, positioning, and detection processes varied for each of the victims due to the influence of the wind on the stability of the UAV. In our case, the wind had an impact because of the size and weight of the UAV, which can be improved upon by using UAVs that are more robust and have greater engine power.

The trials carried out in open spaces did not have any problems with using the GPS georeferencing to obtain the coordinates of the position of the UAV. But, as was to be expected, the trials conducted in closed spaces had problems with the GPS signal, which in turn caused problems with the normal navigation of the UAV. Therefore, a UAV equipped with sensors or three-dimensional cameras is recommended so that it is able to navigate and position itself in closed environments. In any case, it is a line of investigation which can be furthered, with the advantage that the code developed in ROS can be used on other types of UAVs.

The system was designed for navigation in an environment without obstacles. In future studies, a system for navigation around obstacles could be incorporated.

The results obtained from the non-contact system were compared with the traditional contact systems, a finger pulse monitor and a mobile application for the control of respiratory rate, thereby verifying the reliability of the results gathered within the permitted range of ±5 heartbeats among systems.

It is important to mention that at present some proposed systems use UAVs for the search for people [4]. However, none of these systems incorporate the non-contact measurement of heart and respiratory rates of the person [10, 11, 20] to support their assessment and rescue.

5 Conclusions

The development of the present investigation should be considered as the start of the continual improvement of the proposed system and a prototype for wide-scale implementation after making the necessary adjustments to guarantee proper functioning.

As previously mentioned, further work is needed to develop a navigation system that can avoid obstacles within the intervention area. Furthermore, consideration should be given to the possibility of using an infrared camera for low-light settings, and a two-way audio system could be incorporated that allows the rescue team to main audio and visual contact with the victim in real time while the measurement of the heart and respiratory rates is being performed.

The photoplethysmography imaging technique used to carry out the basic triage delivered good results. The biggest drawback lies in determining an appropriate area for analysis, but it is superior to the usual methods that involve contact, as carrying out the measurement in a non-contact manner is much easier when victims are unable to cooperate.

Using a UAV with better features can considerably increase the size of the search area, reduce the execution times of the processes, and ensure greater stability throughout its navigation. The system architecture and design have been developed to be implemented in other robotic systems and to look for better alternatives in the operation and design of its algorithms. The system proposed for the non-contact basic triage can be implemented in other systems that are used in the case of patients who need continual monitoring of their heart and respiratory rates.

Acknowledgments. The research leading to these results has received funding from SmartU-niverCity 2.0 program, funded by "Optimización energética del sistema de recaudo en Unidades de Transporte Urbano".

References

1. Demoraes, F., D'Ercole, R.: Cartografía de las amenazas de origen natural por cantón en Ecuador (2001)
2. Lara Calderon, M.L.: Consecuencias de lecciones no aprendidas que se evidenciaron tras el Terremoto del 16 de abril 2016. TABUGA, comunidad rural ubicada cerca al epicentro (2017)
3. Correa, A.C.: Sistemas robóticos teleoperados. Ciencia E Ingeniería Neogranadina (15), 62–69 (2005)
4. Escobar, J.J..J.M., Ramírez, E.A.D., Hernández, J.C., Matamoros, O.M., Padilla, R.T.: White-donkey: búsqueda de personas con vehículos aéreos no tripulados basada en visión por computadora, vol. 120, pp. 53–63 (2016)
5. Beltrán Huertas, D.C., Bonilla Guerra, J.S.: Robot bio-inspirado para asistencia de búsqueda en situaciones de colapsos estructurales (2018)
6. Bermudez, G., Novoa, K.S., Infante, W.: La robótica en actividades de búsqueda y rescate urbano. Origen, actualidad y perspectivas. Tecnura 8(15), 97–108 (2004)
7. González, R.C.R.: Triage en Emergencias Extrahospitalarias. Monografía de Investigación En Salud. Sevilla: Fundacion INDEX (2014)
8. Arai, M., Tanaka, Y., Hirose, S., Kuwahara, H., Tsukui, S.: Development of 'Souryu-IV' and 'Souryu-V:' serially connected crawler vehicles for in-rubble searching operations. J. Field Robot. 25(1–2), 31–65 (2008)
9. Park, J., Lee, G., Park, J.: Infrared image based human victim recognition for a search and rescue robot. J. Inst. Control Robot. Syst. 22(4), 288–292 (2016)
10. Henderson, R.M.: Congratulations to the Spring 2008 MTT-S undergraduate/pre-graduate scholarship awardees! [Education News]. IEEE Microwave Mag. 9(6), 187–190 (2008). https://doi.org/10.1109/mmm.2008.929730

11. Tsai, T.-H., Shiu, B.-Y., Ho, C.-L., Lin, J.: A vital sign radar receiver with integrated A/D converter and dynamic clutter cancellation. Presented at the 2016 IEEE International Symposium on Radio-Frequency Integration Technology (RFIT), pp. 1–3 (2016). https://doi.org/10.1109/rfit.2016.7578142

12. Rahman, H., Ahmed, M.U., Begum, S.: Non-contact heart rate monitoring using lab color space. In: pHealth (2016)

13. Portilla, K., Santos, V., Trujillo, M.F., Rosales, A.: Non-invasive heart rate monitor applying independent component analysis in videos. Presented at the 2017 International Conference on Information Systems and Computer Science (INCISCOS), pp. 121–127 (2017)

14. Cobos-Torres, J.C., Abderrahim, M.: Measuring heart and breath rates by image photoplethysmography using wavelets technique. IEEE Lat. Am. Trans. **15**(10), 1864–1868 (2017). https://doi.org/10.1109/TLA.2017.8071228

15. Sotelo, V.R.B., Sánchez, J.R.G., Ortigoza, R.S.: Robots Móviles: Evolución y Estado del Arte. Polibits (35), 12–17 (2007)

16. Sobti, A., Arora, C., Balakrishnan, M.: Object detection in real-time systems: going beyond precision. Presented at the 2018 IEEE Winter Conference on Applications of Computer Vision (WACV), pp. 1020–1028 (2018). https://doi.org/10.1109/wacv.2018.00117

17. Kazemi, V., Sullivan, J.: One millisecond face alignment with an ensemble of regression trees. In: Proceedings of the IEEE Conference on Computer Vision and Pattern Recognition (2014)

18. Rajnathsing, H., Li, C.: A neural network based monitoring system for safety in shared work-space human-robot collaboration. Ind. Robot: Int. J. **45**(4), 481–491 (2018)

19. Carrasco, J.L., Jover, L.: Métodos estadísticos para evaluar la concordancia. Med. Clin. **122**(1), 28–34 (2004)

20. Lubecke, V.M., Boric-Lubecke, O., Host-Madsen, A., Fathy, A.E.: Through-the-wall radar life detection and monitoring, pp. 769–772 (2007). https://doi.org/10.1109/mwsym.2007.380053

Aerial Power Lines Measurement Using Computer Vision Through an Unmanned Aerial Vehicle

Luis Gonzalo Lozano Guambaña[1] ⓘ, Juan Carlos Ortega Castro[2] ⓘ,
Diego Javier Morales Jadán[2] ⓘ, Javier Trajano González Redrovan[2] ⓘ,
and Juan-Carlos Cobos-Torres[2(✉)] ⓘ

[1] Grupo Industrial Graiman, Panamericana Norte Km 4,
Cuenca 010150, Ecuador
[2] Catholic University of Cuenca, Av. Américas y G. Torres,
Cuenca 010101, Ecuador
juan.cobos@ucacue.edu.ec

Abstract. Electricity losses in electric power systems are one of the main challenges for the electric power industry worldwide. Although to some extent electricity losses are inevitable, effective actions must be sought to solve this problem, to reduce both technical and non-technical losses. Consequently, this article proposes a system for measuring the caliber of power lines through the use of an unmanned aerial vehicle (UAV) and computer vision. This system will be a tool to decrease technical losses. At present, the caliber measurement is carried out subjectively. An operator or electrical technician by means of his expertise and a visual examination measures the wire gauge without a tool. This causes outdated global information systems, resulting in poor electrical planning. The proposed system focuses on flight control of the UAV, detection of objects using a mask of convolutional neural networks, distance estimation, and caliber measurement of aerial power lines without contact. The system reliability is high, with a concordance correlation (r = 0.97) between the measurements made with a manual meter and the UAV. Therefore, the proposed system improves data quality, reduces time and costs, and minimizes the risk of accidents.

Keywords: Computer vision · UAV · Wire gauge · Aerial power line inspection · Aerial power line measurement

1 Introduction

The electric distribution companies in Ecuador adhere to policies and guidelines established by the Ministry of Electricity and Renewable Energy (MERE). Therefore, the distribution companies must have geographic information systems (GIS) of their electrical networks. This policy is based upon the rationale of using energy resources efficiently. Energy planning encompasses many aspects, such as the electrical, financial, and commercial sectors, among others. This is why georeferenced information takes on a leading role, becoming a powerful tool in energy planning [1]. The Agency

© Springer Nature Switzerland AG 2020
M. Botto-Tobar et al. (Eds.): ICAT 2019, CCIS 1194, pp. 517–530, 2020.
https://doi.org/10.1007/978-3-030-42520-3_41

of Electricity Regulation and Control (ARCONEL) is the entity that controls and supervises these policies and guidelines.

GISs must rely upon trustworthy and timely information to facilitate decision-making in planning, leadership, and management. The information should be made available for the consumption of interested members of the public. The challenge for electric companies lies in gathering quality information. These companies must educate themselves to manage increasingly complex networks. GISs contribute towards this effort by providing information and making external and internal communication possible [2]. A GIS is a computerized system that captures, stores, looks up, analyzes, and shows geospatial data [3]. Therefore, keeping these geospatial data up to date becomes a demanding process in terms of time and resources. Oftentimes, it involves processes that are dangerous to human life. In the particular case of the electrical sector, a process is needed for the inspection, checking, and upgrading of features of electrical lines and structures.

Various methods exist for the inspection of electrical lines, such as manual inspection and the use of helicopters [4] and unmanned aerial vehicles (UAVs) [5]. The process of manual inspection comes with problems like the logistics, costs, and risk of accidents, which of course must be reduced [6]. Manual inspection of electrical lines is prone to subjectivity, as it is dependent on the perception of the technician. The technician often has only practical expertise, and this can cause him to overlook technical issues in components that are not detected at first sight. From a loss assessment point of view, the acquisition and efficient use of the data provided by the information systems should be used to calculate with precision technical and non-technical losses [7]. One of the main causes of technical losses of energy is the inconsistency of geospatial data, such as the wire gauge of overhead conductors.

According to data from the Ecuadorian Agency of Regulation and Control of Electricity [8], one can observe a comparison of the energy consumption and losses in gigawatt hours as well as the consumption percentage of each sector in Ecuador. This comparison, with the cutoff date of March 2019, highlights that technical losses make up 7.25% of the Ecuadorian energy system. On the other hand, the people working in the electrical sector and that intervene directly at electrical energy generation, transmission, and distribution facilities of electrical energy are exposed to working accidents such as falling and electrical risks. These sorts of work accidents can cause bodily harm, disability, or death. The Ecuadorian state created a ministerial agreement [9] to ensure living and working conditions, intending to reduce working risks and occupational diseases. This agreement applies to those working in the electrical area. The goal is to reduce risks by putting policies and procedures into place that companies must follow.

UAVs are a tool that can help human activity, and there are different areas of applications [10]. For example, Goodchild et al. [11] propose the use of UAVs in package delivery so as to reduce $CO2$ emissions, and Kraaijenbrink et al. [12] propose a method to georeference and process thermal images taken of a glacier. As one can see, UAVs are in constant development, in search of new contributions. UAVs in conjunction with technologies such as machine vision create development opportunities and benefits for the inspection process, proposing new applications that allow humans to carry out activities efficiently and that reduce human risk [13].

Different research groups propose methods for electrical line detection. For example, Cerón et al. [14] perform electrical line detection with a histogram of oriented segments (HOS). Hui et al. [15] make use of convolutional neural networks. Likewise, other authors [16, 17] implement the Hough transform. The previously mentioned studies concur that aspects such as the weather, background, vegetation, lighting, noise, and stability of the UAV should be considered to deliver the desired results.

This investigation presents the use of a UAV that autonomously measure the wire gauge of electrical distribution and transmission lines. The study is organized in the following manner: The first section is the introduction. In this section, information has been provided about GISs, the importance of relying upon up-to-date geospatial data, and uses of UAVs mainly in the electrical area through a brief literature review. The definition of the problem and the results are found in the second and third sections respectively. The fourth section is the discussion, and the last section is the conclusion.

2 Material and Method

The structure used for the wire gauge measurement process is described in Fig. 1.

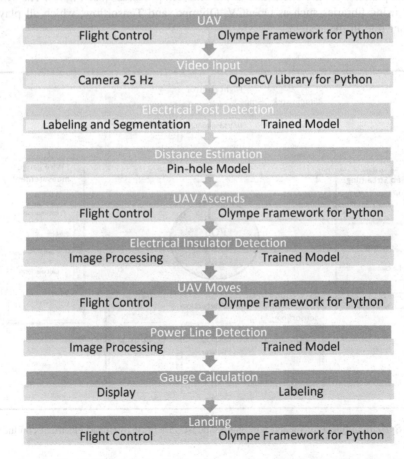

Fig. 1. The structure for the wire gauge measurement process.

The main objective of this investigation is to carry out the wire gauge measurement of an electrical conductor belonging to an overhead electrical network by means of a UAV. UAVs permit access to elevated sites with ease, providing, in this case, efficiency to wire gauge measurement. Furthermore, they help to prevent working accidents or possible falls from ladders or crane baskets. The proposed structure is intended to cover the necessary items in the measurement process. The system consists of different phases that constantly interact. The starting point of the system is the video stream from the onboard camera of the UAV.

2.1 System Architecture

Hereunder the proposed system for the wire gauge measurement of overhead electrical conductors is described. The system is made up of four main processes: video streaming, flight control, object detection, and wire gauge calculation. The Robot Operating System (ROS) robotic software was chosen, which is a collection of frameworks for the development and control of robots. ROS is a powerful alternative as a support platform for the interaction of the different processes (see Fig. 2). The system is built using libraries such as OpenCV, Olympe, and Tensorflow, which all play an important role in the system.

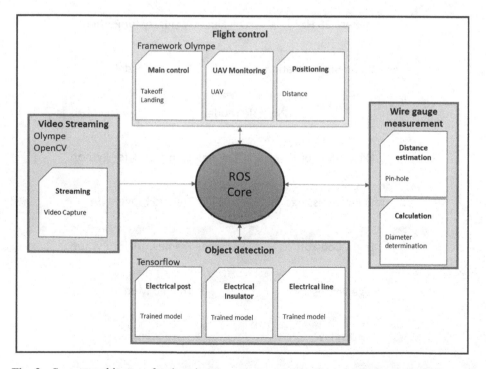

Fig. 2. System architecture for the wire gauge measurement of overhead electrical conductors.

2.2   Uav

The UAV used is called Anafi, of the brand Parrot (see Fig. 3). It is a light and resistant civilian-use drone, weighing 320 grams and boasting a flight time of 25 min. It has a 4K HDR video camera and takes 21 MP photographs. It can resist winds of up to 50 kilometers per hour. It has a transmission range of 4 km. The lens has a focal length of 35 mm. Anafi by Parrot has a library called Olympe that provides an interface to connect the controller using the programming language Python.

Video Resolution

- 4K Cinema 4096 × 2160 24 fps
- 4K UHD 3840 × 2160 24/25/30 fps
- FHD 1920 × 1080 24/25/30/48/50/60 fps

Camera Resolution

- Wide: 21MP (5344 × 4016)/4:3/HFOV 84°
- Rectilinear: 16MP (4608 × 3456)/4:3/HFOV 75.5°.

Fig. 3. UAV Parrot Anafi 4K

2.3   Flight Control and Navigation

The Parrot Anafi UAV has the library Olympe that provides a programming interface for its control from a computer. Command messages can be sent for flight control, to verify the current status, start or stop the video streaming, and record the video stream. Using this library, a script was developed that works within ROS. This framework was selected as it provides the standard operating services such as common-use functionality, sending messages between processes, and package maintenance. In this manner, the necessary processes for the flight control and navigation of the UAV can be processed in various nodes. UAV autonomous navigation is extremely complicated, however. Computation times need to be short. To achieve successful navigation of the UAV, the various pieces of information processed need to have real-time answers.

The flight plan is vertical. The drone should be placed at approximately 50 cm from the post with the electrical lines to be measured (Point A of Fig. 4). Once the flight is started, the UAV should climb to a height of approximately 8.50 m, Point B. While the drone is climbing parallel to the post, it will detect the post to align itself. The condition is that it must keep parallel with the post and at a distance of 50 cm. To keep that distance, the Pin-hole method is used, which enables the calculation of the distance from the post. In this manner, if a tilt in the post is detected, Point B is modified. The destination is reached when the insulators on the post are detected. Once Point B is reached, the drone proceeds to move 5 m to the right, detecting an electrical line. In this same manner as with the post, the UAV detects an electrical line and aligns itself with it. The condition is that it must keep parallel at a distance of 50 cm. Once Position C is reached, the UAV proceeds with the wire gauge measurement through analysis of the images. Once the wire gauge measurement is finished, the UAV should descend and land at its start position, Point A.

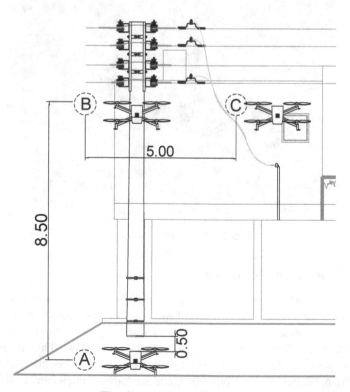

Fig. 4. UAV navigation plan.

2.4   Video Input

Machine vision is based on the capture, processing, and analysis of real-world images. An important part of this investigation is to use the video stream between the UAV and the computer. The streaming starts from the moment the system starts and continues throughout the different phases of the measurement. The video capture is the input point for information into the system. Processing the information from the video will indicate the events and actions to be taken in different phases. Therefore, video streaming and processing must be carried out in parallel with each other.

2.5   Object Detection

The detection phase is performed for three objects in particular: the post, the insulator, and the electrical line (see Fig. 5). By means of Deep Learning algorithms, the UAV through its camera is intended to be able to detect objects. Convolutional neural networks (CNN) algorithms [18] are the main tool that is used in object detection. In our case, Mask R-CNN [19] is used. The detection phase depends on the training of the model, which is carried out before the system is initiated and is essential to the proper functioning of the system. To carry out this training, a dataset of images of the different objects to be detected captured by the camera of the UAV must be acquired and used. The dataset must have data for the training, development, and trials. In the group of training images (1800 images), masks are added to the objects to be detected. The masks are added manually to the dataset. For the training, the open access, pre-trained model mask_rcnn_inception_v2_coco [20] was used. This is facilitated by Tensorflow. As a prerequisite, its parameters must be set so that it will train our model with the objects of interest to us.

Fig. 5. Applying a mask to objects for training

2.6   Distance Estimation

The UAV perceives the real world through image capture. In this phase, the distance between the UAV and the objects it recognizes is estimated. This task is performed using the Pin-hole model, which allows for the estimation of the distance of an object using a camera [21]. The method is exceedingly simple, as the size of the image is related to the distance of the object that is reflected by the focal length (see Fig. 6). The calculation of the distance helps in two ways: Firstly, it allows us to keep the UAV controlled at a distance of 50 cm so that it does not collide with the electrical line and thereby cause a short circuit. Secondly, it allows the relationship between the size of the object in pixels and its real size to be known.

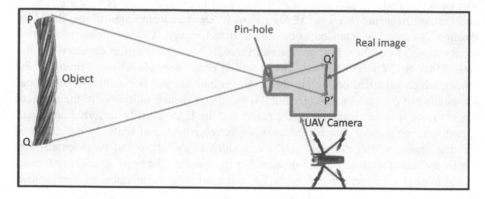

Fig. 6. Pin-hole model.

2.7   Method Used for the Wire Gauge Measurement

Once the UAV is located in front of the electrical conductor, it goes on to determine the wire gauge of the conductor using the Pin-hole method. This is done through a pixel relationship; that is, the number of pixels for each AWG conductor. In Table 1, the specific values in millimeters and pixels for 6, 4, 2, 1/0, and 2/0 wire gauges when the drone is at a distance of 50 cm can be found. These values were determined from the image analysis, as can be seen in Fig. 7.

Table 1. Table of wire gauges and size in pixels from a distance of 50 cm.

AWG conductor	Vertical measurement (Millimeters)	Number of pixels
6	2.125	25
4	2.890	34
2	4.335	51
1/0	5.100	60
2/0	6.290	74

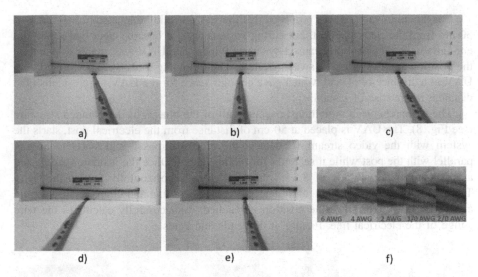

Fig. 7. Photos of the various AWG conductors (a) 6 AWG conductor, (b) 4 AWG conductor, (c) 2 AWG conductor, (d) 1/0 AWG conductor, (e) 2/10 AWG conductor, (f) different AWG conductors.

3 Results

The correct planning and management of the tests provided important feedback to achieve the aim proposed in this investigation. The testing and error-correction plan was carried out in two stages, individual tests and integration tests. In the individual tests, the operation of each proposed level of the system architecture was considered (see Fig. 1), which allowed for the detection and correction of problems in proper time. The integrated tests were carried out to check the functioning of the system when operating with more than one task at the same time. This helped to identify short-comings and to find appropriate solutions.

The functionality tests were performed in different places of the city. In total, the wire gauge of 100 conductors was checked. The chosen places were removed from the city and had little traffic so that the supervisors of the UAV were not at risk. This is appropriate because within the city the electrical posts are only centimeters away from the roadway.

In the tests of the flight control and video streaming capabilities of the UAV, we proved the robustness of the ROS framework, which allowed continuous transmission of data. However, it should be made clear that there were setbacks due to atmospheric issues, such as strong winds, unfavorable climatic conditions, and interference from WiFi signals.

The models trained for the detection of different objects fulfilled the planned aims, correctly detecting the electrical post, the insulator, and the electrical line. This allowed for the distance estimation and for the UAV to be controlled so that it would not collide with the post or electrical line.

For the initial tests, it was proposed that the UAV would be located at a distance of 30 cm from the electrical post, but during various tests, it became clear that it was not enough due to the instability caused by strong winds. The UAV was unable to maintain the position and this caused a swaying effect. The following trials were performed with UAV at a distance of 50 cm, which gave enough space for the UAV to re-establish the distance in the case of instability.

The comprehensive system test executed all of the phases of the planned structure (see Fig. 8). The UAV is placed at 50 cm of distance from the electrical post, starts the system with the video streaming, detects the electrical posts, and starts to ascend parallel with the post while it searches for the insulators. Once the insulator is detected at an approximate altitude of 8.5 m, the UAV travels horizontally 5 m to the right. Throughout the entire movement, it detects the electrical line and aligns itself, keeping a distance of 50 cm. Once the position is reached, it successfully calculates the wire gauge of the electrical line. Finally, it descends and lands.

Fig. 8. Photos taken of the UAV in operation.

To validate the functioning of the system, a set of tests was carried out, where the data from the wire gauge measurements were collected. To prove the precision of the system, the concordance correlation coefficient (CCC) [22] method was used. The method consists of measuring the concordance of two measurements of the same variable. It is a measurement of how well a dataset matches up with a standard measurement. To carry out this method, three data samples were defined: firstly, the data obtained by the measurements carried out by the UAV; secondly, the data obtained from the GIS of Company XXX; thirdly, the data obtained by a manual calibrator. This latter sample is considered the standard measurement. The result obtained by the application of the CCC of the standard measurement compared with the GIS and UAV samples can be observed in Fig. 9.

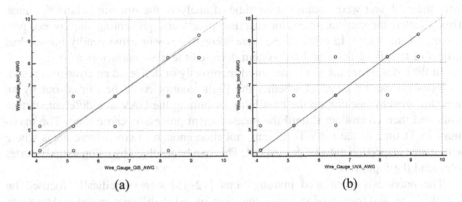

(a)   (b)

Fig. 9. CCC Graph of the samples taken from the calibrator compared with the GIS and UAV, (a) GIS evaluated compared with the calibrator, (b) UAV evaluated compared with the calibrator.

McBride [23] gives the following tips for interpretation: A concordance correlation coefficient of <0.90 is poor, of 0.90 to 0.95 is modest, from 0.95 to 0.99 is substantial, and >0.99 is almost perfect.

Applying this method for the calibrator sample and the GIS sample, the CCC is 0.88, which, based upon the interpretation tips above, is poor. The CCC for the calibrator sample and the UAV sample is 0.97, which is interpreted as substantial.

The Pearson correlation coefficient (p) and Bias correction factor (Cb) obtained from the CCC for the calibrator sample compared with the GIS sample is p = 0.89 and Cb = 0.98. On the other hand, for the calibrator sample compared with the UAV sample results in p = 0.97 and Cb = 0.99.

4 Discussion

Based upon the results obtained, it can be asserted that the wire gauge measurement system has greater precision in capturing information than visual inspection.

As can be appreciated, the result of the proposed method is more efficient and approximates more closely the data obtained by the calibrator. This backs up our investigation, confirming that the GIS information is not up to date and even erroneous. The GIS information provided by technical personnel is based on visual perception and inspection. To rely on correct measurement information, this should be carried out with a manual calibrator. This process takes technical personnel a considerable amount of time and thereby stops them from covering more zones. Furthermore, performing the measurement manually creates a risk of accidents.

On the basis of the evidence, it can be said that the proposed system will minimize the risk of accidents and improve data handling, providing data that are more in line with reality and reducing the time needed to carry out the wire gauge measurement of electrical lines.

The measurements of the UAV that did not match up with the real measurement from the calibrator were because at the time of analysis, the sun was behind the lines. This affected the camera, oversaturating the pixels and preventing the correct processing of the image. In other cases, there were places where the nearby houses had outlets or telephone lines, and the system detected the two conductors as one.

In the cases where the UAV measured incorrectly or delivered an erroneous result, the question can be asked of whether the flight control could be carried out by an operator, thereby opening up the possibility of putting the UAV at a different starting point and then controlling it until the measurement process is carried out. The maximum flight time of the UAV is 25 min, not time enough to cover large areas where wire gauge inspection must be carried out. This can be resolved by using extra batteries to extend the flight time.

The previously mentioned investigations [12–15] were specifically focused on electrical line and transmission tower detection by using different methods. However, no process was carried out to collect information as to the wire gauge of the electrical conductors detected. This investigation provides a step up with the wire gauge measurement of the electrical conductor, information that is necessary for GISs to be maintained up to date.

5 Conclusions

This research work provides a baseline for future investigation, which can provide improvements or look for other areas of application. The executed approach encompasses areas that are under ongoing development, such as computer vision and the civilian use of UAVs.

Taking the results gathered into account, using this research as a tool for the personnel that carry out the measurement of the wire gauge of conductors should be proposed. Additionally, further work and improvements on this line of research are planned, such as incorporating the UAV navigation through GPS, where the operator creates a flight plan, entering the georeferenced information of the various measurement points. The UAV will carry out the route, collect the information, and return to the desired point. Furthermore, the detection of other components is planned to carry out an analysis of the state in which other structures or fittings are found, incorporating a thermal camera to detect hot points. With the information collected, an alert or notification could be sent to carry out preventive or corrective maintenance, in this manner reducing possible energy losses. Another line of research is the facilitation of the UAV to communicate with the GIS to enter or update the information with the use of the mobile network, automatizing the process of entering information so that there is updated information automatically. This will reduce intermediary processes and the time spent on them.

With the work carried out on object detection by means of a UAV, other areas of application can be sought, such as industry, security, health, and even agriculture. It is evident that the field of action is wide.

Acknowledgments. The research leading to these results has received funding from SmartUniverCity 2.0 program, funded by "Optimización energética del sistema de recaudo en Unidades de Transporte Urbano".

References

1. Arancibia, M.E.: El uso de los sistemas de información geográfica -SIG- en la planificación estratégica de los recursos energéticos. Polis (Santiago) **7**(20), 227–238 (2008). https://doi.org/10.4067/S0718-65682008000100012
2. Biehl, M.: Success factors for implementing global information systems. Commun. ACM **50**(1), 52–58 (2007). https://doi.org/10.1145/1188913.1188917
3. Chang, K.: Geographic information system. In: International Encyclopedia of Geography, pp. 1–9 (2017). https://doi.org/10.1002/9781118786352.wbieg0152
4. Avidar, B.: Electronic airborne inspection method for overhead transmission power-lines. In: Proceedings of ESMO 1993, IEEE 6th International Conference on Transmission and Distribution Construction and Live-Line Maintenance, pp. 89–93 (1993). https://doi.org/10.1109/TDCLLM.1993.316255
5. Zhou, G., Yuan, J., Yen, I., Bastani, F.: Robust real-time UAV based power line detection and tracking. In: 2016 IEEE International Conference on Image Processing (ICIP), pp. 744–748 (2016). https://doi.org/10.1109/ICIP.2016.7532456
6. Jones, D.: Power line inspection - a UAV concept, November 2005. https://doi.org/10.1049/ic:20050472
7. Agüero, J.R.: Improving the efficiency of power distribution systems through technical and non-technical losses reduction, May 2012. https://doi.org/10.1109/TDC.2012.6281652
8. Balance Nacional de Energía Eléctrica – ARCONEL. https://www.regulacionelectrica.gob.ec/balance-nacional/. Accessed 12 Sept 2019
9. Ministerio del Trabajo: Reglamento de seguridad del trabajo contra riesgos en instalaciones de energía eléctrica, no. 013, p. 12 (1998)
10. Shakhatreh, H., Sawalmeh, A.H., Al-Fuqaha, A., Dou, Z., Almaita, E., Khalil, I., et al.: Unmanned aerial vehicles (UAVs): a survey on civil applications and key research challenges. IEEE Access **7**, 48572–48634 (2019). https://doi.org/10.1109/ACCESS.2019.2909530
11. Goodchild, A., Toy, J.: Delivery by drone: an evaluation of unmanned aerial vehicle technology in reducing CO_2 emissions in the delivery service industry. Transp. Res. Part D: Transp. Environ. **61**, 58–67 (2018). https://doi.org/10.1016/j.trd.2017.02.017
12. Kraaijenbrink, P.D.A., et al.: Mapping surface temperatures on a debris-covered glacier with an unmanned aerial vehicle. Front. Earth Sci. **6** (2018). https://doi.org/10.3389/feart.2018.00064
13. Kanellakis, C., Nikolakopoulos, G.: Survey on computer vision for UAVs: current developments and trends. J. Intell. Rob. Syst. **87**(1), 141–168 (2017). https://doi.org/10.1007/s10846-017-0483-z
14. Cerón, A., Mondragón, I., Prieto, F.: Onboard visual-based navigation system. Int. J. Adv. Rob. Syst. **15**(2), 1729881418763452 (2018). https://doi.org/10.1177/1729881418763452
15. Hui, X., Bian, J., Zhao, X., Tan, M.: Vision-based autonomous navigation approach for unmanned aerial vehicle transmission-line inspection. Int. J. Adv. Rob. Syst. **15**(1), 172988141775282 (2018). https://doi.org/10.1177/1729881417752821

16. Zhang, J., Liu, L., Wang, B., Chen, X., Wang, Q., Zheng, T.: High speed automatic power line detection and tracking for a UAV-based inspection. In: 2012 International Conference on Industrial Control and Electronics Engineering, pp. 266–269 (2012). https://doi.org/10.1109/ICICEE.2012.77

17. Baker, L., Mills, S., Langlotz, T., Rathbone, C.: Power line detection using Hough transform and line tracing techniques, November 2016. https://doi.org/10.1109/IVCNZ.2016.7804438

18. Du, J.: Understanding of object detection based on CNN family and YOLO. J. Phys: Conf. Ser. **1004**, 012029 (2018). https://doi.org/10.1088/1742-6596/1004/1/012029

19. He, K., Gkioxari, G., Dollár, P., Girshick, R.: Mask R-CNN. arXiv e-prints, arXiv:1703. 06870 (2017). https://doi.org/10.1109/TPAMI.2018.2844175

20. Tensorflow Models. https://github.com/tensorflow/models/blob/master/research/object_detection/g3doc/detection_model_zoo.md. Accessed 18 Sept 2019

21. Megalingam, R.K., Shriram, V., Likhith, B., Rajesh, G., Ghanta, S.: Monocular distance estimation using pinhole camera approximation to avoid vehicle crash and back-over accidents. In: 2016 10th International Conference on Intelligent Systems and Control (ISCO), pp. 1–5 (2016). https://doi.org/10.1109/ISCO.2016.7727017

22. Lin, L.I.-K.: A concordance correlation coefficient to evaluate reproducibility. Biometrics **45**, 255–268 (1989). https://doi.org/10.2307/2532051

23. McBride, G.B.: A proposal for strength-of-agreement criteria for Lin's concordance correlation coefficient. NIWA Client Report: HAM2005-062 (2005)

Developing a Socially-Aware Robot Assistant for Delivery Tasks

Carlos Flores-Vázquez[1,2](✉) [iD], Cecilio Angulo Bahon[2] [iD], Daniel Icaza[1] [iD],
and Juan-Carlos Cobos-Torres[1] [iD]

[1] Catholic University of Cuenca, Av. Américas y G. Torrres,
010101 Cuenca, Ecuador
cfloresv@ucacue.edu.ec
[2] Intelligent Data Science and Artificial Intelligence Research Center,
Polytechnic University of Catalonia BARCELONATECH,
Pau Gargallo 14, Barcelona 08034, Spain
https://www.ucacue.edu.ec/

Abstract. This paper discusses about elements to be considered for developing a Service Robot that performs its task in a social environment. Due to the social focus of the service, not only technical considerations are demanded in order to accomplish with the task, but also the acceptance of use for the people, who interact with all of them. As our particular research topic, we establish a taxonomy to determine the framework for the development of socially-aware robot assistants for serving tasks such as deliveries. This is a general approach to be considered for any service robot being implemented in a social context. This article presents several previous cases of the implementation of service mobile robots, their analysis and the motivation of how to solve their acceptance and use by people. Therefore, under this approach it is very important not to generate false expectations about the capabilities of the robot, because as it is explained in the state of the art analysis that very high unsatisfied expectations lead to leaving the robot unused....

Keywords: Service robotics · Social robotics · Human - robot interaction · Mobile robotics · Technology acceptance

1 Introduction

A *Service Robot* is defined according to the ISO 8373:2012 standard [1] as a robot that performs useful tasks for humans or equipment, excluding industrial automation applications. A *Professional Service Robot* is similarly defined as a robot used for a commercial task, usually operated by a properly trained

Supported by Catholic University of Cuenca - SENESCYT - Polytechnic University of Catalonia BARCELONATECH - GIRVyP Visible Radiation and Prototyping Research Group - Research Project: WAITERS ROBOTS WITH FOCUS ON SOCIAL ROBOTICS.

© Springer Nature Switzerland AG 2020
M. Botto-Tobar et al. (Eds.): ICAT 2019, CCIS 1194, pp. 531–545, 2020.
https://doi.org/10.1007/978-3-030-42520-3_42

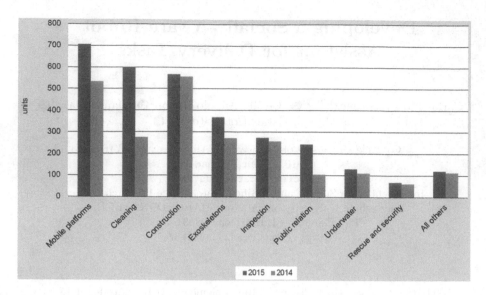

Fig. 1. Service robots for professional use. Sold units 2015 and 2014

operator. Examples are cleaning robots for public places or delivery robots in offices and hospitals. Following the definitions, "a certain degree of autonomy" is required for service robots ranging from partial autonomy, including human-robot interaction, to full autonomy, without operational human-robot intervention. The Executive Summary World of Robotics 2017 Service Robots [2,3], establishes that service robots sales grew in the year 2016 by 24% (59700 units) compared to the year 2015 (48018 units). The projection for the year 2017 estimates a growth of about 17% (78700 units), for the years 2018 to 2020 (397000 units) an annual increase is expected between 20% and 25%.

The International Federation of Robotics (IFR) [4] provides us with a more detailed statistical picture of sales by areas of professional service robotics in units sold during the years 2014 and 2015. As observed in Fig. 1, mobile platforms and robots for cleaning tasks are the two more demanded areas. These figures demonstrate the potential for growth in professional service robotics and a tangible reason for development in this field of robotics.

According to the ISO 8373:2012 standard [1] our proposal about developing a socially-aware robot assistant for serving tasks corresponds to the definition of professional service robots such as in detail restaurants, hotels or any commerce where it is required to move objects from one place to another. Specifically, this proposal is devoted to develop a professional service robot as a delivery assistant. Hence, the main task to be developed by this service robot will be to collaborate with people in the service task by transporting indoors any object (considering the restrictions of the prototype) from one place to another, avoiding obstacles and people on their way.

This service task could seem a simple task to achieve, but the complexity of interacting in social environments is currently one of the most complex problems for robots. A number of previous cases can be found in the literature: from the one developed in China [5] with a total failure of the waiter robots with regard to the expectations of the users, to the experience in the Barcelona-based coffee shop Costa Coffee [6] where successful small demonstrations are working, but it has not been possible to use the proposed robot for prolonged periods.

Our starting hypothesis, based on these past experiences and other similar ones is that the criteria of the user is a fundamental part of the development of this kind of robotic platforms.

The main objective of this work is to define the development of socially-aware robot assistants for serving tasks as delivery in collaboration with a human being from the perspective of social robotics. We defend that this approach will help to overcome past drawbacks in the implementation of service robots in public domains. Moreover, the taxonomy proposed can also be applied to other robotic platforms, contexts and services.

The rest of the article is organized as follows: Sect. 2 presents previous cases of the implementation of waiter robots, Sect. 3 analyzes the social approach in robotics service, finally in Sect. 4 presents the first prototype in terms of its minimum hardware and software requirements and approach that we think should be applied in development of this kind of service robot.

2   Material and Method

Among the conceptualization of service robots [7] similarly to that exposed previously, *Social Robots* are distinguished in three different groups:

- *Assistive Robots* (AR), largely referred to robots which can assist people with physical disabilities through physical interaction.
- *Socially Interactive Robots* (SIR), describing robots for which social interaction plays a key role.
- *Socially Assistive Robots* (SAR) as the intersection of assistive robotics (AR) and socially interactive robotics (SIR).

Our revision of the state of the art will be focused on Social and Service Robots evaluated in tasks in restaurants, that is a common area for Professional Service Robots and Socially Interactive Robots.

The following cases can be listed as pioneering attempts to introduce robots in restaurants (see Fig. 2):

- The Hajime robot restaurant in Thailand was open on April 2010 with a robotic waiter-staff of four robots with a total cost of 150,600 euros [8]. Two of the robots are endowed with one arm, being in charge of taking the dishes off the kitchen and handing them to their most advanced companions, with two members, mobile body and samurai clothing. They move in the restaurant along a rail. Among the reported problems, there is confusion when removing the finished dishes, serve drinks or collect the bill.

(a) Robot Hajime restaurant.   (b) FURO Robot   (c) Dalu Hotpot Robot Restaurant.

Fig. 2. Pioneering restaurant robots.

- On June 2010, the FU-RO restaurant robot (Future Robot Co. Ltd) is opened as a practical restaurant robot with hands-on experience [7]. Its service was implemented to make possible interaction between robot and customers through specialized HRI. The company incorporates several robots with different associated tasks such as greeting robot at an entrance, waiting time robot which provides entertainment, order guide robot for guidance, taking order, and payment robot that help customers pay after meals.
- On December 2010, the Dalu Hotpot Robot Restaurant in China [7] had unveiled seven mechanical custom built robot servers with some resemblance to the classic automatons from 1950s science fiction, and others that are basically wheeled tables.

Currently, from these initial experiences, there is only information and availability of the robot in the FU-RO restaurant. In the other two cases, it is unknown whether they were continuing operating.

In [9] the KeJia robot is introduced. This robot plays the role of a bartender by recognizing and grasping the drink when following the order of people and deliver it to a TurtleBot. Several TurtleBots deliver drinks to people as waiters. The human-robot interaction design for service robots being able of safely and effectively interacting with humans is still a challenging problem in robotics. In the KeJia-TurtleBot system, each robot endows some basic behaviors such as navigation or mobile manipulation. These basic behaviors are combined in several ways to yield more complex social behaviors. The Human-Robot Dialogue module provides the interface for communication between users and the robot. The overall system is illustrated in Fig. 3.

According to this experience, in order to optimize the price-performance relationship, the robotic platform TurtleBot 2 will be considered in our proposal for the implementation and testing of the developed solution (see Fig. 3c). The mobile base is Kobuki and its cost is $2.106.

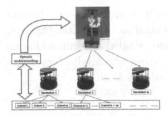 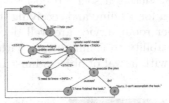 

(a) The architecture of the KeJia-Turtelbot system for bartender-waiter application.

(b) The finite state machine for a simple human-robot dialogue

(c) Turtlebot 2

Fig. 3. The KeJia-TurtleBot overall system.

In [10], a prototype of mobile waiter robot based on Mecanum wheels is introduced. Unlike existing waiter robots, it is able to carry a higher payload. Its development was based on using ROS middleware and implementing SLAM for autonomous navigation. Technologically, the robot is robust in its performance however neither robot human interface, nor the social context are considered. Its cost is about $17.000. The shape of the prototype can be appreciated in Fig. 4.

Fig. 4. URDF (left) and actual robot prototype Beta G (right)

The robot Savioke (see Fig. 5a) was introduced in [11] as a service robot for deliveries and logistics, claiming that they have this robot running in more than 70 hotels. As far as we reviewed the literature, this is the only robot that causes a positive effect considering a social robotics approach. Among its main functions, transportation for sending and receiving items such as towels, toothpaste, newspaper and other room service are well developed, it also has autonomous navigation and the ability to communicate with elevators wirelessly. Nevertheless, the entry and withdrawal of objects must be made by humans.

(a) Savioke Robot.

(b) IEEE   Spectrum;   Robot: TurtleBot; Glass: Jan/Flickr

Fig. 5. Robots from the Kobuki base

In [12], an investigation on how we should give orders to a robot is presented. Specifically, they used a TurtleBot remotely commanded to do their movements (see Fig. 5b) and its voice was generated by a text-to-speech system. This robot is provided with an Artificial Intelligence (AI) system to interact with people. During experimentation the robot played the role of a waiter. The results show that people are overly polite in their order, for example: "Can you Bring Me?" or "If you could X Bring me, that would be great." Instead of direct phrases like "Bring Me X". So the AI system shows that it is still too literal in its interpretation of orders.

2.1   Focusing on Social Robotics

A complete review of social robotics is presented in [13]. Due to the extensiveness of the article and relevance of the information presented, we are exposing it graphically in Fig. 6 and only detailed information is evaluated about sections considered relevant to be applied in our proposal. These areas are highlighted with red color to identify the applicable fields to our project according to this taxonomy.

Design Approaches. Talking about design approaches, the *Functionally designed* approach from [13] will be employed for our proposal. The objective of this approach is to design a robot that outwardly appears to be socially intelligent even if the internal design does not have a basis in science or nature. Hence, design must show the mechanisms, sensations, and traits according to the People's psychology where they understand that a creature is socially intelligent.

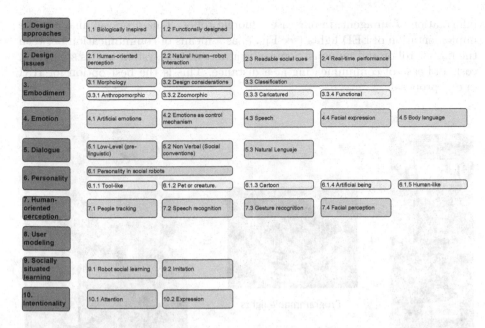

Fig. 6. A proposed taxonomy in social robotics (Color figure online).

Design Issues. We will focus on the *Real-time performance* of the socially interactive robots operating at human interaction rates. Thus, a robot needs to simultaneously exhibit competent behavior, convey attention and intentionality handle social interaction.

Embodiment. It will be considered from two perspectives: *Morphology* because the form and structure of a robot is very important considering that it helps to establish social expectations. *Design Considerations*: since a robot is designed to perform tasks for the human being, then its shape must convey an amount of "productivity" so the user will feel comfortable. Moreover, a robot's design needs to reflect an amount of "robot-ness". This is needed so that the user does not develop detrimentally false expectations of the robot's capabilities.

On the embodiment basis, a robot can be classified as "Functional". The robot's embodiment should first, and foremost, reflect the tasks it must perform. The choice and design of physical features are thus guided purely by operational objectives, especially service robots.

Dialogue. It is an important subject in social robots. The most viable option in our proposal is "Non Verbal (Social conventions)". There exist many non-verbal forms of language, including body positioning, gesturing, and physical action. As most robots have fairly rudimentary capability to recognize and produce speech, non-verbal dialogue is a useful alternative.

At this point it is important to mention the work [14]. The main scope of this work is the idea of expression, which we think about as externalizing hidden information of an agent, in our case a mobile robot. This proposal analyzes the implementation of LED lights (see Fig. 7) as a means of communication between the mobile robot and humans. Some advantages of this system are its simplicity, cost and ease of communicating at a distance. This is the best option for HRI in our proposal.

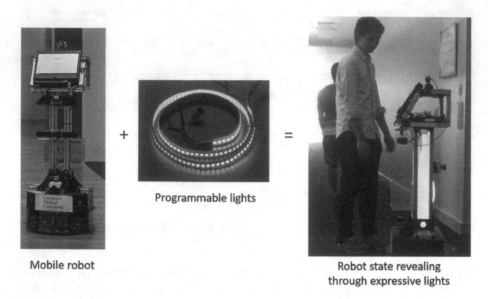

Fig. 7. Augmenting a mobile robot's communication capabilities with a new modality: expressive lights. [14]

Social conventions [13] or norms can also be expressed through non-verbal dialogue. Proxemics, the social use of space, is such one convention. To solve the problems of the proxemics we will use the results tested in [15–17].

Personality. It is organized in social robots into five groups: Tool-like, Pet or creature, Cartoon, Artificial being, Human-like. The "Tool-like" is the one of our interest, that used for robots that operate as *smart appliances*. Because these robots perform service tasks on command, they exhibit traits usually associated with tools.

User Models. They are employed for different purposes. In our proposal user models will be useful for adapting the robot's behavior to suit users with various skills, experience, and knowledge.

Intentionality. It is the last topic to be considered. In detail, "Expression" is understood as the intentionality a robot must exhibit in goal-directed behavior. In our proposal this is considered from its structure and physical aspect for users to understand when seeing the robot which its function and capabilities are.

An important article within the approach of service robotics is [18] the one where they address one of the most important topics for this research proposal such as the non-use of service robots. This research carried out with 70 autonomous robots for 6 months in the houses of different groups of people, explains the reasons why people stop using them or refuse to use them. Among its most relevant conclusions it enunciates:

- Investigating user's motivations and reasons to refuse or abandon the use of a robot providing important insights for the design and acceptance of robots in our society.
- The challenge for robot designers is to create robots that are enjoyable and easy to use to capture users in the short-term and that are also functionally relevant to keep those users in the longer-term.

This reaffirms the users approach as the main participant in the development of a social robot delivery assistant.

3   Results

Based on state of art, we will define the following hardware and software requirements for the developing of the prototype that could be improved based on the user's criteria.

3.1   Hardware

The economic side is one of the aspects that will be considered in this research. The cost of the robot must be accessible to the purchasers, to contemplate the return of their investment in areas such as: publicity, the service it lends and the increase of clients due to this implementation.

Considering that price is one of the major constraints for service robot users, the hardware should be the minimum necessary for the robot to provide functionality and not being excessively expensive. The constituent elements of the robot (see Fig. 8) are listed below.

- Robot Platform: Turtlebot 2 Kobuki base.
- Hard Disk: 500 GB HDD
- Controller: Intel NUC I5
- 3D Sensor: Orbbec Astra
- 2D Sensor: RPLIDAR A3 Lasser Scanner
- Additional battery: 4S2P (4400 mAh)
- Screen and Touch: GeChic 13.3" visualización led-lit Monitor (GeChic 1303i)

For the listed hardware the average cost is $3,000 which is obviously still lower compared to commercial platforms such as iROBOT ROOMBA. However, it is within the feasible costs of a prototype. The selection of the Turtlebot 2 with its Kobuki platform is a low-cost platform that nevertheless meets the minimum characteristics that are necessary to implement our prototype for indoor testing. Among the most important features we mention (Fig. 9):

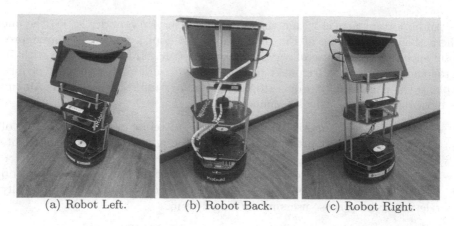

(a) Robot Left.   (b) Robot Back.   (c) Robot Right.

Fig. 8. First prototype: basic structure.

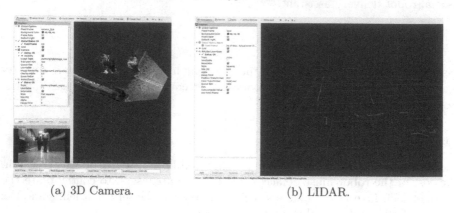

(a) 3D Camera.   (b) LIDAR.

Fig. 9. TEST.

- Maximum linear speed: 70 cm/s.
- Maximum angular speed: 180 deg/s.
- Load capacity: 5 kg (hard floor), 4 kg (carpet).
- Vertical slopes: detects slopes greater than 5 cm in height.
- Maximum unevenness: overcomes unevenness up to 12 mm
- Operating time: 7 h (large battery)
- Charging time: 2.6 h (large battery)

Orbbec Astra is a 3D camera compatible with OpenNI which is a package used in ROS to manage this type of cameras to help during the time of implementation and tuning with SLAM algorithms. The detection ranges of the camera go from 0.4 to 8 m. In the tests performed with people, our results were as follows. While standing the Camera position at 49 cm from the floor, the minimum distance was 0.12 m, the maximum was 5 m and the opening angle was 68°. Probably the obstacle detection with the 3D camera is not enough so we

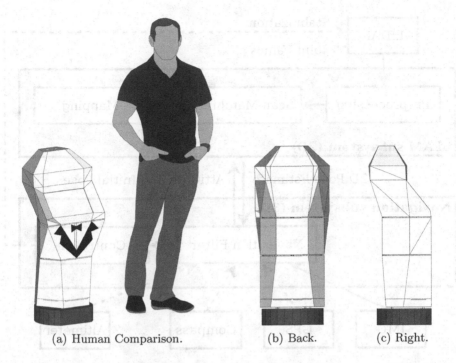

(a) Human Comparison.   (b) Back.   (c) Right.

Fig. 10. First prototype: body Proposal.

considered to include a ROS compatible Lidar sensor 2D that uses the Rplidar package. The main features of this LIDAR sensor are:

- Scanning Angle: 360°.
- Distance range: 10–25 m.
- Scan Speed: 10–20 Hz.
- Angular resolution: 0.3375°, 0.54°.

To perform the following tests with people, the LIDAR was placed 43 cm from the floor. The results allowed us to show that people wearing black clothes were detected at a maximum range of 6 m. The maximum range of detection of people wearing other colors than black was 25 m. In both cases the minimum detection was 0.12 m. The LIDAR was placed 43 cm from the floor.

The design of the Robot Body must consider the taxonomy shown in Fig. 11. In detail 1.2 Functionally Designed, 2.4 Morphology, 3.2 Design Considerations, 3.3.4 Functional, 6.1.1 Tool-like, 10.2 Expression. For all the reasons above, the design illustrated in Fig. 10 does not contemplate elements that generate false expectations about the robot. This implies that as it does not have arms, ears or mouth it is not able to manipulate objects, listen to orders and respond verbally. In this first stage it has only the capacity to move objects from point A (A4 folders, coffee cups and others of the same size) to Point B (in the compartment of its upper part on the screen.) On the screen the robot has two eyes that allows

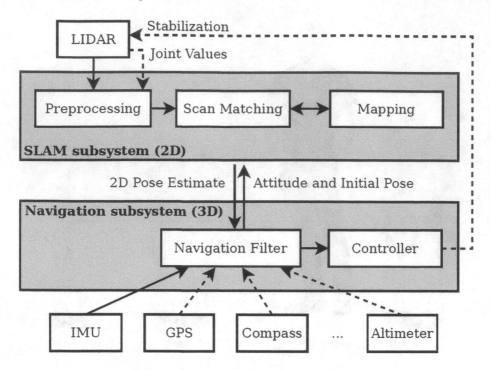

Fig. 11. Mapping and Navigation System. [19]

it to avoid obstacles and fulfill the assigned trajectory. These fulfill the function of communicating with the user and the ability to perceive their environment although this function is performed by the 3D camera and the LIDAR.

Finally as controller we select an Intel NUC I5 is a 4 × 4-inch mini PC, with 8 gigabytes of ram and we include as interface for the user a portable Touch Screen GeChic of 13,3 in.

3.2 Software

While working on robot development, open source platforms are the most popular choice. The problem with these platforms is that they are constantly being tested, improved and generally have compatibility problems with the newer peripherals. Considering all the constraints exposed the following development tools were selected:

- Operating System: Ubuntu 16.04 LTS.
- Framework: ROS Kinetic Kame.
- Additional Packages: OpenNI, Rplidar, Hector Slam, Gmapping.

In the software considerations Ubuntu version with long term support 16.04 was used and not the 18.04 because there are not several ROS packages compatible and necessary for this proposal. ROS Kinetic is the latest version of ROS with supported packages (OpenNI, Rplidar, Hector Slam, Gmapping) for minimum implementation on the TurtleBot 2. The main task for this robot platform is the transport of objects from point A to point B. To perform this task the Robot needs to have a map of the place in which it will move and be able to avoid obstacles whether fixed or mobile as in the case of people. On the basis of the articles [20–23] SLAM (Simultaneous Localization and Mapping) will be applied to carry out our task. This is due to the fact that with SLAM we could make an initial map of the space where the robot will make its displacements and additionally provides information about fixed and mobile obstacles.

The article [19] presents the option that best suits our proposal. This is because it presents a package developed for ROS that integrates a 2D sensor (LIDAR) and another 3D for the generation of maps using SLAM see Fig. 11. The most interesting part of [19] is the integration of the 2D and 3D sensor for SLAM. The 3D estimation applies Kalman Filter, the results of both sensors are not synchronized, the 3D estimation is superimposed on the 2D results.

4 Conclusion

The technological considerations are always the most important for the design of a robot in the work of the engineers. However, this work tries to show that the social considerations are essential to achieve a robot that is accepted and used by the people that are not engineers.

An explicit contribution of this analysis is the taxonomy in graphic form of the social robots. In a specific way, the aspects relevant to the proposal for the development of a robot assistant are analyzed.

Results section proposes a minimum platform for development considering the engineering aspects such as hardware and software.

Articles such as [18] reaffirm that user-focused design is what is needed to overcome the resistance of people to the use of service robots. It is still an outstanding task to establish metrics that allow measuring the performance of a service robot in a social environment.

Within the expected contributions it is proposed to develop a robust robotic platform, which satisfies the expectations of the users. If these expectations are not satisfied the robot will not be used, therefore the objective is not fulfilled.

Another contribution that is expected to be obtained in the research is to obtain an experience in service robotics in social environments, for periods greater than the aforementioned cases.

References

1. International Federation of Robotics: Service robots - definition and classification. In: WR 2016, pp. 9–12 (2012)
2. International Federation of Robotics. Executive Summary World Robotics 2017 Service Robots, pp. 12–19 (2017)

3. Guo, M., Zhang, H., Zhang, G., Huo, J.: La utopía posible: tecnología futurista y calidad de vida. Integr. Comer. **42**, 216–230 (2017)
4. International Federation of Robotics: Executive Summary World Robotics 2016 Service Robots, pp. 1–5 (2016)
5. Thompson, C., Yu, E., CNN: New order? China restaurant debuts robot waiters. http://edition.cnn.com/travel/article/china-robot-waiters/index.html. Accessed 19 Apr 2016
6. RestauracionNews: Costa coffee presenta el primer robot barista del mundo. http://www.restauracionnews.com/costa-coffee-presenta-primer-robot-barista-del-mundo/. Accessed 01 July 2017
7. Pieskä, S., Luimula, M., Jauhiainen, J., Spitz, V.: Social service robots in public and private environments. In: Recent Researches in Circuits, Systems, Multimedia and Automatic Control, pp. 190–196 (2012)
8. ELMUNDO.es: El primer 'camarero-robot'. http://www.elmundo.es/elmundo/2010/07/01/ciencia/1277972988.html. Accessed 01 July 2010
9. Lu, Q., Lu, G., Bai, A., Zhang, D., Chen, X.: An intelligent service system with multiple robots. In: Robot Competition of International Joint Conference on Artificial Intelligence (IJCAI 2013) (2013)
10. Cheong, A., Lau, M.W.S., Foo, E., Hedley, J., Bo, J.W.: Development of a robotic waiter system. IFAC-PapersOnLine **49**(21), 681–686 (2016)
11. Perry, T.S.: CES 2018: delivery robots are full-time employees at a Las Vegas hotel. https://spectrum.ieee.org/view-from-the-valley/robotics/industrial-robots/ces-2018-delivery-robots-are-fulltime-employees-at-a-las/-vegas-hotel?. Accessed 12 Jan 2018
12. Ackerman, E.: How not to order water from a robot waiter. In: IEEE International Conference, and Human Robot Interaction, pp. 1–5 (2018)
13. Fong, T., Nourbakhsh, I., Dautenhahn, K.: A survey of socially interactive robots. Robot. Auton. Syst. **42**(3), 143–166 (2003)
14. Baraka, K., Veloso, M.M.: Mobile service robot state revealing through expressive lights: formalism, design, and evaluation. Int. J. Soc. Robot. **10**, 1–28 (2017)
15. Khambhaita, H., Alami, R.: Assessing the social criteria for human-robot collaborative navigation: a comparison of human-aware navigation planners. In: Proceedings of the IEEE International Symposium on Robot and Human Interactive Communication (RO-MAN), 6p. (2017)
16. Kruse, T., Pandey, A.K., Alami, R., Kirsch, A.: Human-aware robot navigation: a survey. Robot. Auton. Syst. **61**(12), 1726–1743 (2013)
17. Rios-Martinez, J., Spalanzani, A., Laugier, C.: From proxemics theory to socially-aware navigation: a survey. Int. J. Soc. Robot. **7**(2), 137–153 (2015)
18. De Graaf, M., Allouch, S.B., Van Diik, J.: Why do they refuse to use my robot?: reasons for non-use derived from a long-term home study. In: Proceedings of the 2017 ACM/IEEE International Conference on Human-Robot Interaction, pp. 224–233. ACM (2017)
19. Kohlbrecher, S., Von Stryk, O., Meyer, J., Klingauf, U.: A flexible and scalable slam system with full 3D motion estimation. In: Proceedings of the IEEE International Symposium on Safety, Security and Rescue Robotics (SSRR). IEEE, November 2011
20. Omara, H.I.M.A., Sahari, K.S.M.: Indoor mapping using kinect and ROS. In: 2015 International Symposium on Agents, Multi-Agent Systems and Robotics (ISAMSR), pp. 110–116. IEEE (2015)
21. Pajaziti, A.: SLAM-map building and navigation via ROS. Int. J. Intell. Syst. Appl. Eng. **2**(4), 71–75 (2014)

22. Rojas-Fernández, M., Mújica-Vargas, D., Matuz-Cruz, M., López-Borreguero, D.: Performance comparison of 2D SLAM techniques available in ROS using a differential drive robot. In: 2018 International Conference on Electronics, Communications and Computers (CONIELECOMP), pp. 50–58. IEEE (2018)
23. Aagela, H., Al-Nesf, M., Holmes, V.: An Asus_xtion_probased indoor MAPPING using a Raspberry Pi with Turtlebot robot Turtlebot robot. In: 2017 23rd International Conference on Automation and Computing (ICAC), pp. 1–5. IEEE (2017)

Analysis of Normalized Vegetation Index in Castile Coffee Crops, Using Mosaics of Multispectral Images Acquired by Unmanned Aerial Vehicle (UAV)

Julio Mejía Manzano$^{(\boxtimes)}$, Jhon Guerrero Narvaez$^{(\boxtimes)}$,
José Guañarita Castillo$^{(\boxtimes)}$, Diego Rivera Vásquez$^{(\boxtimes)}$,
and Luis Gutiérrez Villada$^{(\boxtimes)}$

Unicomfacauca University, Historical Center Street 4 # 8-30, Popayán, Colombia
{jmejia,jguerrero,joseguanarita,diegorivera,
luisgutierrez}@unicomfacauca.edu.co

Abstract. This paper is aimed at the use of a UAV unmanned aerial vehicle (DJI Phantom 4), and the MicaSense RedEdgeM multispectral camera, for the acquisition of multispectral transported images of the coffee variety of Castilla accessible at Hacienda Los Naranjos in The Cajibío sale located at 1760 m above sea level in the department of Cauca, the gained images were processed using an algorithm developed in the MATLAB® Software to analyze them. The developed routines accept mosaic genres of the images gained through a flight plan that is characterized by having a horizontal and lateral overlap of 75%, the images were gained between February 7 and April 25, 2019 between 11:30 am and 12:00 m, additionally the routines allowed the calculation of vegetation indices such as NDVI, NDRE, GNDVI, and GRVI. That allows us to carry out phenological monitoring in coffee crops, to estimate the vigor or the state of health of the plant.

Keywords: Image processing · Mosaic · Multispectral images · Phenological tracking · Significant differences · Vegetation index

1 Introduction

Using multispectral sensors coupled to UAV has become in recent years a technique for monitoring crops that have improved the use of water resources and fertilizers through the use of precision agriculture techniques, the use of these Techniques in Colombia has focused mainly on the sugarcane sector. Coffee is one of the most widely used agricultural products in the world, it is also Colombia's main agricultural product, and it depends on a significant percentage of the economy and livelihood of a large part of the population. It is produced in over 50 countries and provides a livelihood for over 25 million disadvantaged families worldwide. For consumers, coffee is a beverage that enjoys universal popularity, and sales amount to over 70 billion dollars a year. Coffee is, after oil, the most important commercial product in the world; it beats coal, wheat and sugar [1].

© Springer Nature Switzerland AG 2020
M. Botto-Tobar et al. (Eds.): ICAT 2019, CCIS 1194, pp. 546–559, 2020.
https://doi.org/10.1007/978-3-030-42520-3_43

Precision agriculture comprises the management and administration of the soil and crops taking into account a principle of variability, where it is said that there are inequalities in soil properties within the same land that demand a different treatment according to their conditions. To achieve this differentiation, precision agriculture uses technological tools such as remote sensing sensors, global positioning systems (GPS), geographic information systems (GIS), differentiated rate application technologies (mechatronics, electromechanics) and other technologies. Information and communication (ICT) [2].

The precision agriculture cycle (see Fig. 1) helps to omit the premise of homogeneity that traditional agriculture preserves, in which processes such as land preparation, phytosanitary control, planting, and harvesting are done equally throughout the field, since it seeks to find the spatial and temporal variability within the field to decide on the moment, the location, the necessary quantity and the type of input to be managed, minimizing costs, impact on the environment and maximizing production or productivity [3].

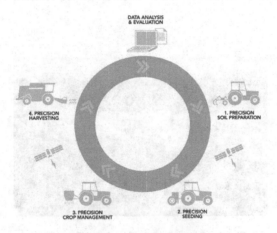

Fig. 1. Precision agriculture cycle [4]

The department of Cauca occupies one of the first positions in coffee production nationwide reaching 9% of the production [5], with about 68,000 hectares of coffee planted and almost 85,000 producers [6], placing it among the 5 first by extension and the number of producers, but this contrasts with the competitiveness data shown by the National Federation of Coffee Growers that rank it below the top 10 departments in competitiveness [7]. This is because of the conditions of the region which position the product as a special coffee, and this is ratified by the information of the national federation of coffee growers of the Cauca [8] who says that the department of the Cauca is a producer of special coffees by designation of origin, showing that coffee growers in the region have vocation and integrity, which regardless of the specific difficulties in the history of production in the region, continually seek alternatives and models of differentiated products to be more competitive.

2   Material and Method

The information acquisition methodology of this paper consists of the acquisition of multispectral aerial images of a coffee crop of the Arabica Coffee type of castle variety, these shots were made between February and April of the year 2019, through the integration of a DJI Phantom 4 Drone and the MicaSense Red Edge M multispectral camera, which allowed us to obtain the necessary information to study vegetation indices in coffee cultivation.

2.1   Description of the Study Crop

The Naranjos farm owned by SupraCafé in the municipality of Cajibío-Cauca, the specific study lot (see Fig. 2) and is identified as lot 3 by TecniCafé, having an approximate age of two and a half years since its planting in August 2016 and is located at an altitude of 1851 m above sea level, the lot has an approximate area of 0.1652 ha^2 and an average of 828 coffee trees of Coffee Arabica variety of Castillo variety.

Fig. 2. Crop of study (Lot 3) of TecniCafé

Lot 3 is divided into 3 zones (see Fig. 3), which presents different states of nutrition based on the nitrogen requirements of the crop where zone 1 has an optimal fertilization of 10 g/m^2 per plant zone 2 it has a low level of fertilization with 5 g/m^2 per plant and zone 3 a high fertilization having a concentration greater than 15 g/m^2 per plant, this is determined in this way since the average nitrogen concentration in coffee of the Cauca is from 10 to 12 g/m^2 per plant, this is achieved through the modification of the fertilizer within each zone which generates a variation in the state of vigor of the plants in the different zones, the treatment performed to the batch 3 can be seen in [9]. These differences in fertilization allow studying and analyzing the behavior of vegetation indices at different stages of fertilization.

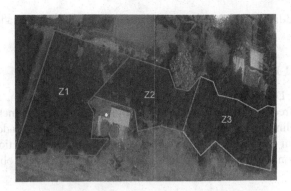

Fig. 3. Crop of study division lot 3.

The variables that will be taken into account when analyzing the information (multispectral aerial images) collected from the study crop (lot 3) as shown in Table 1.

Table 1. Response variables for the study

Variable	Description	Equations
NDVI - Normalized Difference Vegetation Index	Generic index used for leaf cover and plant health	$NDVI = \frac{NIR - RED}{NIR + RED}$
GNDVI - Green Normalized Difference Vegetation Index	Index used to detect areas sensitive to chlorophyll content, measure the rate of photosynthesis and control plant stress	$NDRE = \frac{NIR - REDEDGE}{NIR + REDEDGE}$
NDRE - Red Edge Normalized Difference Vegetation Index	Index sensitive to chlorophyll content in the leaves against the background effects of the soil	$GNDVI = \frac{NIR - GREEN}{NIR + GREEN}$
GRVI - Green proportional vegetation index	It is used to monitor photosynthetically active biomass of plant canopies	$GRVI = \frac{NIR}{GREEN}$

2.2   MicaSense RedEdge-M

The RedEdge-M is a multispectral camera manufactured by the company MicaSense, which manages five spectral bands that allow studies of photogrammetry, cartography and precision agriculture, in a field. The images obtained with the RedEdge-M of its different bands help to construct maps or mosaics of the land, such as the chlorophyll map or other vegetation indices [10], which can also be used to create the digital model of land area.

The RedEdge-M camera allows you to carry out a follow-up study of the crops because when handling the bands Blue, Green, Red, NIR (near-infrared), and Red Edge (red border), it allows you to calculate the maps of the NDVI, GNDVI index, GRVI, NDRE which are the response variables that must be obtained from the coffee study crop (lot 3), in addition, its compact and lightweight design makes it adaptable to small UAV devices.

2.3   Drone DJI Phantom 4

The Phantom 4 Drone, has an OcuSync HD transmission system, which supports automatic dual frequency band switching, is equipped with a 20-inch 1-inch sensor capable of recording 4K/60 fps videos and burst mode at 14 fps. In addition, its Flight Autonomy system includes dual rear view sensors and infrared detection systems for a total of 5 directions of obstacle detection and 4 directions to avoid obstacles [11].

2.4   Flight Plan

Because the crop has an irregular shape measurement of the sides of the land is made and the maximum values of width (70 m) and length (23 m) present in the land are taken to be able to declare a rectangular flight area that covers whole lot 3 study.

For the flight zone, the parameters related to the flight lines, flight height, drone speed, the terrain footprint in the images and the overlap between images are calculated. The first parameter to obtain is the GSD of the MicaSense RedEdge-M camera at a certain height, in this case, we worked with the minimum recommended height (30 m) which guarantees a GSD of 2.1 cm per pixel.

Determined the GSD calculates the footprint of the terrain in the images or the coverage in meters that will cover an image in the terrain.

$$Footprint\ width = \frac{2.1\ cm/pixel * 1280\ pixel}{100} = 26.88\ m, \tag{1}$$

$$High\ footprint = \frac{2.1\ cm/pixel * 960\ pixel}{100} = 20.16\ m, \tag{2}$$

This indicates that each image taken at 30 m with the MicaSense RedEdge-M camera will have a footprint of 26.88 m by 20.16 m of the crop. For the flight lines, the terrain footprint and the desired overlap are taken into account, for this study a standard overlap of 75% was determined, that is, the images will have 75% of information in common, in this case it was noted that the Footprint width covers the width of the flight zone, making it necessary to have a single flight line located in the center (see Fig. 4).

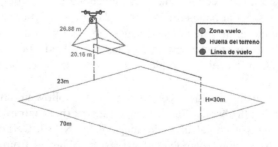

Fig. 4. Diagram of flight lines.

In order to determine the flight speed, the image fingerprint height, the 75% overlap stated above and the image shutter time are taken into account.

$$Base\ in\ the\ Air = 20.16\ m * \left(1 - \left(\frac{75\%}{100}\right)\right) = 5.04\ m, \tag{3}$$

With the value of the distance in the air between image the calculation is made to obtain the speed of the flight, for this process a shutter time of 2 s was determined obtaining.

$$Flight\ speed = \frac{5.04\ m}{2\ seg} = 2.52\ m/seg, \tag{4}$$

3 Results

The method for calculating vegetation indices from multispectral images taken with the MicaSense Red Edge M camera. The indices to be obtained from the study crop (lot 3) are NDVI, GNDVI, GRVI, and NDRE, for this purpose it was developed an algorithm for digital image processing in the MATLAB development environment which allows analyzing and processing the multispectral images of the study and culture taken by the MicaSense Red Edge M camera.

This analysis is composed of three stages, the first is the application of the Mica-Sense Red Edge M camera radiometric calibration model, the construction of panoramas or mosaics from multispectral images and the calculation of vegetation indices.

3.1 Radiometric Calibration

Radiometric calibration or radiometric corrections are those that will directly affect the radiometry of the image. These corrections become necessary due to erroneous values recorded in the image capture due to atmospheric disturbances (atmospheric absorption and dispersion phenomena) or deficiencies in the construction of the sensor [12].

The radiometric calibration process for the images of the MicaSense Edge M network starts by applying the camera's radiometric calibration model to the images of the CRP and the study crop to convert the pixel value images into digital levels (ND) unprocessed to images in absolute spectral brightness values. This model compensates for the black level of the sensor, the sensitivity of the sensor, the gain and exposure settings of the sensor and the vignette effects of the lens. The parameters to implement the radio-metric calibration model are obtained through the metadata see Table 2 saved by the RedEdge M camera in each image.

Table 2. Selected metadata of the image

Metadata	Variable
ImageWidth	y
ImageHeight	x
BitsPerSample	N
RadiometricCalibration	a1, a2, a3
VignettingCenter	cx, cy
VignettingPolynomial	k0, k1, k2, k3, k4, k5
ExposureTime	te
ISOSpeed	is

Equation (5) shows the calibration model of the Red Edge M camera to calculate the spectral brightness "L" from the pixel value "p" [13].

$$L = V(x, y) * \frac{1}{1 + \frac{a2*y}{te} - a3*y} * (p - pBL) * \frac{a1}{\frac{is}{100} * te * 2^N} \tag{5}$$

Where,

- p is the normalized unprocessed pixel value
- pBL is the normalized black level value
- a1, a2, a3 are the radiometric calibration coefficients
- V (x, y) vignette polynomial function for pixel location (x, y).
- is the image exposure time (is/100) is the sensor gain setting
- x, y are the column of pixels and the row number, respectively
- L is the spectral radiation in W/m 2/sr/nm

Having the images in spectral radiance, the images corresponding to the calibrated reflectance panel (CRP) are used to obtain the calibration factor by the band using the following equation

$$Fi = \frac{\rho i}{avg(Li)} \tag{6}$$

- Fi is the reflectance calibration factor for the band i
- ρi is the average reflectance of the CRP for band i, (Blue, Green, Red, and NIR = 0.49, Red Edge = 0.50)
- avg (Li) is the average radiance value of the pixels within the panel for the band i

To complete the calibration process, the factor of each band must be multiplied to the images of the crop in radiance per band, this allows the images to be converted to spectral reflectance values with which the vegetation indices can already be obtained.

3.2 Mosaic Construction

The construction of panoramas or mosaics is used in this type of study because the flight plans result in a set of images that encompasses the entire study crop which necessitates a treatment of these images that allows to build a larger image resolution composed of all these images, thus obtaining a better visualization of the crop.

This process is applied to radiometrically calibrated images and is made between images of their respective band, that is, when using images taken by the MicaSense Red Edge M, five panoramic images of the study culture will be obtained.

The process for the construction of scenarios begins with the detection and extraction of characteristics between pairs of sequential images, that is to say progressively be- tween images X (n) and X (n + 1), followed by X (n + 1) and X (n + 2), until reaching the last image, starting from the fact that "n" is the first image, where the points of interest or coincident characteristics and their location within each image are detected and extracted, to implement it in MATLAB the "detectSURFFeatures" and "extract- Features" commands are used to obtain the points of interest between a pair of images and then the "matchFeatures" command is applied, to compare and select the matching points between a pair of images giving the position of these in each image (see Fig. 5).

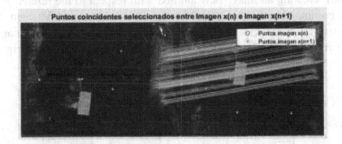

Fig. 5. Selected points of interest.

With the coincident points selected between the images, the transformation that allows the translation, rotation, scale and trimming of image pairs must be obtained, so that the coincident points of the image X (n + 1) overlap and coincide. with the points of interest of the image X (n), allowing to build an image composed of two or several images, this is obtained by means of the "estimateGeometricTransform" command, which finds the transform for each pair of images and when applying them, the panorama is constructed for each set of images (see Fig. 6).

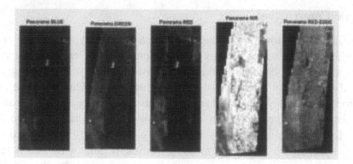

Fig. 6. Panoramas of the study crop by band.

3.3   Index Calculation

The index calculation is done in two steps, the first is the alignment of the pano-branches of the bands that each equation requires for the calculation of indices and the second is the application of this.

To align the scenarios, the MATLAB imregtform and imregister commands are used, these are implemented for alignment of two images where one of them is a reference and the other will be the one that is coupled by means of the obtained transformation, creating an image new aligned with respect to the reference image, the imregt form command estimates a transformation by movement between the panoramas, the imregister command obtains a transformation based on the image intensity register, which allows the alignment of different band panoramas (see Fig. 7).

Fig. 7. Vegetation indices with color scale. (Color figure online)

3.4   Descriptive Analysis

The descriptive analysis refers to the description of the vegetation indices obtained with respect to the nutritional status of the crop all based on what each index perceives.

3.4.1   NDVI Index

The result of the processing of the multi-spectral images of the MicaSense Red Edge M camera for the calculation of the NDVI index of the study culture is indicated (see Fig. 8), it can be seen that this index allows to correctly differentiate the plant material from the cultivation with respect to its environment, giving fidelity to the fact that the process carried out was optimal, in terms of monitoring the three areas of the crop this index does not reach a significant difference between the values of the three areas of the crop, since the values of the zones are between 0.9 and 1, making the whole crop to have a green color, in addition the crop not having detailed grooves always makes the NDVI index fail to identify the soil within the crop.

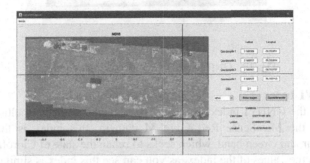

Fig. 8. NDVI panorama (Color figure online)

3.4.2 NDRE Index

The result of the calculation of the NDRE index (see Fig. 9) present in the study crop is presented by means of multispectral image processing taken with the MicaSense Red Edge M camera, this index has a similar functionality to the NDVI, that is, it allows to estimate the state of crop vigor, on the one hand, this index also allows differentiating the study crop from its surroundings, and allows differentiating the treatments presented by the areas of the study crop, since zone 1 and 2 which have less fertilization have the more reddish spots indicating values close to zero, noting the nutritional deficit in these areas.

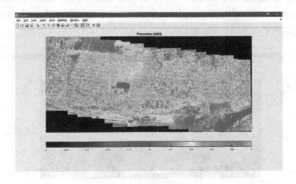

Fig. 9. NDRE panorama

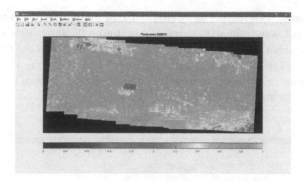

Fig. 10. GNDVI panorama. (Color figure online)

3.4.3 GNDVI Index

The results of the GNDVI index (see Fig. 10) obtained with the multispectral image processing of the MicaSense Red Edge M camera, this index is the relationship between the Nir and Green band, which seeks to relate the state of Clorofila of the plant with respect to the values of the index, as you can see this index is similar to the NDVI allowing the identification of the crop with respect to the soil in addition, it presents a better visualization of the states of the cultivation areas by means of the amount of points or regions of reddish or yellow color within each zone of the crop allowing to identify which zone is in a better state, and in this case it could be said that it is the central zone of the crop or that of the right side that is zones 2 or 3.

3.4.4 GRVI Index

The result of the GRVI index (see Fig. 11) when processing multispectral images taken with the MicaSense Red Edge M camera with the algorithm developed in this research work. This index allows us to better differentiate the study crop from its surroundings than the other indices, and also perceives the different plant species pre- sent in the crop as it is, the trees within the crop, the grass around it, and the objects. Regarding the monitoring of the treatments that each area of the study crop has visually, this index is not efficient since it includes the cultivation in 10 to 20 vales throughout the crop giving a similar color tone in the areas.

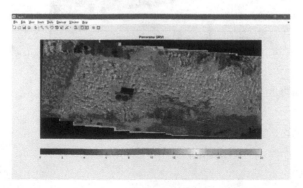

Fig. 11. GRVI panorama (Color figure online)

4 Discussions

In this study, five flights were taken to the study crop with which five scenarios were obtained by index from which the tables were obtained (Tables 3, 4, 5 and 6), which represent the modal values of each zone on their respective flight date.

Table 3. NDVI results.

NDVI	Zone 1	Zone 2	Zone 3
07 02	0,933	0,941	0,934
14 02	0,928	0,938	0,939
21 02	0,932	0,931	0,938
11 04	0,923	0,935	0,925
25 04	0,93	0,935	0,929

Table 4. NDRE results.

NDVI	Zone 1	Zone 2	Zone 3
07 02	0,390	0,408	0,432
14 02	0,403	0,388	0,401
21 02	0,371	0,369	0,356
11 04	0,345	0,396	0,385
25 04	0,393	0,453	0,451

Table 5. GNDVI results.

NDVI	Zone 1	Zone 2	Zone 3
07 02	0,801	0,847	0,800
14 02	0,787	0,810	0,795
21 02	0,775	0,776	0,770
11 04	0,803	0,809	0,803
25 04	0,812	0,804	0,803

Table 6. GRVI results.

NDVI	Zone 1	Zone 2	Zone 3
07 02	4,789	3,418	2,712
14 02	4,487	3,451	7,135
21 02	4,948	6,784	5,735
11 04	4,466	4,771	6,831
25 04	3,818	5,485	7,164

A statistical study was applied to these data, which is composed of a regression and correlation analysis to determine if there is any correlation of behavior between the NDRE, GNDVI and GRVI indices with respect to the NDVI within the study crop. This resulted in the following: For Zone 1, Table 7 was obtained, which indicates a positive (weak) correlation coefficient between the NDRE and GRVI indices with respect to the NDVI and a negative coefficient with the GNDVI index, but all they indicate a low coefficient of determination close to 0.3 indicating that the regression functions of zone 1 have a low correlation.

Table 7. Correlation results Zone 1

Zone 1	NDVI-NDRE	NDVI-GNDVI	NDVI-GRVI
Correlation coefficient	0,585	−0,2578	0,2962
Coefficient of determination	0,3422	0,0664	0,0877
Samples	5	5	5

The Table 8 shows the correlation values of zone 2 of the study crop. This zone has positive correlations between the NDVI, NDRE and GNDVI index, unlike the GRVI index which has a negative correlation, taking into account the coefficient of determination could be said that the best correlation occurs between the NDVI and GNDVI index and there is a correlation close to −1 indicating a direct negative correlation with the GRVI.

Table 8. Correlation results Zone 2

Zone 2	NDVI-NDRE	NDVI-GNDVI	NDVI-GRVI
Correlation coefficient	0,2592	0,9563	−0,9457
Coefficient of determination	0,0671	0,914	0,8943
Samples	5	5	5

For zone 3, Table 9 was obtained, which shows that all the indices have a negative or close to zero correlation coefficient, thus being the area with the lowest correlation be- tween indices, although there is a weak correlation according to the coefficient of de- termination between the NDVI and GNDVI index.

Table 9. Correlation results Zone 3

Zone 3	NDVI-NDRE	NDVI-GNDVI	NDVI-GRVI
Correlation coefficient	−0,2967	−0,6512	−0,203
Coefficient of determination	0,0880	0,4240	0,041
Samples	5	5	5

5 Conclusion

The NDVI index for this case study fails to perceive the different fertilization treatments that each zone has in the study crop, making it not helpful for monitoring the nutritional status of the coffee of Arabian coffees of Castle type, although this may be related to the age of the crop which is approximately 3 years, that is to say in adulthood and that is when the NDVI index loses the sensitivity to the perception of chlorophyll of the plant.

The GRVI index, which makes it possible to differentiate better the Coffea Arabica-type coffee plants of the castle type from the other plant species present in the study crop, making this index the most suitable if we want to detect weeds or plants Invasive coffee crops.

The GNDVI and NDRE indices are efficient to perceive the changes in nutrition in coffee crops of the Coffea arabic type of castle type, since they were the best to perceive the different fertilization treatments present in each area, making them optimal for phenological monitoring to coffee crops in the region.

Taking into account the results of the statistical correlation analysis, it could be said that in general the NDVI, NDRE, GNDVI indices have a behavioral relationship, with the GNDVI having the best correlation with the NDVI since a correlation coefficient was obtained in the zone 2 of 0.9563 and in zone 3 of -0.6512, indicating a strong behavior relationship in the greater area of the crop, regarding the NDRE index a correlation coefficient of 0.585 was obtained in zone 1 which perceives a positive relation- ship and Reliable in this zone, in zone 2 and 3 there is a weak relationship since the coefficients are 0.2592 for zone 2 and -0.2967 for zone 3.

The behavior relationship between the NDVI and GRVI index was only strong in zone 2 of the crop where a coefficient of -0.9457 was obtained, that is, a direct negative correlation, this can occur because each index is implemented for different situations such as the NDVI perceives the state of vigor of a crop and the GRVI allows to identify different species of plants present in a crop.

References

1. Echeverri, D., Buitrago, L., Montes, F., Mejía, I., del González, M.P.: Café para cardiólogos. Revista Colombiana de Cardiología **11**(8), 357–365 (2005)
2. Meneses, V.A.B., Téllez, J.M., Velásquez, D.F.A.: Uso de drones para el análisis de imágenes multiespectrales en agricultura de precisión. Ciencia y tecnología alimentaria (2015)
3. Rivera, A.F.G., Clavijo, F.V., López, A.F.J.: Agricultura de Precisión y Sensores Multiespectrales Aerotransportados (2016)
4. Yost, M.A., Kitchen, N.R., Sudduth, K.A., et al.: Precis. Agric. **20**, 1177 (2019)
5. C&C cauca agenda de competitividad. http://www.cccauca.org.co/public/archivos/documen tos/resultados-resumen.pdf
6. Asoexport. http://www.asoexport.org/productividad-productividad
7. Leiva, F.: La agricultura de precisión: una producción más sostenible y competitiva con visión futurista (2003)
8. Cauca se posiciona como productor de cafés especiales. https://cauca.federaciondecafeteros. org/buenas_noticias/cauca_se_posiciona_como_productor_de_cafes_especiales/
9. Parque tecnológico del café- Tecnicafe. http://www.cauca.gov.co/noticias/parque-tecnolog ico-del-cafe-tecnicafe-una-realidad-al-servicio-de-la-innovacion-de-la
10. MicaSense RedEdge-MTM Multispectral Camera - User Manual (2018)
11. Phantom 4 Manual de usuario (2016). Versión 1.2
12. Romero, M.A., Cazorla, M.: Comparativa de detectores de características visuales y su aplicación al SLAM (2009)
13. Cao, S., Danielson, B., Clare, S., Koenig, S., Campos, C., Sanchez, A.: Radiometric calibration assessments for UAS-borne multispectral cameras: laboratory and field protocols. ISPRS J. Photogramm. Remote Sens., 132–142 (2019)

Is It Possible to Improve the Learning of Children with ASD Through Augmented Reality Mobile Applications?

Mónica R. Romero[1]([⊠]), Estela Macas[2], Ivana Harari[1], and Javier Diaz[1]

[1] LINTI, Laboratorio de Investigación de Nuevas Tecnologías Informáticas, Facultad de Informática, Universidad Nacional de la Plata, Calle 50 y 120, La Plata, Buenos Aires, Argentina
{monica.romerop,iharari,jdiaz}@info.unlp.edu.ar
[2] Facultad de Ciencias y Tecnologías Aplicadas, Escuela de Informática y Multimedia, Universidad Internacional del Ecuador, Quito, Ecuador
esmacasru@uide.edu.ec

Abstract. At present, many researchers and software companies have created a number of mobile applications based on augmented reality that promote learning spaces in children, however, we find few studies where an intervention is carried out with this emerging technology in children with autism. This research sought to verify whether mobile applications can be used in teaching-learning processes in children diagnosed with ASD, the intervention was worked through a multidisciplinary team, which was attended by psychologists, computer engineers, parents and a ASD patient; A curricular strategy was established to verify whether the indicators: cognitive, procedural and communicative after the intervention with RA improve the ability to learn. The experience is described through a case study that shows encouraging results and to some extent promising on the use of new technologies to improve the quality of life of children with ASD.

Keywords: Autism · Mobile application · Teaching · Augmented reality · ICT · ASD

1 Introduction

ASD is a complex neurological disorder that usually lasts a lifetime. Autism is defined as a serious disturbance that affects several areas of development [1], is detected because it affects the development of people who are diagnosed, and is expressed through poor communication, difficulty in establishing social relationships with their environment, and additional language limitation, additionally with problems of imagination and flexibility of thought [2–5].

On the other hand, RA is a technology that allows the user to interact with the physical and real world that surrounds it [6], it is the combination of virtual objects, such as third-dimensional graphics or animations, with real environments [7, 8].

© Springer Nature Switzerland AG 2020
M. Botto-Tobar et al. (Eds.): ICAT 2019, CCIS 1194, pp. 560–571, 2020.
https://doi.org/10.1007/978-3-030-42520-3_44

There are many definitions of RA, in 2010 it is defined as virtual objects or annotations that can be superimposed in the real world, for [9] the RA is a technology capable of complementing perception and interaction with the real world, providing the user a real scenario augmented with additional information generated.

En los últimos años investigadores han generado diversos estudios donde se ha incluido las TIC en la intervención de niños con TEA con resultados prometedores. [10, 11], en este contexto algunos investigadores comentan que estas personas se sienten más confiadas y seguras cuando se utilizan TIC ya que el entorno está controlado y además pueden repetir las actividades propuestas [12, 13].

Augmented reality projects have grown significantly in recent years. obtaining evidence that intervention through augmented reality in activities designed for teaching learning has significantly improved student learning skills in regular schools [6, 8, 14–16]. Through this technology, different learning objects have been created: games, books, teaching material, mobile applications (apps) that are used for different purposes [17–19].

However, from the preliminary systematic review, most studies are based on secondary sources, the reviewed literature lacks research studies on the suitability of these technologies, methodology, quantitative studies and conclusions based on primary sources, [8] Experimental studies aimed at verifying whether this new technology, known as emergent, can improve teaching-learning processes in children with autism are minimal and scarce [20, 21] in this regard, this research conducts a search for mobile applications that contain components of RA to verify, through a multidisciplinary and experimental team if its use in the pedagogical space can be favorable for children with ASD.

The present study is structured as follows: Sect. 2 explains the materials and methods that were used in the execution of the research, structured surveys were conducted for data collection, interviews were performed to the group of informants, and finally conclusions were drawn of direct participation through observation. Section 3 presents the results of the study and last the Sect. 4 presents a critical reflection.

2 Material and Method

This research presents a mixed approach since it uses a qualitative method through interviews led with participants and quantitative by structured surveys [22]. The scope of this work is exploratory, because there are few studies related to evaluations of RA applications where curricular plans are established to measure teaching-learning processes through indicators: cognitive, procedural and communicative [23].

This study is of the micro type since the experimentation is carried out with a single girl, in a controlled space and with a planning previously reviewed by people who treat this disorder. This research is descriptive because it seeks to know in detail about the benefits of the inclusion of RA in the treatment of children with ASD, observing user interactions with applications.

This research focuses on two modalities such as documentary and field, this is because the experimentation must be carried out on specific applications, and on the same environment, and it will also be documentary because the process and the results are supported by a methodology and in the theoretical support of previously conducted research [24]. The fieldwork of this research was carried out in the city of Quito at the Ludic Place Therapeutic Center, located on Seymour Island N41 225 and Floreana Island. This center provides treatment for children with autism spectrum disorder.

The population is composed of the group of people shown in Table 1. Moreover, inclusion criteria of those who participated in the case study to be developed have been included. For this study there are four groups of informants. The informants are those people who will provide us with relevant information: Child with ASD, people close to the child parent, the third group is defined by professionals in treatment of ASD, the people who are in charge of their care and in group 4 professionals in IT engineers in computer science.

Table 1. Research population

Subjects	Inclusion criteria	Quantity
Professionals ASD treatment. Psychologists	Have studies in psychology or educational sciences Have experience of having treated children with ASD Agree with the study to be carried out	2
Patient diagnosed with ASD	Boy or girl diagnosed with ASD Treated by the Ludic Place Center Age between 4–6 years Authorization agreement signed by their legal representatives	1
ICT professionals	Have studies in ICT or related Have experience in the use of RA applications Agree with the study to be carried out Authorization to conduct part of the study interviews	2
Parents	Having a direct relationship with the child diagnosed with ASD Be able to describe the child's behavior Agree with the study to be carried out Authorization to conduct part of the study interviews	2
Total		7

2.1 Case Study

2.1.1 Work Plan

For this purpose, a strategy is established through a discussion group among the professionals who participated in the experiment [25] with the objective of defining a work plan to carry out the intervention through the use of mobile applications based on RA [26].

The work plan was carried out in four sessions. [27, 28] see as an example Fig. 1. The objectives are to have a metric where the use of augmented reality in children with ASD can be critically assessed, [29, 30] additionally there was the participation of A population with 4 groups of informants.

Fig. 1. Work sessions with Girl with mild autism.

Prior to the use of the applications, a small training was given to the psychologists who participated in the experiment. In addition to the first sessions they were to familiarize the girl with the technological equipment that was later used in the intervention, finally, they told the girl how to use the applications. A tablet was used in two sessions and a smartphone in the remaining two sessions; however, the specialist led the use of the application.

2.1.2 Selecting Mobile Applications for ASD

In the collection of information from secondary sources we located a list of 50 mobile applications for students with ASD [6]. Exclusion processes were carried out, which were based on the year of launch of the application, the operating system, if they included RA if they reinforced teaching-learning spaces for children. Once these criteria were applied, the applications were reduced to 18. A work meeting was held with the patient treatment team and they were exposed, at the end according to an expert judgment, which gives having shared with the girl 4 years of therapy two applications were selected for this case study. The selected applications are indicated in the following Table 2.

Table 2. Selection of mobile applications.

App	Name	System O.	Year
AR Animal	AR Animals	Android/iOS	2018
	Alphabet	Android/iOS	2018

2.1.3   AR Animals

AR animals is a mobile application created to teach animals (wild and domestic). Its use is free and just enter the website where they allow you to download a list of pictograms, the application can be downloaded from playstore see as an example Fig. 2.

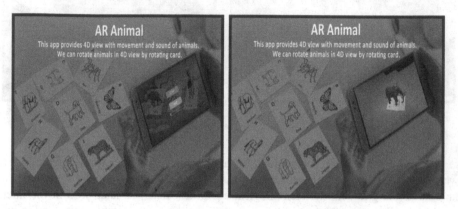

Fig. 2. AR animal application - a graph that represents how through the pictograms and using a smartphone, you can project a chosen animal using augmented reality.

2.1.4   Alphabet FlashCards

This mobile application was created to teach the alphabet and presents with each letter an associated animal. To be able to use it, just download it from Playstore, its use is free, and a pdf with the associated pictograms can be downloaded from the website. Its use is shown in Fig. 3 below.

Fig. 3. Alphabet flashcards App - a graph that represents the use of the mobile application, augmented reality [31].

For the analysis of the experiences, a validated form by an expert opinion (of the Research Group on Educational Technology of the University of Murcia) was used. It is included below in Table 3 of the Activity Description Sheet using augmented reality.

Table 3. Activity description sheet using augmented reality.

Basic reference data	*School*	*Ludic place*
	Web page	https://www.facebook.com/ludicplace/
	City	Quito-Ecuador
	Educational stage	School
	Cycles	Sierra/September to July 2019
Facts about activity	Information sources about the activity	Primary source
	Total number of students involved in the activity completion	1
	Use of new complementary technologies	New technologies Emerging technologies Augmented reality
	Faculty information	Silvia Reyes Paola Navarro Mónica Romero
	Description of the activity: evaluation methodology	
	Objectives	Performs a search of mobile applications containing RA to verify, through a multidisciplinary and experimental team if its use can be favorable for children with ASD
	Contents	Establish a curricular strategy to measure a teaching-learning process through indicators: cognitive, procedural and communicative
	Grouping	Group There is a multidisciplinary population with 4 groups of informants
	Activity completion dates	First half of 2019 January to June
	Places of realization of the activity	Ludic Place Ecuador, Pichincha, Quito, Floreana Island and Seymur

(*continued*)

Table 3. (*continued*)

Analysis of the RA technology involved	RA level		
	Classification of the activity in a subtype of RA	Media description	
		Software	AR animals
		Hardware	Tablet Smart phone

Internet connection (yes, no): YES
Wifi networks: YES
Mobile phone networks: YES

Evaluation	Available data on evaluations of participating staff	The data is found in structured surveys that were conducted through Google forms, as well as a clinical report by the personnel in charge of Ludic Place, there is a signed consent by their legal representatives
	Data available on external evaluations and dissemination of the activity	The project has been socialized with the Research Laboratory of New Information Technologies. LINTI Faculty of Computer Science, National University of La Plata

Source: Activity sheet using RA [32].

2.1.5   Interview

The interviews were carried out with the four groups of informants, this study collected the opinions from several critical points, initially by the professionals who are in charge of caring for the girl with autism, since it is they who play a fundamental role in indicating whether the intervention carried out through mobile applications with RA favored learning in the treatment center.

Last but not least the interviews were designed to extract relevant information from parents who share the longest time with the patient and were present in the work sessions.

3   Result

The evaluation of these sessions were carried out through interviews and observation of the specialists who executed the sessions to identify if the indicators mentioned in the curriculum plan strategy were met by the child with ASD [33].

A record of each section was carried out (work was done in 4 sessions). In each of them several teaching-learning activities were to be completed, for each session 5 activities were defined, the same ones that were fully completed, at the end the 20 proposed activities were evaluated.

Actions were taken for each session so that the girl with autism manages to perform the:

- Identification/Recognition
- Selection,
- Description,
- Visualization
- Imitation

In the different sessions the treating psychologists made a record, the number of RA activities raised and the successes that were evidenced in the experimentation through a scatter plot are shown as variables, the results are shown in Fig. 4.

Number of errors according to the teaching activities proposed RA

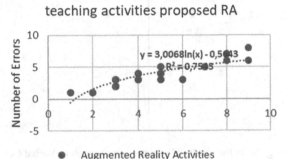

Fig. 4. Below is a graph that represents the number of successes that the girl made according to the activities outlined in the curriculum plan through intervention with the Mobile Application.

In other aspects, a verification of the emotions and reactions of the girl was made after the intervention for this process, several aspects are evaluated, which can be obtained through direct observation and were collected at the end of the planned activities in each of the sessions. This evaluation allows to verify a fundamental aspect such as the behaviors and emotions of the child and are related to verbal communication, non-verbal communication, eye contact, and emotion management.

Moreover, in this space a Likert scale was used, where the people who participate parents and treating psychologists qualify from 1 to 5, where 1 indicates a total disagreement, 2 disagree, 3 neither agree nor disagree, 4 agree and 5 indicate totally agree. For each culminated activity, it was requested to qualify the aforementioned aspects, at the end of the process, scatter plots are made with the collected results. The Fig. 5 shows the results collected.

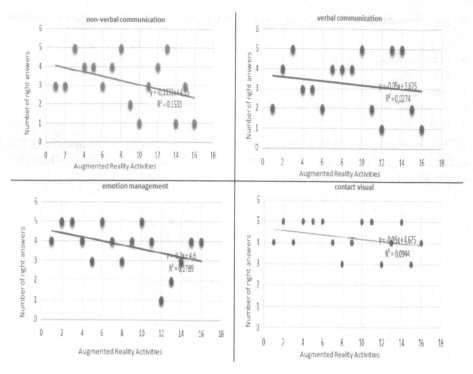

Fig. 5. The following shows the results of various aspects such as Verbal Communication, Non-Verbal Communication, Visual Contact, Emotion Management. On the X axis the 20 teaching activities proposed with RA are shown and on the Y axis the degree of satisfaction determined by the informants is identified

4 Discussion

It is important to consider the presence of a multidisciplinary group for the correct intervention of new technologies or emerging technologies in teaching-learning processes in children with ASD. For this study, work tables and brainstorming were organized where the participation of the treating psychologists was fundamental to guide the intervention process through mobile applications with RA. On the other hand, the parents had a positive impact since they provided all the collaboration including the signing of an informed consent.

From the wide variety of mobile applications that are in the market it is important to make an assessment, defining several aspects in this sense the definition of exclusion criteria can make the number of apps be significantly reduced. From the group of final applications for the process of intervention applications must be socialized with the work group to choose the most suitable, according to the cognitive, social, and academic development information of the patient.

Once the mobile applications have been chosen, it is very important to carry out a technical training to the professionals treating children with ASD. In our case study to the psychologists who participated, since they will be the ones that guide the sessions, so it must be previously defined. The intervention is a space where training, installation, use and any doubts regarding the operation are carried out.

The hardware that is going to be used has an important space since for this experimentation a smartphone was initially introduced and after that a Tablet since the visualization is an important aspect. This strategy that was not initially contemplated made the development of the sessions improve and that the girl can interact better with the app.

In order for the process to be execute correctly, a curricular plan must be adequately considered, where the participation of personnel or expert judgment is essential. Initially, a topic to be addressed must be identified, sub-themes, specific activities.

Future work must be defined and finally the achievements must be stated through indicators. In our case, three were used: cognitive, procedural and communicative that allowed us to evaluate the effectiveness of the intervention.

From the results obtained through experimentation, it can be concluded that the number of successes in the activities raised with the augmented reality intervention were higher than expected, or the average that the student performs without the application of RA, which answers the question posed in this research. It can be indicated that this technology is suitable for teaching and intervention in this type of disorders, the results collected are encouraging.

For the evaluation of different spaces through the direct observation of the participants, referring to the behaviors and emotions that the intervention of mobile applications based on RA, it can be concluded that the most notable improvement is established by eye contact, followed by communication, emotion management and nonverbal communication.

This process went through in an ethical and objective way, with several groups of informants that allow us to collect the results from different points of view after the intervention with RA. However, this strategy to be replicated in other environments, will depend largely on the Autism diagnosis level of the population to be operated or the therapeutic center, since each child with ASD has unique characteristics which makes this work even more complex.

As a future line of research it would be important to expand the study to verify other processes such as multiple intelligences through RA. In Chile there is a mobile application that was developed through a playful space that allows students to reinforce several competences, this initiative was used in regular school, however, it would be important to do it in children with autism.

Augmented reality is an element that can positively influence teaching-learning spaces. Nevertheless, the primary studies carried out, as well as the present research presents limitations in terms of population, so it would be important to replicate the intervention in larger populations and with different severity Autism.

Acknowledgments. We thank the National Secretary of Higher Education Science and Technology SENSCYT - Ecuador for the support provided, in addition to. LINTI Research Laboratory in New Information Technology, University of la Plata, as well as the Ludic Place-Ecuador therapeutic center where the execution of this project was carried out.

References

1. García Guillén, S., Garrote Rojas, D., Jiménez Fernández, S.: Uso de las TIC en el Trastorno de Espectro Autista: aplicaciones. Edmetic **5**(2), 134 (2016)
2. Cabrera, D.: Generalidades sobre el autismo. Rev. Colomb. Psiquiatr. **XXXVI**(1), 208–220 (2007)
3. Lara, J.G.: El autísmo. Historia y clasificaciones. Salud Ment. **35**(3), 257–261 (2012)
4. Cuesta, J.L., Sánchez, S., Orozco, M.L., Valenti, A., Cottini, L.: Trastorno del espectro del autismo: Intervención educativa y formación a lo largo de la vida. Psychol. Soc. Educ. **8**(2), 157–172 (2016)
5. Canal Bedia, R., et al.: Detección y Diagnóstico de Trastornos del Espectro Autista, p. 57 (2007)
6. Cadavieco, J.F., Sevillano, M.A.P., Amador, M.F.M.F.: Realidad aumentada, una evolución de las aplicaciones de los dispositivos móviles (Augmented reality, an evolution of the application of mobile devices). Rev. medios y Educ. (2012)
7. Moralejo, L., Sanz, C., Pesado, P., Baldassarri, S.: Avances en el diseño de una herramienta de autor para la creación de actividades educativas basadas en realidad aumentada. TE Rev. Iberoam. Tecnol. en Educ. y Educ. en Tecnol. (12), 8–14 (2014)
8. Espinosa, C.P.: Realidad aumentada y educación: análisis de experiencias prácticas augmented reality and education: analysis of practical experiencies. Pixel - Bit. Rev. Medios y Educ. (2015)
9. Yuen, S.C.-Y., Yaoyuneyong, G., Johnson, E.: Augmented reality: an overview and five directions for AR in education. J. Educ. Technol. Dev. Exch. **4**(1), 119–140 (2011)
10. Cuesta Gómez, J.L.: Tecnologías De La Información Y La Comunicación: Aplicaciones En El Ámbito De Los Trastornos Del Espectro Del Auti Smo. Rev. Española sobre Discapac. Intelect. **43**(2), 6–25 (2012)
11. Romero, M.R., Diaz, J., Harari, I.: Impact of information and communication technologies on teaching-learning processes in children with special needs autism spectrum disorder, pp. 342–353 (2017)
12. Groba, B., Pereira, J., Munteanu, C.R.: Impacto de las tecnologías de la información y las comunicaciones en el funcionamiento en la vida diaria de niños con trastorno del espectro del autismo, p. 441 (2015)
13. Practicante En el TEACCH Center: Principios y estrategias de intervención educativa en comunicación para personas con autismo: TEACCH. Rev. Logop. Foniatría y Audiol. **27**(4), 173–186 (2007)
14. de Pedro Carracedo, J., Méndez, C.L.M.: Realidad Aumentada: Una Alternativa Metodológica en la Educación Primaria Nicaragüense. IEEE Rita **7**, 102–108 (2012)
15. Basogain, X., Olabe, M., Espinosa, K., Rouèche, C., Olabe, J.C.: Realidad Aumentada en la Educación: una tecnología emergente. Rev. Mex. Tecnol. **2**(3), 14 (2000). http://multimedia.ehu.es
16. De la Horra Villacé, I.: Realidad aumentada, una revolución educativa. Edmetic **6**(1), 9 (2016)
17. Herrera, G., et al.: Pictogram room: Aplicación de tecnologías de interacción natural para el desarrollo del niño con autismo. Anu. Psicol. clínica y la salud = Annu. Clin. Heal. Psychol. (8), 41–46 (2012)
18. Barroso Osuna, J.M., Gallego Pérez, Ó.M., Barroso Osuna, J.M., Óscar, Y., Gallego Pérez, M.: Producción de recursos de aprendizaje apoyados en Realidad Aumentada por parte de estudiantes de magisterio. Learning resource production in augmented reality supported by education students **6**(1), 2254–2259 (2017)

19. De Pedro Carracedo, J., Martínez Méndez, C.L.: Augmented reality: an alternative methodology in primary education in Nicaragua. Artículo **7**, 7 (2012)
20. De, J., Cantero, T., Luis, J., Pérez, S.: Entorno de aprendizaje ubicuo con realidad aumentada y tabletas para estimular la comprensión del espacio tridimensional. RED. Rev. Educ. a Distancia (37), 1–17 (2013)
21. Fracchia, C., Alonso, A., Martins, A.: Realidad Aumentada aplicada a la enseñanza de Ciencias Naturales. Rev. Iberoam. Educ. en Tecnol. y Tecnol. en Educ. (2015)
22. Lévano, A.C.S.: Investigación Cualitativa: Diseños, Evaluación Del Rigor Metodológico Y Retos. Liberabit **13**(2006), 71–78 (2007)
23. Martínez Carazo, P.C.: El método de estudio de caso. Estrategia metodologíca de la investigación científica. Pensam. y Gestión **20**, 165–193 (2006)
24. Jiménez, J.: Atención primaria, principios, organización y métodos en la medicina de familia, pp. 154–165 (2016)
25. Kohler Herrera, J.: Importancia de las estrategias de enseñanza y el plan curricular. Liberabit (11), 25–34 (2005)
26. Enríquez Pigazo, I.: Trastornos del Espectro Autista: estudio de un caso y propuesta de intervención logopédica (2018)
27. Heredero, E.S., Oliva, A.: Experiencias y recursos con las tics para la atención al alumnado con necesidades educativas especiales. Acta Sci. Educ. **36**(2), 279 (2014)
28. Moreno Martínez, N.M., Leiva Olivencia, J.J.: Experiencias formativas de uso didáctico de la realidad aumentada con alumnado del grado de educación primaria en la universidad de Málaga. Edmetic **6**(1), 81 (2016)
29. Ibáñez Bernal, C.: Diseño curricular basado en competencias profesionales: una propuesta desde la psicología interconductual. Rev. Educ. y Desarro. **6**, 45–54 (2007)
30. Crescenzi-Lanna, L., Grané-Oró, M.: An analysis of the interaction design of the best educational apps for children aged zero to eight. Comunicar **24**(46), 77–85 (2016)
31. Andrés Roqueta, C., Benedito, I., Soria Izquierdo, E.: Uso de aplicaciones móviles para la evaluación de la comprensión emocional en niños y niñas con dificultades del desarrollo. Rev. Psicol. y Educ. **12**(1), 7–18 (2017)
32. 3546 @ repositorio.uisek.edu.ec
33. Cruz, O., Ester, E., Guerrero, A., Aguilar, R., García, L., Medina, M.: Evaluación de las funciones cognoscitivas en niños con trastornos del espectro autista. Rev. Neuropsicol. Latinoam. **5**(4), 53–60 (2013)

Open Source System for Identification of Maize Leaf Chlorophyll Contents Based on Multispectral Images

Joe Saverio[1](✉), Allan Alarcón[1](✉), Jonathan Paillacho[1](✉),
Fernanda Calderón[2](✉), and Miguel Realpe[1](✉)

[1] Escuela Superior Politécnica del Litoral, ESPOL, Facultad de Ingeniería
en Electricidad y Computación - FIEC, CIDIS, Guayaquil, Ecuador
{jsaverio,asalarco,jspailla,mrealpe}@espol.edu.ec
[2] Escuela Superior Politécnica del Litoral, ESPOL, Facultad de Ciencias
de la Vida, Guayaquil, Ecuador
mafercal@espol.edu.ec

Abstract. It is important for farmers to know the level of chlorophyll in plants since this depends on the treatment they should give to their crops. There are two common classic methods to get chlorophyll values: from laboratory analysis and electronic devices. Both methods obtain the chlorophyll level of one sample at a time, although they can be destructive. The objective of this research is to develop a system that allows for obtaining the chlorophyll level of plants using multispectral images.

Python programming language and different libraries of that language were used to develop the solution. It was implemented as an image labeling module, a simple linear regression, and a prediction module. The first module was used to create a database that relates the values of the NDVI image with those of chlorophyll, which was then used to obtain a linear regression model for the prediction system to obtain chlorophyll values from the images. The model was trained with 92 images and was obtained a root-mean-square error (RMSE) of 7.27 units CCM (Chlorophyll Content Meter). While the testing was performed using 10 values obtaining a maximum error of 15.5%.

It is concluded that the system is appropriate for chlorophyll contents identification on maize leaves in field tests. However, it can also be adapted for other measurements and crops. The system can be downloaded at [1].

Keywords: Multispectral images · Image labeling · NDVI · Chlorophyll · Linear regression · Open source

1 Introduction

One of the key feature to improve the crop yield and quality is the efficient application of nutrients, to do so is necessary to know the status of the main nutrients in the plants like the Nitrogen (N), which is a essential element for the plant growth [2]. To assess the N in plants exist different direct and indirect methods like plants tissues and chemical analysis, chlorophyll meters and Yara N tester. These methods have a cost in economic and time terms and in some cases they are destructive [2]. Therefore, it is

© Springer Nature Switzerland AG 2020
M. Botto-Tobar et al. (Eds.): ICAT 2019, CCIS 1194, pp. 572–581, 2020.
https://doi.org/10.1007/978-3-030-42520-3_45

necessary to find a system that improves the process of the retrieval of plant biophysical parameters in terms of data collection time, processing time, and operating costs, as well as a non-destructive method of plant analysis in order to increase crop yield.

Throughout the last decades, the advances in the use Unmanned Aerial Vehicle (UAV) become a technique widely used in remote sensing for crop growth assessment and monitoring [3–5]. The biomass determination, leaf area, chlorophyll content, as well as the determination of nutrients in crops at the high spatial resolution, represent a cost-effective alternative to other methods to determine nutritional deficiencies in crops and reduce the data collection period [6].

Studies have shown that through the determination of vegetation indices it is possible to obtain the plant biophysical characteristics due to the indices enhance the vegetation properties of the observations. Due to the multispectral behavior of plants, approximately 90% of the information related to vegetation is found in the red and near-infrared spectral bands [7]. The vegetation indexes are obtained from mathematical estimations where different multispectral bands in the visible (VIS) and near-infrared (NIR) are used as input [8].

One of the most commonly used indexes for crop monitoring applications that include image processing is the Normalized Difference Vegetation Index (NDVI), because it provides information about the presence of biomass in a processed image [9, 10]. This index is used to estimate the quality and development of agricultural fields, making it easier for experts to take decisions in order to increase crop yield. NDVI is commonly used to identify anomalies in crop seasons, yield estimates and distinguish weeds, among other functions.

Recently, machine learning (ML) is an application of artificial intelligence and it has become an attractive technique to quantify and understand the agricultural processes [11]. According to ML the data can be categorized into supervised and unsupervised algorithms, where, linear regression is a supervised learning algorithm that is used in ML and statistics. The regression algorithm establishes a model to adjust the dependency relationship between a specific independent characteristic (value of the independent variable x) and the corresponding value "result" (value of the dependent variable y).

The ML algorithms are usually implemented through the use of open source tools such as OpenCV, MongoDB and Python. The main advantage of open source is the possibility of sharing, modifying and studying the source code of a programmer who shared his work. Also, open source promotes collaboration among users through communities, this allows the rapid and varied development of many programs and tools that wish to be used.

The main objective of this document is to determine the relationship between the values of chlorophyll content in maize plants and the NDVI. To do so, it was conducted a randomized completely block design to the yield of maize at the Experimental Agricultural Farm (Granja Experimental Agricola - GEA) of the Polytechnic University (ESPOL). From the experiment, the concentration of chlorophyll was estimated by using the CCM 200 PLUS sensor, which, measures the energy absorption of the leaf in two regions of wavelength, near-infrared and red spectral bands. On the other hand, the determination of the NDVI and the correlation was carried out through the use of the open source tools for the classification of the multispectral images and the application of the linear regression model.

2 Methodology

A computational system is proposed to analyze the relationship between the multi-spectral images and the biophysical characteristics of the maize crop. First, data collection performed include multispectral image capture and chlorophyll measurements. Then, NDVI and chlorophyll values were processed and stored through de use of open source tools. Finally, a linear regression model is trained in order to predict chlorophyll values using NDVI images (see Fig. 1).

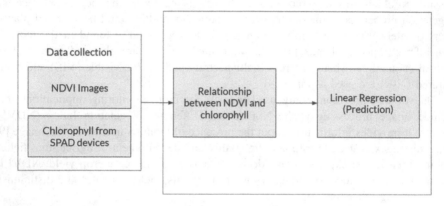

Fig. 1. Proposed methodology.

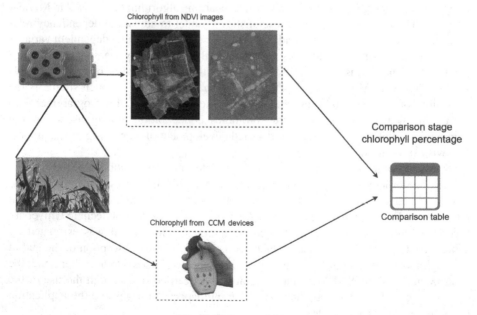

Fig. 2. Data acquisition.

The process for data acquisition is described in Fig. 2. For multispectral images acquisition of the maize crop at GEA, a MicaSense RedEdge-M camera was used. The camera model has 5 lenses, where each one is capable of capturing images in a specific spectral band, such as blue, red, near red, green and near-infrared [12]. These images of the 5 spectral bands were then processed to obtain RGB images and obtain the NDVI values of the maize.

Once the RGB and NDVI images were obtained, a module for associating the NDVI and chlorophyll level obtained with the CCM 200 PLUS (Chlorophyll Content Meter) device was needed [13]. To do this, OpenCV library was used in order to implement an image labeling module that allows the user to draw a region of interest in the NDVI image, and then label it with a chlorophyll value corresponding to that region. The average of the NDVI values into the region of interest will be stored in a database in order to be used later for training a linear regression.

Finally, a linear regression that represents the correlation in the maize crop data was performed using the NDVI images and their corresponding CCM values. Once the linear regression function was obtained, the chlorophyll values were predicted from crop images. Different NDVI values from images of maize leaves were tested.

3   Implementation

3.1   Database

The database structure stores information about the image of the five spectral bands, to facilitate the identification of pixel values inside of the image, as well as the information about the vegetation that is taken through CCM or measurements in the laboratory if necessary. The database was created using MongoDB, there is a collection structure with the following information:

```
{
  "_id": <ObjectId>,
  "name": "filename",
  "extension": "png",
  "NDVI": [
    0: 0.34905,
    1: 0.32853
    ...
    ...
    9: 0.39588
  ],
  "CCM 200 PLUS": 2,
  "LAB": 3
}
```

3.2   Labeling Module

The labeling module is a graphical interface designed to annotate CCM values in the images. The interface shows an NDVI image and allows the user to freely draw regions of interest and then enter the CCM values for each area. In some cases the users showed difficulties to identify the regions on NDVI images, thus an option to simultaneously work with an RGB image of the same data on a split-screen was included. In this case, the interface allows us to manipulate (move, zoom) and to label either of both images and the action would be replicated in the other image, facilitating the identification of vegetation areas (see Fig. 3).

Fig. 3. Labeling module.

3.3   Linear Regression

The NDVI values, approximations of the physical characteristics of plants can be determined. Thus, NDVI values were filtered to leave representing the vegetation. The following equation is used to estimate the index:

$$NDVI = (NIR - Red)/(NIR + Red) \tag{1}$$

Where NIR refers to the near-infrared spectral range, while Red refers to the red spectral range.

It was decided to take NDVI values greater than 0.1 since they represent a range from young vegetation to abundant vegetation, as it can be appreciated below:

According to what they represent [11]

- [−1, −0.1]: Water
- (−0.1, 0.1]: Rocks, sand or snow
- (0.1, 0.4]: Young plantations, shrubs, bushes
- (0.4, 0.9): Abundant vegetation

According to density of the region [11]:

- [−1, 0]: Without vegetation
- (0, 0.15]: Very low density
- (0.15, 0.3]: Low density
- (0.3, 0.45]: Dense
- (0.45, 0.6]: High density
- (0.6, 1]: Very high density

Additionally, data structures for storing NDVI and CCM values were implemented in order to allow compatibility with the types of data used by the Python Scikit-learn library, this library was used to obtain the simple linear regression function between the NDVI values and the theoretical CCM values, the coefficient of linear correlation between both variables and the root of the mean square error.

3.4 Prediction Module

In order to verify the linear regression, a prediction tool was implemented in which the user can upload an image of a crop (five images from Micasense camera) and obtain its chlorophyll values. This was done by selecting any part of the image, so the system provides the chlorophyll value in that selection (see Fig. 4).

Fig. 4. Prediction module.

4   Results

To validate the system, 92 images captured in the maize crop at GEA were used. Each image contains a leaf from a plant in the crop. Figure 5 shows a sample image that was captured one week after seeds were planted.

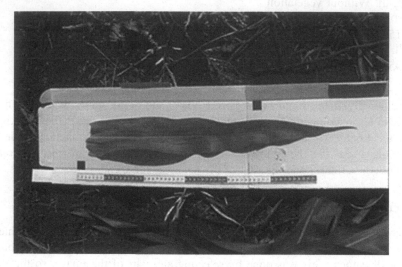

Fig. 5. Sample of RGB leaf image.

4.1   Image Labeling

Image labeling was done on a computer with 8 GB of RAM. From the tests carried out, it was observed that in each of the labeled sheets there are around 150,000 NDVI values, then the average value was used for training. A sample of the information that was stored in the database is shown in Table 1.

Table 1.   Sample values stored in the database.

Image	NDVI min	NDVI max	NDVI avg	CCM (CCI)
NDVI0434	0.57379013	0.77697020	0.69942983	32.2
NDVI0437	0.60665011	0.8095092	0.72107878	37.3
NDVI0440	0.54207837	0.76918220	0.65276326	37.4
NDVI0443	0.78050649	0.92617207	0.87607061	36.6
NDVI0446	0.86651057	0.92999970	0.89974493	38.3

4.2   Training Module Test

In this experiment, the database obtained from the image labeling module was used. The NDVI values belonging to a region of interest were averaged, obtaining 92 NDVI values in total. After training a linear regression (see Fig. 6 and Table 2), the following equation was obtained: y = 43.73495324 x − 2.9578003944927787 and the root of the mean square error (RMSE) was equal to 7.2728 CCM units.

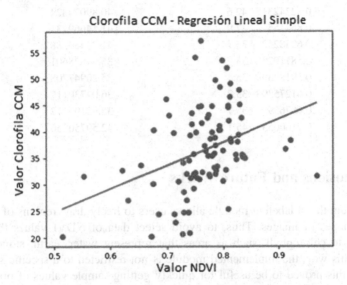

Fig. 6. Linear regression NDVI - CCM

Table 2. Sample values of linear regression.

NDVI	CCM (field measurement)	y (prediction)
0.7012959	34.8	36.78035163
0.8696189	43	40.48576784
0.8162994	41.1	40.35309443
0.69903266	36.3	35.53141767

4.3   Prediction Module Test

Once the linear regression was obtained, the prediction module was tested using crop images (see Fig. 4). After loading an image to the program, any region can be selected, and then the NDVI value of that particular position and the chlorophyll value are shown using the linear regression calculated above.

The following table shows 10 chlorophyll level values obtained using the simple linear regressions mentioned previously and their respective relationship with the NDVI and the actual CCM value (Table 3).

Table 3. Prediction module test.

NDVI	CCM (field measurement)	y (prediction)
0.6682494	38.1	38.44229234
0.5511283	38.1	37.69348461
0.82342315	42.6	40.89074824
0.864155	47.9	41.14669231
0.6238252	35.7	41.24439288
0.5811929	35	35.64828601
0.5918739	35.4	35.20947708
0.61275303	35.2	36.07730117
0.6876595	35.7	35.95191117
0.9160053	45.1	42.59750256

5   Conclusions and Future Works

It is important that a labeling module allows users to freely draw regions of interest in irregular shapes of images. Thus, to avoid select data of NDVI values that do not correspond to chlorophyll, such as areas that represent water, earth, stones, among others. In this way, the implemented module is not restricted to a specific shape (e.g. rectangle), this proved to be useful for quickly getting sample values of pixels inside leaves.

Regarding the simple linear regression results, due to the leaves were in a similar state of health, most of the data were distributed in the same NDVI ranges, where, CCM values are very close to each other, as seen in Fig. 6. This makes the training phase difficult since a diverse range of NDVI values is not obtained as well as CCM values. The linear regression on 10 data tests was obtained a maximum error of 15.5%, with an average of 4.67%.

Future experiments are planned where diverse values of chlorophyll will be obtained by applying different levels of fertilizers to the plants. Additionally, a segmentation preprocess is recommended to be implemented in the labeling module in order to group pixels by colors producing candidate regions, with the objective of reducing the manual labeling time.

References

1. Saverio, J., Alarcón, A.: NDVI-Checking. https://github.com/JoeSvr95/NDVI-Checking
2. Saberioon, M.M., et al.: Assessment of rice leaf chlorophyll content using visible bands at different growth stages at both the leaf and canopy scale. Int. J. Appl. Earth Obs. Geoinf. **32** (1), 35–45 (2014). https://doi.org/10.1016/j.jag.2014.03.018
3. Mogili, U.R., Deepak, B.B.V.L.: Review on application of drone systems in precision agriculture. Procedia Comput. Sci. **133**, 502–509 (2018). https://doi.org/10.1016/j.procs.2018.07.063

4. Schut, A.G.T., et al.: Assessing yield and fertilizer response in heterogeneous smallholder fields with UAVs and satellites. Field Crops Res. **221**, 98–107 (2017). https://doi.org/10.1016/j.fcr.2018.02.018

5. Cao, S., et al.: Radiometric calibration assessments for UAS-borne multispectral cameras: laboratory and field protocols. ISPRS J. Photogram. Remote Sens. **149**, 132–145 (2019). https://doi.org/10.1016/j.isprsjprs.2019.01.016

6. Ampatzidis, Y., et al.: Citrus rootstock evaluation utilizing UAV-based remote sensing and artificial intelligence. Comput. Electron. Agric. **164**, 104900 (2019). https://doi.org/10.1016/j.compag.2019.104900

7. Gilabert, M., Gonzalez-Piqueras, J., García-Haro, J.: Acerca de los índices de vegetación. Revista de teledetección: Revista de la Asociación Española de Teledetección (8), 1133-0953 (1997)

8. Solano, F., Di Fazio, S., Modica, G.: A methodology based on GEOBIA and WorldView-3 imagery to derive vegetation indices at tree crown detail in olive orchards. Int. J. Appl. Earth Obs. Geoinf. **83**, 101912 (2019). https://doi.org/10.1016/j.jag.2019.101912

9. Kyratzis, A., et al.: Investigating correlation among NDVI index derived by unmanned aerial vehicle photography and grain yield under late drought stress conditions. Procedia Env. Sci. **29**, 225–226 (2015). https://doi.org/10.1016/j.proenv.2015.07.284

10. Reyes, J., Godoy, A., Realpe, M.: Uso de software de código abierto para fusión de imágenes agrícolas multiespectrales adquiridas con drones. In: 17th LACCEI International Multi-Conference for Engineering, Education, and Technology: Industry, Innovation, and Infrastructure for Sustainable Cities and Communities, 24–26 July 2019, Jamaica (2019)

11. Liakos, K.G., et al.: Machine learning in agriculture: a review. Sensors (Switz.) **18**(8), 1–29 (2018). https://doi.org/10.3390/s18082674

12. Muñoz-Huerta, R.F., Guevara-Gonzalez, R.G., Contreras-Medina, L.M., Torres-Pacheco, I., Prado-Olivarez, J., Ocampo-Velazquez, R.V.: A review of methods for sensing the nitrogen status in plants: advantages, disadvantages and recent advances. Sensors **8**(13), 10823–10843 (2013)

13. Sainz Rozas, H., Echeverría, H.E.: Relación entre las lecturas del medidor de clorofila (Minolta CCM 200 PLUS 502) en distintos estadios del ciclo del cultivo de maíz y el rendimiento en grano. Revista de la Facultad de Agronomía, p. 103 (1998)

Crowdsensing and Image Processing as a Method for Analysis and Population Count Based on the Classification and Validation of Multimedia

Alexander Mejía[1], Marcelo Olalla[1], Bryan Oscullo[1], Freddy Tapia[1], and Luis Tello-Oquendo[2(✉)]

[1] Computer Science Department, Universidad de las Fuerzas Armadas ESPE, Latacunga, Ecuador
{avmejas, amolalla, bwoscullo, fmtapia}@espe.edu.ec
[2] College of Engineering, Universidad Nacional de Chimborazo, Riobamba Canton, Ecuador
luis.tello@unach.edu.ec

Abstract. The growth of cities and the advancement of technology demands the development of solutions that can help and reduce emerging problems. One of these problems is the control of the population and urban planning based on population count, flow, and density analysis. This paper proposes an approach that could help cities to gather population data in a contextualized manner with the usage of localization, sensor, and weather data. It is based on a method to collect vast quantities of information and a way to validate and analyze gathered data. The data recollection is achieved through crowdsensing and smartphones. Validation and analysis are made with cloud-based image analysis and neural networks. Results show the usefulness and effectiveness of the proposed solution; also, some considerations are presented with the proposal.

Keywords: Analysis · Classification · Population counting · Crowdsensing · Image analysis · Data validation · Georeferencing · Sensor data

1 Introduction

Technological globalization has transformed knowledge and innovation in a short amount of time. This has created new challenges and opportunities based on our necessities. One of these necessities is access to data and its analysis, which has demanded several distributed elements and systems [9]. Also, the rapid growth of population demands that urban cities respond with services and infrastructure efficiently and effectively, which increases the demand for integrated systems like sensors and smartphones [7]. The population growth is a challenge for urban planning and control on high population density zones [1, 13]. This creates the necessities to develop tools and techniques that help determine the number of people in a specific place and under specific conditions. These tools can be a way to mitigate risk and improve urban population based on specific population counts [1]. There are many proposals for the

© Springer Nature Switzerland AG 2020
M. Botto-Tobar et al. (Eds.): ICAT 2019, CCIS 1194, pp. 582–592, 2020.
https://doi.org/10.1007/978-3-030-42520-3_46

determination of population count in a specific place, but they have several limitations [13]. These techniques usually work for population counting, but lack contextualization based on location, weather, time, among others for an in-depth population analysis [1, 13]. This paper proposes an approach to determine the veracity and the results of different tools and applications that could help to determine the population count. The research was conducted in the city of Quito, Ecuador using an artificial intelligence (AI) algorithm for the analysis of images and videos and validation. For doing so, we consider the contextualization of the data with the use of localization, time, and weather.

The remainder of this paper is organized as follows. Section 2 presents the background information regarding the research topic. Section 3 introduces the methodology that we followed to conduct this study. Section 4 details the design and implementation of the proposed architecture. Section 5 presents the results and performance evaluation. Finally, Sect. 6 draws the conclusions and details of the future work planned.

2 Background

Population count and flow analysis, as is proposed in this paper, has different theoretical aspects that are needed to be explained. These concepts are based on the field of data gathering, data analysis with sensors, image analysis, among others.

Crowdsensing. It is a paradigm based on the recollection, sensing, and analysis of data in massive quantities [12]. This works with the usage of a mobile device, sensors, and big crowds as a way of data recollection. Using this paradigm, a possibility opens for the analysis and recollection of vast quantities of data in a practical a relatively easy way [12].

Georeferencing. It is the process of associating the information with a specific geographical location [5]. It is an element that can be found in a wide variety of information systems such as maps, delivery services, and surveillance systems. Georeferencing allows putting information inside a geographical and time context in a precise and clear way. Thanks to this, information can be better specialized and used [5]. To apply to georeference, there is a need to use localization systems or devices like GPS [5].

Population counting. It is the process of finding out the number of people in a specific place or location, its distribution, and density by counting. It is a fundamental aspect in the analysis of population groups [1]. Population counting helps us know if a place presents overcrowding risks, level of attendance to a place or event, or the popularity of a specific location [1]. Using population counting helps with planning and risk mitigation. There is a great variety of different methods to achieve population counting. These methods could be manual or use technological elements like artificial intelligence with neural networks [1].

Convolutional neural networks. They are a type of neural network, which structure is like the primary visual cortex of the human eye [3]. It is structured by four fundamental layers, namely, the image processing layer, the convolutional layer, the subsampling layer (that alternates constantly with the convolutional layer to get to a result), and the classification layer (that is in charge of establishing the results of the neural network). These types of neural networks are commonly used for image analysis [3].

3   Methodology

In this study, we follow a methodology for conducting the research and a methodology for software development.

Regarding the research methodology, we follow the technological investigation. This methodology uses the scientific method to search for new knowledge for its usage and application in the design and improvement of a software product or technological idea. To justify this methodology, based on [4], the main objective of the technical investigation is to gather useful knowledge for the use of it in the solution of problems. These problems usually come from societal needs and the solutions established approach the problem in an economic, social, temporal, social, cultural and geographical problem. All the solution is based on the usage of innovation through technology.

Regarding the software development, we use the SCRUM methodology. It is an agile software methodology, that is used to make software faster without affecting its overall quality. It was selected because it presented the following benefits: (i) It allows agile management of projects in a short schedule, (ii) it helps to adapt quickly to changing requirements, and (iii) it presents a way to track development in a way that is easy to solve problems [8].

4   Design and Implementation

4.1   System Architecture

The system for the validation process and image classification is based on a multi-layered architecture. This architecture divides the whole system into physical and logical layers that allow executing different processes. There are three main layers as illustrated in Fig. 1:

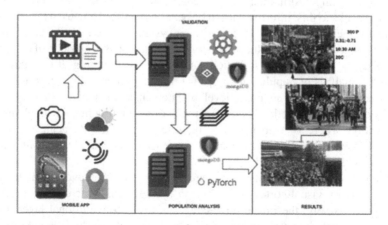

Fig. 1. General system architecture schema.

- Mobile Application (Presentation and User interaction layer),
- validation server (Mid or validation layer),
- population analysis server (Final or analysis layer).

4.2 Analysis and Validation Handling

The process to send data about sensors, images, videos, and the respective validation by sensor data is made in real-time. New entries can be sent at any time without any problem, and the sensor data validation will be done immediately. Image analysis validation is made as soon as a specific quantity of data is captured. This validation is made through cloud-based tools. In this case, the analysis was made for each data lot made. The same process as the image validation is used for population analysis. A neural network is used for this analysis.

4.3 Mobile App

The mobile app is simple, and its main objective is to allow a user to capture images or videos while at the same time gather sensor data from each capture. All captures at sent to a validation server. The app works with the following flow: a user captures a video or image; once the capture is made, immediately the app gathers GPS, light, pressure, time and date; lastly, all the data is sent for analysis. The app has the interfaces shown in Figs. 2 and 3.

Once all the data is sent, the application receives a confirmation message. Inside the app, there is not a validation process being executed. Only the data capture is made. The app was developed for Android mobile devices using Android Studio.

Fig. 2. Mobile application structure. Main Menu

Fig. 3. Mobile application structure. Camera Setup

4.4   Web Services for Validation

The web services are what allows the mobile app to send its captures to the validation server. The service was made using REST. It has two endpoints that manage all the data based on JSON:

- The first endpoint stores a video or image that has been sent from the mobile app. The endpoint verifies if the element that was sent is a video or images and then stores the data in a specific place in the server.
- The second endpoint receives the data from the sensors and complementary data from the mobile app, prepares the weather information from the specified place, and stores it in a database for its future validation and analysis.

The web service was made using. NETs WCF (Windows Communication Foundation).

4.5   Database

The database was the document-based NoSQL database, MongoDB and the data were structured with the following schema:

```
{
    "_id":"5b394986f6259843681c12dc",
    "Validity": "Valid",
    "Temperature_state": "broken clouds",
    "Lux_state": "Valid",
    "Img_name": ["image_201860163339.jpg"],
    "Pressure": 740.6439,
    "Lux": 53.74074,
    "Temperature": 17.1499939,
    "Date_n_time": "Jul 1, 2018 16:33:39",
    "Gps_location": {
        "Type": "point",
        "Location": [-0.317904, 78.49474],
        "Description": "none"
    },
    "Population":0
}
```

This schema allows the whole system to use sensor data, establish capture validation, its location, population, and so on.

4.6   Video Processing

Even though the service and system accept the sending of video, it does not process or analyze video. Instead, the videos are transformed into several images. What is made is the extraction of frames from the video every three seconds, and these frames are stored as images associated with the capture made with the video. This is made with the usage of FFMPEG, a command-line tool for video processing.

4.7   Sensor Validation

The sensor validation is the first action to be made, and it is, basically, a first filter that each capture is passed through. This is the simplest and easiest filter to execute. The validation uses the pressures and light data collected for the capture. The received values are compared with established ranges that are based on real-world pressure and light data. The value for the pressure is accepted when it exists between 337 and 1013 mbar. These ranges correspond to the pressure in mbar encountered at Everest's summit and the pressure found at sea level, respectively [2, 11]. The light data is only accepted when it has a value superior to 10 lux. Any entry with less than 10 lux is denied because it means that the capture is too dark for any validation [6].

4.8   Image Validation

The image validation process is the second and last filter. It uses Microsoft Azure APIs; Computer Vision and Face API and a small knowledge base with different images and critical classifications. The process follows this sequence:

1. The image is sent to Computer vision, and the tags from it are received;
2. The received tags are compared to the knowledge base and then the number of valid, non-valid, and prohibited words are determined.

If an excess of non-valid or prohibited words in the image is found, then the image is rejected. Its status is set as invalid; otherwise, the image is accepted.

4.9   Population Counting and Analysis

Once every validation process is completed, all the data is sent to the analysis server, in which the population is counted. This analysis was conducted by implementing the algorithms in [10, 13]. The process to get the results is the following:

- Load every image to analyze,
- analyze every image on the database,
- once the population is extrapolated from the images, the result is stored in the database.

The analysis was done with PyTorch and the use of a convolutional neural network as depicted in Fig. 1.

5   Results

The purpose of this study is to determine the accuracy of the classification and image validation with the usage of sensor data and image analysis for determining population count and population analysis based on location, time, and weather, among others. For doing so, the following activities were conducted:

- Performance analysis,
- sensor validation and effectiveness analysis,
- image validation and effectiveness analysis,
- effectiveness of the implemented neural network for population counting,
- classification and whole process usefulness.

5.1   Testing Procedure

The tests were conducted in such a way that it is visible the performance and effectiveness of the developed solution. The experiments were made with the use of a smartphone and two servers. The first server filled the role of a middle layer for image validation and had the following characteristics:

- CPU: Intel Xeon E5-2620 2.00 GHz 4 CPUs,
- RAM: 16 GB,
- OS: Windows Server 2012 R2,
- Internet Speed: 40 Mbps Downlink and 30 Mbps Uplink.

The second server was used for the population analysis and had the following characteristics:

- CPU: Intel Xeon E5-2620 2.00 GHz 8 CPUs,
- RAM: 32 GB,
- OS: Ubuntu Server 16.04.

The mentioned characteristics influence the results of all tests. To conduct the experiments, we considered three test groups, that were denominated as batches. Batch 1 consisted of 25 data sets, it was used to validate the images and not for further processing; batch 2 (of 43 data sets) was used to corroborate the validation process, its performance, and the image recognition analysis; batch 3 (30 data sets) was used to validate the population analysis.

5.2 Sensor Validation Results

The sensor validation was adjusted to the parameters that were given. It showed total effectiveness due to a lack of problems during interpretation. It was noticed during the evaluation that a vast majority of captures or data sets were accepted because the batches that were tested were made in the right conditions. When captures were dropped, the lightning conditions of the capture were too dark. The sensor validation works, but it could be ignored. A validation based only on image recognition could be used. Still, the sensor data is useful for classification purposes.

5.3 Image Validation Results

Two aspects were analyzed for the image validation: performance, in terms of processing time, and effectiveness. The performance results were gathered from batch 1. This batch had a total of 25 images, and the test was made in groups of 1, 5, 15, and 25 images. The results achieved were the following:

- 1 image: 8 s;
- 5 images: 48 s;
- 15 images: 1 min 52 s;
- 25 images: 3 min 20 s.

Note that the processing time depends significantly on the internet velocity and image size. Another aspect that was tested is the effectiveness of the image analysis. This evaluation was made with batch 2 using 43 images. The analysis detected a total of three rejected images and 40 accepted ones. Of all the rejected images, just one was a false positive. That image should have been accepted, but the process detected 3 permitted words, 2 not permitted, and 3 prohibited words. Figure 4 depicts a denied image because of the detection of not permitted words and prohibited words. Words like stool, desk, chair, restaurant, room and similar were detected and the algorithm rejected the image. It is necessary to adjust the algorithms to achieve a more accurate evaluation of the image. The success rate of the analyzed batch was 97.67%, but due to the nature of the failure, this percentage could fluctuate.

Fig. 4. A denied image that should be accepted.

5.4   Population Analysis Results

As with the analysis for image validation, there are two perspectives from which we can see the results of this analysis: performance, in terms of resource consumption, and effectiveness. The performance was measured based on batch 2. The procedure is highly dependent on the resources of the server used. The CPU usage was always at 100% and the RAM used approximately 97% of its capacity. Also, the analysis of a single image lasted somewhere around 8 min approximately. This tremendous resource demand and low performance were caused because of the weight of the PyTorch framework. PyTorch would have worked better with a powerful GPU instead of a CPU due to its nature.

Regarding effectiveness, it was found that the population analysis is not that effective and varies significantly based on the condition of the image and the elements in it. A total of 73 images were analyzed from batch 2 and batch 3. The population analysis is not accurate, and there are multiple cases where the estimation of the population is rather exorbitant and far removed from reality as can be seen in Fig. 5. A total of 19 images resulted in exaggerated results. In these images it was evident image errors. It is difficult to know how inexact the analysis is because an empiric observation has shortcomings and does not give exact results, in this case. To get a better result for corroborating the inaccuracy of the analysis, density maps based on the population should be made. Based on the image analysis, the effectiveness was of 73.97%.

With the observations made from the images that present problems, several elements were noticed. The presence of a considerable number of trees in high density can affect the results of the neural network. Cars and vehicles can produce similar problems. The existence of objects with human-like shape creates confusion. A way to fix this is to train the neural network with more significant data samples.

Fig. 5. Image with exaggerated population count result: 1601 people

5.5 Usage of the Results for Classification and Analysis

With all the analysis made and recollected information, there is a way to have relevant information about a place regarding its population based on time and weather. With all the captures, there is the possibility to have information about a specific sector based in a geographical coordinate, population data, date, temperature, pressure, light level, and so forth. There is a way to manage statistics that can be as detailed as a specific geographical point or as a broad city. Because only 73 images were collected and not all the captures presented accurate information, final analysis about a place could not be made. However, the data was enough to see that, with more captures, useful statistics can be gathered about a city and its different places.

6 Conclusions and Future Work

We have presented an approach to count the population and analyze it with the use of crowdsensing and image analysis. A mobile app was developed that can gather images or video referencing while referencing it to sensor data, location and time of capture. To test our proposal, data has been gathered around the city of Quito. After this, all captures were properly validated with the use of cloud-based image analysis and the reference data of each capture. Then, with the use of a convolutional neural network, each capture had its population counted. The whole process allowed us to make a general population analysis contextualized by location, weather conditions, population count, and time. During the testing process, it was determined that the usage of sensor data provided by a mobile phone resulted irrelevant for the validation but useful for the analysis and classification of data. Population counting was not particularly useful, due to a lack of training in the neural network, but still most of the time it was accurate enough. Overall, our results show a useful approach of classifying multimedia for the analysis of population-based on crowdsensing and image analysis. As future work, there exist several research paths that can be taken considering this paper as a baseline. To improve this study, we have planned the following: (i) Using surveillance video cameras for the population analysis processes. This makes the usage of mobile phones for that recollection unnecessary. This work could establish a more natural way to collect and validate data but will be limited to the places where the data can be collected. (ii) Implement other techniques of population counting. This way searching for a way that gives the most accurate and useful results from the analysis. Also, the search for another and better population counting techniques could lead to a process that is resource-efficient and fast, leading to a more dynamic usage of the data.

References

1. Boominathan, L., Kruthiventi, S.S., Babu, R.V.: CrowdNet: a deep convolutional network for dense crowd counting. In: Proceedings of the 24th ACM International Conference on Multimedia, pp. 640–644. ACM (2016)
2. Britannica, E., et al.: Encyclopædia Britannica. University of Chicago, Chicago (1993)
3. Ciresan, D.C., Meier, U., Masci, J., Gambardella, L.M., Schmidhuber, J.: Flexible, high performance convolutional neural networks for image classification. In: Twenty-Second International Joint Conference on Artificial Intelligence (2011)
4. Espinoza Montes, C.: Metodología de investigación tecnológica pensando en sistemas (2014)
5. Hill, L.L.: Georeferencing: The Geographic Associations of Information. MIT Press, Cambridge (2009)
6. Kennedy, J.: Understanding and interpreting lux values (2018). https://docs.microsoft.com/en-us/windows/win32/sensorsapi/understanding-and-interpreting-lux-values
7. Rathore, M.M., Ahmad, A., Paul, A., Rho, S.: Urban planning and building smart cities based on the internet of things using big data analytics. Comput. Netw. **101**, 63–80 (2016)
8. Schwaber, K.: SCRUM development process. In: Sutherland, J., Casanave, C., Miller, J., Patel, P., Hollowell, G. (eds.) Business Object Design and Implementation, pp. 117–134. Springer, London (1997). https://doi.org/10.1007/978-1-4471-0947-1_11
9. Sosa, E.O., et al.: Internet del futuro y ciudades inteligentes. In: XV Workshop de Investigadores en Ciencias de la Computación (2013)
10. Svishwa: Single image crowd counting (CNN-based cascaded multi-task learn- ing of high-level prior and density estimation for crowd counting) (2017). https://github.com/svishwa/crowdcount-cascaded-mtl
11. West, J.B.: Barometric pressures on MT everest: new data and physiological significance. J. Appl. Physiol. **86**(3), 1062–1066 (1999)
12. Yang, D., Xue, G., Fang, X., Tang, J.: Incentive mechanisms for crowdsensing: Crowdsourcing with smartphones. IEEE/ACM Trans. Netw. (TON) **24**(3), 1732–1744 (2016)
13. Zhang, Y., Zhou, D., Chen, S., Gao, S., Ma, Y.: Single-image crowd counting via multi-column convolutional neural network. In: Proceedings of the IEEE Conference on Computer Vision and Pattern Recognition, pp. 589–597 (2016)

Photogrammetry and Augmented Reality to Promote the Religious Cultural Heritage of San Pedro Cathedral in Guayaquil, Ecuador

Joe Llerena-Izquierdo(✉) ⓘ and Luiggi Cedeño-Gonzabay(✉) ⓘ

Universidad Politécnica Salesiana, Guayaquil, Ecuador
jllerena@ups.edu.ec, lcedenog@est.ups.edu.ec

Abstract. This innovative proposal combines the use of the biggest reality and photogrammetry for modeling structures in digital format. It addresses two themes the use of technologies for the restoration of heritage structures in the event of a fortuitous event and the use of a technological tool that allows the dissemination of the religious cultural heritage of the Cathedral of San Pedro in Guayaquil, Ecuador to national and foreign tourists, the interest in knowing a little more about the history of culture and art is encouraged when effective conservation strategies are involved and even more by incorporating the use of technology in smart devices. This allows easy quick access with proper visualization, the use of photogrammetry technique is adopted in several museums around the world and allow tourists to learn about a specific topic in a didactic and interactive way. Another technique is the increasing reality technique, this technique incorporates data in virtual form links on the web audio video, video text or other multimedia through markers to objects serving as a tool that encourages learning more about different topics. The one chosen in this work is the religious cultural heritage. The interface developed in Unity, and the use of the Vuforia development kit, through the mobile application "My Cathedral", allows users to access relevant historical information, visualizing the photogrammetric images in increasing reality, precisely on the most representative objects of the cathedral of San Pedro in Guayaquil.

Keywords: Photogrammetry · Augmented reality · Religious cultural heritage

1 Introduction

The digitization of 3D models in different professional areas is becoming more interesting due to the use of computational techniques to recognize surfaces through a sequence of images and then reconstruct them in an object similar to the real [1, 2]. In medicine [3], with the generation of human, dental or brain prostheses, up to the restoration and healing of cultural heritage or archaeological sites [4]. The use of the technique of digital modeling and reconstruction is increasingly used because it is obtained from the geometry or surface of the object, a representation close to reality and greater level of detail as we perceive it in our daily lives.

Using proper techniques on images becomes an increasingly useful option for the digital reconstruction of structures, buildings or objects that have valuable information

© Springer Nature Switzerland AG 2020
M. Botto-Tobar et al. (Eds.): ICAT 2019, CCIS 1194, pp. 593–606, 2020.
https://doi.org/10.1007/978-3-030-42520-3_47

because of their high historical value [5]. Thus, the reduction in costs [6] compared to traditional practices allows its use to expand [7].

This work presents the elaboration of the 3D modeling applying the technique of photogrammetry for the capture of images and the later processing of four significant structures (religious altars) of the cathedral of San Pedro, in the city of Guayaquil, in Ecuador. In addition to modeling the structures of the altars of "Narcisa de Jesús, Don Bosco, Mary of Schoenstatt, and Jesus of Great Power".

It allows its visualization in increasing reality, by means of the mobile application for intelligent telephone devices with Android platform, called "My Cathedral". Finally, by applying the technique of photogrammetry in the generation of 3D digital models of the most representative altars of the cathedral and presenting them in a mobile application with augmented reality, the promotion and historical, cultural and religious interest by users is expected. The representatives of the Cathedral of the city of Guayaquil can establish actions of restoration before the possible deteriorations of the structures by means of the digital conservation of these objects.

The detailed explanation of photogrammetry and its use in the three-dimensional reconstruction used for the cathedral structures is presented. The development of the generation of the basic photogrammetric model is presented, for its improvement in the 3D Max software. Finally, the results obtained by the method applied in the development of the study are shown. The mobile application and its prototype image with markers can be obtained in GitHub https://github.com/lcedenog/MY_CATHEDRAL.

1.1 Three-Dimensional Reconstruction Based on Photogrammetry

The photogrammetric process allows to obtain information of the physical objects and the medium, in a reliable way, through the registration and measurement of images in photography and their electromagnetic patterns [8]. The orientation of the frames relates the geometric condition of the image and the object projected on the same line where the center of projection is so that they are aligned in the same three-dimensional space [9]. The use of photogrammetry in conjunction with the availability of intelligent devices and new technologies [10, 11], allows low-cost models to be generated with high quality, promoting their use as the main technique for digitalization. Suitable for promoting historical cultural heritage worldwide [12] and orienting it towards preservation or tourism [13, 14].

1.2 Augmented Reality in the Conservation of Cultural Heritage

Augmented reality opens spaces of sensory experiences with visual and audible representation that allow the user to recognize forms of architectural structures or sculptures of high heritage value in a combination between real and digital [15–18].

Experts from different areas make efforts to incorporate techniques and resources supported by new technologies, from the use of drones, intelligent devices and equipment for capturing images in different angles of coverage. The proposals for their work are presented in Table 1.

Preserving cultural heritage digitally for future generations faces several challenges [26], ranging from the use of immersive technologies, adequate means for acquisition

Table 1. Techniques applied by experts in different areas

Techniques	Application	Tools	Resources
3D reconstruction through photogrammetry and augmented reality [19]	Reconstruction of the urban area of Czech Republic	PhotoStruk 3D	Lidar device
Augmented Reality and Gyroscope with programmed flight path [20]	Digital reconstruction in 3D format of Ankara's Roman bath	DJI's Phantom 3 drone y Unity.	Drones, GPS
Augmented Reality and Geolocation [21]	Augmented Reality to publicize the cultural heritage of Indiana University, Bloomington	Vuforia SDK	Smartphone, Camera, GPS
Augmented reality with reflectance transformation image (RTI) [22]	Digital reconstruction of Korean cultural artifacts	Kinect studio	Kinect, device, projector and Tablet
Augmented reality [23]	Preservation of diversity of flora and fauna at the Caloust Gulbenkian Foundation in Lisbon - Portugal	Aurasma – HP Reveals	Smartphone, Camera
Augmented reality and Geolocation [24]	Museum of Industrial Production of Olive Oil in Lesbos (MBEL)	Game-engine middleware software	Smartphone, Camera
Augmented reality with gyroscope [25]	Augmented reality for tracking people based on the sequential model of Monte Carlo	Ubiquitous Augmented Reality	Laptop, VGA Camera

and digitization, and proper planning for the preservation of structures in digital format [27–29]. This allows for the interpretation and dissemination at the social level of the historical and cultural wealth of recognized patrimonies, avoiding oblivion due to the damage or destruction they may cause to works, structures or objects of architectural value.

2   Materials and Methods

The methodology used in this work is quasi-experimental with a quantitative approach and uses the survey as a research technique to a selected target group. The use of augmented reality enables a practical, economic application with high quality results in relation to cost in other methods, such as virtual reality, accessible at a high cost. The use of high-performance equipment with technology for a certain compatibility of them reduces the scope of reaching a large number of users. The use of the augmented reality technique is presented in the development platform of Vuforia in a first version.

The project is developed in four phases, the first is aimed at capturing images to generate the base photogrammetric model, the second phase allows the import of base models in three dimensions, for refinement in the 3D Max software, the third phase develops the mobile application, for mobile operating systems Android and the fourth phase, is conducted the data collection through satisfaction survey to measure the level of user acceptance.

2.1   Capture of the Images and Generation of the Base Photogrammetric Model

The religious structures are in an unopened and poorly lit environment at floor level, making the use of drones impossible. The historical value of the altars makes it impossible for them to be relocated or manipulated to change the place. The first step to make is the capture of images at the height where the structure is for the generation of photogrammetric models bases, for this work are used four of the main altars of St. Peter's Cathedral, the altars are: Don Bosco, Jesus of Great Power, Narcissus of Jesus, Mary of Schoenstatt. From 250 to 300 images are captured per altar at various angles within a visible range that allows a depth to be achieved. Each image is established in intervals that can completely cover a 180° rotation in the frontal direction of the altar (see Fig. 1).

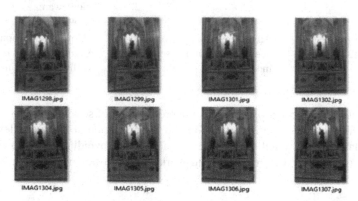

Fig. 1. Captured images for the base photogrammetric model.

In the second step, the profile masks of the structure are created to delimit the area of the base photogrammetric model (see Fig. 2).

Fig. 2. Photogrammetric masks for a delimited base model.

The third step is the generation of the dense point cloud that allows a spatial digital definition of the structure (see Fig. 3).

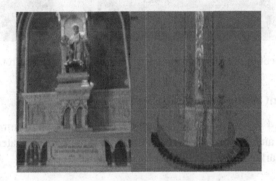

Fig. 3. Creation of the dense cloud of points.

In the fourth step, the mesh that defines the solid body of the 3D model is finally generated (see Fig. 4).

Fig. 4. Creation of the mesh that defines the solid body of the picture.

The generation of the texture of the model and the export of the model are packaged in a standard format file, wave front 3D object file (*.obj).

2.2 Perfecting the Base Models

When the 3D model is obtained, the file extension in obj format is imported into the 3D modeling software Max to correct possible imperfections when generating the final 3D model (see Fig. 5).

Fig. 5. Improvement of the 3D Max model, front view, side view and rotation perspective.

2.3 Development of the Mobile Application

The 3D models (4 models) are imported into the Unity environment, each model represents a scene and this one is related to each altar, they are activated by means of an image target or markers (see Fig. 6).

Fig. 6. Importing the 3D model into the Unity environment.

The creation of Target Image or markers are essential to make the anchorage to a pattern and raise through augmented reality the link to the images of each scene of the 3D model, these markers serve as information required by the record of Vuforia in its information base.

A developer account is required on the Vuforia website and obtain a free development license. This license generates a 380 characters key that must be copied, entering in the configuration of the AR Camera in Vuforia. Subsequently an information base is specified in the target manager of Vuforia called "My Cathedral", database to which we will add information with each of the images created (see Fig. 7).

Fig. 7. Creation of the image target or objective markers.

Vuforia recognizes each of the markers establishing the patterns to distinguish one marker from another and raises the final image in augmented reality. For this within the portal, for Vuforia developers, a score is established for each marker with stars, being 5 stars the highest of the scores. The rapid recognition of the marker depends on these scores as shown below (see Fig. 8).

Fig. 8. Pattern of recognition of Vuforia.

Making use of bookmarks as an access mechanism for augmented reality, it was decided to develop the proposal of the mobile application in the open source Android operating system that has greater acceptance by users in general. The principle of augmented reality is based on virtualizing a real model into a digital one by means of a device capable of presenting or showing objects or models with 3D dimensions and camera detection.

In this way mixes the real world with the digital, Vuforia bases its pattern recognition through Image Target or markers. Establishing in this way an algorithm of evaluation on the characteristics that it contains like pattern and allows to determine how visible it is, with this the speed of load is determined on the basis of the existing 3D models. Table 2 shows the evaluation and characteristics of the Image Target of "My Cathedral".

Table 2. Target Manager of Vuforia used in "My Cathedral".

Name Target	Type	Image	Features	Rating
Welcome	Single image	Logo	345	5 stars
Don Bosco	Single image	Logo	183	4 stars
Narcisa de Jesús	Single image	Logo	185	4 stars
Mary of Schoenstatt	Single image	Logo	187	4 stars
Jesus of Great Power	Single image	Logo	190	4 stars

2.4   Collection of Information

This last phase has a study target group of 21 randomly selected men and women who own a mobile device with an internet connection and are visiting the Cathedral of San Pedro in Guayaquil. Two work scenarios are elaborated, the first scenario consists of carrying out a study that determines the level of knowledge of the population in general about the patrimonial history and religious culture about the altars found in the cathedral of the city, without having used the application "My Cathedral". The second scenario, a satisfaction survey of the application is elaborated, as a research technique, that also allows to understand the knowledge acquired by the users when making use of the mobile application "My Cathedral" for the due improvements of the final application.

3   Results and Discussion

The results of the target group show the percentage of correct and incorrect answers obtained in the surveys before and after the evaluation (see Fig. 9). Incorrect answers are reduced from 70% to 30% while correct answers improve from 10% to 90% with the help of the mobile application.

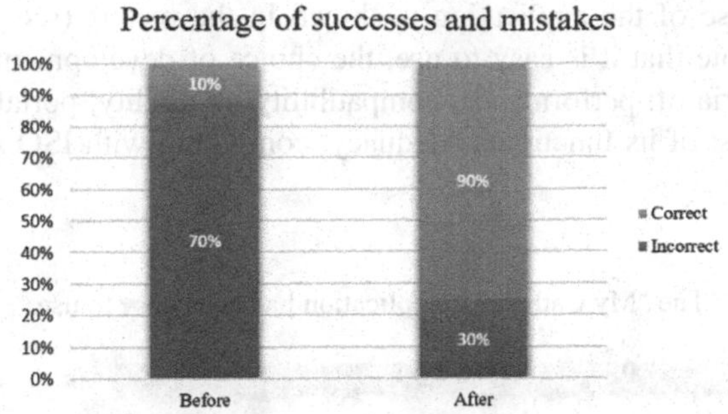

Percentage of successes and mistakes

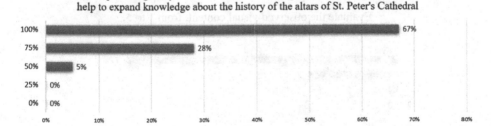

Fig. 9. Evaluation of historical, heritage and religious cultural knowledge, before and after using the "My Cathedral" application.

At first glance, the percentages show that, before the use of the mobile application "My Cathedral", the historical heritage knowledge, as well as cultural and religious knowledge, was lower than the result obtained after using the application. The use of augmented reality in interactive and novel applications facilitates a better way of remembering and transmitting information, in this case of the historical, cultural and religious heritage of the structures taken for the development of this work.

The online survey with a link to the application "My Cathedral" uses the free survey service called "Google Forms". It is addressed to 420 students of the Salesian Polytechnic University, to know the level of acceptance, ease of use and to what extent this set of applied techniques can promote a better dissemination of knowledge of the historical, cultural and religious heritage of the cathedral.

The survey is designed for a confidence level of 96%, with a margin of error of 5%, and is taken as the size of the universe to 100,000 people, which is the monthly average of visitors to the city cathedral.

The final version of the developed application is released, being evaluated by 420 people who installed the application in their intelligent devices. The result determines that 67.10% of those surveyed (see Fig. 10) state that augmented reality helps to broaden knowledge about the history of the altars of St. Peter's Cathedral.

In what percentage do you think that new technologies, such as augmented reality, help to expand knowledge about the history of the altars of St. Peter's Cathedral

Fig. 10. Question 2 of the survey, details the results obtained regarding augmented reality applied to broaden knowledge about the history of the altars.

The ease of use of the application is shown in the results (see Fig. 11), 98% of respondents indicate that it is easy to use, the choice of development tools is initially based on the criteria of: performance, compatibility, reliability, portability and ease of use, due to the cost of its functional adequacy, complying with ISO standards/Quality IEC 25010 [30].

The "My Cathedral" application has been easy to use.

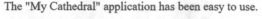

Fig. 11. Question 3, shows the ease of use of the application and the user experience.

99% of the participants (see Fig. 12) show the level of interest generated by the information offered in the mobile application in its various formats.

By using the "My Cathedral" application, you have been interested in the audio, video and image information that have been appearing on the mobile device.

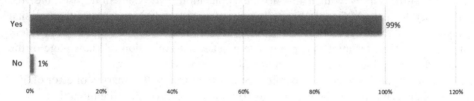

Fig. 12. Question 4 shows the level of interest generated by the information in the mobile application.

The result obtained in the surveys when evaluating the visual quality of the content shown in the application gives 72% of the users the maximum score with reference to the quality of the visualization (see Fig. 13).

Evaluate the observed visual content, from 1 to 5, with 5 being the highest value.

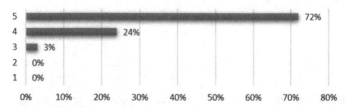

Fig. 13. Question 5, from the survey, shows the result obtained by evaluating the visual quality of the content shown in the application.

42% of the participants indicate that the presentation of the application is complete, 21% adequate, 18% chord, 17% correct (see Fig. 14), giving a higher percentage of conformity of the information that the application brings to the knowledge of the participants.

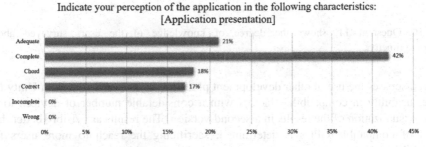

Fig. 14. Question 6 from the survey qualifies the presentation of the application.

The results also present the respondent's degree of interest in knowing the religious cultural heritage of San Pedro Cathedral in Guayaquil. Of the 100% of those surveyed, 68% are open to the possibility of learning more about the cathedral using this novel method so the application meets its objective (Fig. 15).

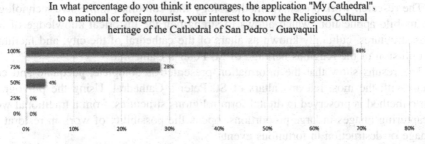

Fig. 15. Question 7 from the survey quantifies the user's interest in learning more about the religious cultural heritage of the cathedral.

78% of respondents indicate that the application "My Cathedral" is an innovative application in our environment, with this the visual and multimedia information available arouses interest in people who use it, thus the use for specific purposes such as: presentation of historical and religious structures, in digital form, preservation of digital replicas in design and form, as well as a digital record of continuous updating in case of fortuitous events or damage to the structure, it is possible by means of applied techniques.

Finally, the result of the degree of knowledge of the users surveyed about photogrammetry is presented, showing that 62% are unaware of this practice and only 38% are informed about it (see Fig. 16.)

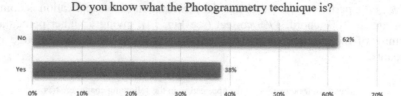

Fig. 16. Question 11, shows the degree of knowledge of the users surveyed about photogrammetry.

It discusses the use of other development platforms such as ARCore with Unity for better usability in compatibility factors with a considerable number of devices for a better visualization of the results in a second version. The results are visibly better, but a list of compatible GPUs is determined, sacrificing the reach to more users for availability and economic costs.

4   Conclusions

The use of increasing reality and the technique of photogrammetry require greater physical resources for their development. The tasks applied for traditional purposes differ in a visible result each time of better quality.

The research work presents a methodology that combines immersive technologies in a mobile application, which allows to strengthen the historical knowledge of religious structures, culturally known as altars of the cathedral of the city, and facilitates the diffusion of the religious heritage of St. Peter's Cathedral.

The results show that the information presented is complete, adequate and consistent with the most relevant altars of St. Peter's Cathedral. Using the photogrammetric method is preserved in digital form religious structures, from a traditional work in capturing images in large proportions, opens the possibility of work in restoration, damage or destruction in fortuitous events.

The increasing reality allows new strategies to promote the historical, cultural and religious heritage of the city's cathedral to national and foreign tourists. According to the level of satisfaction of the participants, this work meets the objectives set with a favorable result. It is demonstrated that the development of the application of augmented reality and the use of photogrammetric technique, improves users' knowledge of the most significant works of the heritage of the cathedral and allows digital preservation of relevant structures. The use of the mobile application "My Cathedral", with these combined technologies does not go unnoticed, opens up possibilities for innovative ways of working.

Acknowledgements. We thank the directors of the Cathedral of San Pedro de Guayaquil, Father Carlos Mena, for the availability and authorization allowed for the accomplishment of this work. To the Salesian Polytechnic University for its unconditional support in the management for the completion of this work. We gratefully acknowledge the support of NVIDIA Corporation with the donation of the Titan X Pascal GPU used for new research.

Referencias

1. Li, G.K., Gao, F., Wang, Z.G.: A photogrammetry-based system for 3D surface reconstruction of prosthetics and orthotics. In: Proceedings of the Annual International Conference of the IEEE Engineering in Medicine and Biology Society, EMBS, pp. 8459–8462 (2011). https://doi.org/10.1109/IEMBS.2011.6092087
2. Remondino, F., Remondino, F.: Heritage recording and 3D modeling with photogrammetry and 3D scanning. Remote Sens. **3**, 1104–1138 (2011). https://doi.org/10.3390/rs3061104
3. Patias, P.: Medical imaging challenges photogrammetry. Virtual Prototyp. Bio Manuf., 45–66 (2008). https://doi.org/10.1007/978-0-387-68831-2
4. Pietroszek, K., Tyson, A., Magalhaes, F.S., Barcenas, C.E.M., Wand, P.: Museum in your living room: recreating the peace corps experience in mixed reality. In: IEEE Games, Entertainment, Media Conference (GEM), pp. 1–4. IEEE (2019). https://doi.org/10.1109/GEM.2019.8811547
5. Berrier, S., Tetzlaff, M., Ludwig, M., Meyer, G.: Improved appearance rendering for photogrammetrically acquired 3D models. In: Digital Heritage, pp. 255–262. IEEE (2015)
6. Gea, G., Gea, S.: Fotogrametría de bajo costo para relevamiento de edificios de valor patrimonial (2009)
7. Warrick, J.A., Ritchie, A.C., Schmidt, K.M., Reid, M.E., Logan, J.: Characterizing the catastrophic 2017 Mud Creek landslide, California, using repeat structure-from-motion (SfM) photogrammetry. Landslides **16**, 1201–1219 (2019). https://doi.org/10.1007/s10346-019-01160-4
8. TS-TOS University: Introduction to photogrammetry, Columbus (2005)
9. Szwoch, M., Bartoszewski, D.: 3D optical reconstruction of building interiors for game development. In: Choraś, M., Choraś, Ryszard S. (eds.) IP&C 2019. AISC, vol. 1062, pp. 114–124. Springer, Cham (2020). https://doi.org/10.1007/978-3-030-31254-1_15
10. Carnevali, L., Lanfranchi, F., Russo, M.: Built information modeling for the 3D reconstruction of modern railway stations, Heritage, vol. 2, no. 3, pp. 2298–2310 (2019)
11. Voinea, G.-D., Girbacia, F., Postelnicu, C.C., Marto, A.: Exploring cultural heritage using augmented reality through Google's Project Tango and ARCore. In: Duguleană, M., Carrozzino, M., Gams, M., Tanea, I. (eds.) VRTCH 2018. CCIS, vol. 904, pp. 93–106. Springer, Cham (2019). https://doi.org/10.1007/978-3-030-05819-7_8
12. Wang, Y., Deng, X., Zhang, K., Lang, Y.: The intangible cultural heritage show mode based on AR technology in museums - take the Li nationality non-material cultural heritage as an example. In: 2018 IEEE 3rd International Conference on Image, Vision and Computing (ICIVC), pp. 936–940. IEEE (2018). https://doi.org/10.1109/ICIVC.2018.8492843
13. Llerena, J., Andina, M., Grijalva, J.: Mobile application to promote the Malecón 2000 tourism using augmented reality and geolocation. In: Proceedings - 3rd International Conference on Information Systems and Computer Science, INCISCOS 2018 (2018). https://doi.org/10.1109/INCISCOS.2018.00038
14. Bec, A., Moyle, B., Timms, K., Schaffer, V., Skavronskaya, L., Little, C.: Management of immersive heritage tourism experiencs: a conceptual model. Tourism Manag. **72**, 117–120 (2019)
15. Rahaman, H., Champion, E., Bekele, M.: From photo to 3D to mixed reality: a complete workflow for cultural heritage visualisation and experience. Elsevier (2019). https://doi.org/10.1016/j.daach.2019.e00102
16. Anderson, E.F., McLoughlin, L., Liarokapis, F., Peters, C., Petridis, P., De Freitas, S.: Developing serious games for cultural heritage: a state-of-the-art review. Virtual Reality **14**(4), 255–275 (2010)

17. Gkion, M., Patoli, M.Z., White, M.: Museum interactive experiences through a 3D reconstruction of the Church of Santa Chiara. In: En IASTED Computer Graphics and Virtual Reallity Conference, Cambridge, United Kingdom (2011)
18. Monaghan, D., O'Sullivan, J., O'Connor, N.E., Kelly, B., Kazmierczak, O., Comer, L.: Low-cost creation of a 3D interactive museum exhibition. In: Proceedings of the 19th ACM international conference on Multimedia, pp. 823–824. ACM (2011)
19. Horníčková, K., et al.: Photostruk–uniting science and humanities for the reconstruction of lost cultural heritage sites and landscape. In: 9th International Conference on Advanced Computer Information Technologies (ACIT), pp. 456–460. IEEE (2019)
20. Unal, M., Bostanci, E., Sertalp, E., Guzel, M.S., Kanwal, N.: Geo-location based augmented reality application for cultural heritage using drones. In: 2nd International Symposium on Multidisciplinary Studies and Innovative Technologies (ISMSIT), pp. 1–4. IEEE (2018)
21. Gao, G., Zhang, Y., Cheng, C., Bu, Y., Shih, P.C.: Designing to enhance student participation in campus heritage using augmented reality. In: 3rd Digital Heritage International Congress (DigitalHERITAGE) held jointly with 24th International Conference on Virtual Systems & Multimedia (VSMM 2018), pp. 1–4. IEEE (2018)
22. Lee, Y.Y., et al.: A SAR-based interactive digital exhibition of Korean cultural artifacts. In: Digital Heritage, pp. 655–658. IEEE (2015)
23. Guimarães, F., Figueiredo, M., Rodrigues, J.: Augmented reality and storytelling in heritage application in public gardens: Caloust gulbenkian foundation garden. In: Digital Heritage, pp. 317–320. IEEE (2015)
24. Chatzidimitris, T., Kavakli, E., Economou, M., Gavalas, D.: Mobile augmented reality edutainment applications for cultural institutions. In: IISA, pp. 1–4. IEEE (2013)
25. Waechter, C.A., Pustka, D., Klinker, G.J.: Vision based people tracking for ubiquitous Augmented Reality applications. In: 8th IEEE International Symposium on Mixed and Augmented Reality, pp. 221–222. IEEE (2009)
26. Caro, J.: Fotogrametría y modelado 3D: un caso práctico para la difusión del patrimonio y su promoción turística (2012)
27. Taupiac, J.-D., Rodriguez, N., Strauss, O.: Immercity: a curation content application in virtual and augmented reality. In: Chen, Jessie Y.C., Fragomeni, G. (eds.) VAMR 2018. LNCS, vol. 10910, pp. 223–234. Springer, Cham (2018). https://doi.org/10.1007/978-3-319-91584-5_18
28. Gonizzi, B., Remondino, F., Palacios, J.F., Visintini, B.: Critical factors and guidelines for 3D surveying and modelling in Cultural Heritage (2014)
29. Gómez Rios, M.D., Paredes Velasco, M.: Augmented reality as a methodology to development of learning in programming. In: Botto-Tobar, M., Pizarro, G., Zúñiga-Prieto, M., D'Armas, M., Zúñiga Sánchez, M. (eds.) CITT 2018. CCIS, vol. 895, pp. 327–340. Springer, Cham (2019). https://doi.org/10.1007/978-3-030-05532-5_24
30. International Organization for Standardization Iso: Iso/Iec 25010:2011: Software Process: Improvement and Practice, vol. 2, no. Resolution 937, pp. 1–25 (2011)

Ar-Math: Augmented Reality Technology Applied for Education

Mateo-Fernando Cordova-Proaño[1], María-Elizabeth Ortega-Camacho[2],
Alejandra González-Castro[2], Mayra-Socorro Gutiérrez-López[2],
Anahí-Montserrat Torres-Tinoco[2] (iD), and Jorge-Luis Pérez-Medina[1](✉) (iD)

[1] Intelligent and Interactive Systems Lab (SI^2 Lab),
Universidad de Las Américas (UDLA), Quito, Ecuador
{mateo.cordova.proano,jorge.perez.medina}@udla.edu.ec
[2] Universidad Tecnológica de Léon, León, GTO, Mexico
eliortegacamacho1997@gmail.com,
{16001541,16001467}@alumnos.utleon.edu.mx, antorres@utleon.edu.mx

Abstract. Currently, technology opens the doors to a new generation of methodologies in the teaching of various subjects. "Serious games" are a tool that allows students to learn while playing, which has several cognitive and social benefits. Augmented reality by its sensory element becomes a tool suitable for application in serious games for young ages. This paper presents a serious game for teaching numerical place value for children up to 6 years, we called as AR-Math.

Keywords: Teaching · Mathematics · Positional value · Serious game · Augmented reality

1 Introduction

The international evaluation PISA [8] develops the evaluation to students in mathematics, science and reading; in mathematics, both Mexico and Ecuador score below average. The educational models are focused on 4 essential competencies in mathematics: the skills and competencies to solve problems independently and autonomously; communicate information regarding mathematics; validate the procedures and results; and finally manage techniques so that they are more efficient. In this last point, there is a concern about the proper use of TICS in the classroom to improve the procedures to solve problems, always promoting autonomous and collaborative study. One of the vulnerable areas is the issue of place value of the number, so this project aims to define an application that favors autonomous learning, so that they use the decimal numbering system to express quantities, to expand and deepen knowledge, way that the understanding and efficient use of mathematical tools is favored.

Likewise, the document "Programas de estudio 2011 guía para el maestro" [18] of the Secretaría de Educación Pública in Mexico, has the purpose of improving the study of mathematics at the elementary school by helping children

© Springer Nature Switzerland AG 2020
M. Botto-Tobar et al. (Eds.): ICAT 2019, CCIS 1194, pp. 607–621, 2020.
https://doi.org/10.1007/978-3-030-42520-3_48

develop ways of thinking, which are most interested in common activities of life or starter's numerical or geometrical problems that are presented in classroom. It also intends to use different resources and techniques to improve procedures to solve problems, always encouraging on the one hand autonomous study but also collaborative work.

With the learning of mathematics, one of the objectives mentions that elementary school children should know how to use the decimal numbering system to express quantities, among many others in order to broaden and deepen knowledge, so as to increase comprehension and the efficient use of mathematical tools. Within the teaching of mathematics, project-based learning stages or problem solving are those that show a more noticeable benefit than others because they apply the use of a relevant tool, as well as the processes that students follow to build knowledge and overcome difficulties that show up in the learning process.

Currently, the learning process seeks improvements in the reasoning, so knowing formulas, algorithms and definitions are important only to use them as tools for solving problems rather than simply memorization. This is the reason why the learning processes in mathematics now takes more time. In this same line it is important that the child is the one who builds his own learning through the resolution of the problem situations by himself, being the teacher only an observer of the process and make questions to reinforce acquired concepts. These situations in the classroom are favorable because students feel more confident to express their ideas freely, reflecting on their acquired knowledge.

The school system in Mexico mentions 4 essential skills in mathematics. These 4 skills are aligned with the 4 competencies of the educational model of mathematics contained in the international evaluation of PISA. This are the same skills that are considered in the technological development of this project [2].

There are educational models focused on the student, in which the only responsible for the learning process is the student. The student creates and thinks for himself so he acquire confidence and discipline, this later helps the student to build other kinds of learning; this is case of the Montessori method. In this model, the teacher, as already mentioned, is only a guide for learning, encourages the student and applies tools in which the child learn in funny ways, leaving aside bad effect and complex processes. In the context of educational innovations, there are efforts to support learning processes in the classroom through the use of "serious games". Currently, technological tools allow software products to be conceived not as interlocutor, but as an instrument of action in a space in which conversations with real or virtual objects take place.

Recent research have prove that digital games can used as learning tools that motivate and generate curiosity resulting in an effective way to optimize learning and student performance [9]. Learning and student engagement due to the use of digital games in daily routines has been confirmed by several independent studies over the years, integrating games in different ages for teaching different assignments.

The present proposal focuses on developing a serious game based on technologies such as Augmented Reality (AR) [7] and Intelligent Algorithms (AI) will helps the learning of the topic of positional value for elementary school children. Thus, the idea of the serious game called Ar-Math is born. Ar-Math stands for augmented reality math. We intended to propose a game where young age kids would learn about positional value using augmented reality as it main component. The task kids are presented is to place tangible targets with numbers printed on them in the correct order so they form a random number. As they see the numbers in thought a camera, they would see animations with each character. The remainder of the paper is organized as follows: Next section presents an analysis of related work. Next, Sect. 3 describes a set of functional and non functional requirements identified for Ar-Math. This section also provides a description of the Ar-Math architecture. Section 4 presents the Ar-Math tool. Section 5 discusses a set of benefits and shortcomings. The conclusions and future work are presented in Sect. 6.

2   Related Work

Serious Games [1,6,16] are applications in which the user enters competition with the computer following a series of specific rules, which uses entertainment to promote learning in the user [19]. This kinds of applications combines video game elements, pedagogy and aspects of human computer interaction engineering.

Even if "serious games" themselves, could be considered as an oxymoron because "Games are inherently fun and not serious" [4]. In the context of serious games the enjoyable component of this works are the most important one [13].

The difference between serious game and video game, is that in a serious game, it is important to measure the learning acquired throughout it. A serious game consider a physical or mental challenge. In order to evaluate the mental challenge, it is necessary to incorporate a set of training and fun rules to reach the pre-established training objectives.

Studies on Serious Games show positive influence, especially on Primary school and Preschool students due to it multi sensory nature [10]. Recent research showed that with Serious Games students are more likely to be engaged to class, scaffolding their learning. Students also showed increased autonomy, independence, motivation and resultant self-esteem. Likewise, the authors have proven that Serious Games Based Learning add value in the geometric thinking learning which includes recognition, visual association, description/analysis, and abstraction/relation) [10]. How technology in education is used have changes considerably over time. In order to enhance education, new paradigms have appeared in classrooms [14,15].

There are endless applications of Augmented Reality. For example, Augmented Reality have been found useful in areas such as Human Computer Interaction (HCI), Computer graphics, vision, architecture, design, collaborative learning, production planning, engineering and gaming [3]. AR is not just a support that potentiate disciplinary learning, user develops skills like information sharing, collaboration and critical thinking [3].

Currently, the Augmented Reality (AR) technology makes it possible to create pedagogical content and digital textbooks. The applications can be used in tablets or smartphones and can be used by teachers and students in the framework of the joint construction of learning objectives. The augmented reality applied in software for education lies in applications to learn history, mathematics, chemistry, biology, anatomy, space, translations and gymnastics. This confirms that the most motivated and committed students always understand the subject better and learn faster using this technology [5,17].

3   The Proposed Solution

In order to specify and design the serious game in a flexible and effective way, we use a user-centered agile development approach (UCD) [12]. To achieve and comply with the use of this agile practice, we must consider compliance with the usability objectives and also ensure that the user experience is satisfactory while adhering to the agile principles of software development [11]. We also, adopt the life cycle of agile development to create the user interfaces, taking as a fundamental aspect the requirements of interaction, the profile of the user and their user context. This agile life cycle will allow us to obtain a usable and accessible interface. Its importance is that it is designed for development environments in which multidisciplinary teams with expertise in user experience (UX) and agile development, want to integrate user-centered design practices (UCD) and agile software development methods (ASD). UCD's approach [12] includes analysis, design and implementation where different concepts, techniques and evaluations are used in the field of human computer interaction (HCI) to guarantee efficiency, effectiveness and quality in the execution of learning tasks by users.

Fig. 1. Sketches made during requirements discovery.

3.1 The Functionalities of Ar-Math

We argue that the Ar-Math software would have the functionalities presented in Tables 1 and 2. Once we started applying the user-centered approach, we were able to discover the requirements of the application while we made some representations of the user interfaces that had developed developments. Figure 1 shows several sketches of the interfaces made. Panel 1 corresponds to the functional requirement "Start module". Panel 2 was used to describe the "Augmented Reality module". Panels 3, 4 and 5 where used to identifies the "Games" and "Exchanges modules".

Table 1. Functional requirements of Ar-Math.

Nro.	Name	Description
1	Start module	The interface has a starting section for the user to choose three sections to perform: Numbers with Augmented Reality, Games or the Exchanges section
2	Augmented Reality module	The application will have a module where the children will access to the AR camera and visualize 3D models of numbers from 0 to 9. When a 3D number is shown in the AR camera the children will hear the name of the number too. The application recognizes all the numbers drawn in the cubes an show them with augmented reality
3	Games module	The application will have a module where the children can play 3D and 2D games that'll help them to learn more about counting numbers and units, tens and hundreds. The child can select different types of exercises
4	Exchanges module	To verify that the participant already dominates the issue of place value, the application contains this section. The application will have a module where the children can interact with an abacus. It is very important that in this section units are exchanged by tens and hundreds by any game
5	Emotion recognition module	The application must use techniques of Computer Vision and Artificial Intelligence to adapt the level of the game based on the emotiotional state of the user

Table 2. No functional requirements of Ar-Math.

Nro.	Description
1	The application needs to be installed in a touch device, such as a smartphone or a tablet with a camera in both sides of the device with at least 3 Megapixels
2	The application will have a design based on cartoon monsters and vibrant colors to bring a fun design for the children
3	The application will visualize the user expressions when they're doing an exercise in the app and depending on the expression (stress, confusion, happiness) the difficulty of the exercises will change
4	The application needs enough storage for it to work correctly in the device

3.2   Possible Usage Scenarios

Figure 2 describes the 4 possible usage scenarios. The first two correspond to the usage of reduce screens such as Smartphones and Tablets. The third configuration contemplates the use of an interactive table. Finally, the fourth configuration foresees a use of a wall screen. These scenarios are described below.

Configuration 1: Using a Smartphone/Tablet. The application when its used on a cell phone allows individualized learning, that is, the user can review the topics and be seeing examples at a low level of complexity because he is only learning.

Fig. 2. Usage scenarios of Ar-Math.

Configuration 2: Using an Interactive Tabletop. In this case the learning that is desired to achieve is more advanced since a didactic material will be used that allows to live the learning, with the interactive tables, it is proposed that the exchange section can have a more sophisticated interaction since it emits sounds if they were resolved right or wrong.

Configuration 3: Using a Wall Screen. This modality of the application is designed in a collaborative learning, that is, the entire group of users can be doing the exercises and the teacher is the one who directs the activity as well as choosing the complexity of the exercises.

3.3   The Architecture

AR-Math is an interactive application based on a video game engine such as Unity version 2019.2.8, in which the user interfaces are shown in next section. Figure 3 presents the architecture of Ar-Math. The user connects to the lessons to learn the numbers and this he does with the Augmented Reality technique either using a cell phone an AR viewfinder or lenses.

AR-Math contains a series of exercises through games to learn the positional value of the number and this is done with programming in c(sharp). The exercises are connected with an artificial intelligence module with which through the recognition of images, it allows the participant's emotions to be identified in order to increase or decrease the complexity of the games.

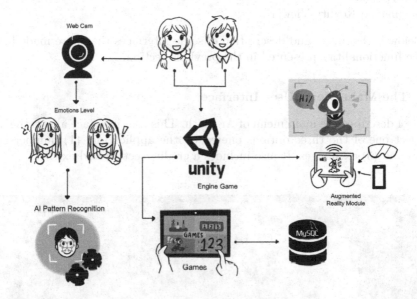

Fig. 3. The architecture of Ar-Math.

AR-Math architecture supports tool development with a development environment, such as Unity[1] or ThreeJS[2]. It also must support software for 3D modeling, and software that allows the development of Augmented Reality such as Vuforia[3] or Zappar[4] markers. For the capture and inference of emotions some computer vision tools defined in python will be used.

---

[1] https://unity.com/.
[2] https://threejs.org/.
[3] https://developer.vuforia.com/.
[4] https://www.zappar.com/.

4 The Tool

The Ar-Math prototype was developed considering software and hardware.

Software:

- Unity version 2019.2.8
- C sharp Language
- Artificial Intelligence Algorithms for Image Recognition in Python

Hardware:

- Webcam
- Smartphones
- Augmented Reality Viewer

Below is the image and description of some interfaces that were made based on the functionalities presented in the previous section.

4.1 The Main Menu User Interface

Figure 4 describes the main menu of Ar-Math. This user interface allows the user to select one of the three options offered by the application: (1) numbers; (2) games and (3) exchanges. It also includes a credits section and exit.

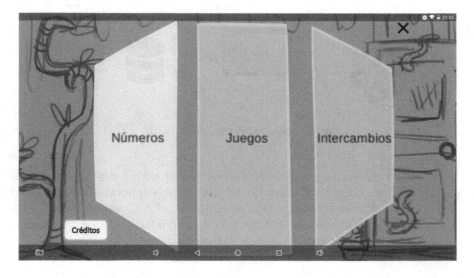

Fig. 4. The main menu of Ar-Math.

Fig. 5. The number user interface of Ar-Math.

4.2 The User Interface for Displaying Numbers Using Augmented Reality

Figure 5 shows the user interface for displaying numbers using Augmented Reality. It allows the user to interact with the augmented reality module using the viewfinder, lenses or smartphone. The user must see a 3-dimensional animation of the selected number. Note that each number represents a character which has the number of eyes associated with the number it represents.

Figure 6 shows an example of how the 3D models were made on Blender[5], because of compatibility with Unity Framework. We can see that the model needs to be as similar as the marker as possible, as shown in the example in Fig. 7. In Unity, each number is link with the corresponding model.

4.3 The User Interface to Play

Figure 8 describes the user interface to play the first game. It allows to the user to choose several games, among them that a character collects coins passed from units, tens to hundreds. This image shows one of the games in the application in which one of the monsters collects coins to move from units to tens and finally hundreds. It is a collection game. The coins will be scattered across the stage. The user must collect the most coins.

Thought the use of animation and augmented reality, numbers could seems "happy" if the user is doing the activity correctly and "sad" if they are wrong. So, kids will be motivated through the game into developing mathematical skills. In order to experience Augmented Reality user just have to place the camera in front of any marker as shown in Fig. 6.

---

[5] https://www.blender.org/.

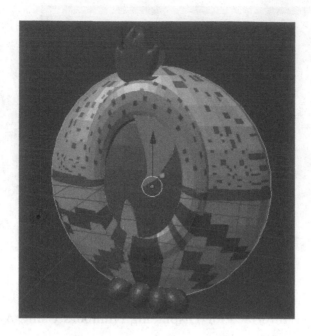

Fig. 6. The 3D Zero model.

Fig. 7. Augmented reality experience.

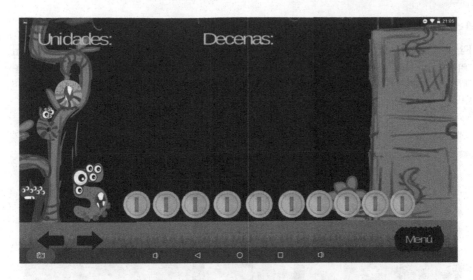

Fig. 8. The game user interface of Ar-Math.

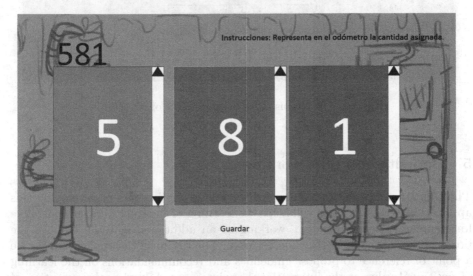

Fig. 9. The odometer game user interface of Ar-Math.

Another game is the odometer in which a quantity is shown to the user. It must write the number that defines the position using the three available options (See Fig. 9). Both games incorporate an incentive sound for each correct answer as well as for each incorrect answer.

4.4   The User Interface for Exchanges

Figure 10 shows the user interface for exchanges. In this interface the user can represent the quantities that are presented on the screen. That is, the user represents quantities in the abacus making exchanges with the tens and hundreds units.

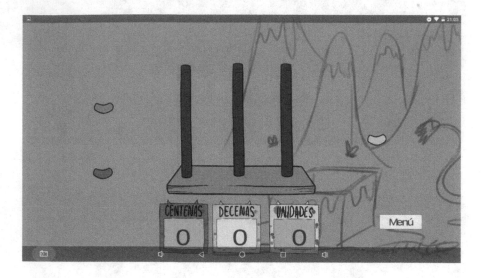

Fig. 10. The abacus game user interface of Ar-Math.

5   Benefits and Shortcomings

The educational model in Mexico, will be one of the benefits of developing this application, a serious game will helps the learning of the topic of positional value for elementary school children, working as an additional material to have the books allow autonomy in the child for learn. The results of the application can be a point of reference to propose guidelines and recommendations on the learning of positional value at the international environment. Likewise, the execution of the project can derive future investigations in other aspects of the learning of mathematics in elementary education levels.

Have a tool that serves as didactic material for learning the topic of place value; demonstrate the use of artificial intelligence in educational applications; provide a comfortable learning space for the user without feeling pressured.

The implementation of Ar-Math allow the generation of employment because it will require the involvement of multidisciplinary teams, among which we can mention educators, designers, multimedia specialists, psychologists, programmers and researchers. Likewise, it is expected to strengthen collaboration ties in different academic and industrial institutions that wish to deep in this topic.

Finally, all the knowledge generated during the execution of this project can be included in new training courses for students in the institutions owners of the project.

The technological tools to be used and the Artificial Intelligence strategies to be implemented, are computational strategies that are acquiring a lot of boom so they become an opportunity to grow the functionalities of the project.

Among the most important techniques of artificial intelligence, is Computer Vision, which is a technique focused on computers can extract information from images, offering solutions to real-world problems. Companies use it since it can help increase production and reduce manufacturing costs. Likewise, the computer vision allows to inspect the production process without fatigue or distractions that are trigger factors of human errors.

Computer vision, serious games and mathematics specifically speaking the positional value of numbers, are issues that have not been implemented for the development of applications. By doing a review on the Internet, there are several alternatives for managing the learning of mathematics, but not a serious game of augmented reality that allows the construction of learning by the user in this particular topic. That is why they will be applied in the proof of concept of this project.

5.1   An Innovative Tool

Innovation Elements of the Project with Respect to the Existing One. There are currently a wide variety of industrial and research applications aimed at facilitating and also improving mathematics learning, some are not based on any educational model, which is applied in this project. The technique of computer vision applied in a serious game that includes Augmented Reality, has not been applied very frequently. One of the most important features for this serious game is the capture and inference of emotions in order to generate a more optimal learning space and confidence to users.

Level of Impact of the Innovation (Economic, Social, Environmental, Scientific, etc. The impact generated by this project has a social impact, since the serious game with augmented reality and inference and inference of emotions, provides support to the second grade teacher to generate a better learning environment, in turn contributes to have children with less frustration when learning about the subject since they have a system that generates a comfortable space for learning and that understands it according to their own scope. It will also have a technological impact since in a serious game various technologies will be implemented, including the Artificial Intelligence itself, which is gaining popularity in ordinary applications today.

5.2   Limitations

Ar-Math, is part of the multimedia area and aims to test concepts using Augmented Reality technology to learn the subject "positional value". Having success in the proof of concept results the subject can be extended to any other. The restrictions will be mainly on the use of the tools, since it will be made use of trial versions of them for the Augmented Reality. However, if the solution obtained is to be marketed in the future, it will be necessary to acquire the licenses such as Vuforia or the Zappar markers.

6   Conclusion and Future Work

In this article we have detailed Ar-Math a serious game developed on the basis of technological principles such as Augmented Reality and Intelligent Algorithms for learning of the topic of positional value for elementary school children. The tool is an innovative mechanism focused on the education of children, in which they seek to understand the operation of the units, tens and hundreds. Ar-Math was conceived from a list of functional and non functional requirements.

As we have seen in the manuscript, Ar-Math allows young children to learn about place value using augmented reality markers as the main component. The markers must be placed in the correct position so that they can form the required number.

We consider conducting usability experiments in real conditions that involve children and teachers. The results will be used to look for new perspectives and evaluate the deficiencies and the performance of the tool. From them we could define new research. Finally, we plan to explore other possibilities of supporting learning activities of positional value. We plan to incorporate the use of tangible objects, such as: cubes to the existing application. This will allow to support other types of interactions where we can study in the participants aspects such as: participation, collaboration, leadership, respect for time and especially the learning of positional value.

References

1. Breuer, J., Bente, G.: Why so serious? On the relation of serious games and learning. J. Comput. Game Culture **4**, 7–24 (2010)
2. Gómez Palacios, M., et al.: Propuesta de aprendizaje de las matemáticas en grupos integrados (2019). Recuperado de: https://bit.ly/2lLa98O
3. Klopfer, E., Perry, J., Squire, K., Jan, M.F.: Collaborative learning through augmented reality role playing. In: Proceedings of th 2005 Conference on Computer Support for Collaborative Learning: Learning 2005: The Next 10 Years! pp. 311–315. International Society of the Learning Sciences (2005)
4. Kuipers, D.A., Wartena, B.O., Dijkstra, A., Prins, J.T., Pierie, J.-P.E.N.: Design for transfer. In: Ma, M., Oliveira, M.F., Petersen, S., Hauge, J.B. (eds.) SGDA 2013. LNCS, vol. 8101, pp. 239–246. Springer, Heidelberg (2013). https://doi.org/10.1007/978-3-642-40790-1_23
5. Mazlan, U.H., Ibrahim, I., Ghazali, N., Zulkifli, Z., Ahmad, S., Ismail, W.S.W.: Applying augmented reality in teaching and learning. In: Computing Research & Innovation (CRINN), October 2017, vol. 2, p. 315 (2017)

6. Michael, D.R., Chen, S.L.: Serious Games: Games that Educate, Train, and Inform. Muska & Lipman/Premier-Trade (2005)
7. Milgram, P., Takemura, H., Utsumi, A., Kishino, F.: Augmented reality: a class of displays on the reality-virtuality continuum. In: Telemanipulator and Telepresence Technologies, vol. 2351, pp. 282–292. International Society for Optics and Photonics (1995)
8. País, F.E.: Resultados de la prueba pisa de la ocde. Este país. Tendencias y opiniones **237**, 61–64 (2011)
9. Papadakis, S.: The use of computer games in classroom environment. Int. J. Teach. Case Stud. **9**(1), 1–25 (2018)
10. Papanastasiou, G.P., Drigas, A.S., Skianis, C.: Serious games in preschool and primary education: benefits and impacts on curriculum course syllabus. Int. J. Emerg. Technol. Learn. **12**(1), 44–56 (2017)
11. Pérez-Medina, J.L., et al.: Usability study of a web-based platform for home motor rehabilitation. IEEE Access **7**, 7932–7947 (2019)
12. Pérez-Medina, J.L., Vanderdonckt, J.: Sketching by cross-surface collaboration. In: Rocha, Á., Ferrás, C., Paredes, M. (eds.) ICITS 2019. AISC, vol. 918, pp. 386–397. Springer, Cham (2019). https://doi.org/10.1007/978-3-030-11890-7_38
13. Ritterfeld, U., Cody, M., Vorderer, P.: Serious Games: Mechanisms and Effects. Routledge, Abingdon (2009)
14. Rodríguez Vizzuett, L.: Proceso de desarrollo de aplicaciones interactivas para el aprendizaje colaborativo a nivel preescolar (2015)
15. Rodríguez-Vizzuett, L., Pérez-Medina, J.L., Muñoz-Arteaga, J., Guerrero-García, J., Álvarez-Rodríguez, F.J.: Towards the definition of a framework for the management of interactive collaborative learning applications for preschoolers. In: Proceedings of the XVI International Conference on Human Computer Interaction, p. 11. ACM (2015)
16. Susi, T., Johannesson, M., Backlund, P.: Serious games: an overview (2007)
17. ThinkMobiles: Usage of augmented reality in education, stay updated in VR/AR and mobile app development (2019). Recuperado de: https://thinkmobiles.com/
18. Vargas García, M., Blanco Lerín, A.: Programas de estudio 2011 guía para el maestro (2011)
19. Zyda, M.: From visual simulation to virtual reality to games. Computer **38**(9), 25–32 (2005)

Mechatronic Prosthesis for Transfemoral Amputation with Intelligent Control Based on Neural Networks

Benalcázar Alexander[1], Comina Mayra[2], Danni De la Cruz[1(✉)], and Tobar Johanna[2]

[1] Department of Electrical and Electronics,
Universidad de las Fuerzas Armadas – ESPE, Sangolquí, Ecuador
drde@espe.edu.ec
[2] Department of Energy Sciences and Mechanics,
Universidad de las Fuerzas Armadas – ESPE, Sangolquí, Ecuador
jbtobar@espe.edu.ec

Abstract. This article presents the design of a low-cost prototype mechatronic prosthesis for a leg with transfemoral amputation. The development for construction of this mechanical prosthesis is detailed in "Low Cost Mechatronics Prototype Prosthesis for Transfermoral Amputation Controlled by Myoelectric Signals" [1]. This prototype was designed and implemented with a mechanical brake of the knee and foot to help people walk. This article expands the development of this prosthesis with the implementation of the electronic part. The prosthesis is activated by signals obtained by inertial sensors (IMU), which capture the angles generated when walking (in thigh, knee and foot) allowing to reproduce the human gait cycle through a controller, based on neural networks, which permits to replicate the movement by the activation of servomotors. In addition, the prosthesis has a constant monitoring system of physiological parameters such as temperature and humidity inside the stump, which can be visualized through an application for smart devices to protect and alert about the patient's wellness.

Keywords: Mechatronics · Prostheses · Gait cycle · Neural networks

1 Introduction

In Ecuador, there were 418001 people who suffered some kind of physical disability in 2017, the 47.07% of these are 196758 people who suffered physical disability [2], in this sector many people with different types of amputations that needed a prosthesis to do their usual activities autonomously were found. Given these crude statistics, Ecuador has not only created different support and inclusion programs for people with disabilities, but has also become a Latin American leader through the "Manuela Espejo" program which started in October 2012; however, the high cost to acquire prostheses has made it difficult for people to access them in a massive way, this problem was the reason to make low-cost prostheses has become a necessity for this social group.

© Springer Nature Switzerland AG 2020
M. Botto-Tobar et al. (Eds.): ICAT 2019, CCIS 1194, pp. 622–636, 2020.
https://doi.org/10.1007/978-3-030-42520-3_49

According to Pablo Parra, a prosthesis is a mechanism that can replace or improve a missing part of the human being [3], which combines anatomical, mechanical and biological knowledge to cover the needs of the patient in a way that improves their quality of life. Currently, thanks to the globalized development of society, prostheses are a sign of the technological breakthrough, proof of this are the Hug Herr bionic prostheses from the Biomechatronic group of the Massachusetts Institute of Technology (MIT) which are made of silicone, titanium, aluminum and carbon. They have their own batteries and are programmed with information downloaded from a computer [4].

For Fermín Martínez from the National Center for Research and Technological Development of Cuernavaca, intelligent control is one of the techniques that has been widely accepted to establish control strategies for prostheses based on standard walking patterns [5]. Above all, neural networks have been used in the control of prostheses to plan the march, whose general purpose is to carry out daily activities in a natural way without affecting the nearby joints, allowing a stable and symmetrical walk.

Finally, for Tad McGeer of Simon Fraser University, one of the methods for training the neural network is the use of the backpropagation algorithm, which is treated as a processor whose output varies with the input [6]. Therefore, the feedback controller responds to an error in tracking by modifying the initial conditions for the sub-sequent steps and so on to eliminate the error. These values that are generated in some way are used to obtain a walking pattern.

2 Materials and Methods

2.1 Design Parameters

In the article titled "Low-cost Mechatronics Prototype Prosthesis for transfermoral amputation controlled by myolectric signals" prepared by Bravo, Comina, Tobar, De la Cruz, Loza and Corella [1]; the construction of the activated prosthesis is developed through myoelectric signals. This article expands its development by performing a gait cycle analysis to obtain parameters that will improve the design and that will serve for the training of the neural network.

To obtain marching patterns, it was necessary to develop a prototype of acquisition and measurement of gait patterns focused on an Ecuadorian biotype for which some interesting data has been collected; as the average height of the male population in Ecuador, which fixed the stature at 1.67 m [7], and the male population for the Metropolitan District of Quito, which determined that the population between 20 and 35 years is approximately 215 420 people [8]. It is worth mentioning that the tests were performed on fifty people without physical or congenital problems related to the gait cycle, who voluntarily participated in this study. The characteristics of the test subjects are detailed in the Table 1:

Table 1. Characteristics of the test subject

Sex	Male
Age range	20–35 (years)
Height range	1.60–1.75 (m)
Weight	Up to 75 (kg)

2.2 Analysis of Gait Cycle

The analysis of the patient's gait cycle was performed to determine the parameters such as: time slow gait, number steps slow gait, step length slow gait, speed slow gait, acceleration slow gait, maximum knee angle, and minimum knee angle. To obtain these values, markers were placed in several points of the patient's leg (Fig. 1), and through an analysis of the photograms during the gait cycle, the values shown in Table 2 were obtained.

Fig. 1. Reference points for gait cycle analysis.

Table 2. Gait cycle variables

Variables	Results
Time slow gait	3 s
Number steps slow gait	5 steps
Step length slow gait	510 mm
Speed slow gait	1.67 m/s
Acceleration slow gait	0.55 m/s^2
Maximum knee angle	65°
Weight	Up to 75 (kg)

2.3 Mechanical Design

In the mechanical design of the prosthesis, the two motors that were placed in the knee and ankle were considered; For proper operation, engine sizing was performed based on the force analysis of Fig. 2.

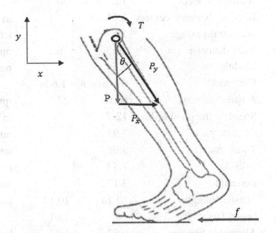

Fig. 2. Load in the human leg

Obtaining, as a result of the analysis, Eq. 1, which considers the patient's mass (m), the acceleration of the slow gait (a), the acceleration of gravity (g), the radius of the knee (R), the angle of the knee (θ), the safety factor (f_s), and the efficiency of the motors (e).

$$T_{fm} = f_s \left(\frac{100}{e}\right) \{m * R * [-a + g * sin(\theta)]\} \tag{1}$$

The necessary torque of the knee motor is 1.18 Nm and the ankle of 0.45 Nm. The design of shafts and gears based on the schemes of Fig. 3 was performed. The material used for the shafts are 303 stainless steel and for the 1060 Aluminum gears.

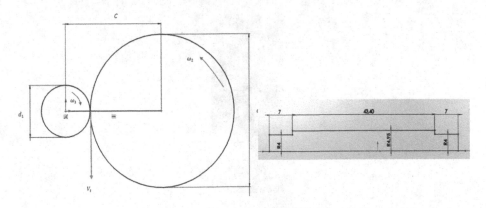

Fig. 3. Axle and gear design

The gear design parameters are shown in Table 3.

Table 3. Gear design variables

Variable	Designation	Gear 1	Gear 2	Unit
N	Number of teeth	12	30	
C	Distance between centers	0.11		m
P_d	Diametral passage	10		teeth/in
d	Pitch diameter	30.5	76.2	mm
m	Module	2.54		mm
F	Face width	0.8 in $< F <$ 1.6 in		
ω	Angular velocity	66.67	27	rpm
V_t	Speed in the passing line	12.7		m/s
p	Circular pitch	7.97		mm
t	Tooth thickness	3.98		mm
a	Adendo	2.54		mm
b	Dedendo	3.175		mm
S_u	Tensile strength	10.14	10.14	ksi
K_v	Speed factor	1	1.01	
σ_{perm}	Allowable bending	0.32	1.33	ksi
W^t	Transmitted load	10.48	8.98	lbf
hp	Power	4.95	4.30	watts
σ_c	Sting resistance	20351.1	11914.5	psi
Z_N	Life factor	$6.91 * 10^7$	$2.79 * 10^7$	rev
S_c	Permissible contact effort	33918.5	19857.5	psi
n_c	Safety factor for wear	1.99	3.39	
n_d	Safety factor for flexion	1.58	2.92	

In the axes a reversible bending effort of 41.76 MPa and a fatigue breakage of $5.21 \cdot 10^{11}$ cycles was obtained. The CAD design of the mechanical parts that make up the prosthesis shown in Fig. 4 was made.

Fig. 4. CAD design of the mechatronic prosthesis

Through the CAE analysis, Von Misses efforts, deformations, unit displacements and safety factors of each part of the prosthesis were obtained as shown in Fig. 5.

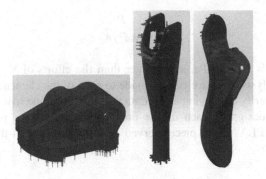

Fig. 5. CAE analysis

The safety factor against fatigue based on finite life was calculated by applying Modified Goodman's equation.

$$n_f = \frac{1}{\frac{\sigma_a}{S_e} + \frac{\sigma_m}{S_{ut}}} \tag{2}$$

These values were used in the mechanical design phase for static load and fatigue analysis, obtaining the safety factors specified in the Table 4.

Table 4. Gear design variables

Parts	Tension (GPa)	Displacement (mm)	Unitary strain	Minimum safety factor
Socket coupling joint	14.5	$1.097 \times 10{-}2$	$1.635 \times 10{-}4$	1.9
Knee coupling joint	16.2	$1.445 \times 10{-}3$	$1.741 \times 10{-}4$	1.7
Knee	2.55	$1.906 \times 10{-}3$	$2.823 \times 10{-}5$	1.7
Steam	11.3	$6.973 \times 10{-}4$	$1.133 \times 10{-}4$	2.5
Ankle	9.58	$1.058 \times 10{-}2$	$9.042 \times 10{-}5$	2.9
Foot	19.9	$1.066 \times 10{-}2$	$2.085 \times 10{-}4$	1.4
Knee axle	10.6	$6.356 \times 10{-}3$	$3.832 \times 10{-}4$	1.62
Foot axle	0.37	$1.083 \times 10{-}5$	$1.44 \times 10{-}6$	1.62

The parts of the prosthesis were made of cast aluminum, a tensile and compression test were performed to determine the resistance of the material to an applied force.

In the tensile test, the specimen bore 120 kg and in the compression test 9000 kg. It was concluded that the 120 kg that support the aluminum test tube is the critical value, the value that was used to calculate the ultimate effort from the Eq. (3).

$$\sigma_u = \frac{P}{\pi d_o^2 / 4} \tag{3}$$

The last effort is 36.55 MPa, it is greater than the efforts of Von Mises that were obtained in the analysis of finite elements made to each of the elements of the prosthesis which allowed to verify employing a real data that the prototype will support the efforts for which it was designed. Each of the prosthesis pieces were printed in 3D, the material used was PLA. These pieces served as a mold to obtain the final aluminum prototype (Fig. 6).

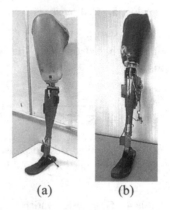

(a)   (b)

Fig. 6. (a) 3D printed prototype, (b) Cast aluminum prototype

2.4   GEPAM System

Sensor Placement and Creating the Database. To determine the location of the inertial sensors, Lee, Ju, Song and Park recommended that the sensors must be aligned to the body segments of interest [9], for which a wired circuit was designed to fix the sensors and have the same point of reference for all subjects, it also facilitates the alignment of the body segments of the leg, thigh and hip respectively. The location of inertial sensors (see Fig. 7) were located taking as reference the Z-axis to determine the flexion angle of the joints, and the Y-axis to determine the roll angle.

For the creation of the database, tests were carried out individually, under conditions described in a protocol where each sample obtained from a test subject is stored inside a register for 25 s where the angles generated by the inertial sensors placed in the thigh, knee and foot when walking.

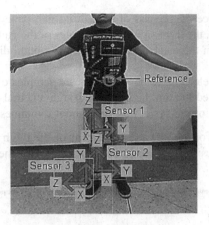

Fig. 7. Location of inertial sensors

For the experimentation, 50 tests were carried out on people without physical or congenital problems related to the gait cycle, who voluntarily participated in this study. During the test an average of 20 steps were performed where 50 samples were obtained, generating approximately a total of 15,000 gait patterns. In the example of gait patterns obtained (see Fig. 8), the generated gait patterns follow a similar behavior for each step and this varies from subject to subject, it is unique and depends on several physiological factors of each person.

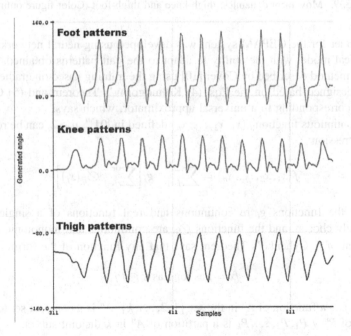

Fig. 8. Example of walking patterns obtained

Controller Design. Once the process of measuring gait patterns has been completed, before neuronal training, the most significant samples of the gait patterns obtained from each individual stored in the database are selected to obtain distinctive characteristics for each one of them.

Subsequently, after selecting the most significant samples of the study, we proceed to purify the data obtained to train the system through a neural network, using filtering techniques and division of the system by movement dynamics (see Fig. 9):

- Upward dynamics: For the upward movement of the thigh (orange circles).
- Descending dynamics: For the downward movement of the thigh (blue circles).

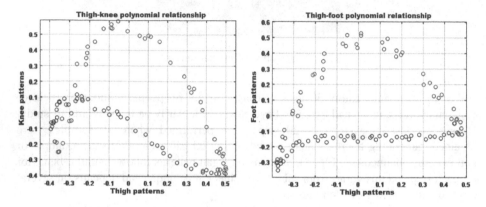

Fig. 9. Movement dynamics: thigh-knee and thigh-foot (Color figure online)

The controller for the GEPAM system was developed using neural networks to find a mathematical model with the ability to adapt to the gait patterns obtained, the cross-validation method described by Cover [10] using the training based on gradient descent has been designed based on the Eqs. (4) Kolmogorov's Theorem and (5) Cybenko's Theorem, corresponding to a universal approximator, which says:

Any continuous function $f(x_1, x_2, \ldots, x_n)$ defined in $[01]^n$, $n \geq 2$, can be represented by the expression:

$$f(x_1, x_2, \ldots, x_n) = \sum_{i=1}^{2n+1} g_i \left[\sum_{j=1}^{n} \varnothing_{ij}(x_j) \right] \tag{4}$$

Where the functions g_i re continuous and real functions of a single variable, appropriately chosen, and the functions $\varnothing_{ij}$ are continuous and monotonic increasing independent of f. Cybenko's Theorem says that any function of the form:

$$fx = j \text{ as long as } x \in Pj \tag{5}$$

Where f is a function of A^n in the set $\{1, 2, \ldots, k\}$, A^n is a compact set (closed and bounded) of R^n y $P_1, P_2, \ldots, P_k$ is a partition of A^n in k disjoint subsets.

Equations (4) and (5) establish that a multilayer perceptron with a single hidden intermediate layer (see Fig. 10.) can uniformly approximate any multivariate function with the desired degree of precision.

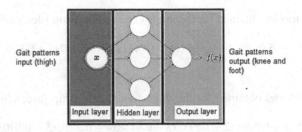

Fig. 10. Neural network architecture

For this study a comparison was made between the four types of continuous functions (polynomial, exponential, trigonometric and logarithmic) to determine the best possible model to train the neural network, the selection criteria applied will be the minimum error.

To perform the network training, a polynomial function has been used, since after some tests it was determined that it is the most suitable model, as shown in the result of the pattern training: Thigh-knee and Thigh-foot (see Fig. 11), the training of the network is adjusted in a good way to the trend guided by the walking patterns and also has the training error and validation lower compared to other models, both in the thigh-leg and thigh-foot relationship.

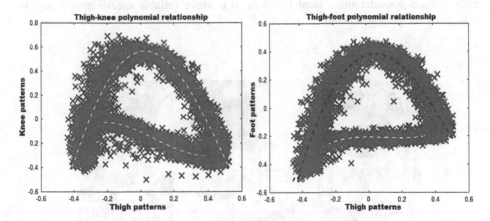

Fig. 11. Result of pattern training: thigh-knee and thigh-foot

Polynomial model obtained for the thigh-knee relationship (ascending dynamics):

$$-0.1072 - 0.6318x - 0.6682x2 + 2.7520x3 - 0.7690x4 + 0.5196x5 - 0.1589x6 + 0.0773x7 \tag{6}$$

Polynomial model obtained for the thigh-knee relationship (descending dynamics):

$$0.5615 + 0.2788x - 5.0298x2 + 0.6878x3 + 1.7431x4 + 0.2907x5 + 0.8702x6 + 0.2096x7 \tag{7}$$

Polynomial model obtained for the thigh-foot relationship (ascending dynamics):

$$-0.1161 + 0.0160x - 0.5091x2 + 1.5935x3 - 0.4158x4 + 0.2410x5 - 0.0703x6 + 0.0230x7 \tag{8}$$

Polynomial model obtained for the thigh-foot relationship (descending dynamics):

$$0.4660 + 0.4324x - 3.8903x2 + 0.8937x3 + 1.0291x4 + 0.5267x5 + 0.5027x6 + 0.2064x7 \tag{9}$$

3 Results

3.1 Functional Test of the Prosthesis in a Patient

To evaluate the response of the GEPAM system before carrying out the functional tests, certain considerations must be taken to achieve reliable measurements and to safeguard the patient's health:

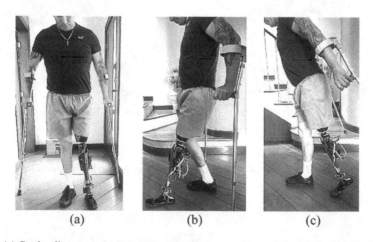

(a)   (b)   (c)

Fig. 12. (a) Static alignment (b) Dynamic alignment: extension (c) Dynamic alignment: bending

- A static alignment test is performed that consists of verifying the height, rotation and inclination of the prosthesis so that it can achieve a stable walk with full support on a flat surface (see Fig. 12(a)).
- A dynamic alignment test is performed, where the patient begins to take their first steps and verify if the length of the step and the support of the prosthesis is adequate. (see Fig. 12(b)&(c)).
- The running cycle tests were performed on a flat surface, at normal speed and with the support of canes to safeguard the safety of the patient.
- During the calibration phase, certain parameters of the GEPAM system were modified by the patient's suggestions, in order to provide comfort and confidence when making a step. Once the considerations were verified, the running tests were performed obtaining the results shown in Fig. 13:

Support phase

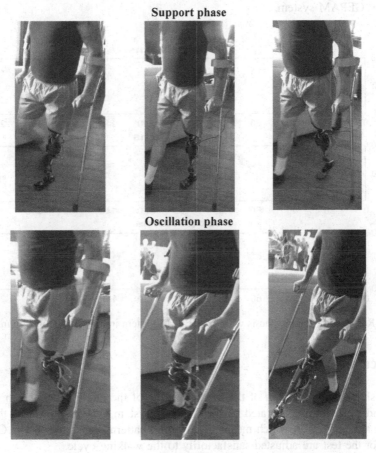

Oscillation phase

Fig. 13. Walk cycle test performed with a patient

Given the response of the prosthesis with the GEPAM system in a patient (see Fig. 14), we can see that the response of the neural network to an input walking pattern, the red signal represents the patterns generated by the GEPAM system, while the blue signal represents the pattern generated by the inertial sensors in the prosthesis, concluding that the test is performed satisfactorily, however, there are considerations such as:

- The GEPAM system has been modified for this experimentation, since the patient's walking speed is lower than the speed of the system, two pause zones have been introduced, where the system stops generating patterns and keeps the last generated patterns, this with the desire to provide security and confidence to the patient when making the step.
- The prosthesis has certain mechanical limitations that prevent it from fully replicating the gait cycle generated by the system; however, the response of the mechanism has been satisfactory, adequately replicating the movements generated by the GEPAM system.

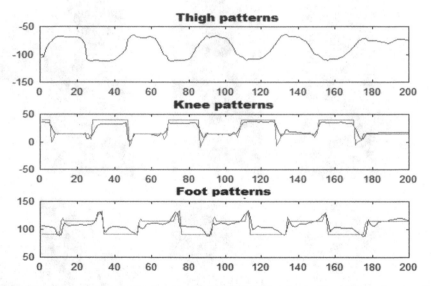

Fig. 14. Response of the prosthesis with the GEPAM system in a patient (Color figure online)

4   Discussions

Table 5 shows a comparison of the measurements of the angle generated in the test bench and the patterns generated by the GEPAM system, which shows that the error variation around 4.31%, which means that the gait patterns generated by the GEPAM system for the test are adjusted satisfactorily to the walking cycle.

Table 5. Results prostheses in test bench vs. GEPAM system

Knee			Foot		
Prosthesis	GEPAM	Error	Prosthesis	GEPAM	Error
13.69	12.95	6.96%	67.69	63.92	5.90%
21.25	20.35	4.82%	74.13	70.08	5.78%
29.69	27.84	6.65%	88.31	86.89	1.63%
38.75	38.96	0.54%	97.81	100	2.19%
40.38	37.98	6.32%	99.56	100	0.44%
43.50	46.20	5.84%	102.75	96.54	6.43%
48.63	47.14	3.73%	104.88	100	4.88%
50.81	49.70	2.23%	114.44	119.97	4.61%

Table 6 shows a comparison of the measurements of the angle generated by the patient and the patterns generated by the GEPAM system, which shows that the error variation around 4.65%, which means that the gait patterns generated by the GEPAM system for the test performed with the patient are adjusted satisfactorily to the walking cycle.

Table 6. Results prostheses in patient vs. GEPAM system

Knee			Foot		
Prosthesis	GEPAM	Error	Prosthesis	GEPAM	Error
14.87	15	0.87%	87.075	91	4.31%
15.44	15	2.93%	89.392	91	1.77%
15.19	15	3.93%	93.641	91	2.90%
36.94	40	4.65%	105.942	113.536	3.93%
38.62	40	3.95%	112.667	115	2.03%
38.19	40	4.53%	116.867	115	1.62%
39.06	40	2.35%	117.525	115	2.20%

5 Conclusions

The control system for a leg prosthesis for transfemoral amputation using a gait pat-tern generator system (GEPAM), satisfactorily reproduces the gait cycle with an efficiency of 95% and a delay of 0.25 s, by activating inertial sensors and servo-motors; it also has a monitoring system if too constantly monitors the relative humidity and temper-ature within the socket of the prosthesis, by a mobile application that alerts the user to abnormal operating ranges, in order to safeguard the health of patient.

In addition, compared to the system activated by myoelectric signals, the perfor-mance of the GEPAM system has a lower error variation corresponding to 4.65% compared to 5%, obtaining a model that adjusts satisfactorily to the running cycle generating adequate patterns of movement.

References

1. Bravo, X., Comina, M., Tobar, J., De la Cruz, D., Corella, J.: Low cost mechatronics prototype prosthesis for transfermoral amputation controlled by myoelectric signals. In: Engineering and Computer Science (2017)
2. Consejo Nacional para la Igualdad de Discapacidades CONADIS. Ministerio de Salud Pública del Ecuador. Obtenido de (2017). http://www.consejodiscapacidades.gob.ec/estadistica/index.html
3. Parra, P.: Bioingeniería: "Prótesis". Universidad Politécnica Salesiana, Cuenca (2010)
4. Tardón, L.: El diseñador biónico Hugh Herr, Premio Princesa de Asturias de Investigación Científica. El Mundo, pp. 12–13 (2016)
5. Martínez, F.: Diseño de prótesis transfemoral activa. Centro Nacional de Investigación y Desarrollo Tecnológico, Cuernavaca (2013)
6. McGeer, T.: Passive Dynamic Walking. Simon Fraser University, Burnaby (2005)
7. Organización Mundial de la Salud. Ecuador está en el tercer puesto de los países con menor estatura en Sudamérica. El Tiempo (2016)
8. Central Intelligence Agency. The World FactBook. https://www.cia.gov/library/publications/the-world-factbook/geos/ec.html.Accessed12Mar2018
9. Lee, M., Ju, H., Song, J., Park, C.: Kinematic model-based pedestrian dead reckoning for heading correction and lower body motion tracking. Sensors **15**, 28129–28153 (2015)
10. Cover, T.: Methodologies of Pattern Recognition. Watanabe, Satoshi (1969)

Construction of a Computer Vision Test Platform: VISART for Facial Recognition in Social Robotics

Edwin Rodríguez, Christian Gutiérrez, Cristian Ochoa, Freddy Trávez,
Luis Escobar$^{(\boxtimes)}$, and David Loza

Department of Energy and Mechanics, Universidad de las Fuerzas Armadas - ESPE,
170121 Sangolquí, Pichincha, Ecuador
{rsedwin,cagutierrez6,ceochoa,fstravez,lfescobar,dcloza}@espe.edu.ec
https://decem.espe.edu.ec/

Abstract. Robotics has undoubtedly found its way deeper into every day human tasks up to the point where now they even share workspaces with people. In this context, social robotics has increased its field of action, one important part is the advances in aspects that make up the Human-Machine Interaction. This article reports the advance that an investigation team at Universidad de las Fuerzas Armadas ESPE has done to create a test platform for Computer Vision algorithms. This approach consists of a 3-DOF articulated robotic head with anthropomorphic characteristics, considering the guidelines established by different authors for social robotics, besides it provides affordable, flexible and scalable hardware resources to facilitate the implementation, testing, analysis and development of different stereo artificial vision algorithms. The system, called Visart, is intended to work in more natural situations than independent cameras, therefore, work deeply in Human-Robot Interaction. As the first report of the Visart prototype, some basic tests with the platform were performed, they consist of face detection, facial and expression recognition, object tracking, and object distance calculation. The results obtained were consistent, opening a door for further research focusing on the comparison of more computer vision algorithms to examine their performance in real scenarios and evaluate them more naturally with our prototype for Human-Robot Interaction.

Keywords: HRI · Computer Vision · Human-Robot Interaction · Anthropomorphic robot · Social robotics

1 Introduction

Since the end of the XX century, robots have found their way into the lives of humans in diverse tasks, being originally intended to partially or completely replace workers in simple repetitive duties [5,11]. The last decade of global competition to improve and automate processes, technological advancement of robotics has greatly increased and opened up the market for new applications

© Springer Nature Switzerland AG 2020
M. Botto-Tobar et al. (Eds.): ICAT 2019, CCIS 1194, pp. 637–651, 2020.
https://doi.org/10.1007/978-3-030-42520-3_50

[12]. Most of these include robots and humans working or interacting together in some other way, requiring a series of new challenges to be solved. Human-Robot Interaction (HRI) corresponds to a field of study dedicated to understanding, designing and evaluating robotic systems used in interaction with humans [9], which generally involves environments where conditions are unpredictable and constantly changing, unlike industrial workspaces where the different factors are under control [2,20,21].

Unlike HMI (Human Machine Interaction) and HCI (Human-Computer Interaction), HRI focuses on a more direct and close interaction between the whole robotic system and the user. This forces to improve robotics applications on how they can be useful in human daily life [17,29]. This is why a complete understanding of how the user interacts with robot for long periods of time is utmost important - even more when humans and robots have interacted since 1940's [7].

Communication with robots has become more complex, resembling relationships between two humans. Consequently, the dialogue process must consider many aspects - one of them being Linguistic Competence that considers the ability to understand each participant [7]. Face to Face communication is a pillar of the dialogue action and it occurs at a time scale in the order of 40 ms, a high level of reliability in facial recognition (primitive sensory perception) by the robot is necessary considering the uncertainty present at this time scale [17,29].

Charles Darwin acknowledged facial expression as one of the most important means for human beings communication [17,29]. Accordingly, facial recognition makes up one of the main capabilities for Face to Face communication that a robot can be equipped to understand and adapt to changes in its environment, and thereby improve performance in its interaction, is Computer Vision [2,14, 21]. Examples of Facial Recognition implementation can be found in prototypes such as: Kismet Robot [32], SCARINO [23], Maki Social Robot, SAM Robot, UPS, TEA Robot and NAR Robot, ESPE [22].

2   Social Robotics Guidelines

In recent years of robotics advancement, the areas of interest have changed their focus from the analysis in the robot environment towards the study of social and emotional aspects of interaction with humans [16]. For most people, interaction with robots is easier to carry out when the robot exhibits social behavior in some way and if it has anthropomorphic characteristics [20]. Based on the guidelines proposed by [15] for the development of robots oriented towards interaction with people, the strategies to address the following recommendations for the implementation of the robotic system are proposed as:

2.1   Appearance

The appearance of the robot is one of the main features, because it defines the degree of user acceptance, likelihood to interact with the system and, therefore, the success of the robot in the desired application. [18] determined that for different applications, people prioritize the robot's appearance rather than its functionality, to which the robot's behavior is also taken into account to perform seemly natural actions. The reason for this is that cognitive expectations that a human shows in front of a machine go hand in hand with its human-like appearance [10,15]. To avoid venturing into the field of realism, but into the field of credibility, the robot's appearance must also be consistent with the purpose and capabilities of the robot. It may be unnecessary to provide the robot with unmotivated anthropomorphic features such as an excess of facial expressions that seek to create an intelligent social entity instead of a social robot tool [4,30]. Managing to implement just the right amount of social traits in a machine can help to maintain user expectations closer to those that the robot may be able to fulfill.

2.2 Affective Interactions and Empathy

One technique to strengthen relationships between a human and a robot is to demonstrate emotions and pro-social actions. With this approach, the distress and disagreement in the user can be greatly reduced [25]. Since the trajectory of the study of the appearance of a human-like robot is already considerably extensive, in recent years there have been a significant number of studies focused on how to express emotions and other nonverbal behaviors in social robots [15]. An area that is still poorly explored within this field is the ability to identify emotions and empathize with the user [26]. Empathy plays such an important role in human relationships that its implementation in social robots has proven effective enough to improve the user's perception of the robot as well as its interaction over extended periods [27].

2.3 Memory and Adaptation

Although advances in intelligence and robotic memory, consequently their direct benefits, are still under constant study, it can be anticipated that coherence in the actions of interaction of the robotic system will be improved [34]. Memory is a skill that the robot must have to establish interaction with different users. Therefore, the robot's behavior will be limited not only to predefined events, but also to learning and experiences previously obtained and stored in itself. Cloud Robotics is a technique to provide robustness in terms of memory and computing in robotic systems [13]. This perspective has benefits as Big Data, Cloud Computing, Collective Robot Learning, and Human Computation [13]. So, the system's scalability is ensured by using processing resources on cloud modules. Even simple references from users that the robot can remember increase the user's feeling of trust [15] because it is a greater step towards a more realistic personality. The analysis of behavior and social interactions has benefited from the advances in the different methods for the automatic measurement and monitoring, computer vision methods being one of the main tools [28]. The principal aspects that artificial vision usually analyzes for recognizing and sensing the environment are shapes, sizes, location of objects, color, lighting, texture, and composition. A correct implementation of Computer Vision can help the actions and behavior of the robot easily meet the expectations of the user and thus, achieve an interactive environment where the human and robot participate in more natural and intuitive interactions [32]. The reason for this is that, by equipping the robotic system with a Computer Vision tool, it improves its environmental perception capabilities, giving it a more independent personality and functionality in addition to the potential cognitive abilities that can be achieved. Therefore, providing the robot that will be used for social interaction with a computer vision system for recognition and monitoring of objects becomes a necessity to meet the need conferring it with greater autonomy, giving it the power to decide to carry out more than its programmed activities.

3   Design and Construction

Visart design constraints were based on the social robotics guidelines and the dynamics of its Human-Robot Interaction intents. Through, the Quality Function Deployment matrix (QFD) [31] the most important technical characteristics of the system were obtained and cited as follows:

- Open source platform development
- Use of artificial vision in the final structure of the prototype
- Rapid prototyping manufacturing
- Movement using 3 DOF mechanism
- Compatibility with computer vision algorithms
- Use of HMI for results validation.

3.1   Mechanical Subsystem

Taking the appearance considerations from the social robotics guidelines, the development of the platform will be limited to the construction of a robotic head. This head has to resemble a human in a minimalist way, avoiding the Mori valley. The robot will show its emotions through its eyes to take advantage of human abilities to interpret intentions based on one's actions. Additionally, the robot has to be able to move in order to operate alongside the algorithms for tracking objects, faces, and colors [4]. This is why, a 3-DOF mechanism will be designed for simulating a simplified version of the human neck (flexion, extension, rotation). The mechanical design involves the analysis of the kinematic chain of the 3-DOF mechanism (Fig. 1), according to fatigue design criteria established by [3]. In addition, the selection of actuators M_x, M_y and M_z (Fig. 1) are also detailed. Through the use of Computer-Aided Design (CAD) and Computer-Aided Engineering (CAE) software, mechanical properties were verified. Tensile, Von Misses stress and deformation values were found through software simulation.

Visart's mechanical design also includes kinematic analysis for the ball type joint, according to Denavit Hartenberg criteria [6]. Kinematic chains were

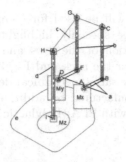

Fig. 1. Visart 3-DOF mechanism. General scheme.

described according to Fig. 2, a total of 7 links, 3 rotational joints, 3 universal joints, and 2 spherical joints were defined. The result is a 3-DOF mechanism. The head's rotational and translational matrix were obtained and shown in Eqs. 3 and 1 respectively. Subsequently, the dynamic analysis of the system was carried out, starting from a kinematic analysis and using the Euler-Lagrange method to obtain the Lagrangian of the system, resulting in Eq. 2.

$$
P_c = \begin{bmatrix} -19.5c(\phi) - 10s(\phi) - 40c(q_2)s(\phi) \\ 10c(\phi) - 19.5s(\phi) + 40cos(q_2)c(\phi) \\ 40s(q_2) \end{bmatrix} \tag{1}
$$

$$
\begin{bmatrix} 0.024372 & -0.0806804 \\ -0.0806804 & 0.024372 \end{bmatrix} \begin{bmatrix} \ddot{\theta}_i \\ \ddot{\theta}_d \end{bmatrix} = \begin{bmatrix} \tau_i \\ \tau_d \end{bmatrix} \tag{2}
$$

$$
R_c = \begin{bmatrix} c(\phi) & -c(q_2)s(\phi) & s(q_2)s(\phi) \\ s(\phi) & c(\phi)c(q_2) & -s(q_2)c(\phi) \\ 0 & s(q_2) & c(q_2) \end{bmatrix} \tag{3}
$$

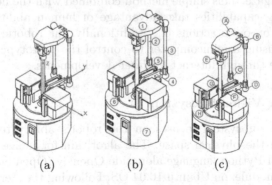

(a)   (b)   (c)

Fig. 2. Visart 3-DOF kinematic analysis. Links and joints distribution. (a) Coordinate system. (b) Links (c) Rotational and universal joints.

Since the robotic system will be used for computer vision algorithms, the visual sensory system will be specially highlighted, which will be achieved through the shape and position of the eyes and cameras. The final system structure is shown in Fig. 2 for the neck and Fig. 3 for the head. A total of 32 components are included on Visart's mechanical design. Most elements, including body and face, were manufactured by additive manufacturing using FDM (Fused Deposition Modeling) with PLA (Polylactic acid or polylactide) as prime material.

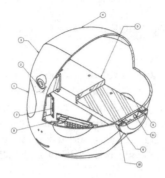

Fig. 3. Visart's head assembly. 1: Glasses, 2: Camera, 3: Front Head Case, 4: Back Head Case, 5: Head Bracket, 6: Ear Base, 7: Led Matrix, 8: Ear Protection, 10: Ears, 11: Ear LED ring.

3.2   Electronic Subsystem

In order to show empathy, one of the most effective tools are the robot's eyes [20]. With this consideration, Visart was built with an 8×8 LED matrix as an interactive component of the robot to show its emotions and transmit expressions through its gaze. This simple method combined with the head's movement and face tracking capabilities take advantage of human abilities to interpret intentions based on one's actions [4]. Additionally, the robotic platform has a modular design using a minicomputer to control the prototype, this computer can communicate through internet for later developments.

3.3   Computer Vision Subsystem

Computer Vision subsystem focuses on the robot's ability to extract visual information from the physical space. The algorithm for image processing was programmed with Python language alongside OpenCv (Open Source Computer Vision Library) module, on Ubuntu 16.04 OS. Following the stereo vision guidelines two parallel web cams located on the top of the head were used (Fig. 3).

For the mathematical modeling of stereo vision camera arrangement, the cameras were considered parallel to b distance (Fig. 4: a). The origin of the coordinate system was set on the left camera. Using homogeneous matrices, any point projection within the robot's visual field can be calculated. For a projected point M (Fig. 4: b) on the stereo system, at a focal length f, according to the pin-hole model, the matrices shown in Eq. 4 were obtained.

$$
\begin{bmatrix} x_i \\ y_i \\ z_i \end{bmatrix}_{left} = \begin{bmatrix} fk_x & 0 & fC_x & hfC_x \\ 0 & 1 & 0 & 0 \\ 0 & k_z & fC_z & hfC_z \end{bmatrix} \begin{bmatrix} x_m \\ y_m \\ z_m \\ 1 \end{bmatrix}
\tag{4}
$$

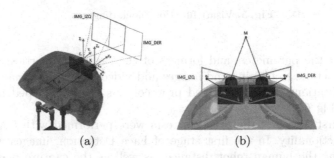

(a)   (b)

Fig. 4. Visart stereo vision field space for mathematical modeling.

4 Validation Methodology

Once all Visart robot subsystems have been implemented, the test platform is set-up as indicated in the block diagram shown in Fig. 5. The computer vision algorithms will be loaded on the Intel NUC Personal Computer kit, equipped with an Intel Core i5-4250U processor with a 2.6 GHz speed, integrated with an Intel HD Graphics 5000 2 GB graphics processor and 6 GB of RAM. The 2 cameras of the robot will send the captured images directly to the computer for processing. After the analysis, the program will send the respective movement and expression commands to the control board. This platform will be in charge of writing the commands received by the computer in the Visart actuators.

With the complete platform set-up, the cameras are first calibrated. For this task, the ChessBoard calibration method will be used to align the lines captured in the images between both cameras and with the real world. The algorithm code is based on the work of [1]. Once the image calibration of both cameras is achieved, the image is trimmed to repair the resulting distortion, so that the actual coordinates of the detected objects can be obtained.

For platform validation tests, a test protocol is established to verify the operation and capabilities of the system to use computer vision algorithms. Based on [8,33] consideration of ISO/IEC 19794-5:2005, the following protocol also takes

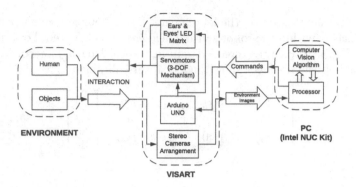

Fig. 5. Visart function block diagram

into account the parameters and formats of the scenes, properties, attributes, and best practices to obtain good images and video of faces for analysis in biometric applications. The established protocol was then developed in 4 stages detailed in Fig. 6:

In the first 2 stages, the platform tests were performed with 5 subjects to analyze functionality. In the first stage of Face Detection, images were taken while varying the human robot distance, as well as the capture planes (front, right and left side) and light incidence in order to check the ability to recognize human faces in different ambient. Fluorescent lights were used to control the surrounding light in indoors. For the registration process of a user, 15 images of his face in different orientations are stored. Likewise, the facial expression recognition algorithm -limited to joy, surprise, sadness, and anger- was trained with a database of 100 images per expression using the Python Tensorflow tool. The tests were performed with all the individuals in the same scene.

For the last two test stages, objects of different colors (green, red, blue and yellow) were used. The objective was to recognize and track the object along Visart visual field. For this task 6 × 6 cm, 3D-printed PLA cards were used. Finally, in order to analyze the bifocal setup of computer vision of the robot, Kolmogorov and Zabih-GC's graph-cutting algorithm [24] was tested for the generation of disparity maps, showing a grayscale pattern in several planes of the image allowing 3D reconstruction of its environment. The test consisted of an object placed at different distances from the platform (5, 10, 20, 40, 80, 120, 160, 200 cm) and then estimated through the computer vision algorithm.

5 Results

Indoor tests were carried out in a controlled environment. A controlled beam of light directed to the user's face was emitted by fluorescent lamps. Under these conditions, all the tests detailed in Sect. 4 were performed. Outdoor tests, on the other hand, brought out unpredictable outcomes due to the environmental conditions. Each stage test resulted as follows:

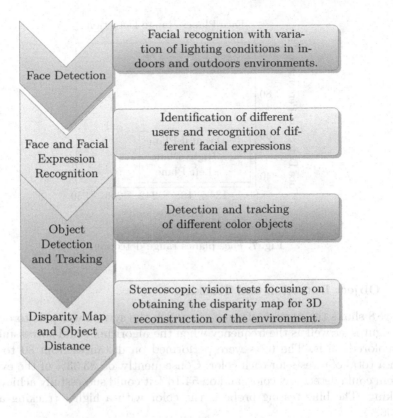

Fig. 6. Validation methodology stages.

5.1 Face Detection

Figure 7 shows the data obtained by varying the distance from the user to Visart, and trying several face positions. The values shown below correspond to the means of the experiments. The data were taken from 6 ranges, starting at 80 cm up to 350 with a 54 cm increase. For frontal face plane faces are detected 80% of the times, while right face planes 56.67% and left face plane 53.33% of the times. For distances greater than 296 cm the detection percentage is zero.

5.2 Face and Facial Expression Recognition

The experiment started with the 15 images database for each user, and was performed in a dark room at 100% illumination -with the fluorescent lamp pointed towards the user- and 0% illumination -without the fluorescent lamp-. In addition, recognition for several subjects in the same image frame was tested. In all cases for registered users, the system could recognize them 100% of the times with 100% illumination at close range. As a result, the system was able to recognize expressions such as happiness, sadness, and surprise.

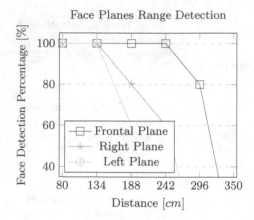

Fig. 7. Face planes range detection.

5.3   Object Detection and Tracking

Figure 8 shows the portion of events in which the system was able to detect the test figures, as well as the frequency when the algorithm could successfully track the colored cards. The tests were performed on distances from 80 to 350 cm, with a total of 6 tests for each color. Consequently, on 83.33% of the events the system could detect any color, and on 54.16% it could successfully achieve object tracking. The blue testing probe is the color with a higher tracking accuracy (66.66%).

5.4   Disparity Map and Object Distance

Through the disparity map generation algorithm, a grayscale-image was obtained. This allowed 3D reconstruction of the robot environment though using both camera images. Therefore, it was possible to compute the estimated distance from any object to the robot and then calculate measurement errors. Deviation values are shown in Fig. 9. The best results were obtained from distances from 20 to 120 cm with an error lesser than 5%.

6   Discussion

The results obtained show that facial recognition is effective on face to face interaction among the user and Visart. However, the platform face recognition ability is greatly reduced when Visart captures only right or left face profile. This issue may be due to the vision algorithm used or the image quality that is influenced by the environmental brightness or cameras' resolution. Besides, Visart can successfully recognize 3 basic facial expressions [29] used during people interaction. As well as the ability to track color testing probes. In the latest instance, the repetitiveness is determined by the corresponding color. Experiments with the

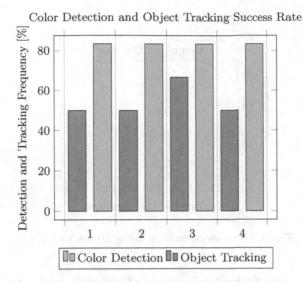

Fig. 8. Visart success rate for color detection and object tracking. Column 1: Green card, 2: Red card, 3: Blue card and, 4: Yellow card. (Color figure online)

blue testing probe got the highest tracking accuracy rate. This may be related to the probe color opacity and the type of material used to manufacture the cards and the 3D, including the printing parameters such as infill and layer height. The contrast of the color and the environment is also a concern considering that the color detection algorithm interprets the information as grayscale.

7 Future Work

A robust validation process is necessary, in order to obtain a greater amount of data to identify the platform behavior. For this purpose works like [23] and [19] will be taken as a reference. In order to improve the platform's functionality, tests in more diverse environments are planned to be done including the integration an analysis of several computer vision algorithms. This will lead to making the best choice for every test stage before taking the platform into real life scenarios.

Since educational robotics has also advanced at the same pace as other areas of social robotics [30], more tools such as speech recognition can be implemented into VISART to fulfill different functions that involve interaction with young students either as a tutor or class assistant. Because of this, it is important to carry out a validation process of social interaction with students of different ages and identify the strengths and weaknesses of the platform. This will also promote the use of new educational tools and methodologies that encourage curiosity and technology development by students; and even the identification of possible obstacles and risks in the use of robots for teaching.

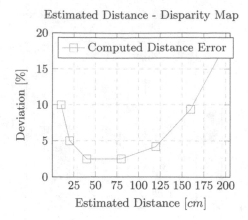

Fig. 9. Distance estimation using Stereo Vision.

Another great area to develop is the use of VISART as a interaction tool for management of Cyber Physical Systems in elderly care, home automation and as office assistant.

8   Conclusion

The Visart platform was proven useful for face detection in any of the planes up to 134 cm - between user and robot. However, for longer distances only frontal plane facial recognition is guaranteed. Considering that 242 cm is the maximum reliable detection distance, the system validates itself for use in social interaction as most social actions take place within the 100 cm range. In this length, facial expressions such as happiness, sadness and surprise could also be recognized successfully.

Object tracking capabilities of the system are significantly encouraging with its 80% accuracy in recognition obtained. The object in some cases could not be tracked due to several factors such as the reflection index of the test cards, the different light conditions consequence of the robot's movement, the image processing algorithm and even the cameras used.

The platform is based on open-source software and has the capability for local data processing. These features provide the operating conditions to mitigate network connection problems - such as low-latency responses, quality of service and downtime.

All these regards contribute to Visart being a platform that brings together the necessary characteristics in both appearance and functionality, that ensure an effective HRI in social robotics [9,15,23]. Therefore, it is possible to take the next step towards testing its effectiveness with more real people interactions against the use of various algorithms for face detection and facial recognition, object detection and monitoring, and proximity estimation in the interaction environment.

References

1. Armea, A.: Calculating a depth map from a stereo camera with OpenCV (2017). https://albertarmea.com/post/opencv-stereo-camera/
2. Azevedo, H., Belo, J.P.R., Romero, R.A.F.: Cognitive and robotic systems: speeding up integration and results. In: 2017 Latin American Robotics Symposium (LARS) and 2017 Brazilian Symposium on Robotics (SBR), Curitiba, pp. 1–6. IEEE (2017). https://doi.org/10.1109/SBR-LARS-R.2017.8215337
3. Budynas, R.G., Keith, N.: Diseño en ingeniería mecánica de Shigley, 8th edn. McGrawHill, Mexico (2008)
4. Duffy, B.R.: Anthropomorphism and the social robot. Robot. Auton. Syst. **42**(3–4), 177–190 (2003). https://doi.org/10.1016/S0921-8890(02)00374-3. https://link inghub.elsevier.com/retrieve/pii/S0921889002003743
5. Faust, R.A.: Robotics in Surgery: History, Current and Future Applications. Nova Publishers, New York (2007). Google-Books-ID: p70afWyqcrMC
6. Fernández, L., Sotomayor, L.: Análisis cinemático inverso y directo del robot paralelo (2016)
7. Fong, T., Thorpe, C., Baur, C.: Collaboration, dialogue, human-robot interaction. In: Jarvis, R.A., Zelinsky, A. (eds.) Robotics Research. Springer Tracts in Advanced Robotics, vol. 6, p. 10. Springer, Heidelberg (2003). https://doi.org/10.1007/3-540-36460-9_17
8. Gomez Camacho, M., Barrera, Y.V.: Prototipo de un sistema biométrico integrado para la identificación de personas - fase I (2016). https://doi.org/10.13140/rg.2.1.3267.2881
9. Goodrich, M.A., Schultz, A.C.: Human-robot interaction: a survey. Found. Trends Hum. Comput. Interact. **1**(3), 203–275 (2007). https://doi.org/10.1561/1100000005
10. Hayashi, K., Shiomi, M., Kanda, T., Hagita, N.: Who is appropriate? A robot, human and mascot perform three troublesome tasks. In: 19th International Symposium in Robot and Human Interactive Communication, Viareggio, Italy, pp. 348–354. IEEE (2010). https://doi.org/10.1109/ROMAN.2010.5598661
11. Hockstein, N.G., Gourin, C.G., Faust, R.A., Terris, D.J.: A history of robots: from science fiction to surgical robotics. J. Robot. Surg. **1**(2), 113–118 (2007). https://doi.org/10.1007/s11701-007-0021-2
12. IFR: World Robotics Survey: Industrial robots are conquering the world (2015). https://ifr.org/news/world-robotics-survey-industrial-robots-are-conquering-the-world-/
13. Kehoe, B., Patil, S., Abbeel, P., Goldberg, K.: A survey of research on cloud robotics and automation. IEEE Trans. Autom. Sci. Eng. **12**(2), 398–409 (2015). https://doi.org/10.1109/TASE.2014.2376492
14. Krishnan, A., Hs, A.: Design and implementation of robotic vision for face recognition. In: 2017 International Conference on Circuits, Controls, and Communications (CCUBE), Bangalore, pp. 135–138. IEEE (2017). https://doi.org/10.1109/CCUBE.2017.8394148
15. Leite, I., Martinho, C., Paiva, A.: Social robots for long-term interaction: a survey. Int. J. Social Robot. **5**(2), 291–308 (2013). https://doi.org/10.1007/s12369-013-0178-y

16. Lindblom, J., Andreasson, R.: Current challenges for UX evaluation of human-robot interaction. In: Schlick, C., Trzcieliński, S. (eds.) Advances in Ergonomics of Manufacturing: Managing the Enterprise of the Future. Advances in Intelligent Systems and Computing, vol. 490, pp. 267–277. Springer, Cham (2016). https://doi.org/10.1007/978-3-319-41697-7_24

17. Littlewort, G., et al.: Towards social robots: automatic evaluation of human-robot interaction by facial expression classification. In: Thrun, S., Saul, L.K., Schölkopf, B. (eds.) Advances in Neural Information Processing Systems 16, pp. 1563–1570. MIT Press, Cambridge (2004)

18. Lohse, M., Hegel, F., Wrede, B.: Domestic applications for social robots: an online survey on the influence of appearance and capabilities. J. Phys. Agents (JoPha) 2(2), 21–32 (2008). https://doi.org/10.14198/JoPha.2008.2.2.04

19. Martinez, B., Valstar, M.F., Jiang, B., Pantic, M.: Automatic analysis of facial actions: a survey. IEEE Trans. Affect. Comput. 10(3), 325–347 (2019). https://doi.org/10.1109/TAFFC.2017.2731763. https://ieeexplore.ieee.org/document/7990582/

20. McGinn, C.: Why do robots need a head? The role of social interfaces on service robots. Int. J. Social Robot. (2019). https://doi.org/10.1007/s12369-019-00564-5

21. Nagai, K., Sakabe, H., Ohka, M.: Finger direction recognition toward human-and-robot cooperative tasks. In: 2017 International Symposium on Micro-NanoMechatronics and Human Science (MHS), Nagoya, pp. 1–3. IEEE (2017). https://doi.org/10.1109/MHS.2017.8305242

22. Ochoa, C., Trávez, F.: Diseno y construcción de una plataforma robótica para pruebas técnicas sobre visión artificial que permita el aprendizaje y desarrollo del software en robots interactivos. Ph.D. thesis, Universidad de las Fuerzas Armadas ESPE, Sangolquí, Ecuador (2018)

23. Pinillos, R., Marcos, S., Feliz, R., Zalama, E., Gómez-García-Bermejo, J.: Long-term assessment of a service robot in a hotel environment. Robot. Auton. Syst. 79, 40–57 (2016). https://doi.org/10.1016/j.robot.2016.01.014. https://linkinghub.elsevier.com/retrieve/pii/S0921889015300440

24. Pinto Sánchez-Matamoros, L.F., Carrasco, J.A.P.: Análisis de la aplicación de algoritmos de K-means y Continuous Max-Flow a la segmentación de imágenes en color. Dep. Teoría de la Señal y Comunicaciones Escuela Técnica Superior de Ingeniería Universidad de Sevilla, p. 80 (2015)

25. Preston, S.D., de Waal, F.B.M.: Empathy: its ultimate and proximate bases. Behav. Brain Sci. 25(1), 1–20 (2002). https://doi.org/10.1017/S0140525X02000018

26. Reyes Fernández, B.: Empathy and moral development: implications for caring and justice. Actualidades en Psicología 20(107), 141 (2011). https://doi.org/10.15517/ap.v20i107.40

27. Riek, L.D., Robinson, P.: When my robot smiles at me: enabling human-robot rapport via real-time head gesture mimicry. J. Multimodal User Interfaces 3(1–2), 99–108 (2010)

28. Robie, A.A., Seagraves, K.M., Egnor, S.E.R., Branson, K.: Machine vision methods for analyzing social interactions. J. Exp. Biol. 220(1), 25–34 (2017). https://doi.org/10.1242/jeb.142281

29. Stewart, M., Gwen, B., Ian, L., Javier, F., Movellan, R.: Real time face detection and facial expression recognition: development and applications to human computer interaction, Madison, Wisconsin, USA, p. 53. IEEE (2003)

30. Tapus, A., Maja, M., Scassellatti, B.: The grand challenges in socially assistive robotics. IEEE Robot. Autom. Mag. 14(1) (2007)

31. Tontini, G.: Integrating the Kano model and QFD for designing new products. Total Qual. Manag. Bus. Excell. **18**(6), 599–612 (2007). https://doi.org/10.1080/14783360701349351
32. Wiese, E., Metta, G., Wykowska, A.: Robots as intentional agents: using neuroscientific methods to make robots appear more social. Front. Psychol. **8**, 1663 (2017). https://doi.org/10.3389/fpsyg.2017.01663. http://journal.frontiersin.org/article/10.3389/fpsyg.2017.01663/full
33. Yi, D., Liu, R., Chu, R.F., Lei, Z., Li, S.Z.: Face matching between near infrared and visible light images. In: Lee, S.-W., Li, S.Z. (eds.) ICB 2007. LNCS, vol. 4642, pp. 523–530. Springer, Heidelberg (2007). https://doi.org/10.1007/978-3-540-74549-5_55
34. Yii Lim, M., Aylett, R.: Human-like memory retrieval mechanisms for social companions. In: AAMAS, Taipei (2011)

Face Recognition Systems in Math Classroom Through Computer Vision Traditional Techniques

Luis Granda$^{(\boxtimes)}$, Luis Barba-Guaman$^{(\boxtimes)}$, Pablo Torres-Carrión$^{(\boxtimes)}$ (iD),
and Jorge Cordero$^{(\boxtimes)}$

Grupo de Investigación Inteligencia Artificial Aplicada,
Universidad Técnica Particular de Loja, Loja, Ecuador
{lfgranda,lrbarba,pvtorres,jmcordero}@utpl.edu.ec

Abstract. The methods and techniques of detection of the human face and facial recognition have presented a great impulse in recent years, thanks to the advance in areas such as artificial vision and machine learning. Although deep neural network techniques are in vogue, traditional techniques allow you to create applications that do not consume many resources from computing devices. In this research, we present a facial recognition system that implements the Eigenfaces method, developed in C # of Microsoft Visual Studio and open-source video processing libraries such as OpenCV as EmguCV. The application is divided into two sections: the first called register where, through an integrated camera, images of the user's face or other means such as video and stored images are captured, and the second section is known as recognition where the user is compared with all the records of the data set, indicating whether this is registered and the recognition percentage. The project was implemented with a universe of the size of twenty-five users, of which six are men (24%) and nineteen are women (76%), developing tests for five weeks.

Keywords: Facial recognition · Computer vision · Eigenfaces · Image processing

1 Introduction

Facial recognition is a very active research area in the field of computer vision, the main basis is the correspondence of the detected face with the user's record in the training base. The facial recognition process implements algorithms that detect the facial area, process the obtained image, then the extraction of facial features is performcd, here we obtain the fundamental information of the facial features and intrinsic characteristics of each face (eyes, nose, chin, eyebrows, mouth, etc.), which generates a unique bio-metric pattern for each individual.

These algorithms have different accuracy rates, in different conditions such as lighting, distance, movement, camera quality, background, and contrast. Generally, facial recognition uses the PCA algorithm (Principal Component Analysis), and the Eigenfaces algorithm, the combination of these allows to recognize of an unknown test image, compares it with the training images stored in the database and obtains

© Springer Nature Switzerland AG 2020
M. Botto-Tobar et al. (Eds.): ICAT 2019, CCIS 1194, pp. 652–664, 2020.
https://doi.org/10.1007/978-3-030-42520-3_51

information about the person recognized [1]. The detection and recognition of human faces play an important role in areas such as video surveillance, security, and facial image data management [2]. Facial recognition systems and similar biometric systems (3D scanning of the face, geometric analysis of the face, etc.) present a similar methodology. The process begins with the detection of the face, here the face area is identified. In a second stage the intrinsic characteristics of the face are examined and extracted; Focusing on those features that allow isolating the face of the individual within a set of images. Finally, the verification of the person is carried out. See Fig. 1.

Fig. 1. General process of facial recognition.

Facial detection and recognition involve a process of locating one or more faces in images or videos. To do this, it relies on a face detection algorithm that searches for rectangular regions within the image, or failing that, regions that contain objects similar to those of a training set previously established, this function performs the process of scanning the image at different scales with the objective of recognizing faces with different dimensions. Facial recognition is a very active and essential research area within computer vision, whose fundamental basis is the correspondence of the detected face with the user's record in the training base. To carry out facial recognition there are multiple algorithms; in the present investigation methods based on linear appearance were implemented, as shown in Fig. 2.

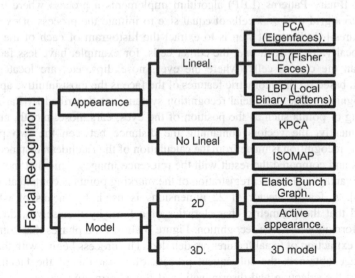

Fig. 2. Facial recognition methods [3].

2 Methods Based on Linear Appearance

2.1 Haar Cascade Face Method

The process of detection of people is implemented natively in the OpenCV library for its free use. The OpenCV library provides various classifiers for the detection of body regions in its extension [13, 14], detailed in Table 1.

Table 1. Haar cascade classifiers for a region of the body.

Detection region	Name of classifier
Body	"haarcascade_fullbody"
Nose	"haarcascade_mcs_nose"
Mouth	"haarcascade_mcs_mouth"
Eyes	"haarcascade_eye_tree_eyeglasses"
Face	"haarcascade_frontalface_alt_tree"
	"haarcascade_frontalface_alt"

This system implements the "haarcascade_frontalface_default.xml" file for the face detection process. This classifier is fundamental in the initial stage of the detection process, established in the learning-machine, the cascade function is trained with a set of valid and invalid photographs of the test subjects, detecting the required areas, in the case of our system faces.

2.2 LBP Faces Method

The Local Binary Patterns (LBP) algorithm implements a process where the face is divided into a grid of 7×7 cells of equal size to initiate the process of extraction of facial features [4]. The next step is to extract the histogram of each of the cells and assign a local value to each one; the corner cells, for example, have less facial information than the central cells (where the eyes, nose, lips, etc. are located). Facial recognition based on the geometric features of the face is the most intuitive approach to facial recognition. The first facial recognition system was described by Parker [5], in the marking of points such as the position of the eyes, ears, nose, mouth, etc.

Subsequently, the vectors containing the distance between the two points are plotted. The recognition is made from the calculation of the Euclidean distance between the vectors and compared the result with the reference images. This first method has a great disadvantage: The exact registration of the marking points is complicated, even at present [6]. A feature vector of 22 dimensions is used, by means of which it is determined that the geometric characteristics of a face, by themselves, do not have enough information for face recognition. Figure 3 shows the phases that comprise the process of extraction of facial features are detailed. The process begins with the resizing of the image obtained, the subsequent process of extraction of the facial features, ending with the selection and discrimination of the determining values.

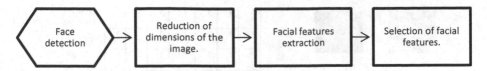

Fig. 3. Process of facial features extraction

- **Face detection:** The image of the student's face to be identified in the set of users is captured.
- **Reduction of dimensions of the image:** Here begin the processes such as: adaptation of the image of the face to the gray scale and obtaining the area of interest.
- **Facial features extraction:** The information of the facial features and intrinsic characteristics of each face (eyes, nose, chin, eyebrows, mouth, etc.) is obtained, which generates a unique biometric pattern for everyone.
- **Selection of facial feature:** The information obtained is compared with the data set. The results obtained reflect the percentage of similarity of the student to be identified.

$$LBP(X_c, Y_c) = \sum_{p=0}^{P-1} 2^p S(i_p - i_c) \tag{1}$$

Where:

(X_c, Y_c), it's the central pixel of the neighborhood.

(i_c), it's the pixel intensity in gray scale.

(i_p), it's the intensity of the contiguous pixel in gray scale.

The function S defined by:

$$S(x) = \left\{ \begin{array}{l} 1, x \geq 0 \\ 0, x \leq 0 \end{array} \right\} \tag{2}$$

$(S(x))$, it's face details.

This algorithm allows you to capture a lot of details in the images, generates a histogram of the image in grayscale and obtains the information in the most efficient way. The histogram can be considered a texture descriptor [4].

Figure 4 shows how to reduce an matrix (3 × 3) of pixels. The process starts by determining the central pixel, then the value of the intensity of that pixel is calculated; finally, the neighboring pixels in the matrix are analyzed to determine the threshold through binary values. A binary result is obtained after the concatenation that will determine the label of the initial matrix.

Fig. 4. Local binary patterns original process [7].

DataSet: This prototype provides an approximate 75% of the photographs as training data for the algorithm, and the remaining 25% of the data allow to evaluate facial recognition.

Gray scale: It allows to calculate an LBP for each channel, or the conversion of the image in gray scale.

Central pixel: The pixel will be the object of the analysis, additionally a comparative order must be established. See (3) [8].

$$if\ \{(value\ of\ neighbor \geq\ Central\ value\ ) = 1\}else = 0 \tag{3}$$

2.3   Fisher Faces Method

The Fisherfaces method implements the Fisher Linear Discriminant (FDL) for the face classification process and reducing the dimensions of the image; aspects such as the light reflected in the faces and the different positions of the face are considered. The core of the FDL algorithm allows the combination of characteristics that better separates between classes, maximizing the relationship between the classes and the dispersion within them, instead of maximizing the total dispersion. In the representation of dimensions, similar classes must be grouped closely, while different classes must be separated from each other at the longest possible distance [9].

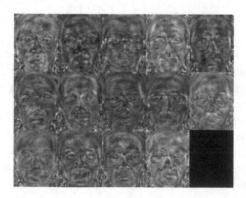

Fig. 5. Examples from Extended Yale Face Database images.

The performance of the Fisherfaces algorithm depends largely on the input data [10]. If only photographs with well-lit scenes are entered as input data and in the recognition stage the process is performed with a bad lighting scenario, the method is most likely to find incorrect components by varying the predominant features of the face. See Fig. 5.

Let X be a random vector with samples taken from c classes [10]:

$$X = \{X_1, X_2, \ldots, X_c\} \tag{4}$$

$$X_i = \{X_1, X_2, \ldots, X_n\} \tag{5}$$

The dispersion matrix S_B y S_w are calculated by the following algorithm:

$$S_B = \sum_{i=1}^{c} N_i(\mu_i - \mu)(\mu_i - \mu)^T \tag{6}$$

$$S_W = \sum_{i=1}^{c} \sum_{x_i \in X_i} N_i(\mu_i - \mu)(\mu_i - \mu)^T \tag{7}$$

Where, μ is total average:

$$\mu = \frac{1}{N} \sum_{i=1}^{N} x_i \tag{8}$$

And, μ_i is the middle of the class, $i \in \{1, \ldots, c\}$:

$$\mu_i = \frac{1}{|X_i|} \sum_{x_i \subset X_i} x_j \tag{9}$$

The Fisher's classical algorithm search a W projection, that maximizes the class separability:

$$W_{opt} = \arg\max_w \frac{|W^T S_B W|}{|W^T S_W W|} \tag{10}$$

The optimization problems can be described like this:

$$W_{pca} = \arg\max_w |W^T S_T W| \tag{11}$$

$$W_{fld} = \arg\max_w \frac{|W^T W_{pca}^T S_B W_{pca} W|}{|W^T W_{pca}^T S_w W_{pca} W|} \tag{12}$$

The transformation matrix W, which projects a sample in the dimensional space $(c-1)$, is describe by:

$$W = W_{fld}^T W_{pca}^T \tag{13}$$

2.4   Eigenfaces Method

This algorithm implements a set of eigenvectors for the recognition of the human face and it was elaborated by Sirovich and Kirby [11]. The word "eigen" means a set of eigenvectors. It is important to mention that Eigenfaces algorithm can be applied to a different field than face recognition (it can be applied to any data set or objects of the same type that need to be compared) [2]. For facial recognition to be effective many images of each person are required.

Stages of the Eigenfaces Method
The Fig. 6 synthesizes the four stages that comprise the algorithm process [12].

```
A.-Training        D.- Matrix
data set.          creation

B.- Gray           C.-
scale              Histogram.
```

Fig. 6. Eigenfaces algorithm process.

Eigenfaces Algorithm Process
In the identification process was proposed to use Euclidean distance between two images, represented by the Eigenfaces algorithm, as shown in Table 2.

Table 2. Euclidean distance between two images.

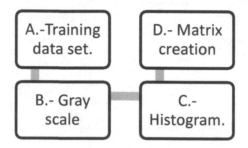

	source		Compare to image	
d (		,		) = (0.00)
d (		,		) = (0.08)
d (		,		) = (9.54)

Identification Through Eigenface Algorithm

The identification algorithm performs a linear projection of the image space to a space features of smaller dimensions. From the vectors obtained as a result of the implementation of Eq. 14.

$$QF = 36\% \; de \; Eigenface_1 + (-8 \; eigenface_2) + 21\% \; Eigenface_n \qquad (14)$$

Where:

- (QF), it is Query Face.
- $(eigenface_1)$ it is face (1).
- $(eigenface_2)$ it is face (2).
- $(eigenface_n)$ it is face (n) (Fig. 7).

Fig. 7. The first seven faces were obtained by implementing the Eigenfaces algorithm with images from the Yale Face extended database.

3 Methodology

3.1 Feature Dataset for Identification

These are the detailed features of the prototype system:

- Subjects have different facial details (with a mustache, or not, use of glasses or not).
- The images have 256 levels in the grayscale for each pixel.
- All the photographs in the dataset contain only the image of the face.
- The images of the faces have a variety of facial expressions.
- Improvements in the graphical user interface section with more intuitive windows and interfaces.
- Improvements in the facial recognition process of the subjects, with a lower error rate and more robust and stable software.
- Each photograph has a standard resolution of 100 * 100 pixels, with a density of 96 pixels per inch (Table 3).

Table 3. Facial Recognition level in Classroom.

#	Person	Detection face level	Images
1	Alma	High	
2	Andrea	Low	
3	Carolina	Low	
4	Cristina	High	
5	Elena	Middle	
6	Katty	Low	
7	Lorena	Low	
8	Manuel	High	

Facial detection and recognition involve locating one or more faces in images or videos. In this sense, it is essential to look for rectangular regions within the image or, failing that, regions that contain objects similar to those of a set of training data. This function performs the process of scanning the image at different scales with the aim of recognizing faces with different dimensions.

3.2   Register Systems

In Fig. 8 the user interface is shown. This allows to capture and save the images of the users to be identified.

Fig. 8. Stage of images captures.

3.3   Training and Recognition Stage

The first stage is to have all registered users. Next comes the recognition stage, using the webcam or a video previously generated and stored on the computer. The system compares all the records stored in the data set, then the system relates users with similar characteristics, discarding users who do not comply with the similarities; finally the output, the system already presents a response to the end-user with the identity of the subject. See Fig. 10. This implementation requires parameters of the EigenObjectRecognizer object, making comparisons between these parameters to determine the identity of the subject (Fig. 9).

Fig. 9. Facial recognition of students: Alma (80%) and Manuel (93%).

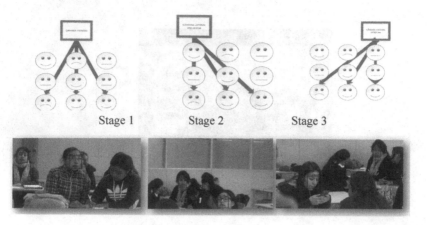

Stage 1   Stage 2   Stage 3

Fig. 10. Test stage in the classroom.

The Table 4 shows the accuracy in the test stage. Table 5 shows the "false positive" in the test stage.

Table 4. Accuracy and time face recognition.

Eigenfaces		LBP Faces		Fisherfaces	
Accuracy (%)	Time (ms)	Accuracy (%)	Time (ms)	Accuracy (%)	Time (ms)
97	187.15	87	648.10	94	257.23

Table 5. False positive test.

Algorithm name	Images for students	Recall (False positive)		
		Stage 1	Stage 2	Stage 3
LBP faces	4	0–4%	35%	47%
	10	0–4%	9%	21%
Eigenfaces	4	0–4%	23%	27%
	10	0–4%	11%	16%
Fisherfaces	4	0–4%	28%	34%
	10	0–4%	17%	34%

4 Discussion

The analyzed algorithms Eigenfaces, LBP Faces, and Fisherfaces are traditional algorithms that allow images to be analyzed quite quickly in face detection. The face detection technique through the eigenfaces algorithm with its training and classification phases has a high accuracy of 97% and a processing time of 187.15 ms. However, the images have to be frontal and with good lighting, so sometimes it does not correctly detect the face. In contrast, the Fisherfaces algorithm is less sensitive to the variation in lighting. In this case, the images must have different lighting conditions and angles on the faces. The values of this algorithm were with an accuracy of 94% and a processing time of 257.23 ms fairly good. Finally, the LBP Faces algorithm has a good accuracy of 87%, but the time of 648.10 ms. In the recognition process, it is the highest of the analyzed algorithms. It can also be verified that when using more images in the training of the algorithms, greater characteristics can be used for later comparison.

5 Conclusions

The user recognition is immediate, considered as response time in real-time, while more student images have to be analyzed, the greater the computational time required for the comparison. This paper proposes and analyzes a people detect method using a traditional algorithm such as LBP faces, Eigenface and Fisherface. Many experiments based on this system have been performed, and the results demonstrate that Eigenfaces algorithm was faster and more efficient during the development of the training tests of the data set, this justifies its implementation as the standard method of detection for the development of the system, and this method entails a higher computational cost.

The Eigenfaces algorithm presents a better performance for images with the variation of both lighting and position of the student's face (see Table 5). By using a greater number of photographs for the recognition dataset, it leads to a better performance of the algorithm when measuring performance through the recall metric (false positives). Future research will be used with embedded systems such as Raspberry Pi or Nvidia Nano Jetson to carry out the process of emotion analysis and present statistics to teachers in the classroom.

References

1. Gao, Y., Qi, Y.: Robust visual similarity retrieval in single model face databases. Pattern Recogn. **38**(7), 1009–1020 (2005)
2. Kaushik, S., Dubey, R.B., Madan, A.: Study of face recognition techniques. Int. J. Adv. Comput. Res. (2014)
3. Lu, X.: Image analysis for face recognition. Pers. Notes **5**, 1–37 (2003)
4. Pietikäinen, M., Hadid, A., Zhao, G., Ahonen, T.: Computer Vision Using Local Binary Patterns, vol. 40. Springer, London (2011). https://doi.org/10.1007/978-0-85729-748-8_14
5. Parker, R.E.: Picture processing during recognition. J. Exp. Psychol. Hum. Percept. Perform. **4**(2), 284–293 (1978)
6. Cama Castillo, Y.A.: Prototipo computacional para la detección y clasificación de expresiones faciales mediante la extracción de patrones locales binarios, pp. 1–94 (2015)
7. Ahonen, T., Hadid, A., Pietikäinen, M.: Face description with local binary patterns: application to face recognition. IEEE Trans. Pattern Anal. Mach. Intell. **28**(12), 2037–2041 (2006)
8. Huang, D., Shan, C., Ardabilian, M., Wang, Y., Chen, L.: Local binary patterns and its application to facial image analysis: a survey. IEEE Trans. Syst. Man Cybern. Part C Appl. Rev. **41**(6), 765–781 (2011)

9. Smiatacz, M.: Eigenfaces, Fisherfaces, Laplacianfaces, Marginfaces - how to face the face verification task. Adv. Intell. Syst. Comput. **226**, 187–196 (2013)
10. Face Recognition with OpenCV – OpenCV 2.4.13.6 documentation. https://docs.opencv.org/2.4/modules/contrib/doc/facerec/facerec_tutorial.htmllast. Accessed 01 Sept 2019
11. Sirovich, L., Kirby, M.: Low-dimensional procedure for the characterization of human faces. J. Opt. Soc. Am. A **4**(3), 519 (1987)
12. Gomathi, E., Baskaran, K.: Recognition of faces using improved principal component analysis. In: The 2nd International Conference on Machine Learning and Computing, ICMLC 2010, pp. 198–201 (2010)
13. Barba Guamán, L.: Utilización de métodos de visión artificial para PC como apoyo en la automoción. Dissertation, ETSI_Sistemas_Infor (2015)
14. Rosebrock, A.: Deep Learning for Computer Vision with Python: Starter Bundle. PyImageSearch (2017)

Correction to: Applied Technologies

Miguel Botto-Tobar(iD), Marcelo Zambrano Vizuete(iD),
Pablo Torres-Carrión(iD), Sergio Montes León(iD),
Guillermo Pizarro Vásquez(iD), and Benjamin Durakovic(iD)

Correction to:
M. Botto-Tobar et al. (Eds.): *Applied Technologies*,
CCIS 1194, https://doi.org/10.1007/978-3-030-42520-3

In the originally published version of the paper on p. 39 the authorship information was incorrect. The names and sequence of the authors have been corrected as "Pablo Torres-Carrión, Ruth Reátegui, Priscila Valdiviezo, Byron Bustamante and Silvia Vaca".

In the originally published version of the paper on p. 473, the author's full name was incorrect. The author's name has been changed to "Ramon Alcarria".

The updated version of these chapters can be found at
https://doi.org/10.1007/978-3-030-42520-3_4
https://doi.org/10.1007/978-3-030-42520-3_38

© Springer Nature Switzerland AG 2020
M. Botto-Tobar et al. (Eds.): ICAT 2019, CCIS 1194, p. C1, 2020.
https://doi.org/10.1007/978-3-030-42520-3_52

Author Index